Physics Programs

Physics Programs

A Manual of Computer Exercises for Students of Physics and Engineering

Edited by
A. D. Boardman
Department of Pure and Applied Physics
University of Salford

A Wiley–Interscience Publication

JOHN WILEY & SONS
Chichester · New York · Brisbane · Toronto

Copyright © 1980 by John Wiley & Sons Ltd.

All rights reserved.

No part of this book may be reproduced by any means, nor transmitted, nor translated into a machine language without the written permission of the publisher.

British Library Cataloguing in Publication Data:

Physics programs
  1. Physics—Data processing—Problems,
exercises, etc. 2. Engineering—Data
processing—Problems, exercises, etc.
I. Boardman, A. D.
530′.028′54044   QC52   79-41214

ISBN 0 471 27674 X

Typeset in Northern Ireland at The Universities Press (Belfast) Ltd. and printed at Pitman Press, Bath

Contributors

G. C. AERS	Theoretical Physics, Daresbury Laboratory, Science Research Council, Daresbury, Warrington, WA4 4AD, U.K.
A. D. BOARDMAN	Department of Pure and Applied Physics, University of Salford, Salford, M5 4WT, U.K.
M. E. S. CHAPMAN	Mullard Hazel Grove (Associated Semiconductor Manufacturers) Ltd., Bramhall Moor Lane, Stockport, U.K.
R. D. CLARKE	Department of Mathematics, Manchester Polytechnic, John Dalton Faculty of Technology, Chester Street, Manchester, U.K.
M. I. DARBY	Department of Pure and Applied Physics, University of Salford, Salford, M5 4WT, U.K.
R. GERBER	Department of Pure and Applied Physics, University of Salford, Salford, M5 4WT, U.K.
P. J. GRUNDY	Department of Pure and Applied Physics, University of Salford, Salford, M5 4WT, U.K.
B. W. JAMES	Department of Pure and Applied Physics, University of Salford, Salford, M5 4WT, U.K.
G. J. KEELER	Department of Pure and Applied Physics, University of Salford, Salford, M5 4WT, U.K.
G. S. KEEN	V. G. Data Systems Ltd., 2 Tribune Avenue, Broadheath, Manchester, U.K.
D. J. MARTIN	Department of Pure and Applied Physics, University of Salford, Salford, M5 4WT, U.K.
J. A. E. TEMPLE	Theoretical Physics Division, A.E.R.E., Harwell, Oxon, U.K.
P. A. YOUNG	Department of Pure and Applied Physics, University of Salford, Salford, M5 4WT, U.K.

Contents

Preface ix

OPTICS

1. **Ray Tracing and Lens Aberrations** 1
   P. A. Young
2. **Attenuated Total Reflection Analysis of Surface Polaritons** . 47
   G. C. Aers and A. D. Boardman
3. **Computer-generated Holograms** 79
   A. D. Boardman and M. E. S. Chapman

MAGNETISM

4. **Calculation of the Fields near Permanent Magnets** 125
   M. I. Darby
5. **Particle Capture in High Gradient Magnetic Separation** . . . 149
   R. Gerber
6. **Magnetization in the Crystal Field System Praseodymium** . . 187
   J. A. G. Temple

SOLID STATE AND QUANTUM PHYSICS

7. **Elastic Waves in Crystalline Solids** 221
   B. W. James
8. **A Computer-assisted Tutorial in Time-independent Non-degenerate Perturbation Theory** 239
   D. J. Martin
9. **Simulation of Phonon Dispersion Curves and Density of States** 269
   G. J. Keeler

10. **Electron Energy Bands in a One-Dimensional Periodic Potential** 301
    R. D. Clarke and D. J. Martin

APPLIED PHYSICS

11. **Computer Simulation of Hot Electron Behaviour in Semiconductors using Monte Carlo Methods** 355
    A. D. Boardman

12. **Modelling of the Thermal Conductivity of Unidirectional Composite Materials** 411
    G. S. Keen and B. W. James

13. **Computational Study of Diffraction by Microcrystalline and Amorphous Bodies** 461
    P. J. Grundy

Index 481

Preface

The computer is ubiquitous in all fields of endeavour but none more so than in the pure sciences and engineering. Learning how to use a computer to break through the barrier presented by the traditional use of idealized solutions to problems in the teaching of physics, engineering, or other sciences is a task confronting us all. Indeed there seems to be a danger that, because few problems of any real practical interest can actually be studied in a classroom context, science seems to be only an academic exercise in the mind of the student. This impression, unfortunately, can lead to a serious loss of enthusiasm for the subject.

Using computers is a way to lift this restraint, making it unnecessary to restrict classroom discussion to simple models that do not really do justice to any branch of the subject. The same can also be said of experimental studies. It is often too costly, or impossible, for a student to perform certain types of experiments yet a lot can be learned from 'experiments' done with a good computer simulation program.

It is obvious that there is a need for material that contributes towards the fulfilment of these aims. It is with this need in mind that the present volume has been prepared. The material is broadly based and ranges from fairly easy to quite demanding. At the basic level of running the computer programs as classroom demonstrations the material should be accessible to a wide audience.

The chapters are loosely grouped into optics, magnetism, solid-state physics, and modelling. Each chapter is self-contained with enough theory given for the whole chapter, and the associated computer program (s), given at the end of the chapter, to be understood. The programs are copied directly from fully working source texts on the computer. They can be used, without understanding the coding, in the exercises or classroom demonstrations, except where system variations require attention. For long-term project work the student may wish to write a new program incorporating as much of the present work as is easily understood. The program published in this book could then be used as a benchmark while the student incorporates modifications into the new program in attempting the more advanced exercises.

All the computer programs use STANDARD FORTRAN but several chapters use graphics. The straightforward graph plotting calls should have instantly recognized equivalents in any establishment as, indeed, should the calls from the main general-purpose graphics package, called GINO-F. This acronym stands for Graphical Input/Output—Fortran developed by the Computer-Aided Design Centre, Cambridge University. Some of the programs that use these instructions are clearly commented as to their use. All the programs, with the exception of that in Chapter 2, were developed at the University of Salford. The program in Chapter 2 was developed at the Theoretical Physics Institute, University of Alberta, Canada.

Some of the authors make use of a general-purpose library, widely referred to in the U.K. as the NAG library. NAG (Numerical Algorithms Group) was originally formed to develop a numerical library for use on the British ICL machines. It is expected, however, that other countries have similar facilities, so it is simply a matter of identifying the routine required. If any difficulty is encountered please write directly to me. Also, it is expected that we can make available tapes or some other recorded version of these programs at a reasonable cost.

It is hoped that this handbook will appeal to the professor and the student alike. The former may welcome the acquisition of some striking demonstration material while the student will have a package that can be used at a leisurely pace, or at one that will stretch his intellectual resources. As far as the British scene is concerned second-year and final-year honours students should gain a great deal from it, both in exercises and long-term project work. Some of the material, such as the ray tracing, the computer-assisted tutorial in quantum mechanics, and the computer modelling of heat flow, is quite suitable material for first-year students. Some of the work is, admittedly, quite difficult and may be more suitable at graduate course level. It is, therefore, expected to be of use at many undergraduate and graduate levels in North American institutions.

I should like to thank all the contributors for their cooperation with my editorial demands, for keeping to a fairly tight schedule, and for submitting material of such high quality with such enthusiasm. It is also a pleasure to thank Dr. Larmouth, Director of the Computing Laboratory, on behalf of all his staff for their constant support.

Finally, I wish to thank my wife Mary who has done much of the secretarial work and is a constant source of encouragement when the hours are long and the tasks seem to be endless.

Salford   A. D. BOARDMAN
May 1979

PART 1

Optics

Physics Programs
Edited by A. D. Boardman
© 1980 John Wiley & Sons Ltd.

CHAPTER 1

Ray Tracing and Lens Aberrations

P. A. YOUNG

1. INTRODUCTION

A great deal of our knowledge of the physical and biological world comes from our use of microscopes, telescopes, cameras, and other optical devices that use light waves to form images of greater brightness or detail than we can obtain from our eyes alone. A basic part of the design of such optical systems is the tracing of rays through them and the determination of their deviations from perfect imagery, the so-called aberrations.

The two concepts that are important in discussing the propagation of light and the formation of images are the wavefront and the ray. A wavefront is defined as the locus of points which the light has taken the same time to reach, and the ray as the direction in which the light energy is travelling. In isotropic materials, such as glass, the rays are perpendicular to the wavefronts.

Because light is a wave-motion it can be diffracted, and the concepts of ray and wavefront breakdown in situations in which diffraction is important. These include points at which light waves converge to form images. Diffraction, in fact, makes it impossible to realize a ray physically, by, for example, passing light through smaller and smaller pinholes set so as to define a direction of energy travel; nevertheless, it remains a useful idealization and it is the propagation of these ideal rays, and associated waves, that is the province of geometrical optics.

1.1 Fermat's principle

A fundamental link between the wavefront and the ray is provided by Fermat's principle of least time that, in words, is

the path taken by a light ray is such that the time of travel is a minimum.

Subsequent work has shown that, although a minimum is often involved, a better statement is that the time should be stationary (maximum, minimum, or inflexion point). In the notation of the calculus of variations this is written as

$$\delta \int dt = 0, \qquad (1)$$

where t is the time and the integral is taken between suitable limits. Now if the speed v of the wave, i.e. ds/dt, and the refractive index $n = c/v$ is introduced then equation (1) has the form

$$\delta \int n \cdot ds = 0 \qquad (2)$$

where c, the constant velocity of light *in vacuo*, has been deleted. The quantity $\int n \cdot ds$ is called the optical path along the ray. For regions of constant refractive index $\int n \, ds$ is ns which is the familiar rule that

Optical path = refractive index × geometrical path

It follows from Fermat's principle that light rays obey the observed laws of geometrical optics,[1] and in particular for refraction, if a ray (called the incident ray) in a medium of refractive index n strikes a surface, at an angle I to the normal, that separates it from a medium of refractive index n' then it continues as a refracted ray at an angle I' to the normal such that
(a) the incident ray, the refracted ray, and the normal lie in one plane;
(b) the angles and refractive indices obey Snell's law, viz.:

$$n \sin I = n' \sin I'. \qquad (3)$$

2. RAY TRACING IN THE PARAXIAL APPROXIMATION

It is a consequence of the wave nature of light that just as rays are a physical impossibility so also is the ideal optical system defined as one in which all rays leaving a single object point converge on (or appear to diverge from) a unique image point, and even within the realm of geometrical optics the quasi-ideal system (ignoring diffraction) can only be realized in a few cases, of which the plane mirror is the simplest example.[2] The situation, however, is not as bad as it seems because sufficiently close approximations to ideal systems can be obtained as to be practically useful, and it is the closeness to ideal imagery that is specified by the aberrations. Furthermore, these may be determined by ray tracing using the laws of geometrical optics.

However, before any detailed design is carried out to find the exact form and nature of the deviations from perfect imagery, as revealed by the actual paths of the rays, it is useful to use what is known as the paraxial or Gaussian approximation in which all rays are assumed to be close to the axis of a system and all angles are assumed to be small. These assumptions lead to the paraxial equations which can be used, for instance, to determine: (1) the system focal length; (2) the position of the ideal, or Gaussian, image from which the deviations can be measured; (3) an estimate of the size of the aberration of the image as measured by the difference in optical path between the actual ray and the Gaussian image ray.

Assume that the system has a unique axis of symmetry and that a paraxial ray is described by the two parameters u and y, as shown in Figure 1a; u is the angle the ray makes with the axis, and y is the distance from the axis of a point on the ray, usually at one of the optical surfaces. (Note that for small u and y, no distinction is made between the position of a ray intercept on a surface or on its tangent plane.)

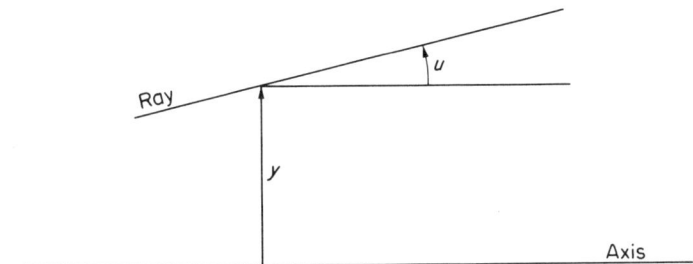

Figure 1a. Ray parameters

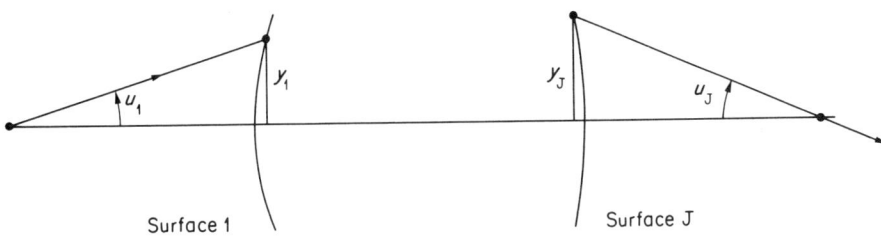

Figure 1b. Initial and final ray parameters

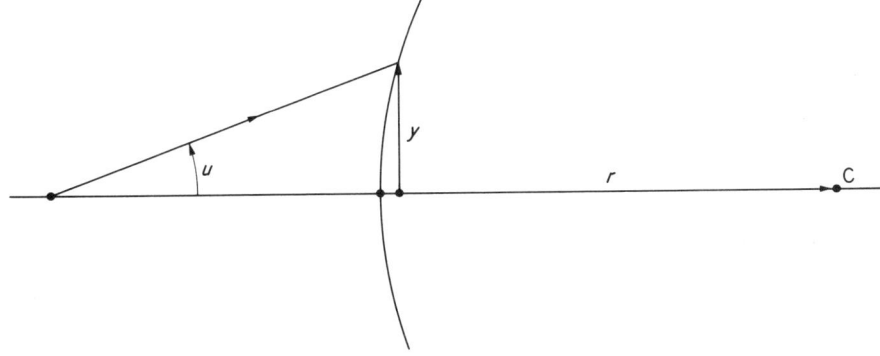

Figure 1c. Positive parameters

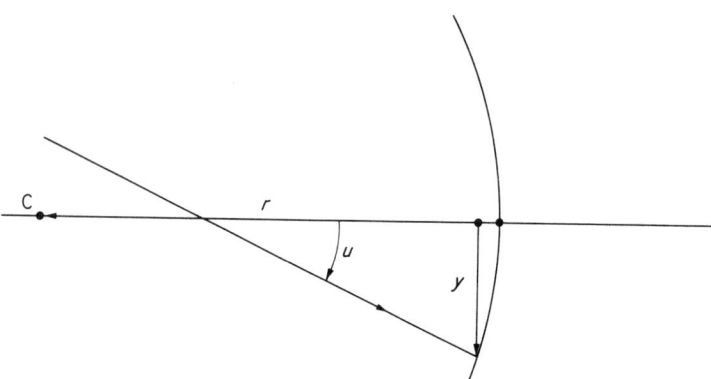

Figure 1d. Negative parameters

The tracing of a ray means, as shown in Figure 1b, the determination of the angle u_J and height y_J at which the ray leaves the final (Jth) surface of the system, given the angle u_1 and height y_1 at which it enters the first surface.

The rays are assumed to traverse the system from left to right and the normal Cartesian conventions on the signs of distances and angles apply. The radii r of surfaces are positive if they are convex to the left. Figures 1c and 1d, in which C is the centre of curvature, show situations in which all the quantities are positive or negative respectively. The procedure for ray tracing is broken down into two parts, these are refraction at a surface and transfer from one surface to the next.

2.1 Refraction and transfer

The sth surface AB of radius r_s dividing two regions of refractive indices n_s and n'_s is shown in Figure 2a. Suppose that a ray makes axial angles u_s and

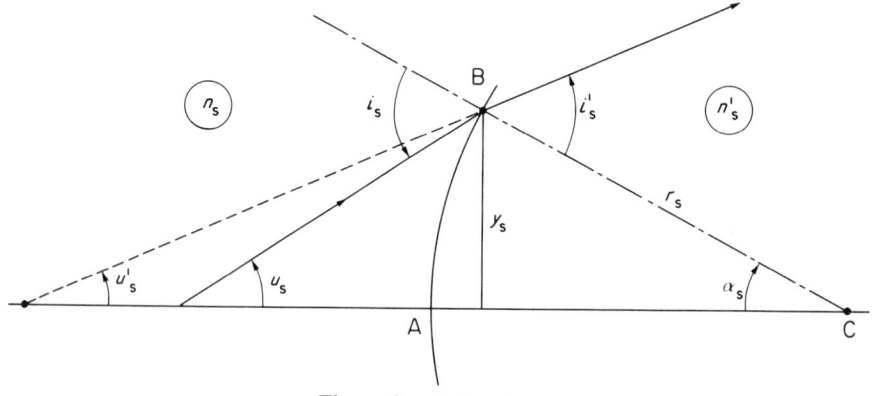

Figure 2a. Refraction

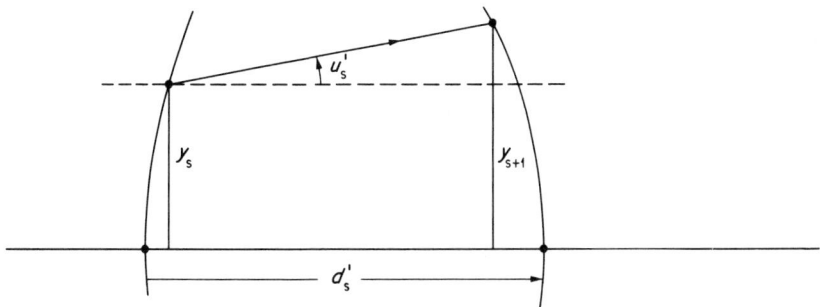

Figure 2b. Transfer

u'_s before and after refraction respectively, and have corresponding angles of incidence and refraction of i_s and i'_s. Furthermore, suppose that C is the centre of curvature of the surface and that the incident height y_s subtends an angle of α_s at the centre. (It is a consequence of the sign convention on distances that α_s is *positive*.) Then, from the diagram,

$$i_s = \alpha_s + u_s, \qquad (4)$$
$$i'_s = \alpha_s + u'_s.$$

Now for small angles, Snell's law, equation (3), becomes $n'i' = ni$ so that

$$n'_s(\alpha_s + u'_s) = n_s(\alpha_s + u_s). \qquad (5)$$

Also, for small angles, $\alpha_s = y_s c_s$ where $c_s = 1/r_s$ is the curvature of the surface. (Note that $c_s \to 0$ when $r_s \to \infty$ and can thus be used numerically for plane surfaces.) Hence, on substituting for α_s we find

$$n'_s u'_s = n_s u_s - y_s K_s. \qquad (6)$$

where K_s, the power of the surface, is

$$K_s = (n'_s - n_s) c_s. \qquad (7)$$

After a ray leaves the surfaces at angle u'_s and height y_s, it proceeds to the $s+1$ surface, a distance d'_s away along the axis, and intercepts it at height y_{s+1}. In Figure 2b, it is seen, for small angles and heights, that

$$y_{s+1} = y_s + d'_s u'_s. \qquad (8)$$

2.2 Ray tracing procedure

A given ray is traced through a system by successive use of equations (6) and (8), noting that at each surface $n_{s+1} = n'_s$ and $u_{s+1} = u'_s$. The trace is started in one of two ways:
(1) An initial axial point O on the object, a distance l_1 from the first

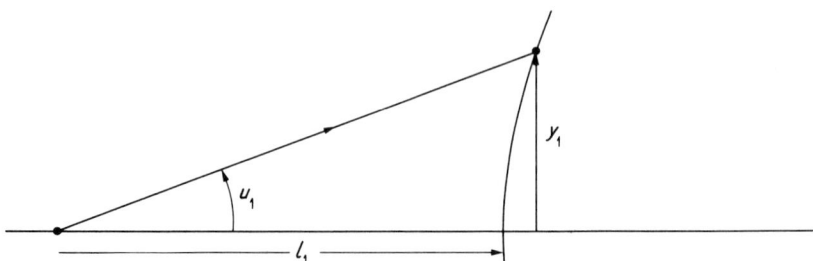

Figure 2c. Starting with a ray at a given incidence height

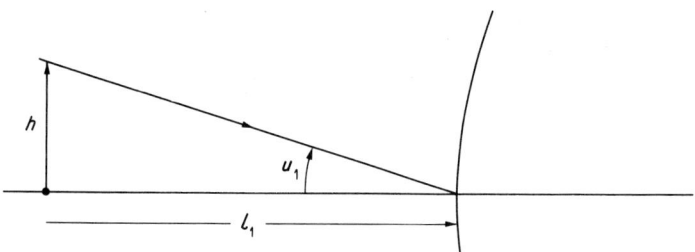

Figure 2d. Starting with a ray from a known object height

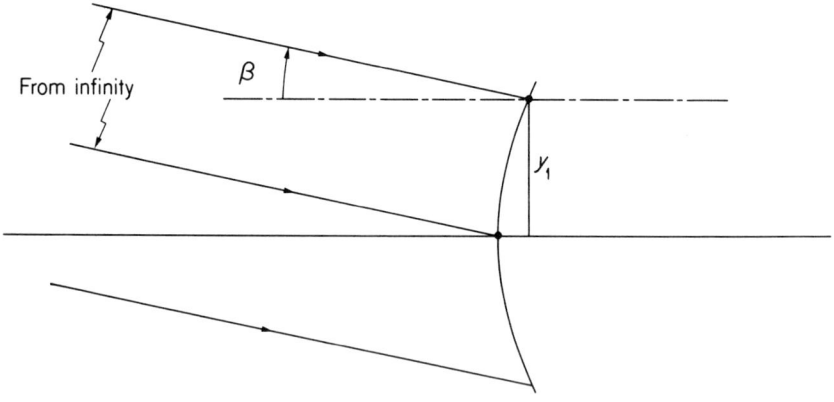

Figure 2e. Rays from infinity, field angle β

surface, is chosen together with an incidence height of y_1 as shown in Figure 2c. Then

$$u_1 = y_1/l_1, \tag{9}$$

and the trace starts with a refraction.

(2) An object of height h is selected at a distance l_1 (Figure 2d) and a ray through the centre of the first surface is chosen, then

$$\begin{aligned} y_1 &= h, \\ u_1 &= -h/l_1, \end{aligned} \tag{10}$$

and the trace starts with a transfer.

Note that for objects at infinity h and l_1 simultaneously tend to infinity but the field angle $\beta = -h_1/l_1$ remains constant, as shown in Figure 2e. If a ray at height y_1 on the first surface is chosen, then with $u_1 = \beta$, the trace starts with a refraction.

3. STOPS AND PUPILS

It should be pointed out that the 'surface' referred to above can be simply a circular hole for which $n' = n$ and $c = 0$. Such apertures, or stops, are often placed in optical systems to limit the extent of the beams passing through, and also to control the aberrations. Amongst the various stops and lens apertures in an optical system there will be one, or its image, which seen from the object side subtends the smallest angle at the axial object point: this is known as the entrance pupil and it limits the maximum angle that the rays can make with the axis and still pass through the system.

If the pupil is an image of a stop the corresponding stop is called the aperture stop, if the pupil is real it is itself the aperture stop. The image of the entrance pupil as seen from the image side is the exit pupil. For an off-axis object point the pupil will still, to a large extent, control the angular

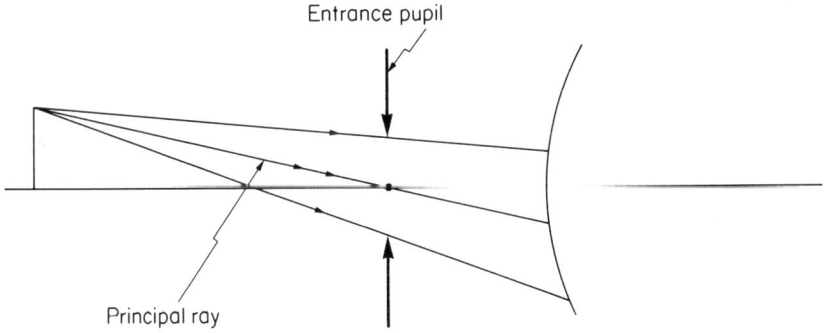

Figure 3a. Entrance pupil and principal ray for a real pupil

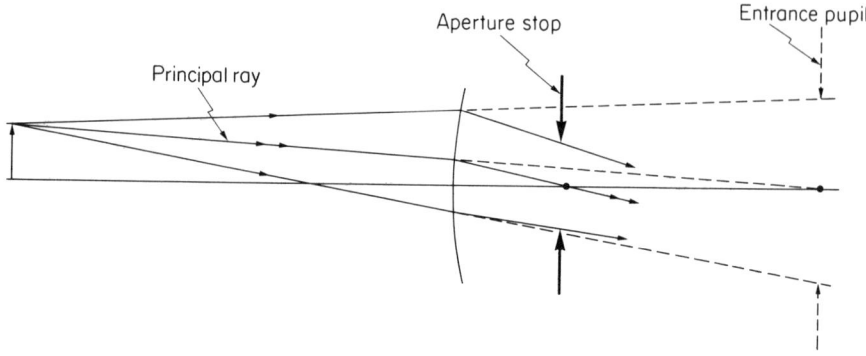

Figure 3b. Entrance pupil and principal ray for a virtual pupil

aperture of the rays that pass through the system. The ray through the centre of the entrance pupil, that defines a cone of rays that can pass through the system, is an important ray and is known as the principal ray. Figures 3a and 3b show stops, pupils, and principal rays in an optical system; the principal ray is indicated by double arrows.

4. FOCAL LENGTH

If the initial angle u_1 is zero and the initial height y_1 is finite then one can determine the focal length. After passing through the system the ray will (except in what are known as telescopic systems) leave the final surface at a finite angle, u'_J, and height y_J and pass through the focal point F'. The point on the axis directly below the intersection point of the initial ray and the final ray is the principal point P', distance p' from the last surface, whilst the distance from the last surface to F' is known as the back focal length, bfl'.

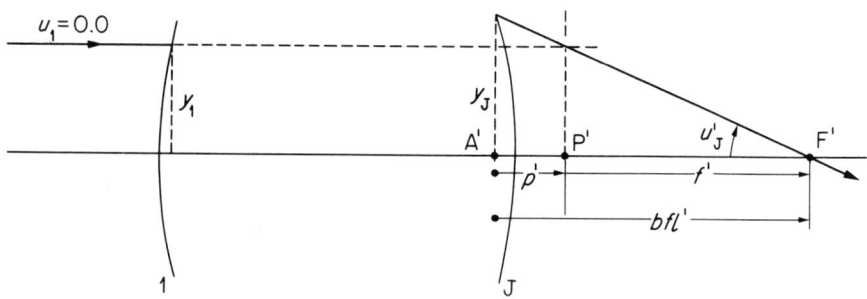

Figure 3c. Focal parameters, P', F', p', f', bfl'

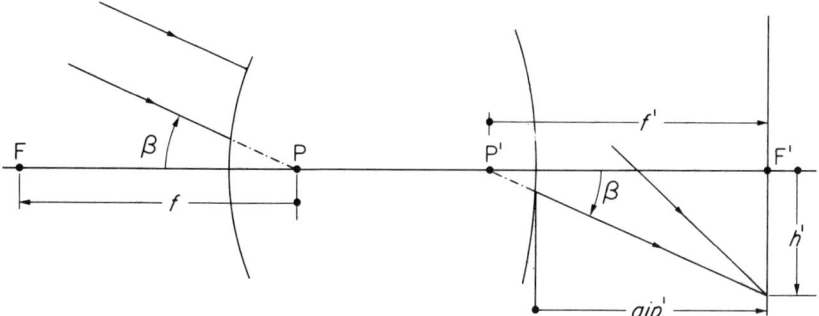

Figure 3d. Gaussian image plane and size for an infinite object

In Figure 3c, it is seen that

focal length: $\quad f' = -y_1/u'_J,$ (11)

back focal length: $\quad bfl' = -y_J/u'_J,$ (12)

position of principal point: $\quad p' = bfl' - f'.$ (13)

If a ray be traced from infinity back through the lens ($u'_J = 0$) then similar points F and P and distances f, p, and ffl (front focal length) are defined on the object side. In systems in which, as is usually the case, $n_1 = n'_J$ the focal lengths f and f' are numerically equal and the principal points, P and P' are also the so-called nodal points such that a ray directed towards P on the object side leaves the system on the image side as if directed away from P'. This is shown in Figure 3d.

4.1 Gaussian image

If the position of the Gaussian, or paraxial, image with respect to the last, Jth, surface of the system, is gip' and the size of the image is h' then
(1) for an object effectively at infinity, as shown in Figure 3d, the image distance is

$$gip' = bfl' \qquad (14)$$

and its size is

$$h' = f'\beta; \qquad (15)$$

(2) for an object at a finite distance l, and of size h, the image position is found by tracing a paraxial ray from the axial object at any non-zero angle u_1 and compatible $y_1 = lu_1$. The image position is then given, as shown in Figure 3e, by

$$gip' = y_J/u'_J \qquad (16)$$

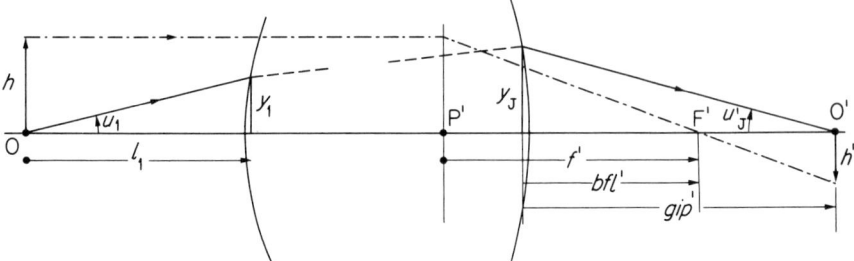

Figure 3e. Gaussian image plane and size for a finite object

The size of the image can be seen, from Figure 3e, to be given by

$$\frac{-h'}{F'O'} = \frac{h}{P'F'}$$

so that

$$h' = -(gip' - bfl')h/f'. \tag{17}$$

5. LENS ABERRATIONS

In a perfect optical system all rays leaving a point object, O, converge on (or diverge from) a point image, O'. If we apply Fermat's principle to the rays travelling from the object to the image via the system then any ray takes a minimum time so that all rays from O to O' must take the same time; this fact is more conveniently stated as: the optical path along all the rays from object to image are equal. A suitable way to measure the defects of any real optical system is therefore in terms of the differences in the optical paths of

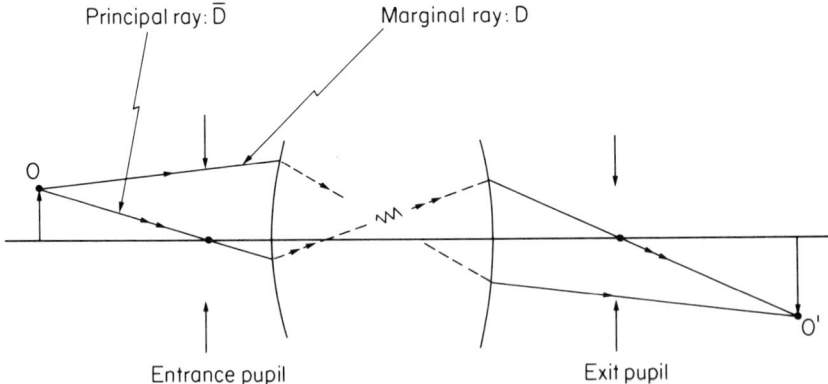

Figure 4a. Principal ray and marginal ray

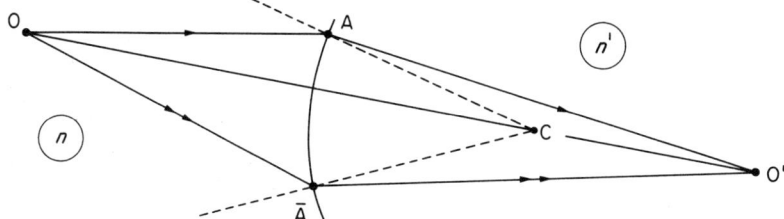

Figure 4b. Wave aberration due to refraction

the rays, which is done by comparing the optical path of any given ray with that of a reference ray, the latter being chosen as the principal ray. This is shown in Figure 4a and the conventional choice of a measure W, of the aberration, is

$$W = \bar{D} - D, \tag{18}$$

where $\bar{D}$ is the total optical path of the principal ray and D is the total optical path of the given ray.

5.1 Aberration due to one refraction

In Figure 4b O is an object and O' is its image in a surface separating media of refractive indices n and n'. $O\bar{A}O'$ is a principal ray and OAO' is a given ray that, since it is usually taken to be a ray at the edge or margin of the pupil, is called a marginal ray. Then

$$\begin{aligned} \bar{D} &= n \cdot O\bar{A} + n' \cdot \bar{A}O', \\ D &= n \cdot OA + n' \cdot AO', \end{aligned} \tag{19}$$

and the difference in optical path, defined by equation (18), is

$$\begin{aligned} W &= n'(\bar{A}O' - AO') + n(O\bar{A} - OA) \\ &= n'(\bar{A}O' - AO') - n(\bar{A}O - AO) \\ &= \Delta\{n(\bar{A}O - AO)\}, \end{aligned} \tag{20}$$

where Δ means take the difference of the value in the expression after and before refraction.

5.2 Spherical aberration

The most important aberration, which is present even for axial objects, is spherical aberration. It is also the one that is most readily calculated from equation (20). Spherical aberration takes the general form of producing differing focusing positions for different incident heights, as shown in Figures 5a and 5b. The focus F_G is the Gaussian focus (F') and is the one that is

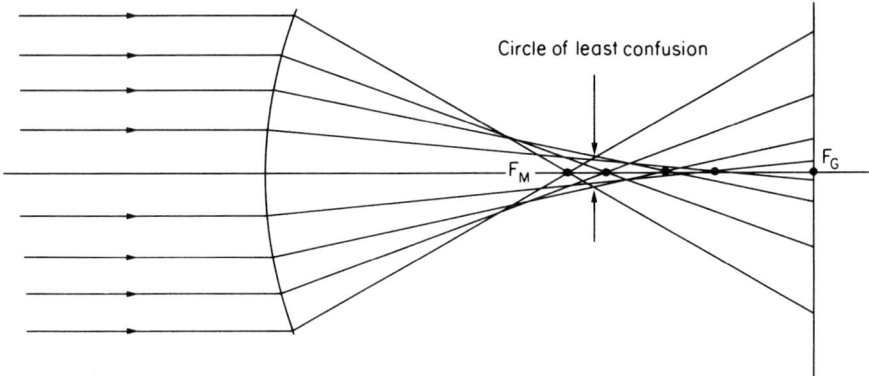

Figure 5a. Spherical aberration

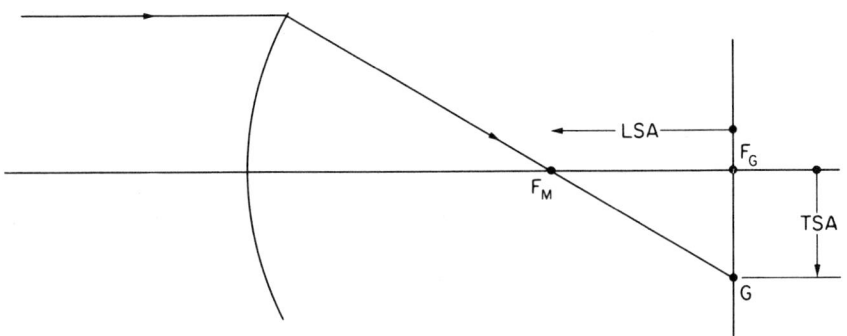

Figure 5b. Longitudinal and transverse spherical aberration

obtained from the paraxial ray tracing equations. The distance along the axis between the marginal focus F_M and the Gaussian focus F_G is the longitudinal spherical aberration, LSA. The distance between F_G and the marginal ray intercept G on the Gaussian image plane is the transverse spherical aberration, TSA. Similar results are obtained for off-axis objects, as shown in Figure 5c. Suppose Figure 4b is redrawn explicitly for an axial object, as shown in Figure 5d, such that $\bar{A}A$ is a surface, centre C, radius r, separating media of refractive indices n and n', α is the angle subtended at the centre by $\bar{A}A$, and i is the angle of incidence of a marginal ray OA, from an axial object O. H is the foot of the perpendicular from A to the axis $O\bar{A}C$,

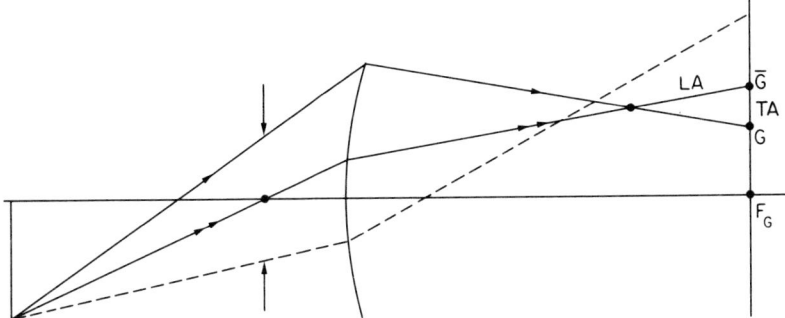

Figure 5c. Longitudinal and transverse aberration for an off-axis object

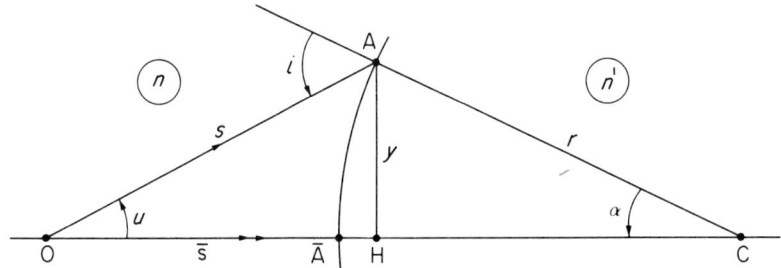

Figure 5d. Calculation of wavefront aberration

$HA = y$, $\bar{A}O = \bar{s}$ and $AO = s$. Then, from Figure 5d,

$$AH = r(1 - \cos \alpha),$$

and

$$OH = \bar{s} + r(1 - \cos \alpha) = s \cos u.$$

Therefore

$$\bar{A}O - AO = \bar{s} - s = s(\cos u - 1) + r(\cos \alpha - 1)$$

Now

$$y = s \sin u = r \sin \alpha,$$

hence

$$\begin{aligned}
\bar{s} - s &= y[(\cos u - 1)/\sin u + (\cos \alpha - 1)/\sin \alpha] \\
&= -y(\tan u/2 + \tan \alpha/2) \\
&= -y[\sin(u/2)\cos(\alpha/2) + \sin(\alpha/2)\cos(u/2)]\sec(\alpha/2)\sec(u/2) \\
&= -y \sin\left(\frac{u + \alpha}{2}\right)\sec(\alpha/2)\sec(u/2).
\end{aligned}$$

Thus, since

$$i = u + \alpha,$$
$$\bar{s} - s = -y \sin i/2 \sec \alpha/2 \sec u/2$$
$$= -\tfrac{1}{2} y \sin i \sec \alpha/2 \sec u/2 \sec i/2.$$

On expanding the sec terms to third order, but retaining $\sin i$ for convenience since $n \sin i$ is an invariant quantity, we obtain

$$\bar{s} - s = -\tfrac{1}{2} y \sin i \left(1 + \frac{\alpha^2}{8}\right)\left(1 + \frac{u^2}{8}\right)\left(1 + \frac{i^2}{8}\right) + \ldots$$
$$= -\tfrac{1}{2} y \sin i - \tfrac{1}{16} y \sin i (\alpha^2 + u^2 + i^2) + \ldots.$$

We then find, from equation (20), that

$$W = -\tfrac{1}{2}\Delta(yn \sin i) - \tfrac{1}{16}\Delta(yn \sin i (\alpha^2 + u^2 + i^2) \ldots). \tag{21}$$

Now y, α, and $n \sin i$ are constant across a surface, so that, for instance

$$\Delta(yn \sin i) = yn' \sin i' - yn \sin i = 0$$

and similarly for terms including α. Equation (21) therefore becomes

$$W = -\tfrac{1}{16} y(n \sin i)\Delta(u^2 + i^2)$$
$$= -\tfrac{1}{16} y(n \sin i)\Delta[(i - u)^2 + 2ui]$$
$$= -\tfrac{1}{16} y(n \sin i)\Delta(\alpha^2 + 2ui)$$
$$= -\tfrac{1}{8} y(n \sin i)\Delta(ui)$$
$$= -\tfrac{1}{8} y(n \sin i)\Delta\left[\frac{u}{n}(n \sin i)\frac{i}{\sin i}\right]$$
$$= -\tfrac{1}{8} y(n \sin i)^2 \Delta\left(\frac{u}{n}\right)(1 + \tfrac{1}{6}\sin^2 i + \ldots),$$

where we have used

$$i = \sin i + \tfrac{1}{6}(\sin^3 i) + \ldots.$$

Finally, expanding to third order in i gives, for the first term $S1$ of W,

$$S1 = -\tfrac{1}{8} y A^2 \Delta(u/n), \tag{22}$$

in which $A = ni$, is the paraxial value of the Snell refraction invariant. (Note, it is the convention to denote by S_I—in which S stands for Seidel, one of the earliest investigators of lens aberrations[3]—the value $-yA^2\Delta(u/n)$, so that $S1 = \tfrac{1}{8} S_I$.) As a consequence of refraction each surface therefore contributes to the wavefront spherical aberration an amount equal to $S1$ and it can be shown that the total aberration to third order for a system of surfaces is equal to the sum of the contributions at each surface as the transfer process produces no new aberration.[4]

During a computer calculation, the value of S1 is determined at each surface from the known values of n, c, u, and y, using the identities

$$\Delta(u/n) = u'/n' - u/n,$$

and

$$A = ni = n(u + \alpha) = nu + nyc.$$

6. BENDING

After an initial design has been set up, and its aberration calculated, it can be altered so as to minimize the aberration. The simplest way to do this is to leave the glass types and separations unchanged, but to alter the curvatures of the surfaces in a systematic way, because the aberration varies quite strongly with this parameter. The easiest way to do this would be to add a constant amount Δc to each curvature, so that positively curved surfaces become more curved and negatively curved surfaces less so, and then to recalculate the aberration. However, it is found that if this is done then the focal length changes and a way to avoid this, as first shown by Hopkins,[5] is to alter the curvature in such a way that the angle α is changed by a constant amount $\Delta \alpha$, or in other words to make Δc at each surface inversely proportional to the height y so that $\Delta \alpha = y \Delta c$ is constant. The fact that this leaves the focal length unchanged will now be demonstrated.

As shown earlier, equation (6), the equation for refraction at each surface can be written as

$$n'_s u'_s - n_s u_s = -(n'_s - n_s) \alpha_s.$$

Writing this out in full for each surface we have

$$n'_1 u'_1 - n_1 u_1 = (n_1 - n'_1) \alpha_1,$$
$$n'_2 u'_2 - n_2 u_2 = (n_2 - n'_2) \alpha_2,$$
$$n'_J u'_J - n_J u_J = (n_J - n'_J) \alpha_J,$$

so that on adding, and using the fact that

$$n_2 = n'_1, \qquad u_2 = u'_1, \ldots$$

we find that

$$n'_J u'_J - n_1 u_1 = n_1 \alpha_1 - n'_J \alpha_J + \sum n_I (\alpha_J - \alpha_{J-1}).$$

It is usual in an optical system for the initial and final refractive indices to be the same, that is (for air),

$$n'_J = n_1 = 1.0$$

so that

$$u'_J - u_1 = (\alpha_1 - \alpha_J) + \sum n_J (\alpha_J - \alpha_{J-1}) \qquad (23)$$

If, in equation (23), each value of α is altered by the same amount $\Delta\alpha$ it is seen that the right-hand side remains unchanged, hence for a given u_1 the final angle u'_f is unaltered and it follows, from equation (11), that the focal length of the system stays at its initial value.

6.1 Effect of bending

The effect of bending is to make $S1$ vary as a parabolic function of $\Delta\alpha$ (see Figure 8c) that may or may not pass through the $\Delta\alpha$ axis. In the former case the choice of lens shape will depend on the presence of other aberrations, especially the asymmetric aberration known as coma which is often a minimum near the turning point (not zero) value of $S1$. In the latter case the numerically minimum value of $S1$ may be brought closer to zero by altering the refractive indices or separations, but this cannot always be done in a systematic manner—as can bending—and is outside the scope of this chapter.

7. FINITE RAY TRACING

So far we have considered only the paraxial approximation that has allowed us to determine the focal length, the Gaussian image plane, and a measure of the wavefront spherical aberration and its variation with lens shape.

When a design has approached a sufficient minimum in aberration—which may require changes in refractive indices and element separations as well as changes in curvatures—then it is time to trace some actual rays through the system and determine how these are spread out over the image plane.

To trace the actual paths of rays means that we must determine the true heights Y and angles U at which a ray strikes a surface rather than their paraxial approximations y and u.

The procedure for doing this which is called finite ray tracing makes use of Snell's law and coordinate geometry.

7.1 Finite ray tracing in two dimensions

Consider an optical system that has an axis of symmetry OZ and with the OYZ plane containing the off-axis object point. The plane OYZ is called the meridian plane and it is a consequence of Snell's law, and the axial symmetry, that a ray which leaves the object in this plane will remain in it. The process of tracing such a ray thus involves only two-dimensional geometry which simplifies the analysis to a certain extent, as we shall now see. Figure 6a shows a ray $\tilde{P}P$ leaving a surface at $\tilde{P}$, with known coordinates $(\tilde{Y}, \tilde{Z})$, and proceeding at a known angle U with the axis to the next surface, centre C, radius r, distance t along the axis, which it intersects at P whose coordinates are (Y, Z) with respect to axes with origin at O.

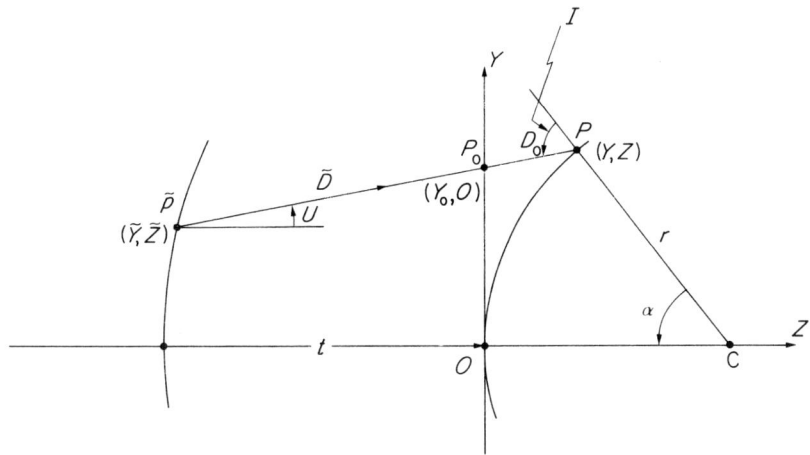

Figure 6a. Parameters for tracing a finite ray

Suppose the ray intersects the tangent plane, (at O), at point P_0 whose coordinates are $(Y_0, 0)$ and adopt the definitions

$$\tilde{P}P = D, \tag{24}$$
$$\tilde{P}P_0 = \tilde{D}, \tag{25}$$
$$P_0 P = D_0. \tag{26}$$

The quantities to be determined are the coordinates (Y, Z) and the angle I that the ray $\tilde{P}P$ makes with the normal PC. If the direction cosines of the ray are

$$M = \sin U, \tag{27}$$
$$N = \cos U, \tag{28}$$

then from the figure it is seen that

$$\tilde{P}P_0 \sin U = \tilde{D}M = Y_0 - \tilde{Y}, \tag{29}$$
$$\tilde{P}P_0 \cos U = \tilde{D}N = t - \tilde{Z}. \tag{30}$$

Hence the equations

$$\tilde{D} = (t - \tilde{Z})/N, \tag{31}$$
$$Y_0 = \tilde{Y} + (t - \tilde{Z})M/N, \tag{32}$$

determine $\tilde{D}$ and Y_0 in terms of the known ray parameters. $\tilde{Z}$ is measured with respect to the origin of the first curved surface.

D_0 and hence (Y, Z) are found in the following way. As before

$$MD_0 = Y - Y_0, \qquad (33)$$
$$ND_0 = Z, \qquad (34)$$

and a third equation is provided by the equation of the surface OP, viz.:

$$Y^2 + (Z-r)^2 = r^2,$$

or, with $c = 1/r$,

$$c(Y^2 + Z^2) - 2Z = 0. \qquad (35)$$

Substitution of equations (33) and (34) into equation (35) leads to

$$c(MD_0 + Y_0)^2 + c(ND_0)^2 - 2(ND_0) = 0,$$

which, since

$$M^2 + N^2 = 1$$

is also

$$cD_0^2 - 2(N - cMY_0)D_0 + cY_0^2 = 0.$$

Now, in accordance with standard practice,[6] we use

$$F = cY_0^2, \qquad (36)$$
$$G = N - cMY_0, \qquad (37)$$

which are known quantities, so that D_0 is simply a root of the quadratic equation

$$cD_0^2 - 2GD_0 + F = 0, \qquad (38)$$

i.e.

$$D_0 = (G - \sqrt{G^2 - cF})/c, \qquad (39)$$

where the negative sign is taken since D_0 has to be the lesser of the two possible values—as shown in Figure 6b. Equation (39), because the limit $c \to 0$ can cause numerical trouble, has a better computational form after rationalization with the numerator. Thus

$$D_0 = \frac{G - \sqrt{G^2 - cF}}{c} \frac{G + \sqrt{G^2 - cF}}{G + \sqrt{G^2 - cF}},$$

or

$$D_0 = F/(G + \sqrt{G^2 - cF}). \qquad (40)$$

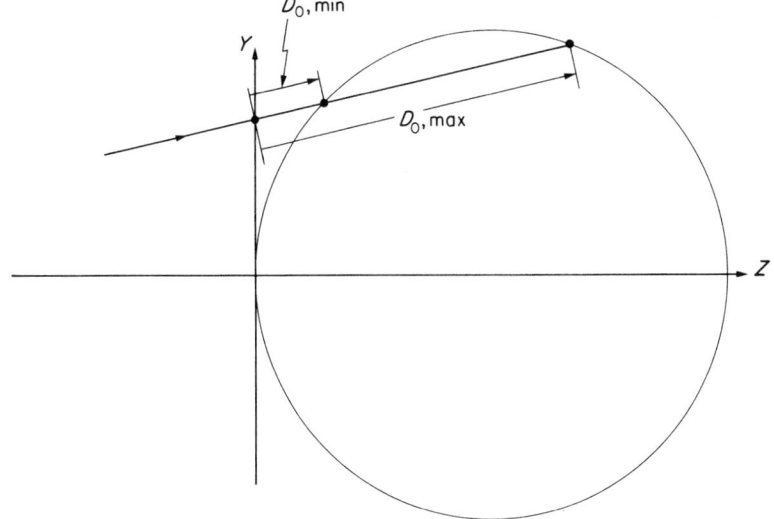

Figure 6b. Maximum and minimum values of D_0

Also, from Figure 6a,
$$I = U + \alpha,$$
therefore
$$\cos I = \cos U \cos \alpha - \sin U \sin \alpha,$$
or
$$\cos I = N \cos \alpha - M \sin \alpha,$$
whilst, from Figure 6a,
$$\cos \alpha = (r - Z)/r = 1 - cZ, \tag{41}$$
$$\sin \alpha = Y/r = cY, \tag{42}$$
hence
$$\cos I = N - NcZ - McY$$
$$= N - cN(ND_0) - cM(Y_0 + MD_0),$$
using equations (33) and (34). Therefore
$$\cos I = (N - cMY_0) - cD_0$$
$$= G - cD_0$$
$$= \sqrt{G^2 - cF}. \tag{43}$$

using equation (39). We have found D_0 and cos I, and Y and Z follow from equations (33) and (34), so the description of the transfer from one surface to the next is complete. We now need to determine the refracted ray.

Before and after refraction,
$$I = U + \alpha,$$
$$I' = U' + \alpha,$$
hence
$$\sin U = \sin I \cos \alpha - \cos I \sin \alpha,$$
$$\sin U' = \sin I' \cos \alpha - \cos I' \sin \alpha.$$

Therefore
$$n' \sin U' - n \sin U = (n' \sin I' - n \sin I)\cos \alpha - (n' \cos I' - n \cos I)\sin \alpha$$
$$= -(n' \cos I' - n \cos I)cY,$$
using Snell's law and equation (42). This means that
$$n' \sin U' = n \sin U - YK, \tag{44}$$
where the power K is given by
$$K = (n' \cos I' - n \cos I)c \tag{45}$$
and $n' \cos I'$ is determined from Snell's law and the known value of cos I (equation (32)). Similarly, by considering
$$\cos U = \cos I \cos \alpha - \sin I \sin \alpha$$
one can show that
$$n' \cos U' = n \cos U - ZK + (n' \cos I' - n \cos I) \tag{46}$$
In terms of M and N equations (44) and (46) may be written as
$$n'M' = nM - YK \tag{47}$$
$$n'N' = nN - ZK + (n' \cos I' - N \cos I) \tag{48}$$
or alternatively, in two dimensions, N' may be determined from M' by using the fact that
$$M'^2 + N'^2 = 1$$
which can be used as a check on the values of M' and N'.

The derivation of the equations required for tracing a finite ray through an optical system is now complete. However, it is interesting to note that equations (33) and (29) may be combined as
$$Y = Y_0 + MD_0 = \tilde{Y} + M\tilde{D} + MD_0 = \tilde{Y} + MD$$
$$= \tilde{Y} + D \sin U$$

which reduces to the paraxial equation (8), viz.:

$$y_+ = y + d'u'$$

when y and u are small, and that equation (44) reduces, similarly, to the corresponding paraxial equation (6) viz.:

$$n'u' = nu - Ky.$$

8. COMPUTER PROGRAMS

Two programs and typical outputs are given below, one S1BEND to calculate the wavefront spherical aberration $S1$ and its variation with the bending parameter $\Delta\alpha$, and the other RAYTRC to calculate the transverse aberration on the Gaussian image plane.

The comments in the program are designed to be self-explanatory; in both programs a distinction between objects at a finite distance and objects at infinity is made by means of the logical variable FINITE. In S1BEND it is assumed that the position of a finite object is specified by the incident ray angle $U(1)$ and its height $Y(1)$ on the first surface as these are required in both cases—finite, or infinite; in RAYTRC, however, the image height must be calculated and also the initial incidence angle varies, so it is convenient to specify a finite object by its size and distance, cf. equation (10). The data required for the programs are listed in the comments. Some specific data for a typical optical system are considered in section 8.1. Besides looking at the effects of bending the S1BEND program may also be used to investigate the variation of $S1$ with angle $U(1)$ and incident height $Y(1)$ by setting the bending parameters to zero and running the program for the required incident ray parameters.

8.1 Example of data

A suitable optical system on which to try the programs is provided by the TESSAR type photographic lens, such as the one specified by O'Neill,[7] of which a diagram is shown in Figure 7 and the data are given in Table 1.

The focal length of the lens is given as (approximately) 50 mm, but further details are not given, so we shall assume therefore that it is designed to cover a 35 mm film and that it operates at a maximum relative aperture, or stop number, of f/2.8, (i.e. focal length/aperture = 2.8). Using these figures we see that the maximum semi-angle of the rays entering the lens is approximately $\tan^{-1}(17.5/50)$, i.e. 20°, whilst the aperture is given by 50/2.8 or approximately 18 mm, the semi-aperture is therefore 9 mm. The initial

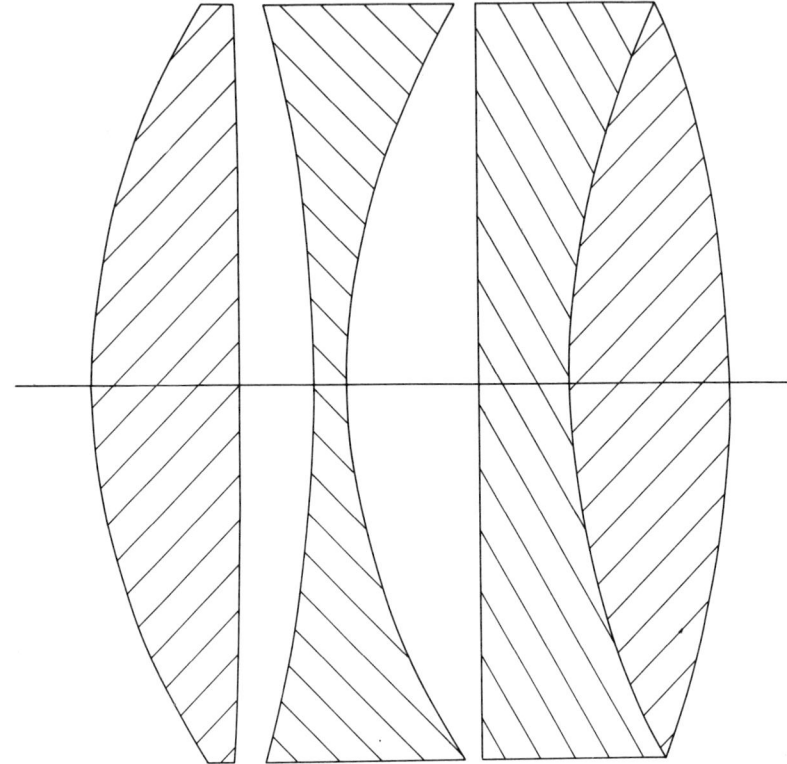

Figure 7. Diagram of the TESSAR lens

Table 1. Data for the TESSAR lens

n	c (cm^{-1})	d (cm)
1.0		
	0.61425	
1.6116		0.357
	−0.036271	
1.0		0.189
	−0.28927	
1.6053		0.081
	0.63221	
1.0		0.325
	0	
1.5123		0.217
	0.52083	
1.6116		0.396
	−0.41667	
1.0		

data for an infinite object are thus:

Logical parameter	FINITE = FALSE
Incidence height	$Y(1) = 0.9$ cm
Field angle	$U(1) = -20°$

8.2 Results from S1BEND

The aberration program S1BEND, using the initial data given in section 8.1 and zero bending (i.e. DALPH0 = 0.0, NALPH0 = 0) gives the following results

Focal length	$f' = 5.08$ cm
Spherical aberration	$S1 = -3.76 \times 10^{-3}$ cm

The focal length agrees with the quoted value of ≈ 5.0 cm, but we shall now have to consider the significance of the $S1$ value.

The first point to remember is that $S1$ is the difference in optical path between the marginal ray (at 20° and 0.9 cm) and the principal ray and that in an ideal optical system this should be zero. Now, superficially, $S1$ is quite small and indeed it is only about 0.075% of the focal length. (Note that the negative value means that the wave at the margin is behind the ideal spherical wave through the centre of the exit pupil.) The closeness to perfection, however, is not measured absolutely nor by comparison with the focal length but rather in terms of the wavelength of light.

Lord Rayleigh[8] was one of the first to discuss the influence of aberrations on optical image formation, and he showed that if a system is defocused then the image is expected to show no appreciable deterioration if the optical paths of the rays reaching the image region differ by less than one-quarter of a wavelength. This result is now called the Rayleigh limit and is frequently used quite generally, although it is only strictly true for a defocusing error. Subsequent work has shown that the limit can be raised and that a wavelength error of 0.95λ is tolerable for spherical aberration if some defocusing is introduced, because a better image can be found away from the Gaussian image, as shown in Figure 5a where the so-called circle of least confusion shows a sharper concentration of rays than the Gaussian image. Furthermore, if higher order spherical aberration terms are considered, that were neglected in section 5.2, then the limit can be raised to approximately 6λ.[9]

The magnitude of $S1$, i.e. 3.76×10^{-3} cm can now be looked at again in the light of these comments. For a wavelength of 590 mm, $S1$ amounts to 64λ which is on extremely large value, compared with the required value of about one wavelength. It should be said in mitigation, however, that this

Table 2. Variation of S1 with field angle and aperture

Y(1)	U(1) −20°	−10°	0°
0.9	−37.6	−11.6	9.49
0.75	−20.8	−8.31	4.57
0.625	−9.65	−5.57	2.21
0.5	−1.11	−3.17	0.90
0.375	+4.78	−1.30	0.28
0.25	+8.10	−0.07	0.056
0.125	+9.04	+0.51	0.003
0	+7.84	+0.49	0
−0.125	+4.84	+0.03	0.003
−0.25	+0.48	−0.64	0.056
−0.375	−4.75	−1.20	0.28
−0.5	−10.2	−1.25	0.90
−0.625	−15.3	−0.30	2.21
−0.75	−19.2	+2.22	4.57
−0.9	−21.1	+8.26	9.49
		$\times 10^{-4}$ cm	

value of S1 corresponds with the extreme edge of the field and the largest aperture. A more realistic appreciation of the extent of the aberration is obtained by repeating the calculation for other values of $U(1)$ and $Y(1)$.

Table 2 contains the results obtained with S1BEND by setting the bending parameter NALPHA equal to zero and using incident heights that scan the aperture, and incident angles of $-20°$, $-10°$, and $0°$. The results are shown graphically in Figure 8a from which it can be seen that the aberration falls rapidly with incidence height and approaches a sufficiently small value for apertures less than 0.625 (f/4), especially for angular fields between 0° and 10°. These curves show, in effect, the shape of the wavefront as a function of aperture and it is interesting to note, for example, that the spherical aberration for $(-20°, 0.75)$ is exactly $16\times$ that for $(-10°, 0.375)$. This follows from the fact that at each surface S1 is proportional to y, A^2, and u; in fact, for a given axial intersection point, we have: (1) $u \propto y$ and (2) $A = ni = n(u + yc) \propto y$ so that, in all, we find for each surface

$$S1 \propto y^4.$$

The proportionality does not, however, hold for the first surface as u and y can be chosen independently there, and to show the fourth-power dependence we need to halve both the initial angle of incidence and the incidence height. The curve for $U(1) = 0°$, however, follows exactly a fourth-power curve.

Typical results that show the effect of bending the lens are given, in part,

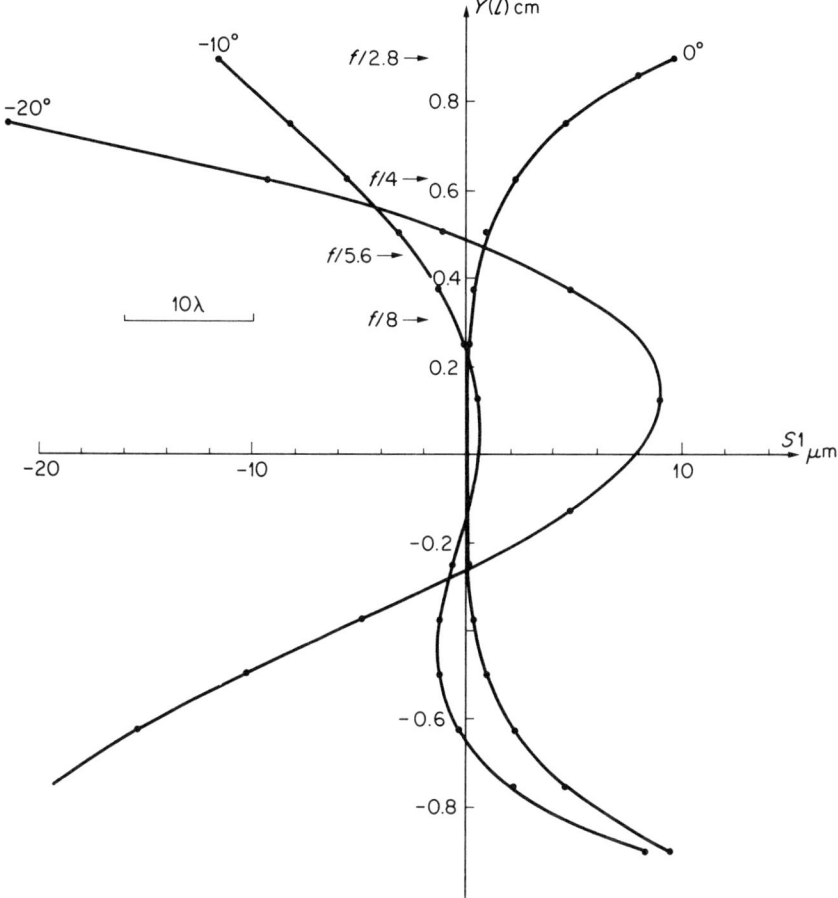

Figure 8a. Variation of $S1$ with aperture $Y(1)$ for various angles $U(1)$

in Table 3, and illustrated in Figure 8b. It is seen that the curves are roughly parabolic and this is usually the case. It is also seen that $S1$ can be exactly zero for some values of $U(1)$ and $Y(1)$, but that these do not occur at the same value of the bending parameter. From an examination of these curves and of those in Figure 8a it would appear that the TESSAR lens, as specified, is reasonably well corrected for spherical aberration but that it might be worth examining a lens as specified by $\Delta\alpha = -0.17$, $c(1) = 0.44758$ in order to see if the curves are flatter.

It is not wise to push the conclusions too far as we have considered only the effect of spherical aberration, whereas other aberrations, especially coma and astigmatism, become important as we move to larger angular fields and the minimization of these may be the more important criterion. Coma is

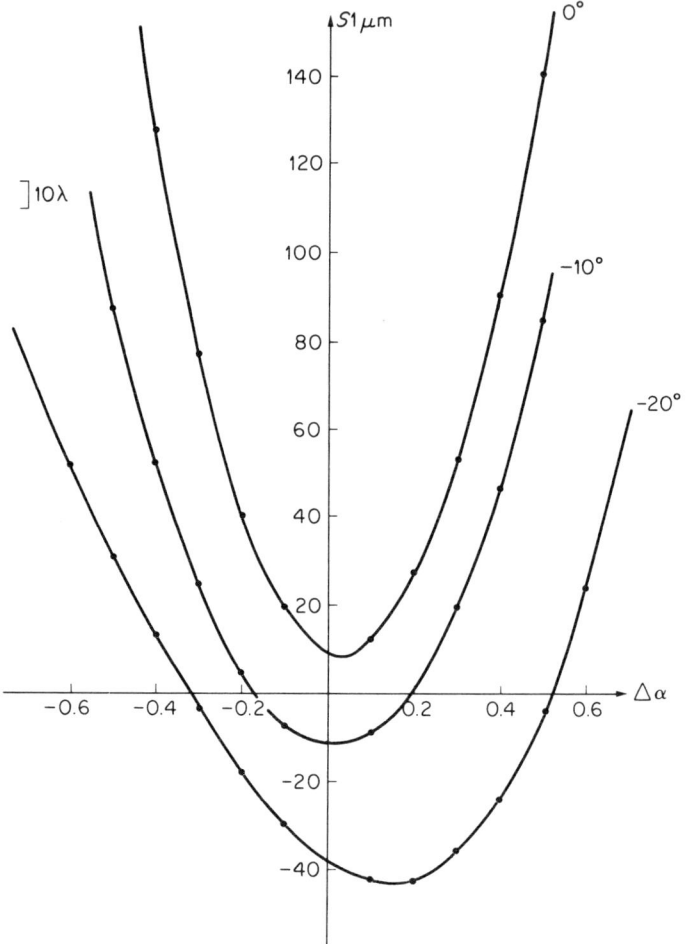

Figure 8b. Variation of $S1$ with bending parameter $\Delta\alpha$

usually a minimum near the turning point of the $S1$ parabola and indeed it can be shown that for the TESSAR it is near zero at $\Delta\alpha = 0$, ($U(1) = -20°$, $Y(1) = 0.9$).

8.3 Results from RAYTRC

Although the curves for $S1$ give a fair idea of the wavefront errors of the lens it is the pattern of light on the image plane that is of final interest, and the magnitude of the deviations from perfect imagery is realized by seeing

RAY TRACING AND LENS ABERRATIONS

Table 3. Typical results from a lens bending

```
NUMBER OF SURFACES            7

CURVATURES OF SURFACES
    0.61425E 00   -0.36271E-01   -0.28927E 00    0.63221E 00    0.00000E 00
    0.52083E 00   -0.41667E 00
REFRACTIVE INDICES
    1.00000        1.61160        1.00000        1.60530        1.00000
    1.51230        1.61160        1.00000
SEPARATIONS OF SURFACES
    0.35700        0.18900        0.08100        0.32500        0.21700
    0.39600


FIELD ANGLE IN DEGREES         0.0000
INCIDENCE HEIGHT               0.9000


INTERVALS OF ALPHA             0.1000
NUMBER OF VALUES OF ALPHA      13



        DALPHA= -0.6000

    FOCAL LENGTH              5.07986
    BACK FOCAL LENGTH         5.72729

C 1=-0.524157E-01   C 2=-0.694984E 00   C 3=-0.976354E 00
C 4=-0.151389E 00   C 5=-0.692691E 00   C 6=-0.213361E 00
C 7=-0.911931E 00


    SPHERICAL ABERRATION      0.28279E-01



        DALPHA= -0.5000

    FOCAL LENGTH              5.07986
    BACK FOCAL LENGTH         5.50726

C 1= 0.586954E-01   C 2=-0.593528E 00   C 3=-0.871236E 00
C 4=-0.338594E-01   C 5=-0.588523E 00   C 6=-0.107366E 00
C 7= 0.015075E 00


    SPHERICAL ABERRATION      0.19605E-01
```

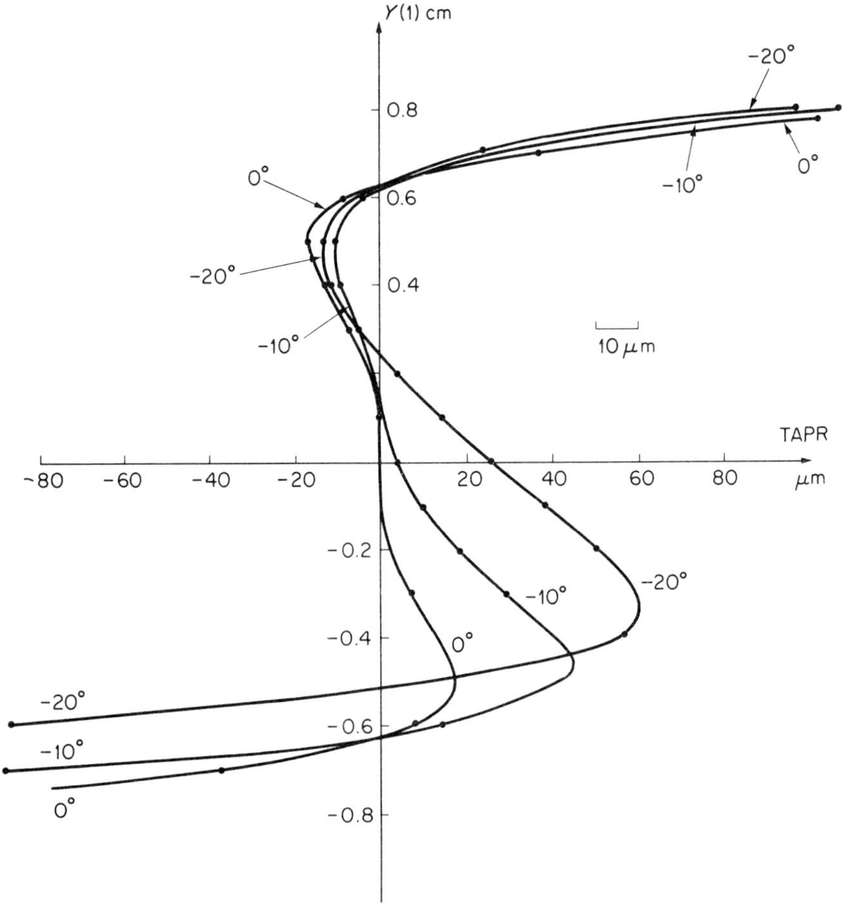

Figure 8c. Variation of the transverse aberration with aperture Y(1)

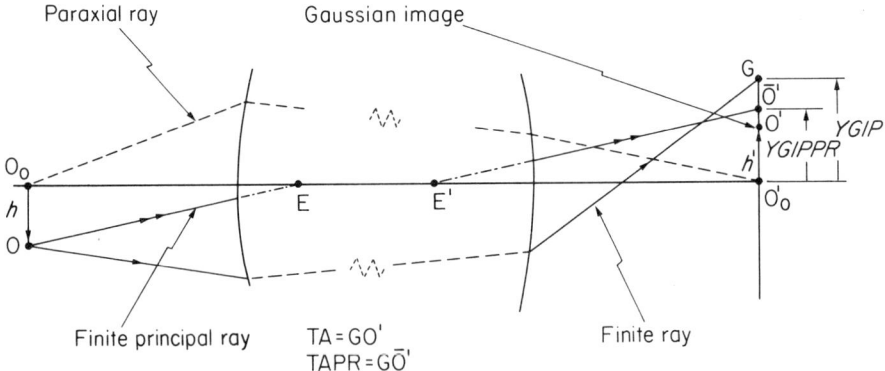

Figure 8d. Definition of transverse aberration quantities

how far the ray intersections on the image plane differ from each other and how far they deviate from the ideal image point upon which, for a perfect system, they should converge.

The program RAYTRC outputs, as a function of incident height Y(1), the transverse aberrations TA and TAPR, where

$$TA = YGIP - YPXL,$$

$$TAPR = YGIP - YGIPPR,$$

YGIP is the ray intersection on the Gaussian image plane, TA is the transverse aberration with respect to the Gaussian image of height YPXL (YPXL = h' of equation (15)), and TAPR is the transverse aberration with respect to the finite principal ray which intersects the Gaussian image plane at YGIPPR. If the finite principal ray does not intersect the image plane at the paraxial image point then the system suffers from distortion. The variables are illustrated in Figure 8d, in which O_0O is an object of height h, O'_0O' is the paraxial image of height h', $\bar{O}'$ is the finite principal ray intersection, and G is the finite ray intersection.

A typical output is shown at the end of the chapter and the results are plotted in Figure 8c. A comparison of the transverse aberrations with respect to the Gaussian image, TA, and with respect to the principal ray, TAPR, shows that they differ by about 630 μm giving a distortion of 0.63/17.5 or 3.5 per cent, which is just about acceptable at the edge of the field.

The spread about the principal ray image, i.e. the range of values of TAPR, is small between about $Y(1) = +0.65$ and $Y(1) = -0.5$ for the $-10°$ field, but only between $Y(1) = +0.65$ and $Y(1) = 0$ for the $-20°$ field. Actually as the eye, working at its least distance of distinct vision, can resolve distances of the order of 100 μm, then allowing for a magnification of 10×, ray deviations of 10 μm or less would be acceptable in the image. The system is therefore usable at f/4 provided a stop is introduced, probably near surface number 4 and made small enough to cut out the $-20°$ field rays entering the lower part of the lens. Note that no allowance is made in the program for the finite sizes of the lens elements other than the possibility that the rays miss the lens surface entirely. A more realistic trace would need the insertion of lens diameters.

9. SUGGESTIONS FOR FURTHER STUDY

The Tessar lens is a fairly complicated optical system that needs a great deal of analysis to realize its limitations and capabilities; however, it is instructive to carry out the investigation of some simpler systems using S1BEND and RAYTRC.

9.1 The single lens

It is shown, in most elementary optics texts,[10] that a single lens has minimum spherical aberration if:
(a) for an infinite object, it is nearly convexo-plane,
(b) for a finite object, it is biconvex.

These situations are illustrated in Figures 9a and 9b and the two results are readily checked by starting with a biconvex lens and bending it. Suitable starting values are given in Table 4.

It is also of interest to find how the $S1 \sim \Delta\alpha$ parabola shifts vertically as the refractive index is varied between 1.5 and 1.7, which corresponds to the range of readily available glasses.

9.2 The achromatic lens

One of the basic defects of all refracting systems is that, when they are used with white light, the refractive index varies with wavelength and hence the aberration also varies; however, the variation of $S1$ is of second order

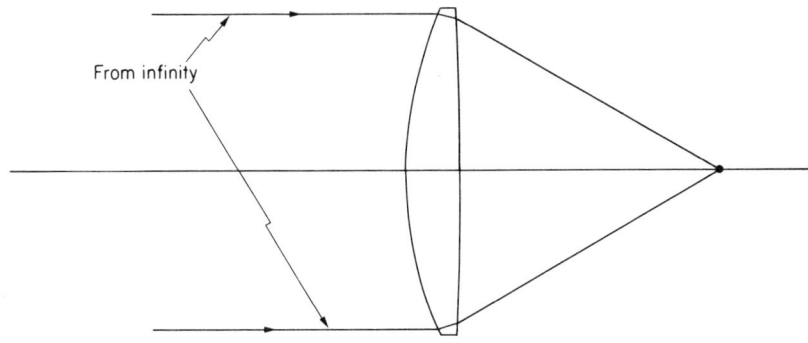

Figure 9a. Almost convexo-plane lens with minimum spherical aberration for object at infinity

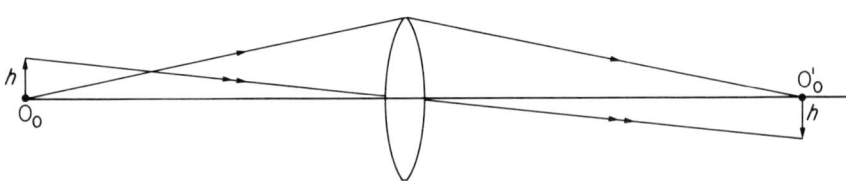

Figure 9b. Bi-convex lens with minimum spherical aberration for object imaged at unit magnification

Table 4. Data for a single lens

n	c	d
1.0		
	0.1	
1.5		0.2
	−0.1	
1.0		
Finite objects	YOBJ	1.0
	DOBJ	10.0
	Y(1)	1.0
Infinite objects	Y(1)	1.0
	U(1)	0.0
Use $\Delta\alpha = 0.05$		

compared with the variation in the position of the Gaussian image plane (and focal length) and of the size of the Gaussian image that are, respectively, the longitudinal and transverse chromatic aberrations.

The simplest way to investigate the chromatic effects is, of course, to recalculate f', $S1$, and TA using different values of n. The range of variation of n over the visible spectrum is of the order of a few per cent and is specified by the so-called V-value, where

$$V = (n_D - 1)/(n_F - n_C)$$

and C and F refer to particular red and blue lines in the spectrum of hydrogen and D to a yellow line in the spectrum of sodium, viz.:

Line	Colour	Wavelength (nm)
C	Red	656.3
D	Yellow	589.3
F	Blue	486.1

To find n_F and n_C from the given values of n_D and V we can assume that the refractive index obeys Cauchy's equation:

$$n = A + B/\lambda^2$$

from which it is easily shown that

$$B = (n_D - 1)\lambda_F^2 \lambda_C^2 / V(\lambda_C^2 - \lambda_F^2)$$

and

$$A = n_D - B/\lambda_D^2.$$

The values of V for the glasses used in the Tessar lens are given in Table 5, together with the calculated values of A, B, n_F, and n_C. A recalculation of

Table 5. Refractive index constants and values

n_D	V	A	B	n_F	n_C
1.6116	58.54	1.5959	5.4688	1.6190	1.6085
1.6053	38.03	1.5813	8.3315	1.6166	1.6007
1.5123	56.35	1.4986	4.7589 $\times 10^{-15}$	1.5187	1.5096

Table 6. Data for a telescope object glass*

n	c (cm^{-1})	d (cm)
1.0		
	0.072516	
1.6203		0.290
	−0.16388	
1.5728		0.590
	0.0079681	
1.0		
	LOGICAL = FALSE	
	$Y(1) = 3.00$	
	$U(1) = -3.00°$	

* From reference 7, page 41.

Table 7. Data for a microscope objective

n	c (cm^{-1})	d (cm)
1.0		
	1.96154	
1.572		0.08
	−2.69564	
1.620		0.025
	0.0	
1.0		
	LOGICAL = TRUE	
	$Y(1) = 0.125$ cm	
	$U(1) = 2.38°$	
	$YOBJ = 0.15$ cm	
	$DOBJ = 3.00$ cm	

the Tessar using these values can be carried out to see how it responds to different colours. A good lens should show a variation in focal length of less than 1 part in 2000.

Most elementary optics texts also discuss the realization of achromatism by combining two lenses made of glasses of high and low V-values respectively, and thus as further simple examples one can try the lenses specified in

Tables 6 and 7. These are both cemented achromatic doublets of the type first made by Dollond in 1759, which Newton concluded in 1668 were impossible to produce. The former is an object glass for a telescope and the latter is an objective for a microscope. It is interesting to note that these lenses have minimum spherical aberration when the external shape of the lens agrees with the conditions given in section 9.1.

9.3 The concave mirror

The simplest optical system of all is a single mirror. We have not so far considered reflecting systems, but they can be taken into account by using the fact that when light is reflected the beam reverses direction and Snell's law may be written as

$$\sin I' = -\sin I,$$

that is equivalent to putting

$$n' = -n \text{ or in air, } n = 1, n' = -1.$$

The spherical aberration of a concave mirror may thus be obtained by using the data

$$NS = 1$$
$$N(1) = 1.0$$
$$N(2) = -1.0$$
$$D(2) = 0.0$$

and other parameters $C(1)$, $Y(1)$ as required.

9.4 Other aberrations

The program S1BEND calculates only the amount of spherical aberration $S1$ but, as mentioned in section 8.2, there are other aberrations that may be present, of which the most important are coma, $S2$, and astigmatism, $S3$. Coma appears, as shown in Figure 5c, as a difference in focusing position, on the principal ray, for rays above and below the principal ray, whilst astigmatism is a difference in focus for rays in the meridian plane and those in a plane perpendicular to the meridian. Both aberrations can be calculated if the parameters for the paraxial principal ray are known, in particular if $\bar{A}$ (corresponding to A for the paraxial marginal ray) is known at each surface.

The expressions for wavefront coma $S2$ and astigmatism $S3$ are

$$S2 = \tfrac{1}{2}A\bar{A}y\Delta(u/n)$$
$$S3 = \tfrac{1}{2}\bar{A}^2 y\Delta(u/n).$$

The proofs of these expressions are given by Welford,[11] and they may be incorporated into the program by adding the principal ray quantities $\bar{y}$ and $\bar{u}$ to the main program and to the subroutine PARAXL (as COMMON variables) and calculating at each surface

$$\bar{A} = n\bar{u} + n\bar{y}c$$

and thence $S2$ and $S3$. For any object the initial value of $\bar{y}(1)$ is zero at the entrance pupil, whose position may need calculating, as is done in RAYTRAC. For a finite object $\bar{u}(1) = -\text{YOBJ/DOBJ}$, for an infinite object $\bar{u}(1) = u(1)$.

REFERENCES

1. H. A. Buchdahl, *An Introduction to Hamiltonian Optics* (University Press, Cambridge), 3 (1970).
2. M. Born and E. Wolf, *Principles of Optics*, 3rd ed. (Pergamon Press, Oxford), 147 (1965).
3. Born and Wolf, *Principles*, 203, also L. Seidel, *Astr. Nachr.*, **43,** 289, 305, 321 (1856).
4. W. T. Welford, *Aberrations of the Symmetrical Optical System* (Academic Press, London), 111 (1974).
5. H. H. Hopkins, *Wave Theory of Aberrations* (Clarendon Press, Oxford), (1950).
6. Welford, *Aberrations*, 49.
7. E. L. O'Neill, *Introduction to Statistical Optics* (Addison-Wesley, Mass.), 43 (1963).
8. Lord Rayleigh, *Phil. Mag.* **8,** 403 (1879).
9. J. M. Palmer, *Lens Aberration Data* (Hilger, Bristol), 56 *et seq.* (1971).
10. F. A. Jenkins and H. E. White, *Fundamentals of Optics*, 4th edn., (McGraw-Hill, New York), 153 (1976).
11. Welford, *Aberrations*, Ch. 7.

S1BEND: PROGRAM TO CALCULATE SPHERICAL ABERRATION

```
C THIS PROGRAM CALCULATES THE FOCAL-LENGTH AND THE INITIAL WAVEFRONT
C  SPHERICAL ABERRATION S1 OF AN OPTICAL SYSTEM. IT ALSO CALCULATES
C  VIA THE SUBROUTINE ALPHA0 THE INITIAL VALUES OF ALPHA (ALPHA=N*I)
C  AND THEN VARIES ALPHA BY +OR- NALPHA*DALPH0 AND RECALCULATES
C  THE ABERRATION
C AS AN ALTERNATIVE IF NALPHA IS SET EQUAL TO ZERO THEN THE VARIATION
C  OF S1 WITH INCIDENCE HEIGHT Y(1) IS DETERMINED FOR VALUES OF Y(1)
C  THAT DIFFER BY SUCCESSIVE INTERVALS OF DELY,WHERE DELY IS SET EQUAL
C  TO THE VALUE OF DALPH0
C THE DATA SHOULD BE READ IN THE FOLLOWING ORDER:
C          FINITE          (FINITE IS TRUE FOR A FINITE OBJECT,
C                           FINITE IS FALSE FOR AN INFINITE OBJECT)
C          NS              (THE NUMBER OF SURFACES)
C          C(J) J=1,NS     (THE SURFACE CURVATURES)
C          N(J) J=1,NS+1   (THE REFRACTIVE INDICES)
C          D(J) J=2,NS     (THE SURFACE SEPARATIONS)
C          Y(1)            (THE INITIAL INCIDENCE HEIGHT)
C          U(1)            (THE INITIAL INCIDENCE ANGLE,IN DEGREES,
C                           U(1) IS POSITIVE FOR A FINITE OBJECT
C                           U(1) IS NEGATIVE FOR AN INFINITE OBJECT)
C          DALPH0,NALPHA   (DALPH0 IS THE INTERVAL BETWEEN SUCCESSIVE
C                           VALUES OF ALPHA AND 2*NALPHA+1 IS THE
C                           TOTAL NUMBER OF VALUES OF ALPHA)
C THE PROGRAM USES THE FOLLOWING SUBROUTINES:
C          ALPHA0  TO CALCULATE THE INITIAL VALUES OF ALPHA
C          FOCUS   TO CALCULATE THE FOCAL LENGTH AND THE
C                   NEW VALUES OF C(J)
C          ABERRN  TO CALCULATE THE WAVEFRONT SPHERICAL ABERRATION
       LOGICAL FINITE
       REAL N
       COMMON NS,N,C,D,U,Y,DALPHA,NALPHA,ALPHA
       DIMENSION N(20),C(20),D(20),U(20),Y(20),ALPHA(20)
       WRITE(2,303)
       READ(1,100) FINITE
       READ(1,101) NS
       WRITE(2,201) NS
       NSP=NS+1
C READ AND WRITE THE SYSTEM PARAMETERS
       READ(1,102) (C(J),J=1,NS)
       WRITE(2,209)
       WRITE(2,202) (C(J),J=1,NS)
       READ(1,102) (N(J),J=1,NSP)
       WRITE(2,210)
       WRITE(2,205) (N(J),J=1,NSP)
       READ(1,102) (D(J),J=2,NS)
       WRITE(2,211)
       WRITE(2,205) (D(J),J=2,NS)
       WRITE(2,301)
C READ AND WRITE THE RAY PARAMETERS
       READ(1,102) Y(1)
       READ(1,102) U(1)
       IF (.NOT.FINITE) GOTO 2
       WRITE(2,204) U(1),Y(1)
       GOTO 5
     2 CONTINUE
       WRITE(2,200) U(1),Y(1)
```

S1BEND: PROGRAM TO CALCULATE SPHERICAL ABERRATION

```
    5 CONTINUE
      IF(ABS(Y(1)).LE.0.0001) Y(1)=0.0001
      PI=3.1415926536
      U(1)=U(1)*PI/180.0
C READ AND WRITE THE VARIATION DALPH0 AND THE NUMBER NALPHA
C  OF VARIATIONS ON EACH SIDE OF THE STARTING POINT
      READ(1,103) DALPH0,NALPHA
      IF(NALPHA.EQ.0) GOTO 23
      NALPH2=2*NALPHA+1
      WRITE(2,207) DALPH0,NALPH2
      CALL ALPHA0
       DO 20 NP=1,NALPH2
      NPM=NP-NALPHA-1
      DALPHA=DALPH0*FLOAT(NPM)
      WRITE(2,208) DALPHA
C NOW CHANGE ALL THE VALUES OF ALPHA
         DO 21 J=1,NS
      ALPHA(J)=ALPHA(J)+DALPHA
   21    CONTINUE
      CALL FOCUS
      CALL ABERRN
C NOW CHANGE ALPHA BACK AGAIN
         DO 22 J=1,NS
      ALPHA(J)=ALPHA(J)-DALPHA
   22    CONTINUE
   20 CONTINUE
      GOTO 29
C IF NALPHA=0 CALCULATE THE VARIATION OF S1 WITH Y(1)
   23 IF(Y(1).LT.0.0) Y(1)=-Y(1)
C STORE THE STARTING VALUE OF Y(1)
      Y1=Y(1)
      CALL FOCUS
      WRITE(2,206)
      DELY=DALPH0
      Y(1)=Y(1)+DELY
   24 Y(1)=Y(1)-DELY
      IF(ABS(Y(1)).LT.0.0001) GOTO 25
      IF(Y(1).LT.0.0) GOTO 27
      CALL ABERRN
      GOTO 24
   25 Y(1)=0.0
      CALL ABERRN
   26 Y(1)=Y(1)-DELY
      CALL ABERRN
      IF(ABS(Y(1)+Y1).LT.0.0001) GOTO 29
      GOTO 26
   27 Y(1)=Y(1)+DELY
      Y(1)=-Y(1)
      CALL ABERRN
   28 Y(1)=Y(1)-DELY
      CALL ABERRN
      IF(ABS(Y(1)+Y1).LT.0.0001) GOTO 29
      GOTO 28
   29 CONTINUE
      STOP
  100 FORMAT(L5)
```

S1BEND: PROGRAM TO CALCULATE SPHERICAL ABERRATION

```
  101 FORMAT(I2)
  102 FORMAT(5F10.0)
  103 FORMAT(F8.0,I2)
  200 FORMAT(23H FIELD ANGLE IN DEGREES,F16.4/1X,16HINCIDENCE HEIGHT,F22
     8.4//)
  201 FORMAT(1X,19H NUMBER OF SURFACES,8X,I2/)
  202 FORMAT(E16.5,4E14.5)
  204 FORMAT(27H INCIDENCE ANGLE IN DEGREES,F12.4/17H INCIDENCE HEIGHT
     8,F22.4/)
  205 FORMAT(F12.5,4F14.5)
  206 FORMAT(6X,8HINCIDENT,8X,9HSPHERICAL/6X,8H HEIGHT ,8X,10HABERRATION
     8/)
  207 FORMAT(19H INTERVALS OF ALPHA,F20.4/26H NUMBER OF VALUES OF ALPHA,
     8I8///)
  208 FORMAT(8X,7HDALPHA=,F8.4/)
  209 FORMAT(1X,23H CURVATURES OF SURFACES)
  210 FORMAT(1X,19H REFRACTIVE INDICES)
  211 FORMAT(1X,24H SEPARATIONS OF SURFACES)
  301 FORMAT(1X/)
  303 FORMAT(1X///)
      END
            SUBROUTINE ALPHA0
C THIS SUBROUTINE SETS UP THE INITIAL VALUES OF ALPHA
      REAL K,N
      COMMON NS,N,C,D,U,Y,DALPHA,NALPHA,ALPHA
      DIMENSION N(20),C(20),D(20),U(20),Y(20),ALPHA(20),UF(20),YF(20)
      UF(1)=0.0
      YF(1)=Y(1)
      J=1
   30 ALPHA(J)=YF(J)*C(J)
      IF (J.EQ.NS) GOTO 31
      K=(N(J+1)-N(J))*C(J)
      UF(J+1)=(N(J)*UF(J)-YF(J)*K)/N(J+1)
      YF(J+1)=YF(J)+D(J+1)*UF(J+1)
      J=J+1
      IF (J.LE.NS) GOTO 30
   31 CONTINUE
      RETURN
      END
            SUBROUTINE FOCUS
C THIS SUBROUTINE CALCULATES THE FOCAL LENGTH FL AND
C   THE NEW VALUES OF C(J)
      REAL K,N
      COMMON NS,N,C,D,U,Y,DALPHA,NALPHA,ALPHA
      DIMENSION N(20),C(20),D(20),U(20),Y(20),ALPHA(20)
      IF(NALPHA.EQ.0) GOTO 41
      C(1)=ALPHA(1)/Y(1)
   41 K=(N(2)-N(1))*C(1)
      U(2)=-Y(1)*K/N(2)
        DO 40 J=2,NS
      Y(J)=Y(J-1)+D(J)*U(J)
      IF(NALPHA.NE.0)  C(J)=ALPHA(J)/Y(J)
      K=(N(J+1)-N(J))*C(J)
      U(J+1)=(N(J)*U(J)-Y(J)*K)/N(J+1)
   40   CONTINUE
      FL=-Y(1)/U(NS+1)
```

S1BEND: PROGRAM TO CALCULATE SPHERICAL ABERRATION   PAGE -

```
      BFL=-Y(NS)/U(NS+1)
      WRITE(2,400) FL
      WRITE(2,401) BFL
  400 FORMAT(3X,13H FOCAL LENGTH,F22.5)
  401 FORMAT(3X,18H BACK FOCAL LENGTH,F17.5/)
      RETURN
      END
           SUBROUTINE ABERRN
C THIS SUBROUTINE PERFORMS A PARAXIAL RAYTRACE AND
C  CALCULATES THE WAVEFRONT SPHERICAL ABERRATION
      REAL K,N,NNP
      COMMON NS,N C,D,U,Y,DALPHA,NALPHA,ALPHA
      DIMENSION N(20),C(20),D(20),U(20),Y(20),ALPHA(20)
      DATA CC/1HC/
      S1=0.0
      J=1
C REFRACT
   50 CONTINUE
      K=(N(J+1)-N(J))*C(J)
      U(J+1)=(N(J)*U(J)-Y(J)*K)/N(J+1)
C CALCULATE THE ABERRATION
      A=N(J)*Y(J)*C(J)+N(J)*U(J)
      NNP=N(J+1)*N(J)
      DELUN=(N(J)*U(J+1)-N(J+1)*U(J))/NNP
      S1=S1-0.125*A*A*Y(J)*DELUN
      IF(J.EQ.NS) GOTO 55
C TRANSFER TO THE NEXT SURFACE
      Y(J+1)=Y(J)+D(J+1)*U(J+1)
      J=J+1
      IF(J.LE.NS) GOTO 50
   55 CONTINUE
      IF(NALPHA.EQ.0) GOTO 57
      WRITE(2,500) (CC,J,C(J),J=1,NS)
      WRITE(2,501)
      GOTO 58
   57 WRITE(2,503) Y(1),S1
      GOTO 59
   58 WRITE(2,502) S1
   59 CONTINUE
      RETURN
  500 FORMAT(3(1H ,A1,I2,1H=,E13.6,2X))
  501 FORMAT(1H /)
  502 FORMAT(3X,21H SPHERICAL ABERRATION,E18.5///)
  503 FORMAT(F13.4,E20.5)
      END
```

TYPICAL RESULTS OF A CALCULATION OF S1 AS A FUNCTION OF Y(1)

NUMBER OF SURFACES   7

CURVATURES OF SURFACES
   0.61425E 00   -0.36271E-01   -0.28927E 00   0.63221E 00   0.00000E 20
   0.52083E 00   -0.41667E 00
REFRACTIVE INDICES
   1.00000   1.61160   1.00000   1.60530   1.00000
   1.51230   1.61160   1.00000
SEPARATIONS OF SURFACES
   0.35700   0.18900   0.08100   0.32500   0.21700
   0.39600

FIELD ANGLE IN DEGREES   -20.0000
INCIDENCE HEIGHT   0.9000

   FOCAL LENGTH   5.07906
   BACK FOCAL LENGTH   4.40715

```
   INCIDENT         SPHERICAL
   HEIGHT          ABERRATION

    0.9000        -0.37623E-02
    0.8000        -0.26029E-02
    0.7000        -0.16042E-02
    0.6000        -0.77258E-03
    0.5000        -0.11107E-03
    0.4000         0.38084E-03
    0.3000         0.70713E-03
    0.2000         0.87522E-03
    0.1000         0.89604E-03
    0.0000         0.78394E-03
   -0.1000         0.55680E-03
   -0.2000         0.23593E-03
   -0.3000        -0.15398E-03
   -0.4000        -0.58436E-03
   -0.5000        -0.10239E-02
   -0.6000        -0.14369E-02
   -0.7000        -0.17851E-02
   -0.8000        -0.20262E-02
   -0.9000        -0.21146E-02
```

RAYTRC: PROGRAM TO PERFORM A FINITE RAY TRACE

```
C THIS PROGRAM TRACES A FINITE MERIDIAN RAY THROUGH AN
C OPTICAL SYSTEM AND CALCULATES ITS TRANVERSE ABERRATION
C ON THE GAUSSIAN IMAGE PLANE AS
C     1) TA    WITH RESPECT TO THE GAUSSIAN IMAGE
C     2) TAPR  WITH RESPECT TO THE PRINCIPAL RAY
C THE PROGRAM USES THE SUBROUTINES:
C     FOCUS    TO FIND THE FOCAL LENGTH FL AND THE BACK FOCAL
C              LENGTH BFL
C     PARAXL   TO FIND,FOR A FINITE OBJECT,THE GAUSSIAN IMAGE
C              PLANE GIP AND THE PARAXIAL IMAGE SIZE YPXL
C              (FOR AN INFINITE OBJECT GIP=BFL AND
C                              YPXL=FL*FIELD ANGLE)
C     PUPIL    TO FIND THE EXIT PUPIL DISTANCE EXPP TO THE RIGHT
C              OF THE FIRST SURFACE AND THE INCIDENT HEIGHT OF
C              THE PRINCIPAL RAY ON THE FIRST SURFACE
C     TRACE    TO TRACE THE FINITE RAYS
C THE DATA SHOULD BE READ IN THE FOLLOWING ORDER:
C     FINITE         (FINITE IS TRUE FOR A FINITE OBJECT,
C                     FINITE IS FALSE FOR AN INFINITE OBJECT)
C     NS             (THE NUMBER OF SURFACES)
C     C(J) J=1,NS    (THE SURFACE CURVATURES)
C     N(J) J=1,NS+1  (THE REFRACTIVES INDICES)
C     D(J) J=2,NS    (THE SURFACE SEPARATIONS)
C     Y(1)           (THE MAXIMUM  INCIDENCE HEIGHT ON THE
C                     FIRST SURFACE)
C     U(1)           (THE FIELD ANGLE IN DEGREES-INCLUDE FOR AN
C                     INFINITE OBJECT-OMIT FOR A FINITE OBJECT)
C     YOBJ,DOBJ      (THE OBJECT HEIGHT AND ITS DISTANCE
C                     TO THE LEFT OF THE FIRST SURFACE
C                     -OMIT FOR AN INFINITE OBJECT)
C     NR             (THE NUMBER OF RAYS TO BE TRACED)
C     DAP,JP         (THE DISTANCE DAP OF THE APERTURE STOP
C                     TO THE RIGHT OF SURFACE NUMBER JP)
      REAL N,K
      COMMON NS,N,C,D,U,Y,FL,BFL,GIP,YGIP,YPXL,YOBJ,DOBJ,EXPP
      DIMENSION N(21),C(20),D(19),U(21),Y(20)
      LOGICAL FINITE
      PI=3.1415926536
      READ(1,100) FINITE
      READ(1,101) NS
      WRITE(2,201) NS
C READ AND WRITE THE SYSTEM PARAMETERS
      NSP=NS+1
      READ(1,102) (C(J),J=1,NS)
      READ(1,102) (N(J),J=1,NSP)
      READ(1,102) (D(J),J=2,NS)
      WRITE(2,202) (C(J),J=1,NS)
      WRITE(2,203) (N(J),J=1,NSP)
      WRITE(2,204) (D(J),J=2,NS)
C READ THE MAXIMUM INCIDENT HEIGHT Y(1) ON THE TANGENT
C PLANE AT THE FIRST SURFACE
      READ(1,102) Y(1)
      IF (FINITE) GOTO 10
C FOR AN INFINITE OBJECT READ AND WRITE THE FIELD ANGLE U(1) IN DEGREES
      READ(1,102) U(1)
      WRITE(2,206) U(1)
```

RAYTRC: PROGRAM TO PERFORM A FINITE RAY TRACE   PAGE -

```
      U(1)=U(1)*PI/180.0
      GOTO 20
C FOR A FINITE OBJECT,READ THE OBJECT HEIGHT YOBJ AND
C ITS DISTANCE DOBJ  FROM THE FIRST SURFACE
   10 READ(1,102) YOBJ,DOBJ
      WRITE(2,205) YOBJ,DOBJ
C READ THE NUMBER NR OF RAYS TO BE TRACED
   20 READ(1,101) NR
      NRM=NR-1
      DELY=2.0*Y(1)/FLOAT(NRM)
C CALCULATE THE FOCAL LENGTH
      CALL FOCUS
      GIP=BFL
      IF(.NOT.FINITE) YPXL=FL*U(1)
      IF(FINITE) CALL PARAXL
C TO FIND THE PRINCIPAL RAY HEIGHT AT THE FIRST SURFACE
C FIRST STORE THE INITIAL VALUE OF Y(1) AS Y0
      Y0=Y(1)
      CALL PUPIL
      Y(1)=-EXPP*U(1)
      IF(FINITE) Y(1)=YOBJ*EXPP/(DOBJ+EXPP)
      IF(FINITE) U(1)=-YOBJ/(DOBJ+EXPP)
C TRACE THE PRINCIPAL RAY
      CALL TRACE
      YGIPPR=YGIP
C CALCULATE THE TRANSVERSE ABERRATIONS TA AND TAPR FOR
C  VALUES OF Y(1) THAT SCAN THE APERTURE IN NR STEPS STARTING
C  WITH THE INITIAL VALUE OF Y(1)
C FIRST RETRIEVE THE INITIAL VALUE OF Y(1) FROM Y0
      Y(1)=Y0
      WRITE(2,301)
      Y(1)=Y(1)+DELY
         DO 25 I=1,NR
      Y(1)=Y(1)-DELY
      IF(FINITE) U(1)=ATAN((Y(1)-YOBJ)/DOBJ)*PI/180.0
      CALL TRACE
      TA=YGIP-YPXL
      TAPR=YGIP-YGIPPR
      WRITE(2,207) Y(1),TA,TAPR
   25 CONTINUE
      STOP
  100 FORMAT(L5)
  101 FORMAT(I2)
  102 FORMAT(5F10.0)
  201 FORMAT(1H ///6X,18HNUMBER OF SURFACES,I5/)
  202 FORMAT(11X,10HCURVATURES/5(5F12.5/))
  203 FORMAT(11X,18HREFRACTIVE INDICES/5(5F12.5/))
  204 FORMAT(11X,11HSEPARATIONS/5(5F12.5/))
  205 FORMAT(1H /6X,13HOBJECT HEIGHT,F20.3/6X,15HOBJECT DISTANCE,F18.3/)
  206 FORMAT(1H /6X,11HFIELD ANGLE,F22.3,9H DEGREES/)
  207 FORMAT(1X,F14.4,E19.4,E18.4)
  301 FORMAT(1H /8X,8HINCIDENT,8X,10HTRANSVERSE,8X,10HT.A.  FROM/8X,
     *8H HEIGHT ,8X,10HABERRATION,6X,14H PRINCIPAL RAY/)
      END
            SUBROUTINE PARAXL
C THIS SUBROUTINE TRACES A PARAXIAL RAY FROM AN AXIAL POINT
```

RAYTRC: PROGRAM TO PERFORM A FINITE RAY TRACE

```
C  ON A FINITE OBJECT AND CALCULATES THE POSITION GIP AND
C  THE SIZE YPXL OF THE GAUSSIAN IMAGE
      REAL K,N
      COMMON NS,N,C,D,U,Y,FL,BFL,GIP,YGIP,YPXL,YOBJ,DOBJ
      DIMENSION N(21),C(20),D(19),U(21),Y(20)
      U(1)=Y(1)/DOBJ
         DO 30 J=1,NS
      K=(N(J+1)-N(J))*C(J)
      U(J+1)=(N(J)*U(J)-Y(J)*K)/N(J+1)
      IF (J.EQ.NS) GOTO 35
      Y(J+1)=Y(J)+D(J+1)*U(J+1)
   30    CONTINUE
   35 GIP=-Y(NS)/U(NS+1)
      YPXL=-(GIP-BFL)*YOBJ/FL
      RETURN
      END
              SUBROUTINE FOCUS
C THIS SUBROUTINE CALCULATES THE FOCAL LENGTH FL
C  AND THE BACK FOCAL LENGTH BFL
      REAL K,N
      COMMON NS,N,C,D,U,Y,FL,BFL
      DIMENSION  C(20),N(21),D(19),U(21),Y(20)
      K=(N(2)-N(1))*C(1)
      IF(Y(1).EQ.0.0) Y(1)=1.0
      U(2)=-Y(1)*K/N(2)
         DO 40 J=2,NS
      Y(J)=Y(J-1)+D(J)*U(J)
      K=(N(J+1)-N(J))*C(J)
      U(J+1)=(N(J)*U(J)-Y(J)*K)/N(J+1)
   40    CONTINUE
      FL=-Y(1)/U(NS+1)
      BFL=-Y(NS)/U(NS+1)
      WRITE(2,401) FL
      WRITE(2,402) BFL
  401 FORMAT(1H0,5X,12HFOCAL LENGTH,F22.4)
  402 FORMAT(6X,17HBACK FOCAL LENGTH,F17.4/)
      RETURN
      END
              SUBROUTINE PUPIL
C THIS SUBROUTINE CALCULATES THE POSITION OF THE EXIT PUPIL
C  FROM THE KNOWN POSITION OF THE APERTURE STOP
      REAL KP,NP,N
      COMMON NS,N,C,D,U,Y,FL,BFL,GIP,YGIP,YPXL,YOBJ,DOBJ,EXPP
      DIMENSION C(20),N(21),D(19),U(21),Y(20),CP(20),NP(21),DP(20),
     *UP(21),YP(20)
C READ AND WRITE THE DISTANCE DAP OF THE APERTURE STOP
C  TO THE RIGHT OF SURFACE NUMBER JP
      READ(1,103) DAP,JP
      WRITE(2,209) DAP,JP
C SET UP VARIABLES FOR A REVERSE RAY TRACE FROM THE STOP
C  NOTE THAT THE SIGNS OF THE CURVATURES ARE REVERSED
      M=0
   50 M=M+1
      JP2M=JP+2-M
      NP(M)=N(JP2M)
      IF(M.EQ.JP+1) GOTO 55
```

RAY TRACING AND LENS ABERRATIONS 43

RAYTRC: PROGRAM TO PERFORM A FINITE RAY TRACE

```
      JP1M=JP+1-M
      CP(M)=-C(JP1M)
      IF(M.EQ.JP) GOTO 50
      DP(M+1)=D(JP1M)
      GOTO 50
   55 CONTINUE
C PERFORM A PARAXIAL RAYTRACE TO FIND THE EXIT PUPIL
      IF (ABS(DAP).GE.1.0E-4) GOTO 60
      UP(1)=0.1
      YP(1)=0.0
      GOTO 65
   60 YP(1)=0.1
      UP(1)=YP(1)/DAP
   65 CONTINUE
      DO 70 J=1,JP
      KP=(NP(J+1)-NP(J))*CP(J)
      UP(J+1)=(NP(J)*UP(J)-YP(J)*KP)/NP(J+1)
      IF(J.EQ.JP) GOTO 75
      YP(J+1)=YP(J)+DP(J+1)*UP(J+1)
   70 CONTINUE
   75 EXPP=YP(JP)/UP(JP+1)
C EXPP IS THE EXIT PUPIL PLANE
  103 FORMAT(F8.0,I2)
  209 FORMAT(6X,22HAPERTURE STOP DISTANCE,F8.4,22H TO THE RIGHT OF SURFA
     *,6HCE NO.,I2/)
      RETURN
      END
            SUBROUTINE TRACE
      REAL N,K
      COMMON NS,N,C,D,U,Y,FL,BFL,GIP,YGIP,YPXL,YOBJ,DOBJ
      DIMENSION N(21),C(20),D(19),U(21),Y(20)
C SET UP THE INITIAL VALUES AT THE FIRST SURFACE
      DD=0.0
      Z=0.0
      UU=U(1)
      YY=Y(1)
C DEFINE DIRECTION COSINES DCM AND DCN
      DCM=SIN(UU)
      DCN=COS(UU)
      J=1
C SET UP DUMMY VARIABLES FOR THE LOOP
   85 CC=C(J)
      RN=N(J)
      RNP=N(J+1)
C        TRANSFER
      YY=YY+(DD-Z)*DCM/DCN
      F=CC*YY*YY
      G=DCN-CC*YY*DCM
      COSISQ=G*G-CC*F
      IF(ABS(Y(1)).LE.1.0E-8.AND.ABS(DCM).LE.1.0E-8) GOTO 86
      IF(COSISQ.GE.0.0.AND.COSISQ.LE.1.0) GOTO 87
      YPXL=0.0
      YGIP=9999.0
      WRITE(2,220) J,Y(1)
      GOTO 80
   86 YPXL=0.0
```

RAYTRC: PROGRAM TO PERFORM A FINITE RAY TRACE

```
      YGIP=0.0
      GOTO 80
   87 COSI=SQRT(COSISQ)
      IF(F.EQ.0.0) D0=0.0
      IF(F.EQ.0.0) GOTO 88
      D0=F/(G+COSI)
   88 YY=YY+D0*DCM
      Z=D0*DCN
C     REFRACT
      SINI=SQRT(1.0-COSI*COSI)
      SINIP=RN*SINI/RNP
      IF(SINIP.LE.1.0) GOTO 90
      YPXL=0.0
      YGIP=9999.0
      WRITE(2,221) J,Y(1)
      GOTO 80
   90 COSIP=SQRT(1.0-SINIP*SINIP)
      K=RNP*COSIP-RN*COSI
      DCM=(RN*DCM-YY*K*CC)/RNP
      DCN=(RN*DCN-Z*K*CC+K)/RNP
      IF(J.EQ.NS) GOTO 95
      J=J+1
      DD=D(J)
      GOTO 85
   95 YGIP=YY+(GIP-Z)*DCM/DCN
   80 RETURN
  220 FORMAT(/1X,38HCOSISQ IS OUT OF BOUNDS AT SURFACE NO.,I2,4X,5HY(1)=
     *,F8.4)
  221 FORMAT(/1X,38HCRITICAL ANGLE EXCEEDED AT SURFACE NO.,I2,4X,5HY(1)=
     *,F8.4)
      END
```

TYPICAL RESULTS FROM RAYTRC

```
NUMBER OF SURFACES    7
    CURVATURES
0.61425    -0.03627   -0.28927    0.63221    0.00000
0.52083    -0.41667
    REFRACTIVE INDICES
1.00000    1.61160    1.00000    1.60530    1.00000
1.51230    1.61160    1.00000
    SEPARATIONS
0.35700    0.18900    0.08100    0.32500    0.21700
0.39600

FIELD ANGLE              -20.000  DEGREES

FOCAL LENGTH              5.0799
BACK FOCAL LENGTH         4.4072

APERTURE STOP DISTANCE  0.1000 TO THE RIGHT OF SURFACE NO. 4


    INCIDENT      TRANSVERSE      T.A.  FROM
    HEIGHT        ABERRATION      PRINCIPAL RAY

    0.9000        -0.3759E-01     0.2549E-01
    0.8000        -0.5342E-01     0.9660E-02
    0.7000        -0.6067E-01     0.2415E-02
    0.6000        -0.6361E-01    -0.5249E-03
    0.5000        -0.6441E-01    -0.1333E-02
    0.4000        -0.6421E-01    -0.1126E-02
    0.3000        -0.6354E-01    -0.4608E-03
    0.2000        -0.6266E-01     0.4163E-03
    0.1000        -0.6166E-01     0.1423E-02
   -0.0000        -0.6053E-01     0.2551E-02
   -0.1000        -0.5929E-01     0.3787E-02
   -0.2000        -0.5805E-01     0.5028E-02
   -0.3000        -0.5714E-01     0.5936E-02
   -0.4000        -0.5741E-01     0.5673E-02
   -0.5000        -0.6078E-01     0.2301E-02
   -0.6000        -0.7179E-01    -0.8710E-02
   -0.7000        -0.1018E 00    -0.3877E-01
   -0.8000        -0.1839E 00    -0.1208E 00

CRITICAL ANGLE EXCEEDED AT SURFACE NO. 4    Y(1)= -0.9000
   -0.9000         0.9999E 04     0.1000E 05
```

TYPICAL RESULTS FROM RAYTRC

NUMBER OF SURFACES   7

CURVATURES
0.61425   -0.03627   -0.28927   0.63221   0.00000
0.52083   -0.41667
REFRACTIVE INDICES
1.00000   1.61160   1.00000   1.60530   1.00000
1.51230   1.61160   1.00000
SEPARATIONS
0.35700   0.18900   0.08100   0.32500   0.21700
0.39600

FIELD ANGLE   0.000 DEGREES

FOCAL LENGTH   5.0799
BACK FOCAL LENGTH   4.4072

APERTURE STOP DISTANCE 0.1000 TO THE RIGHT OF SURFACE NO. 4

INCIDENT HEIGHT	TRANSVERSE ABERRATION	T.A. FROM PRINCIPAL RAY
0.9000	0.5122E-01	0.5122E-01
0.8000	0.1705E-01	0.1705E-01
0.7000	0.3689E-02	0.3689E-02
0.6000	-0.8162E-03	-0.8162E-03
0.5000	-0.1685E-02	-0.1685E-02
0.4000	-0.1283E-02	-0.1283E-02
0.3000	-0.6604E-03	-0.6604E-03
0.2000	-0.2184E-03	-0.2184E-03
0.1000	-0.2888E-04	-0.2888E-04
-0.0000	0.0000E 00	0.0000E 00
-0.1000	0.2888E-04	0.2888E-04
-0.2000	0.2184E-03	0.2184E-03
-0.3000	0.6604E-03	0.6604E-03
-0.4000	0.1283E-02	0.1283E-02
-0.5000	0.1685E-02	0.1685E-02
-0.6000	0.8162E-03	0.8162E-03
-0.7000	-0.3689E-02	-0.3689E-02
-0.8000	-0.1705E-01	-0.1705E-01
-0.9000	-0.5122E-01	-0.5122E-01

Physics Programs
Edited by A. D. Boardman
© 1980 John Wiley & Sons Ltd.

CHAPTER 2

Attenuated Total Reflection Analysis of Surface Polaritons

G. C. AERS and A. D. BOARDMAN

1. INTRODUCTION

The internal degrees of freedom of a medium are, generally, excited by the passage through it of an electromagnetic wave. In fact, because of the medium, the electromagnetic wave becomes a new type of wave in which the original electromagnetic field is modified by the induced polarization of the medium. This new coupled mode of excitation is known as a polariton,[1] the exact nature of which is further specified according to which elementary excitation is involved. For example, a photon coupled to the elementary excitation of an electron plasma (as in the case of a metal or semiconductor) is called a plasmon-polariton. A photon coupled to the lattice vibrations in a crystal is called a phonon-polariton, and so on.

The study of such coupled modes provides useful information about the characteristic quantities used to describe the medium. One such important quantity is the dielectric tensor function ε which relates the displacement vector $\mathbf{D}$ in the medium to the electric field $\mathbf{E}$. This relationship may be written, for an isotropic medium, as $\mathbf{D} = \varepsilon_o \varepsilon \mathbf{E}$, where ε_0 is the permittivity of free space and ε is now a scalar function of frequency, whose structure can be examined experimentally by using optical techniques.

Particularly useful excitations, in this connection, are surface polaritons;[2] that is excitations which propagate along the boundaries of dielectric media and whose associated fields decay exponentially with distance from a boundary in the direction of the normal (see Figure 1 for isotropic media). Surface polaritons also serve as a sensitive probe of the structure of material surfaces since they are so closely associated with them.

One of the most frequently examined excitations is the surface plasmon-polariton which is a TM wave (magnetic field in the plane of the surface and normal to the propagation direction) and the remainder of this chapter will be devoted to this particular case, although the arguments can be, in principle, easily extended to describe other types of polaritons.

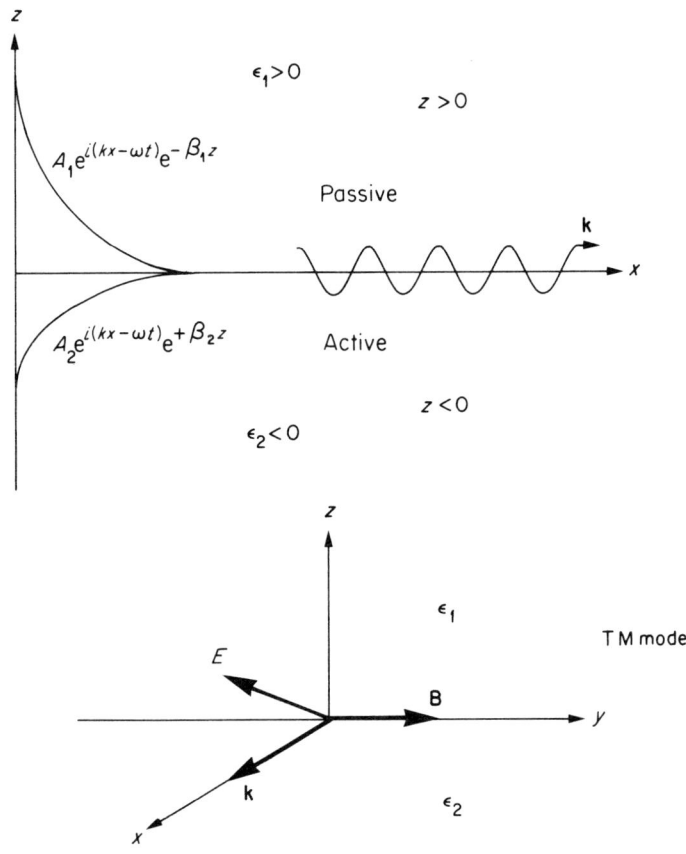

Figure 1. Schematic diagram of the variation of the field amplitudes associated with a surface wave propagating at the boundary between two media. The amplitudes have an oscillatory behaviour in the direction of propagation but die away exponentially in a direction normal to the boundary

Figure 1 shows a surface polariton mode that is a surface guided mode, of the TM type, between two semi-infinite isotropic media. It is not too difficult to show that, for plasmon systems such as metals, it is indeed the TM mode that propagates. Conversely, for magnon systems, such as ferromagnetic insulators, it is the TE mode that is a surface mode. The form of a surface wave is $A_1 \exp[i(kx - \omega t)] \exp(-\beta_1 z)$ for $z > 0$ and $A_2 \exp[i(kx - \omega t)] \exp(\beta_2 z)$ for $z < 0$. The relationship between angular frequency ω and wave number k is usually called the dispersion equation. This name arises because such an equation determines how a wave packet (pulse) will spread out.

Here, β_1 and β_2 are determined by Maxwell's equations under the TM mode assumption $\mathbf{E} = (E_x, 0, E_z)$, $\mathbf{B} = (0, B_y, 0)$. For a medium with a dielectric function ε the relevant components of Maxwell's equations are

$$\frac{\partial E_x}{\partial z} - ikE_z = i\omega B_y, \qquad \frac{\partial B_y}{\partial z} = \frac{i\omega}{c^2} \varepsilon E_x,$$

$$ikE_x + \frac{\partial E_z}{\partial z} = 0, \qquad kB_y = \frac{-\omega}{c^2} \varepsilon E_z, \tag{1}$$

from which it follows that, if $E_x = A_1 e^{\pm \beta z}$,

$$E_z = \pm i \frac{k}{\beta} A_1 e^{\pm \beta z}, \qquad B_y = \pm i \frac{\omega \varepsilon}{\beta c^2} A_1 e^{\pm \beta z}, \tag{2}$$

where

$$\beta^2 = k^2 - \varepsilon \omega^2 / c^2.$$

Hence

$$\beta_1 = \left(k^2 - \varepsilon_1 \frac{\omega^2}{c^2}\right)^{\frac{1}{2}}, \qquad \beta_2 = \left(k^2 - \varepsilon_2 \frac{\omega^2}{c^2}\right)^{\frac{1}{2}}. \tag{3}$$

The boundary conditions at the interface between the two media are that E_x and B_y are continuous. This leads immediately to

$$\frac{\beta_1}{\beta_2} = -\frac{\varepsilon_1}{\varepsilon_2} > 0, \tag{4}$$

and

$$k^2 = \frac{\omega^2}{c^2} \frac{\varepsilon_1 \varepsilon_2}{\varepsilon_1 + \varepsilon_2} > 0. \tag{5}$$

Equation (5) is the dispersion equation of the surface waves. Equations (4) and (5) also show that the existence of such waves requires ε_1 and ε_2 to be of opposite sign and their sum to be negative. This can be understood as follows: equation (4) can be satisfied only if $\varepsilon_1 > 0$, $\varepsilon_2 < 0$, or vice versa; if this is so then $\varepsilon_1 \varepsilon_2 < 0$, hence $\varepsilon_1 + \varepsilon_2 < 0$, for equation (5) to be satisfied.

A medium with a positive dielectric function is often termed a 'passive medium', and commonly used examples of such media include air, vacuum, glass, or similar media whose dielectric functions are normally fairly constant over the frequency range of interest. The dielectric function for the passive medium will be labelled ε_1 for this work and has the value unity for air or vacuum, 2.25 for typical glasses, and 11.683 for silicon, say. A medium with a negative dielectric function is called an 'active' medium and include metals and semiconductors at frequencies below the plasma frequency ω_p. This fact has led to the study of metals through the simple, but surprisingly good, model in which the dielectric function of a free electron gas is used in the form

$$\varepsilon_2(\omega) = \varepsilon_L\left(1 - \frac{\omega_p^2}{\omega(\omega + i\nu)}\right), \qquad \omega_p^2 = \frac{Ne^2}{\varepsilon_L \varepsilon_0 m^*}, \qquad (6)$$

where ε_L is the high-frequency dielectric constant which, while it is unity for metals, is usually between 10 and 20 for semiconductors; N is the free electron density; e is the electronic charge; m^* is the effective mass, ε_0 is the permittivity of free space; and ν is a damping term describing electron–photon collisions, etc. and is quite small, compared to ω, in the optical range of frequencies. If ν is included the dielectric function of an active medium is a complex quantity with a large negative real part, provided $\omega < \omega_p$. $\varepsilon(\omega)$ is, approximately, real and negative for $\omega < \omega_p$. The plasma frequency for metals is typically $\sim 10^{15}$–10^{16} and in semiconductors $\sim 10^{13}$–10^{14}, therefore surface plasmon-polaritons might be expected to be generated at infrared frequencies for semiconductors and at optical frequencies for metals.

For real materials ν ought to be included in the dielectric function and, as a consequence, equation (5) may then be interpreted in two different ways that correspond to a certain choice in experimental procedure. Firstly the wave number may be assumed to be real so that equation (5) may be solved for a complex frequency, corresponding to a wave that is temporally damped. Alternatively, the frequency may be taken as real so that equation (5) yields a complex wave-number, corresponding to spatial damping. In either case the surface wave is not a true, long-lived, normal mode and leaks energy into the active medium. This can be seen from the fact that β_1 and β_2 are complex and hence the fields may have a small propagating component normal to the surface.

The dispersion curves obtained by these two choices are radically different, as can be seen from Figure 2. The curve obtained for real wave number and complex frequency is, actually, almost indistinguishable from the curve that would be calculated with $\nu = 0$. In the simple model used here, this curve has a non-dispersive ($d\omega/dk = 0$) large wave number limit given by

$$\omega = \omega_p\left(\frac{\varepsilon_L}{\varepsilon_L + \varepsilon_1}\right)^{\frac{1}{2}}. \qquad (7)$$

However, the curve obtained with real frequency and complex wave numbers, although similar at low frequencies, does not have such a limit, instead it exhibits a 'bend-back', at a critical frequency, to lower wave numbers. This difference is a fundamental property of 'leaky' or damped waves and, as will be shown later, is closely related to the experimental procedure used

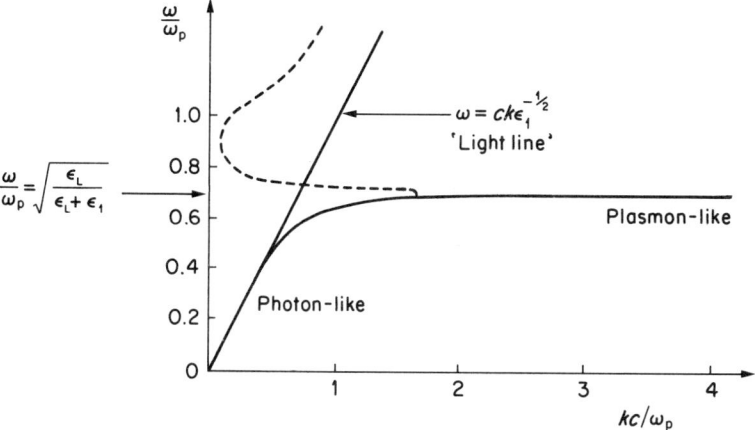

Figure 2. Surface plasmon-polariton dispersion curve for a free electron metal with finite damping plotted for: (a) real k, complex ω (solid curve), (b) real ω, complex k (broken curve). The curves are plotted in dimensionless units

in the measurement of the dispersion curve. Note that, for small k, both curves of Figure 2 are asymptotic to what is termed the 'light line', i.e. the straight-line dispersion relationship of electromagnetic waves in the passive medium. This is given by $\omega = ck\varepsilon_1^{-\frac{1}{2}}$ and represents the propagation of light in the passive dielectric.

Several techniques are available in the experimental study of surface plasmon-polaritons, and their use is dictated by the frequency and wave number range in which the excitation is being examined. For example, at short wavelengths (that is large wave number values) where the surface polariton is most 'plasmon-like', the most suitable probe of the dispersion equation is an experiment involving electrons, such as thin-film electron diffraction.[3] Conversely, at longer wavelengths (that is small wave number values) where the surface polariton is most 'photon-like', the most suitable probe is an experiment involving light of the appropriate frequency, and it is one such technique that is to be discussed here, from a computational point of view. It is known as attenuated total reflection (ATR).

ATR involves the use of the evanescent wave that is set up at a medium–air interface when light in a high refractive index medium, such as glass, suffers total internal reflection.[4] It is the reduction of the reflected wave due to absorption that is called ATR. If the weakening occurs by some other means it is usually called frustrated total internal reflection (FTIR). The names, however, have sometimes been interchanged. Rather remarkably, FTIR can be traced back to Newton,[5] and the knowledge and use of ATR has been widespread since the early 1960s due to pioneering work by

Fahrenfort and Harrick.[4,6] In spite of all this work, however, the fundamental idea of using ATR to generate surface polaritons was first published by Otto[7] as late as 1968.

An inspection of Figure 2 reveals that, in the photon-like region, where it might be thought that surface polaritons could be stimulated by incident electromagnetic waves, the dispersion curve, for either choice of solution, which we discussed earlier, lies *below* the light line. Unfortunately, for a plane electromagnetic wave, incident on the interface between the ε_1 and ε_2 of Figure 1, at some angle θ_1, the relationship between the component of wave number parallel to the surface is

$$k = \frac{\omega}{c} \varepsilon_1^{\frac{1}{2}} \sin \theta_1. \tag{8}$$

This corresponds to a line *steeper* than the light line of Figure 2 and, hence, to the region *above* the light line. Indeed, only under the impractical condition of grazing incidence ($\theta_1 = 90°$) does k approach a surface-wave value. The awkward conclusion is that surface polariton waves cannot be generated by propagating plane electromagnetic waves, in a semi-infinite passive medium, on to an interface separating it from a semi-infinite active medium. If surface modes are to be generated, then an ATR system, as first proposed by Otto (Figure 3a) and subsequently by Kretschmann[8] (Figure 3b), must be used.

The first system, proposed by Otto, and shown in Figure 3a will be called the prism–air–medium system (PAM). This consists of a prism (usually hemicylindrical) with a dielectric constant ε_1 separated from a thick sample of the active medium by a small air (vacuum) gap of dielectric constant $\varepsilon_2 < \varepsilon_1$. Light incident, through the prism on to the prism–air interface, at angles greater than the critical angle θ_c is normally totally reflected back out through the prism. However, under total internal reflection conditions, there will always be an exponentially decaying evanescent field extending into the air gap. The totally reflected wave has a component of wave number parallel to the surface given by equation (8). If the dispersion curve for surface polaritons on a metal–air interface is now considered, as in Figure 4, it can be seen that the curve, in the small k region, lies only *below* the air or vacuum light line ($\omega = ck$). Now the evanescent wave, created by total reflection in the prism, actually exists in the range $\theta_c \leq \theta_1 \leq 90°$ for which

$$\frac{1}{\sqrt{\varepsilon_1}} < \sin \theta_1 < 1. \tag{9}$$

Hence the range of surface wave number associated with the evanescent wave is

$$\frac{\omega}{c} < k < \sqrt{\varepsilon_1} \frac{\omega}{c}, \tag{10}$$

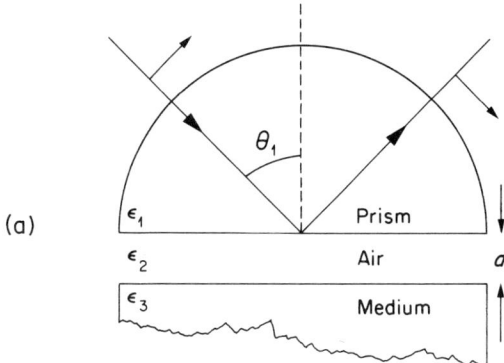

Figure 3a. PAM ATR configuration with an air gap of thickness d separating a hemicylindrical prism from a bulk sample of the active medium. Plane-polarized light is incident at an angle θ_1

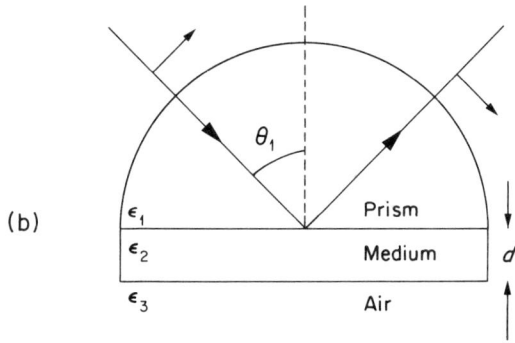

Figure 3b. PMA ATR configuration with an active film of thickness d deposited on the base of the hemicylindrical prism

and corresponds to a region *above* the prism light line but *below* the air/vacuum light line. As indicated on Figure 4, it is clear that, because the evanescent wave has a larger k than the corresponding vacuum wave vector, a range of k of the air/metal surface polariton becomes directly accessible by using an ATR system.

From a practical point of view it is evident that, provided the air gap is sufficiently small, then the evanescent field from the prism can reach the air–medium interface and a surface plasmon-polariton can be excited. Energy from the incident wave is then used to stimulate a surface wave resulting in a weakening of the reflected intensity returned by the prism,

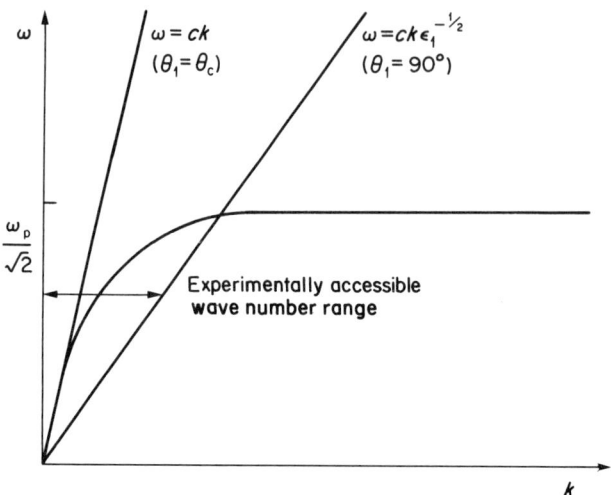

Figure 4. Schematic illustration of the way in which the ATR configuration enables a portion of the surface plasmon-polariton dispersion curve for a medium–air interface to be observed

hence the name 'attenuated total reflection'. By tracing the locus of the minimum in the reflected intensity, the surface polariton dispersion curve may be obtained.

There is also a second system, developed by Kretschmann, and shown in Figure 3b. This will be called the prism–medium–air system (PMA) and consists of a thin film of an active medium deposited onto the base of a hemicylindrical prism. The surface wave at the medium–air interface in the PMA case is excited in just the same way as the PAM case, except that the fields from the prism must now penetrate the film to reach the medium–air interface. In both cases the excitation of a surface plasmon-polariton occurs at the second interface and would be expected to be observed as a sharp minimum in the reflected intensity of (TM) p-polarised light. Also, in each case, (a) the prism light line corresponds to grazing incidence, (b) the air light line corresponds to the prism-air critical angle θ_c, (c) intermediate angles correspond to straight lines of different slope, passing through the origin of Figure 4.

A typical ATR experiment would be to measure, as a function of real frequency, the locus of real angles at which the reflected intensity is a minimum, or to measure, as a function of a real angle, the locus of real frequencies at which a minimum occurs. The experimental dispersion curves, obtained by these approaches, turn out to be significantly different,[9] corresponding, essentially, to the fact that a choice is made as to whether the resulting surface wave is temporally or spatially damped. This difference is exactly analogous to the choice of theoretical solutions of the dispersion equation described earlier.[10] Measurements with a fixed real angle and scanned real frequency correspond to a theoretical solution of the dispersion equation with a real wave number and a complex frequency, and produce the dispersion curve shown in Figure 2. The imaginary part of the eigenfrequency obtained from the solution of the dispersion equation has a direct interpretation as the linewidth in the frequency scan of the ATR experiment. It should be re-emphasized here that the ATR experiment clearly involves real frequencies and angles in contrast with the theoretical solutions of the dispersion equation. The choice of a fixed real frequency and scanned real angle corresponds to solving the dispersion equation with a real frequency and a complex wave number and results in the folded back curve. The imaginary part of the wave number obtained from the solution of the dispersion equation is interpreted as a linewidth in the angle scan of the ATR experiment. In either case it is clear that the air gap or film width is a critical factor since, if they are too large, there is insufficient coupling of energy into the surface wave and the resulting minimum in the reflected intensity is very weak, or, if they are too small, then the medium/air surface mode will be perturbed by the presence of the second interface (thin-film effects) and the measured dispersion curve will not be that of a free surface wave at a single air–medium interface.

A detailed picture of what is expected experimentally can be built up from a computational model and it is the purpose of this chapter to develop and exploit a model of this kind. For example, it is possible to find the optimum values of the variables (frequency, prism dielectric constant, gap, or thin-film width) for observation of the surface polariton. Moreover, since ATR, like many other optical problems, is adequately theoretically described by simple classical considerations (at least at the frequencies of interest) and leads in many cases to theoretical results which fit experiment very well,[11] a satisfying scope is afforded to the student to examine the behaviour of real systems in the form of computer experiments. In this regard the results of this chapter are best used graphically and also in an 'on-line' interactive fashion so that students may rapidly vary the parameters as they wish, in order to observe the behaviour of the system. The program is therefore designed for use on computer systems with visual display unit (VDU) graphical output capabilities, although simple modifications could be effected to make it suitable for running as a batch program leading to 'hard copy' graphs. In either case, a graph plotting package is really required in addition to this program. If this is not possible, listing the results file will enable the student to plot the graphs by hand.

The exact form of the graphical output in this case is a plot of the reflected intensity of incident p-polarized light, either as a function of incident angle for a fixed frequency or vice-versa. If more sophisticated graphics software is available, some minor modifications to the program could enable the plotting of a 'reflectivity surface', that is, a three-dimensional plot of reflected intensity both as a function of frequency and angle. This gives very good insight into some of the processes involved in ATR experiments[9,12] since the dispersion curve can be seen directly as a curved valley in the surface.

2. ATR MODEL

Consider the simple model of a two-interface system shown in Figure 5. This model, with a suitable choice of the dielectric functions ε_1, ε_2, and ε_3, may be used to describe both PAM and PMA systems of Figures 3a and 3b. For p-polarised light, of unit electric field amplitude, incident through medium 1 at angle θ_1 to the surface normal, the electric field in each medium may be expressed in the form

Medium 1:	Incident	$\exp[i(\alpha_1 \sin \theta_1 x - \alpha_1 \cos \theta_1 z - \omega t)]$	(11)
	Reflected	$r \exp[i(\alpha_1 \sin \theta_1 x + \alpha_1 \cos \theta_1 z - \omega t)]$,	(12)
Medium 2:	Incident	$E_2^t \exp[i(\alpha_2 \sin \theta_2 x - \alpha_2 \cos \theta_2 z - \omega t)]$,	(13)
	Reflected	$E_2^r \exp[i(\alpha_2 \sin \theta_2 x + \alpha_2 \cos \theta_2 z - \omega t)]$,	(14)
Medium 3:	Transmitted	$E_3^t \exp[i(\alpha_3 \sin \theta_3 x - \alpha_3 \cos \theta_3 z - \omega t)]$,	(15)

where r, E_2^t, E_2^r, and E_3^t are reflected and transmitted electric field amplitudes, α_1, α_2, and α_3 are the propagation constants in each medium, and θ_1, θ_2, and θ_3 are the angles subtended by the incident and reflected waves to the surface normal.

The boundary conditions appropriate in this model are the continuity of E_x, the tangential component of the electric field in the plane of incidence, and the continuity of D_z, the normal component of the displacement vector **D**. D_z is given in an isotropic medium by $D_z = \varepsilon_0 \varepsilon E_z$, where ε is the dielectric function of the medium and E_z is the normal component of the electric field. Application of these boundary conditions at each interface yields the four equations.

$$\cos \theta_1 + r \cos \theta_1 = E_2^t \cos \theta_2 + E_2^r \cos \theta_2, \quad (16)$$

$$\varepsilon_1 \sin \theta_1 - r\varepsilon_1 \sin \theta_1 = E_2^t \varepsilon_2 \sin \theta_2 - E_2^r \varepsilon_2 \sin \theta_2, \quad (17)$$

$$E_2^t \cos \theta_2 \exp(i\alpha_2 \cos \theta_2 d) + E_2^r \cos \theta_2 \exp(-i\alpha_2 \cos \theta_2 d)$$
$$= E_3^t \cos \theta_3 \exp(i\alpha_3 \cos \theta_3 d, \quad (18)$$

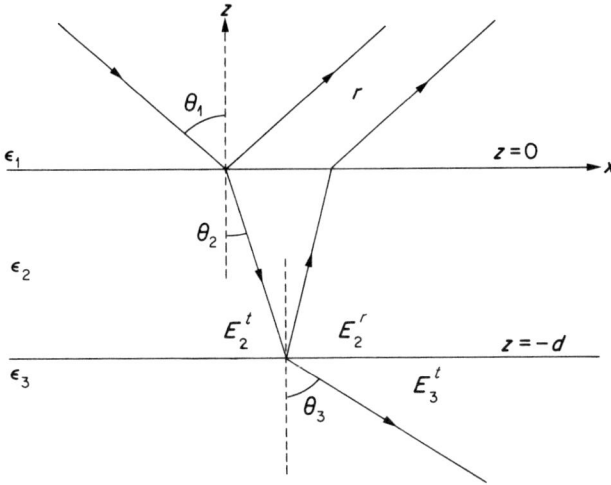

Figure 5. Illustration of the two-interface model for an ATR system. By suitably assigning ε_2 and ε_3 either the PMA or PAM configuration can be treated

$$E_2^t \varepsilon_2 \sin \theta_2 \exp(i\alpha_2 \cos \theta_2 d) - E_2^r \varepsilon_2 \sin \theta_2 \exp(-i\alpha_2 \cos \theta_2 d)$$
$$= E_3^t \varepsilon_3 \sin \theta_3 \exp(i\alpha_3 \cos \theta_3 d), \quad (19)$$

where the exponential terms common to all the equations have been suppressed and d is the thickness of medium 2.

These equations can be manipulated to give an expression for the reflected amplitude inside medium 1 (the prism) of the standard form

$$r = \frac{r_{12} + r_{23} \exp(2i\alpha_2 \cos \theta_2 d)}{1 + r_{12} r_{23} \exp(2i\alpha_2 \cos \theta_2 d)}, \quad (20)$$

where r_{12} and r_{23} are the Fresnel coefficients of the two interfaces and are given by

$$r_{12} = \frac{\varepsilon_2^{\frac{1}{2}} \cos \theta_1 - \varepsilon_1^{\frac{1}{2}} \cos \theta_2}{\varepsilon_2^{\frac{1}{2}} \cos \theta_1 + \varepsilon_1^{\frac{1}{2}} \cos \theta_2}, \quad (21)$$

$$r_{23} = \frac{\varepsilon_3^{\frac{1}{2}} \cos \theta_2 - \varepsilon_2^{\frac{1}{2}} \cos \theta_3}{\varepsilon_3^{\frac{1}{2}} \cos \theta_2 + \varepsilon_2^{\frac{1}{2}} \cos \theta_3}. \quad (22)$$

Here, α_2 is given by $\varepsilon_2^{\frac{1}{2}} \omega/c$ and the angles, using Snell's law, satisfy

$$\cos \theta_2 = \left(1 - \frac{\varepsilon_1}{\varepsilon_2} \sin^2 \theta_1\right)^{\frac{1}{2}}, \quad (23)$$

$$\cos \theta_3 = \left(1 - \frac{\varepsilon_1}{\varepsilon_3} \sin^2 \theta_1\right)^{\frac{1}{2}}. \quad (24)$$

For the PMA case $\varepsilon_1 = n^2$, where n is the refractive index of the prism, $\varepsilon_3 = 1$ and ε_2 is given by equation (6). In the PAM case ε_2 and ε_3 are interchanged. In either case, since the dielectric function of the active medium is complex and surface waves are generated for incident angles $\theta_1 > \theta_c$, the angles θ_2 and θ_3 and the quatities r, r_{12}, and r_{23} are complex variables, even though the input variables ω and θ_1 and the reflected intensity $R_p = |r|^2$ are real quantities. This situation cannot be simplified by setting the damping term $\nu = 0$ since, as pointed out earlier, the surface wave is leaky. This means that if an absorption mechanism were not provided, all of the energy from the surface wave would eventually leak back into the prism. Thus, setting $\nu = 0$ leads to a value for R_p of unity at all frequencies and angles $\theta_1 > \theta_c$.

In both PMA and PAM configurations a surface wave is generated at the medium–air interface. Hence an examination of the behaviour of the Fresnel coefficient r_{23} should reveal something about the generation mechanism of surface waves. If the reflectivity $R_{23} = |r_{23}|^2$ is considered further in, say, the PMA case then for $\theta_1 > \theta_c$

$$R_{23} = |r_{23}|^2 = \left| \frac{(n^2 \sin^2 \theta_1 - \varepsilon_2)^{\frac{1}{2}} - \varepsilon_2 (n^2 \sin^2 \theta_1 - 1)^{\frac{1}{2}}}{(n^2 \sin^2 \theta_1 - \varepsilon_2)^{\frac{1}{2}} + \varepsilon_2 (n^2 \sin^2 \theta_1 - 1)^{\frac{1}{2}}} \right|^2, \qquad (25)$$

where it should be remembered that ε_2 for metals, e.g. sodium, aluminium, silver, or gold has, at optical frequencies, a large *negative* real part. A surface wave is generated at $\theta_1 = \theta_p$ with a wave number given by equation (8) so that, by equation (5),

$$n \sin \theta_p = \mathrm{Re} \left[\frac{\varepsilon_2}{\varepsilon_2 + 1} \right]^{\frac{1}{2}}, \qquad (26)$$

where Re denotes the real part. At $\theta_1 = \theta_p$ the denominator of r_{23} tends to zero and $r_{23} \to \infty$. The physical meaning of this is clear because it is an example of a resonance in which a finite-amplitude electromagnetic wave can be created at the surface without any incident power.[13] If, for the moment, we consider a single interface between two semi-infinite media labelled 2 (metal) and 3 (air), or vice versa, then r_{23} is the conventional reflection coefficient of an incident plane electromagnetic wave. Now r_{23} can be regarded as a ratio N/D so that $N = 0$ corresponds to the familiar Brewster effect $r_{23} = 0$ while, neglecting the imaginary part of ε_2 for the moment, $D = 0$ corresponds to the surface plasmon-polariton resonance under discussion here. This fascinating connection allows us to regard the Brewster effect as a surface-wave phenomenon. The $N = 0$ case corresponds to unbound, radiative surface waves (Brewster modes) while the $D = 0$ case corresponds to bound, non-radiative surface waves (called Fano modes in honour of Fano who published an early pioneering paper on this topic[14]).

Naturally, finite damping makes ε_2 have a small imaginary part and in a real system r_{23} will simply become very large as θ_1 approaches θ_p.

Since the ATR system is a coupled two-surface system, the dispersion relation of surface polaritons on a free metal/air surface is not strictly observed. Instead it is a form disturbed by the presence of the prism and corresponds to the denominator of equation (20), being zero, i.e.

$$1 + r_{12} r_{23} \exp(2i\alpha_2 \cos\theta_2 d) = 0. \tag{27}$$

It should be noted, however, that in the limit, when the two interfaces are sufficiently far apart (uncoupled) to be treated separately, the single interface result is obtained. It may seem strange that the launching of a surface wave, corresponding to a pole in r is expected to be seen as a minimum in $R_p = |r|^2$. This can be understood, however, by remembering that a direct computer solution of equation (27) would involve complex quantities and therefore a solution in terms of complex frequency and real wave number of vice versa must be sought. If instead of solving equation (27) the response in the R_p, of an ATR calculation, is observed by varying *real* frequency and *real* angle, it cannot be expected that a true solution of (27) will be found. In fact the ATR response actually corresponds to a projection, on to the real frequency-angle plane of a multidimensional complex space, of the solution of the dispersion equation, and is thus not observed as a pole in R_p but as a sharp minimum.[15] The fact that a wave with real frequency, incident at a real angle, can be used to launch a leaky wave satisfying a complex dispersion relation can also be understood in terms of a 'forced' resonance[14,16] phenomenon.

3. ATR COMPUTER PROGRAM

The core of an ATR program of the kind indicated in section 1, is a very simple calculation. Once the choice of ATR configuration has been made, the system parameters such as d, ε_1, etc. specified, and the dielectric functions appropriately assigned to each of the three media in the model, the process consists simply of a loop, either over frequency or angle, in which $R_p = |r|^2$ is calculated from equation (20) using the same short subroutine for both loops. The results may then be written to a file for subsequent reading at a later stage, and also stored in arrays for the purposes of graphical output. When the loop is completed, a graph may then be plotted, of R_p as a function of the chosen variable. On completion of this graph, the student may then wish to repeat the operation with different parameters to optimize the conditions for observation of the surface-wave minimum, or having done so, to carry out a set of scans to obtain the dispersion curve. The program, therefore, must return to some convenient point for recalculation.

Another important point in programs of this kind is that some students may need more complete prompting than others, to ensure that sensible parameters are chosen for experiment. Students should therefore have a choice of short prompting for their input of variables, or of more explanatory prompting that they can subsequently switch off as their experience grows.

There are also some more detailed points in the calculation that make the program more convenient for the student but, since they involve further choices of action, complicate the simple basic operation.

For example, when choosing a scan of frequency for fixed angle, the student is constrained to calculating the dielectric function of the active medium from equation (6), using the free electron model, since this quantity is frequency dependent. However, in the case of a scan over angle for fixed frequency, the dielectric function, being independent of angle in this simple model, is a constant across the scan. This means that it is possible either to calculate the dielectric function from (6) or to use directly experimental values of the optical constants n_r and n_i, that is the real and imaginary parts of the refractive index, knowing the simple relation $\varepsilon = (n_r + in_i)^2$. If the free electron model is used, the required data are the high-frequency dielectric constant ε_L (unity for all metals), the plasma frequency ω_p, and the damping constant ν. Although the value of this last parameter is not critical, it must be there. For most cases it is about 1 per cent of the plasma frequency.

To understand some of the other technical aspects of the program it is useful to examine the flow diagram of Figure 6 containing the whole program structure except for the choices in prompting options that would overcomplicate the diagram.

After the usual declarations, specification of the arrays and opening of input, output, and data-file channels, the program offers a simple explanation of the ATR technique in a few lines. The user is then asked if detailed input prompting is required. Depending on the answer, a parameter is set that controls the form of prompting for the rest of the program. At the start of each new calculation the program returns to this point so that the users can decide each time if their memories need refreshing about typical values for certain parameters. Following this, an identifying name for the active medium is requested, of up to 12 letters. This is merely for convenient identification of the data stored in the data file, and to label the graph.

The data for the program is put in at this stage, starting with the ATR configuration required. Depending on the response to this, either the air gap or film thickness in nanometres is requested, together with the refractive index of the prism. A further choice is then required of a fixed angle (varying frequency) or fixed frequency (varying angle) scan of the reflected intensity.

If a fixed angle scan is chosen, the model is constrained to the free

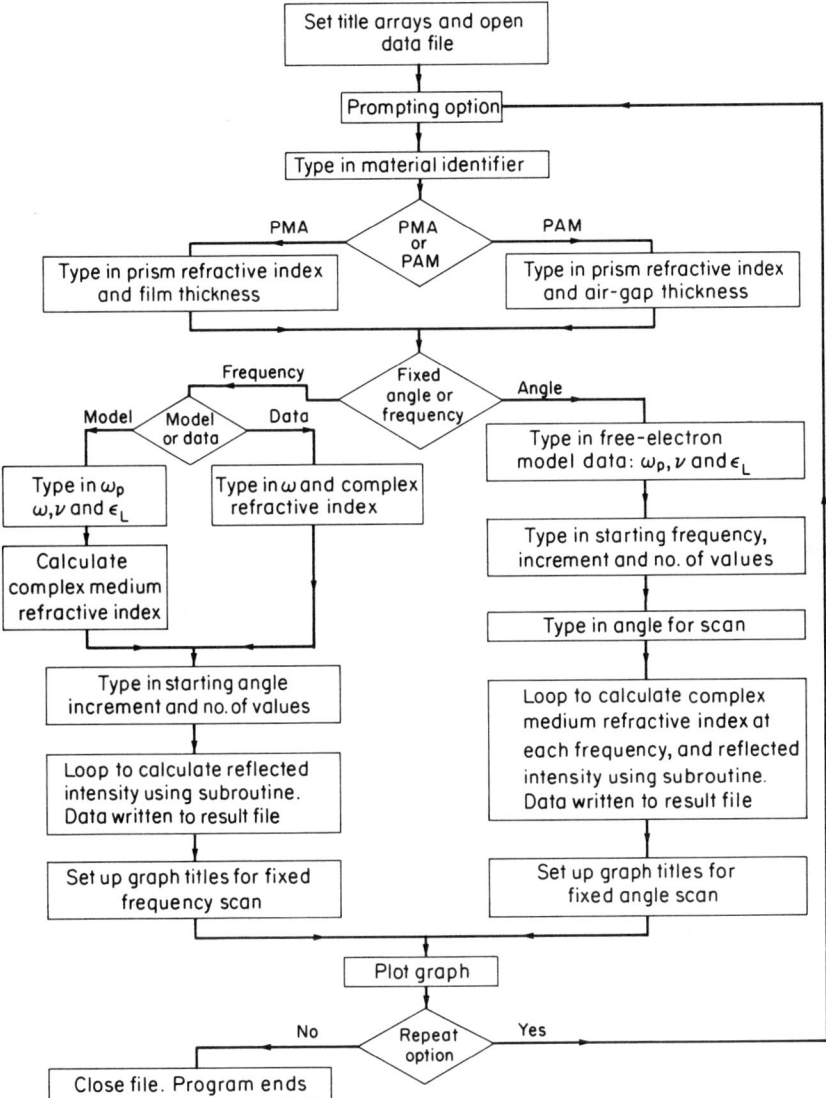

Figure 6. Flow diagram for the ATR program, illustrating the options available to a user

electron model (since the dielectric function of the active medium is frequency dependent). In this model, the plasma frequency, high-frequency dielectric constant, and damping parameter are required as constants for the scan. In addition the user then specifies the range and precision of the frequency scan in terms of a starting frequency, an increment, and the number of values required. For graphical purposes this must be limited, since for most computer systems, arrays of definite length have to be specified in graphics work. In the program given here, the maximum number of points is 200; however, this can be increased simply by changing the array size at the head of the program.

At this stage the program requests the angle at which the scan is to be carried out and then enters a frequency scan loop in which the dielectric function is calculated, and the subroutine is called that calculates $R_p = |r|^2$. The frequency value and R_p at this frequency are then stored in the graphics arrays and also output to the data files, together with the calculated values of the squares of the Fresnel coefficients $R_{12} = |r_{12}|^2$ and $R_{23} = |r_{23}|^2$. In addition, if required by the particular graphics package used, the maximum and minimum values of ω and R_p are calculated at this stage, and a suitable title specified.

If we now return to the point at which the direction of scan was chosen and consider the alternative possibility of a fixed frequency (varying angle) scan, this choice leads to a more complicated procedure since the dielectric function may be calculated either from the free electron model or from experimental values of the optical constants. If the free electron model is chosen, the plasma frequency, high-frequency dielectric constant, and damping parameter are required as before, together with the frequency at which the scan is to be carried out. If the experimental approach is used, the program requires the scan frequency as before, but then requires only values for the optical constants, that is the real and imaginary parts of the refractive index. Whichever approach is used, the remaining data required are the starting value of angle for the scan, the angle increment, and the number of points in the scan. The program then enters the angle scan loop and uses the same subroutine to calculate R_p. The data are stored as before and the program then converges to the same point for either choice of fixed variable.

At this stage a suitable graphics package (appropriate to the system) is called to display the reflected intensity R_p as a function of the scanned variable. On completion of this graph, the user is given the option of terminating the program or of selecting new parameters. If he chooses the former, the program closes the data file and terminates. It should be pointed out that at the start of the program the data file is emptied, so that the data file should be read, if required, immediately after terminating the program.

If the user wishes to perform further calculations, the program returns to the point at which the level of prompting is specified. Each subsequent calculation adds a new table to the data file so that the accumulated results are in a convenient neatly labelled form.

Several features of this program will vary between computer systems; however, these points could be quite simply modified. For example, input data in this program are handled mostly by a free-format subroutine FREAD. The advantage of this is that the use of free format reduces the possibility of format errors by the user. Most systems have a package equivalent to FREAD, but if one is not available a conventional FORTRAN format is adequate. Similarly the input–output and data file assignments will be different. In this case the package FTNCMD should be replaced by the appropriate subroutines. In particular the command

CALL FTNCMD ('$EMPTY RESULT OK', 16)

simply empties the data file.

The length of the logical variables used in the character input will depend on the system, and the function EQUC(Y1, Y2) merely checks if the two variables Y1 and Y2 are the same. This function can clearly be replaced by a conventional statement or equivalent function. The logical variables Y1, Y2, Y3, and Y4 are test variables against which are checked the responses of the user to questions involving a choice of action. Y1 = Y(yes) is used to check yes or no answers. Y2 = A(angle) is used to check the response to a choice of angle or frequency scan. Y3 = D(data) checks the response to a choice of experimental data or the free electron model in the case of an angle scan. Y4 = M(PMA systems) identifies the ATR configuration. The variables Y5 and Y6 (2) are used to receive the responses to the questions and are then checked against these 'constants' for the choice of action. In most cases the first letter of the response is sufficient to delineate the choice. However, in the case of the choice between the PMA and PAM systems it is the second letter which is used and hence Y6 has two elements; the second of which is checked against Y4.

The remaining system-specific aspects of the program are graphical ones. The core of the graphics is the external package which plots the reflected intensity R_p as a function of the chosen variable. In this case, this package is the subroutine CGPL, and the required parameters will be broadly the same in any package. However, the associated subroutines used in this program may be discarded if not required or replaced by equivalent packages. In order to make this task easier, there follows a brief explanation of the purpose of each routine in order of appearance in the program.

Firstly, a title array of some kind is usually required by most packages. This is contained in the array T. In addition, since the abscissa of the plot will depend on the choice of variable to be scanned, the title for this must be assigned after the data arrays X and Y are filled. Consequently, two alternative titles, contained in the variables TA1 to TA4 and TF1 to TF4, are specified at the beginning of the program and assigned to T at the appropriate stage.

Early in the program, the routine AUX093 is used to specify the device upon which the graph is to be plotted. This is a common requirement and most systems have equivalent routines. Note that by a suitable choice here, hard copies of graphs could be obtained although this has not been done here, since a visual display is the immediate purpose of this work.

The graph plotting package CGPL used here requires scale factors to define the units of the X and Y axes. These are calculated from SCX and SCY which in turn use the routine SLE to obtain a convenient scale. In many systems a scale factor is not required so that SCX, SCY, and the routine SLE or its equivalent may often be left out of the program. However, the maximum and minimum values of the data points, XMAX, YMAX, XMIN, and YMIN are usually required.

Finally the routine PLOT closes the graphics package data file. An equivalent is usually required to avoid 'overplotting' errors whereby some points from each graph are plotted at the next call to the graph plotting routine.

The remainder of the variables involved in CGPL are simply those involved in the choice of symbols, form of curve, etc. It has been found that the cross is usually the most suitable symbol, and since most curve-fitting procedures are inclined to 'overshoot' on rapidly varying functions such as R_p, it is generally better to plot only individual data points rather than a smooth curve.

The modifications outlined above should not present any serious problems to anyone who is reasonably familiar with their own graphics facilities.

4. TYPICAL RESULTS

As an example of the kind of results obtained from this program, data for sodium (a good free electron metal, which has a complex refractive index $0.044 + i2.42$ at $\omega = 3.2 \times 10^{15}$ rad s^{-1})[*] was used. The program was run for a fixed frequency scan, and a fixed angle scan. Table 1 contains a truncated copy of the data file showing the interesting parts of a fixed frequency scan using the PMA configuration. The graph obtained with this data is shown in Figure 7.

It can be seen that the reflected intensity in the fourth column of the table exhibits a sharp peak at 42° (the critical angle for a prism–air interface). This feature is typical of the PMA configuration and is due to the fact that the final medium in the system (ε_3) is air. An examination of the third column containing $R_{23} = |r_{23}|^2$ reveals a very large peak near 47° corresponding to the launching of a surface wave associated with the medium–air interface. Note that the reflected intensity shows a minimum very near to this angle, the shift being due to the fact that perturbation occurs because we really have a coupled system, even in an optimum parameter situation. The degree to which this perturbation occurs should now be studied as a function of the film thickness to determine the optimum value of d consistent with a reasonable reduction from unity of R_p.[15]

---

[*] All frequencies used in the program are angular frequencies whose proper units are rad s^{-1}. In the program we have used the tidier, but not strictly correct, label Hz.

TABLE 1

```
SUBJECT MEDIUM IS SODIUM

         PRISM-MEDIUM-AIR CALCULATION
         -----------------------------

PRISM REFRACTIVE INDEX=  0.1500E+01
FILM THICKNESS=  0.4000E+02 NANOMETRES

         FIXED FREQUENCY CALCULATION
         ---------------------------

         CALCULATION USING REFRACTIVE INDEX DATA
         ---------------------------------------

SCAN FREQUENCY=  0.3200E+16
GIVEN COMPLEX REFRACTIVE INDEX=   0.4400E-01 +J  0.2420E+01

   ANGLE(DEG)        R12            R23             R
     0.0          0.9680E+00     0.9747E+00     0.6787E+00
     0.1000E+01   0.9680E+00     0.9747E+00     0.6787E+00
     0.2000E+01   0.9679E+00     0.9746E+00     0.6785E+00
         .            .              .              .
         .            .              .              .
     0.3600E+02   0.9610E+00     0.9603E+00     0.6260E+00
     0.3700E+02   0.9607E+00     0.9600E+00     0.6282E+00
     0.3800E+02   0.9604E+00     0.9602E+00     0.6339E+00
     0.3900E+02   0.9601E+00     0.9613E+00     0.6460E+00
     0.4000E+02   0.9599E+00     0.9643E+00     0.6715E+00
     0.4100E+02   0.9596E+00     0.9719E+00     0.7318E+00
     0.4200E+02   0.9594E+00     0.2179E+01     0.9776E+00
     0.4300E+02   0.9591E+00     0.8131E+01     0.9760E+00
     0.4400E+02   0.9589E+00     0.2232E+02     0.9573E+00
     0.4500E+02   0.9587E+00     0.6757E+02     0.9022E+00
     0.4600E+02   0.9585E+00     0.3204E+03     0.7930E+00
     0.4700E+02   0.9583E+00     0.8271E+04     0.6940E+00
     0.4800E+02   0.9582E+00     0.5893E+03     0.6898E+00
     0.4900E+02   0.9581E+00     0.1718E+03     0.7333E+00
     0.5000E+02   0.9580E+00     0.8778E+02     0.7764E+00
     0.5100E+02   0.9579E+00     0.5627E+02     0.8092E+00
     0.5200E+02   0.9578E+00     0.4070E+02     0.8333E+00
         .            .              .              .
         .            .              .              .
     0.8500E+02   0.9891E+00     0.6518E+01     0.9850E+00
     0.8600E+02   0.9912E+00     0.6488E+01     0.9880E+00
     0.8700E+02   0.9934E+00     0.6465E+01     0.9909E+00
     0.8800E+02   0.9956E+00     0.6449E+01     0.9939E+00
     0.8900E+02   0.9978E+00     0.6439E+01     0.9970E+00
     0.9000E+02   0.1000E+01     0.6436E+01     0.1000E+01
----------------------------------------------------------------
----------------------------------------------------------------
```

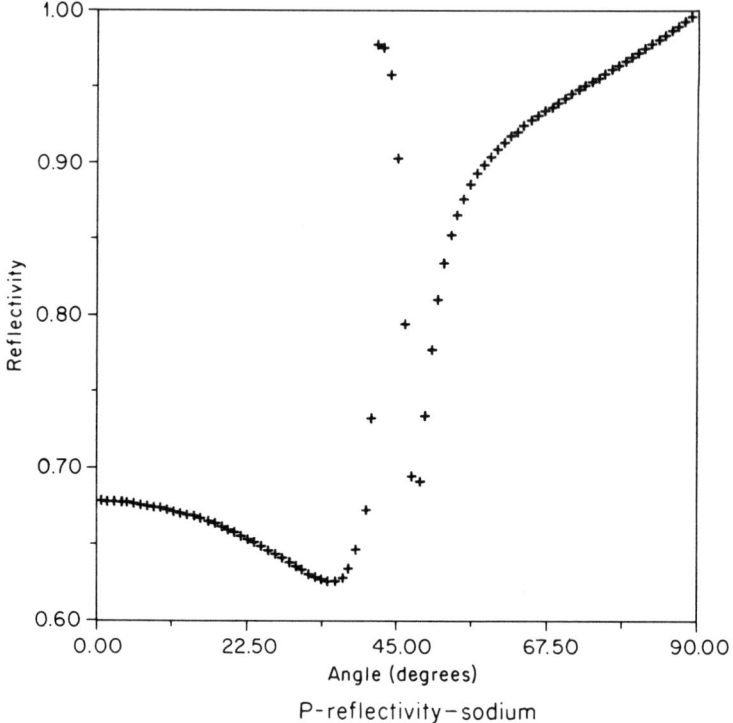

Figure 7. Plot of reflected intensity R_p as a function of angle of incidence for sodium data using the PMA configuration. Angular frequency of scan is 3.2×10^{15} rad s^{-1} ($\approx 0.4 \times \omega_p$)

As a second example, a truncated set of results is displayed in Table 2 for a fixed angle scan in the PAM configuration using a free electron model with data appropriate to sodium. The graph obtained with this data and configuration is shown in Figure 8. This scan at 60° shows a deep minimum in the reflected intensity at a frequency a little above $\omega/\omega_p = 0.5$. This corresponds to a 'pole' in R_{23} that is at a slightly higher frequency, the difference again being due to the perturbation by the prism–air interface. Note that $R_{12} = |r_{12}|^2$ is unity for all frequencies since the scan is made at an angle greater than $\theta_c = 42°$, the critical angle for the prism–air interface.

These examples vividly illustrate the principle of attenuated total reflection. Other metals can be considered, of course, e.g. silver with a complex dielectric constant $\varepsilon = -18.3 + i0.4$ at the He–Ne laser[17] frequency $\omega = 2.979 \times 10^{15}$ rad s^{-1}, or aluminium[18] with complex refractive index $\sqrt{\varepsilon} = 1.53 + i5.65$ at $\omega = 2.218 \times 10^{15}$ rad s^{-1}. Semiconductors can also be investigated such as GaAs or, more popularly, InSb[19] for which $\varepsilon_L = 15.68$. Computational experiments can be made with either experimental refractive

TABLE 2

SUBJECT MEDIUM IS SODIUM

PRISM-AIR-MEDIUM CALCULATION

PRISM REFRACTIVE INDEX= 0.1500E+01
AIR GAP THICKNESS= 0.2000E+03 NANOMETRES

FIXED ANGLE CALCULATION

FREE ELECTRON MODEL CALCULATION

PLASMA FREQUENCY= 0.8230E+16
DAMPING TERM= 0.1236E+15
HIGH FREQ. DIELECTRIC CONSTANT= 0.1000E+01
ANGLE FOR ATR SCAN= 0.6000E+02 DEGREES

FREQ.	R12	R23	R
0.3500E+16	0.1000E+01	0.2443E+02	0.9788E+00
0.3525E+16	0.1000E+01	0.2577E+02	0.9782E+00
0.3550E+16	0.1000E+01	0.2722E+02	0.9776E+00
0.3575E+16	0.1000E+01	0.2879E+02	0.9769E+00
0.3600E+16	0.1000E+01	0.3051E+02	0.9762E+00
0.3625E+16	0.1000E+01	0.3239E+02	0.9754E+00
0.3650E+16	0.1000E+01	0.3444E+02	0.9745E+00
0.3675E+16	0.1000E+01	0.3671E+02	0.9736E+00
0.3700E+16	0.1000E+01	0.3920E+02	0.9725E+00
0.3725E+16	0.1000E+01	0.4196E+02	0.9713E+00
0.3750E+16	0.1000E+01	0.4503E+02	0.9700E+00
0.3775E+16	0.1000E+01	0.4845E+02	0.9686E+00
0.3800E+16	0.1000E+01	0.5227E+02	0.9669E+00
0.3825E+16	0.1000E+01	0.5657E+02	0.9651E+00
.	.	.	.
.	.	.	.
0.4275E+16	0.1000E+01	0.7457E+03	0.7099E+00
0.4300E+16	0.1000E+01	0.1004E+04	0.6296E+00
0.4325E+16	0.1000E+01	0.1405E+04	0.5221E+00
0.4350E+16	0.1000E+01	0.2044E+04	0.3892E+00
0.4375E+16	0.1000E+01	0.3050E+04	0.2591E+00
0.4400E+16	0.1000E+01	0.4376E+04	0.1972E+00
0.4425E+16	0.1000E+01	0.5227E+04	0.2545E+00
0.4450E+16	0.1000E+01	0.4617E+04	0.3957E+00
0.4475E+16	0.1000E+01	0.3282E+04	0.5464E+00
0.4500E+16	0.1000E+01	0.2193E+04	0.6677E+00
0.4525E+16	0.1000E+01	0.1491E+04	0.7556E+00
.	.	.	.
.	.	.	.
0.4900E+16	0.1000E+01	0.8518E+02	0.9860E+00
0.4925E+16	0.1000E+01	0.7682E+02	0.9877E+00
0.4950E+16	0.1000E+01	0.6961E+02	0.9890E+00
0.4975E+16	0.1000E+01	0.6336E+02	0.9902E+00
0.5000E+16	0.1000E+01	0.5790E+02	0.9913E+00

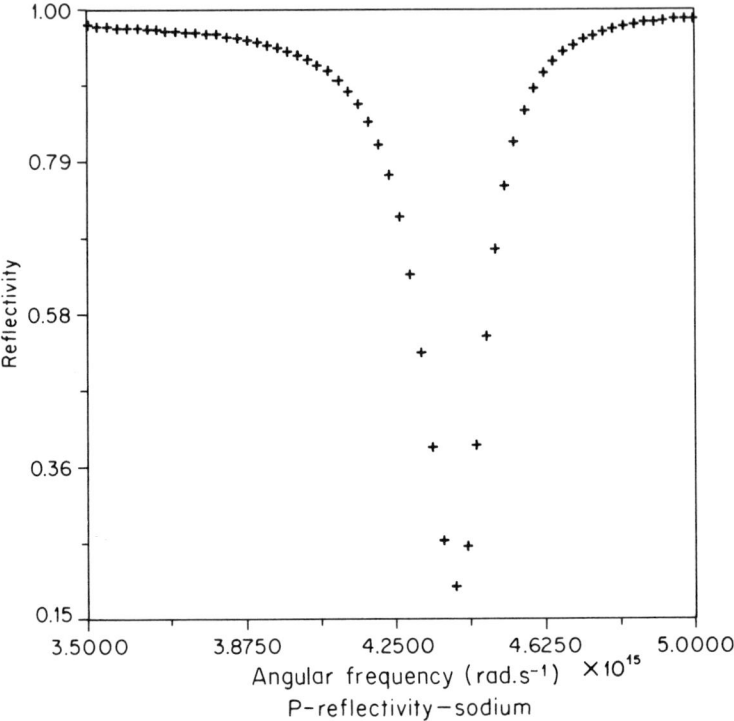

Figure 8. Plot of reflected intensity R_p as a function of incident angular frequency for sodium using a free electron model for the PAM configuration. Angle of scan is 60°

index data or model parameters. For example, in the case of semiconductors, ω_p can be varied as the electron density is varied and the free electron model can be easily modified to incorporate phonon terms as ω_p is lowered. The variation of the results with the size of damping parameter is also an interesting one as is the variation of θ_{min}, the angle of the minimum in R_p, with air gap thickness (PAM), or film thickness (PMA). Only for certain values of d will θ_{min} become θ_p and, in the case of PMA, double-valued results will appear at larger d.

Finally, as can be appreciated, there is ample scope for experimentation and any results obtained will require interesting, and often subtle, physics for a satisfactory interpretation to be found.

REFERENCES

1. E. Burstein, *Polaritons*, Eds. E. Burstein and F. de Martini (Pergamon, New York), 1 (1974).
2. E. Burstein, A. Hartstein, J. Schoenwald, A. A. Maradudin, D. L. Mills, and R. F. Wallis, *Polaritons*, Eds. E. Burstein and F. de Martini, (Pergamon, New

York), 89 (1974); K. L. Kliewer and R. Fuchs, *Advances in Chemical Physics*, **17,** 356 (1974); E. N. Economou and K. L. Ngai, *Advances in Chemical Physics*, **17,** 265 (1974); D. L. Mills and E. Burstein, *Rep. on Prog. in Phys.*, **37,** 817 (1974); C. Kittel, *Introduction to Solid State Physics*, 5th edn. (Wiley, New York); A. D. Boardman, *Progress in Surface Science* (to be published).
3. C. Kunz, *Z. Phys.*, **196,** 311 (1966).
4. P. A. Wilks and T. Hirschfeld, *App. Spectro. Rev.*, **1,** 99 (1967). N. J. Harrick, *Internal Reflection Spectroscopy* (Wiley, New York, 1967).
5. I. Newton, *Optiks II*, Book 8, 97 (1817).
6. J. Fahrenfort, *Spectrochim. Acta.*, **17,** 698 (1961); N. J. Harrick, *Phys. Rev.*, **125,** 1165 (1962).
7. A. Otto, *Z. Phys.*, **216,** 398 (1968).
8. E. Kretschmann, *Z. Phys.*, **241,** 313 (1971).
9. R. W. Alexander, G. S. Kovener and R. J. Bell, *Phys. Rev. B*, **14,** 1458 (1976).
10. G. Borstel, E. Schuller, and H. J. Falge, *Phys. Stat. Sol.* (b), **76,** 759 (1976).
11. G. Borstel, and H. J. Falge, *Phys. Stat. Sol.* (*b*), **83,** 11 (1977).
12. G. C. Aers and A. D. Boardman, *UKSC Conference on Computer Simulation*, (IPC Science and Technology Press Ltd., Guildford, U.K.), 320 (1978).
13. M. Cardona, *Am. J. Phys.*, **39,** 1277 (1971).
14. U. Fano, *J. Opt. Soc. Am.*, **31,** 213 (1941).
15. G. C. Aers, A. D. Boardman, and P. Clark, *Phys. Stat. Sol.* (b), 171 (1978).
16. A. Hessel and A. A. Oliner, *Applied Optics*, **4,** 1275 (1965).
17. H. J. Simon, D. E. Mitchell, and J. G. Watson, *Am. J. Phys.*, **43,** 630 (1975).
18. A. Otto, 'Festkorperprobleme XIV', in *Advances in Solid State Physics*, Eds. O. Madelung and H. J. Queisser (Pergamon, Oxford, 1974).
19. B. Fischer, N. Marschall, and H. J. Queisser, *Surf. Science*, **34,** 50 (1973).

```
C       CALCULATION OF REFLECTED INTENSITY OF P POLARISED LIGHT FROM
C       THE SURFACE OF A DIELECTRIC MEDIUM.
C       ----------------------------------------------------------------
        LOGICAL*1 Y1,Y2,Y3,Y4,Y5,Y6(2),EQUC
        DIMENSION X(200),Y(200),Z(200),T(20)
        COMPLEX J,R2,R3
C
C       GRAPHICS TITLES AND INPUT PARAMETERS.
C
        DATA Y1/'Y'/,Y2/'A'/,Y3/'D'/,Y4/'M'/
       1  ,T(1)/'     '/,T(2)/'     '/,T(3)/'P-RE'/,T(4)/'FLEC'/,
       2   T(5)/'TIVI'/,T(6)/'TY--'/,T(10)/'    '/,T(11)/'    '/,T(12)
       3  /'    '/,T(17)/'REFL'/,T(18)/'ECTI'/,T(19)/'VITY'/,T(20)/
       4  '    '/,TA1/'FREQ'/,TA2/'UENC'/,TA3/'Y (RD'/,TA4/'S-1)  '/
       5  ,TF1/'ANGL'/,TF2/'E (D'/,TF3/'EGRE'/,TF4/'ES) '/
C
C         INPUT/OUTPUT CHANNELS ASSIGNED.
C
        CALL FTNCMD('$EMPTY RESULT OK',16)
        CALL FTNCMD('ASSIGN 7=RESULT',15)
        CALL FTNCMD('EQUATE 5=SCARDS',15)
        J=CMPLX(0.0,1.0)
        WRITE(6,1)
C
C       OPTION FOR DETAILS OF ATR METHOD.
C
        CALL FREAD(5,'STRING:',Y5,1)
        IF(.NOT.EQUC(Y5,Y1))GOTO 100
        WRITE(6,2)
        WRITE(6,3)
100     WRITE(6,4)
        CALL FREAD(5,'I*4:',NC)
C
C       SET DEVICE CODE FOR GRAPHICS OUTPUT.
C
        CALL AUX093(NC)
110     WRITE(6,5)
C
C       OPTION OF DETAILED INPUT PROMPTING.
C
        CALL FREAD(5,'STRING:',Y5,1)
        K=0
        IF(EQUC(Y5,Y1))K=1
C
C       MATERIAL IDENTIFIER REQUIRED.
C
        WRITE(6,6)
        READ(5,7)(T(I),I=7,9)
        WRITE(7,8)(T(I),I=7,9)
C
C       CHOICE OF ATR CONFIGURATIONS
C
```

```
      WRITE(6,9)
      CALL FREAD(5,'STRING:',Y6,2)
      IF(EQUC(Y6(2),Y4))GOTO 140
C
C     PRISM-AIR MEDIUM CALCULATION CHOSEN
C
      WRITE(7,10)
      IF(K.EQ.0)GOTO 120
      WRITE(6,10)
      WRITE(6,11)
C
C     INPUT DATA FOR PRISM-AIR-MEDIUM CALCULATION
C
120   WRITE(6,12)
      CALL FREAD(5,'R:',RP)
      IF(K.EQ.0)GOTO 130
      WRITE(6,13)
130   WRITE(6,14)
      CALL FREAD(5,'R:',D)
      WRITE(7,15)RP,D
      D=D*1.0E-9
      GOTO 170
140   WRITE(7,16)
C
C     PRISM-MEDIUM-AIR CALCULATION CHOSEN.
C
      IF(K.EQ.0)GOTO 150
      WRITE(6,16)
      WRITE(6,11)
150   WRITE(6,12)
C
C     INPUT DATA FOR PRISM-MEDIUM-AIR CALCULATION.
C
      CALL FREAD(5,'R:',RP)
      IF(K.EQ.0)GOTO 160
      WRITE(6,17)
160   WRITE(6,18)
      CALL FREAD(5,'R:',D)
      WRITE(7,19)RP,D
      D=D*1.0E-9
C
C     SETS PARAMETERS TO CALCULATE MAXIMUM AND MINIMUM VALUES OF
C     PLOTTED VARIABLES.
C
170   XMAX=-1.0E60
      YMAX=-1.0E60
      XMIN=1.0E60
      YMIN=1.0E60
      WRITE(6,20)
C
C     CHOICE OF FIXED ANGLE OR FIXED FREQUENCY SCAN.
C
      CALL FREAD(5,'STRING:',Y5,1)
```

```
      IF(EQUC(Y5,Y2))GOTO 260
      WRITE(7,21)
      IF(K.EQ.0)GOTO 180
C
C     FIXED FREQUENCY CALCULATION CHOSEN.
C
      WRITE(6,21)
      WRITE(6,22)
C
C     CHOICE OF FREE ELECTRON MODEL OR EXPERIMENTAL DATA FOR
C     REFRACTIVE INDEX.
C
180   WRITE(6,23)
      CALL FREAD(5,'STRING:',Y5,1)
      IF(EQUC(Y5,Y3))GOTO 200
      WRITE(7,24)
      IF(K.EQ.0)GOTO 190
      WRITE(6,24)
C
C     FREE ELECTRON MODEL CHOSEN.
C
      WRITE(6,25)
C
C     INPUT DATA FOR FREE ELECTRON MODEL CALCULATION.
C
190   WRITE(6,26)
      CALL FREAD(5,'R:',WP)
      WRITE(6,27)
      CALL FREAD(5,'R:',W)
      WRITE(6,28)
      CALL FREAD(5,'R:',DP)
      WRITE(6,29)
      CALL FREAD(5,'R:',EL)
      WRITE(7,30)WP,W,DP,EL
C
C     CALCULATION OF REAL AND IMAGINARY PARTS OF REFRACTIVE INDEX.
C
      RR=REAL(CSQRT(EL-EL*WP*WP/(W*(W+J*DP))))
      RI=AIMAG(CSQRT(EL-EL*WP*WP/(W*(W+J*DP))))
      WRITE(7,31)RR,RI
      GOTO 220
200   WRITE(7,32)
C
C     CALCULATION USING REFRACTIVE INDEX DATA CHOSEN.
C
      IF(K.EQ.0)GOTO 210
      WRITE(6,32)
210   WRITE(6,27)
C
C     INPUT DATA FOR EXPERIMENTAL REFRACTIVE INDEX.
C
      CALL FREAD(5,'R:',W)
      WRITE(6,33)
```

```
      CALL FREAD(5,"R:',RR)
      WRITE(6,34)
      CALL FREAD(5,'R:',RI)
      WRITE(7,35)W,RR,RI
220   WRITE(6,36)
C
C     INPUT PARAMETERS FOR ANGLE SCAN.
C
      CALL FREAD(5,'R:',TH1)
      WRITE(6,37)
      CALL FREAD(5,'R:',THGAP)
      WRITE(6,38)
      CALL FREAD(5,"I*4:',NTH)
      WRITE(7,39)
      IF(EQUC(Y6(2),Y4))GOTO 230
C
C     PRISM-AIR-MEDIUM CONFIGURATION.
C
      R2=CMPLX(1.0,0.0)
      R3=CMPLX(RR,RI)
      GOTO 240
C
C     PRISM-MEDIUM-AIR CONFIGURATION.
C
230   R2=CMPLX(RR,RI)
      R3=CMPLX(1.0,0.0)
240   DO 250 I=1,NTH
C
C     CALLS SUBROUTINE TO CALCULATE REFLECTED INTENSITY.
C
      CALL REF(TH1,RP,R2,R3,W,D,R12,R23,R)
C
C     CALCULATES MAXIMUM AND MINIMUM VALUES OF VARIABLES.
C
      X(I)=TH1
      Y(I)=R
      IF(XMAX.LT.X(I))XMAX=X(I)
      IF(XMIN.GT.X(I))XMIN=X(I)
      IF(YMAX.LT.Y(I))YMAX=Y(I)
      IF(YMIN.GT.Y(I))YMIN=Y(I)
C
C     OUTPUTS RESULTS TO DATA FILE.
C
      WRITE(7,40)TH1,R12,R23,R
      TH1=TH1+THGAP
250   CONTINUE
      NP=NTH
C
C     GRAPH TITLES FOR ANGLE SCAN.
C
      T(13)=TF1
      T(14)=TF2
      T(15)=TF3
```

```
              T(16)=TF4
              GOTO 320
        260   WRITE(7,41)
        C
        C     FIXED ANGLE CALCULATION CHOSEN.
        C
              WRITE(7,24)
              IF(K.EQ.0)GOTO 270
              WRITE(6,41)
              WRITE(6,25)
        C
        C     INPUT DATA FOR FREE ELECTRON MODEL CALCULATION.
        C
        270    WRITE(6,26)
              CALL FREAD(5,'R:',WP)
              WRITE(6,28)
              CALL FREAD(5,'R:',DP)
              WRITE(6,29)
              CALL FREAD(5,'P:',EL)
              WRITE(7,42)WP,DP,EL
        C
        C     INPUT PARAMETERS FOR FREQUENCY SCAN.
        C
              WRITE(6,43)
              CALL FREAD(5,'R:',W)
              WRITE(6,44)
              CALL FREAD(5,'R:',WGAP)
              WRITE(6,45)
              CALL FREAD(5,'I*4:',NF)
              IF(K.EQ.0)GOTO 280
              WRITE(6,46)
        280   WRITE(6,47)
              CALL FREAD(5,'R:',TH1)
              WRITE(7,48)TH1
              DO 310 I=1,NF
              RR=REAL(CSQRT(EL-EL*WP*WP/(W*(W+J*DP))))
              RI=AIMAG(CSQRT(EL-EL*WP*WP/(W*(W+J*DP))))
              IF(EQUC(Y6(2),Y4)) GOTO 290
        C
        C     PRISM-AIR-MEDIUM CONFIGURATION.
        C
              R2=CMPLX(1.0,0.0)
              R3=CMPLX(RR,RI)
              GOTO 300
        C
        C     PRISM-MEDIUM-AIR CONFIGURATION.
        C
        290   R2=CMPLX(RR,RI)
              R3=CMPLX(1.0,0.0)
        C
        C     CALLS SUBROUTINE TO CALCULATE REFLECTED INTENSITY.
        C
        300   CALL REF(TH1,RP,R2,R3,W,D,R12,R23,R)
```

```
C
C      CALCULATES MAXIMUM AND MINIMUM VALUES OF VARIABLES.
C
       X(I)=W
       Y(I)=R
       IF(XMAX.LT.X(I))XMAX=X(I)
       IF(XMIN.GT.X(I))XMIN=X(I)
       IF(YMAX.LT.Y(I))YMAX=Y(I)
       IF(YMIN.GT.Y(I))YMIN=Y(I)
C
C      OUTPUTS RESULTS TO DATA FILE
C
       WRITE(7,40)W,R12,R23,R
       W=W+WGAP
310    CONTINUE
       NP=NF
C
C      GRAPH TITLES FOR FREQUENCY SCAN.
C
       T(13)=TA1
       T(14)=TA2
       T(15)=TA3
       T(16)=TA4
320    WRITE(7,49)
C
C      PLOTS GRAPH OF REFLECTED INTENSITY AS FUNCTION OF CHOSEN
C      VARIABLE.
C
       CALL SLE(XMAX,XMIN)
       CALL SLE(YMAX,YMIN)
       SCX=ABS(XMAX-XMIN)/4.0
       SCY=ABS(YMAX-YMIN)/4.0
       CALL CGPL(X,Y,Z,NP,64,1,1,1,1,XMIN,SCX,4.0,
     1 YMIN,SCY,4.0,T,8)
       CALL CGPL(X,Y,Z,NP,132,1,1,1,1,XMIN,SCX,4.0,
     1 YMIN,SCY,4.0,T,8)
       CALL PLOT(0.0,0.0,999)
       WRITE(6,50)
C
C      OPTION TO REPEAT PROGRAM.
C
       CALL FREAD(5,'STRING:',Y5,1)
       IF(EQUC(Y5,Y1))GOTO 110
C
C      --------------------------------------------------------------
C
1      FORMAT(22X,'ATR CALCULATION',/,22X,15(1H-),//,'DO YOU WANT',
     1 ' ANY INFORMATION ABOUT ATR METHOD?-TYPE YES OR NO')
2      FORMAT('PROGRAM CALCULATES AND DISPLAYS GRAPHICALLY THE ',/
     1 ,'REFLECTED INTENSITY FROM THE SURFACE OF A MATERIAL',/,
     2 'IRADIATED WITH P-POLARISED LIGHT:',//,
     3 12X,'(A)  AS A FUNCTION OF FREQUENCY FOR A FIXED ANGLE.',
     4 //,12X,'(B)  AS A FUNCTION OF ANGLE FOR A FIXED FREQUENCY.'
```

```
     5   ,//,'AND FOR TWO ATTENUATED TOTAL REFLECTION(ATR)',
     6   ' CONFIGURATIONS:',//,12X,'(1)   PRISM-MEDIUM-AIR      PMA',
     7   //,12X,'(2)   PRISM-AIR-MEDIUM      PAM',//
     8   ,'THE PAM SYSTEM CONSISTS OF A SLAB OF THE MEDIUM',/
     9   ,'UNDER INVESTIGATION, SEPARATED FROM THE INPUT PRISM',/
     1   ,'BY A SMALL AIR GAP.THE PMA SYSTEM CONSISTS OF A',/
     2   ,'THIN FILM OF THE MEDIUM DEPOSITED DIRECTLY ONTO THE',/
     3   ,'PRISM SURFACE.',/
     4   ,'   IN BOTH CASES THE EXPONENTIALLY DECAYING FIELD',/
     5   ,'FROM THE TOTALLY INTERNALLY REFLECTED INPUT LIGHT',/
     6   ,'RAY HAS A SUFFICIENTLY LARGE COMPONENT OF MOMENTUM',/
     7   ,'IN THE PLANE OF THE MEDIUM SURFACE AND IN THE ',/
     8   ,'DIRECTION OF PROPAGATION , TO LAUNCH A SURFACE WAVE.',/
     9   ,'THIS IS SEEN AS A SHARP MINIMUM IN THE REFLECTED',/
     1   ,'INTENSITY (WHICH WOULD OTHERWISE BE UNITY).')
3     FORMAT('   NOTE THAT A RAY INCIDENT DIRECTLY ONTO THE SURF',
     1   'ACE',/,'OF THE MEDIUM HAS INSUFFICIENT MOMENTUM IN THE ',/
     2   ,'DIRECTION OF PROPAGATION TO LAUNCH A SURFACE WAVE.',/)
4     FORMAT('TYPE DEVICE CODE-4010+4013=1',/
     1   ,18X,'4014+4015=2')
5     FORMAT('DO YOU WANT FULL INPUT PROMPTING? TYPE YES OR NO')
6     FORMAT('TYPE IDENTIFIER FOR MATERIAL: UP TO 12 LETTERS')
7     FORMAT(3A4)
8     FORMAT('SUBJECT MEDIUM IS ',3A4,/)
9     FORMAT('WHICH ATR CONFIGURATION DO YOU WANT? TYPE PMA ',
     1   'OR PAM')
10    FORMAT(10X,'PRISM-AIR-MEDIUM CALCULATION',/,10X,28(1H-),/)
11    FORMAT('IN THE CASE OF METALS FOR WHICH THE SURFACE PLASMON'/
     1   ,'FREQUENCY IS USUALLY IN THE OPTICAL REGION, PRISMS',/
     2   ,'OF REFRACTIVE INDEX SIMILAR TO GLASS(ABOUT 1.5) ARE',/
     3   ,'USED.FOR SEMICONDUCTORS THE REGION OF INTEREST IS',/
     4   ,'USUALLY THE INFRA-RED WHERE SILICON IS USED.IN THIS',/
     5   ,'CASE THE REFRACTIVE INDEX IS 3.418.',/)
12    FORMAT('TYPE REFRACTIVE INDEX OF PRISM')
13    FORMAT('IN PAM CALCULATIONS THE AIR GAP THICKNESS  D, IS',/
     1   ,'CRITICAL AND TYPICALLY THE ORDER OF 200 OR 300 ',
     2   'NANOMETRES',/)
14    FORMAT('TYPE AIR GAP THICKNESS (IN NANOMETRES)')
15    FORMAT('PRISM REFRACTIVE INDEX=',E12.4,/
     1   ,'AIR GAP THICKNESS=',E12.4,' NANOMETRES',/)
16    FORMAT(10X,'PRISM-MEDIUM-AIR CALCULATION',/,10X,28(1H-),/)
17    FORMAT('IN PMA CALCULATIONS THE FILM THICKNESS D, IS',/
     1   ,'CRITICAL AND TYPICALLY THE ORDER OF 20 TO 50 NANOMETRES',/)
18    FORMAT('TYPE FILM THICKNESS (IN NANOMETRES)')
19    FORMAT('PRISM REFRACTIVE INDEX=',E12.4,/
     1   ,'FILM THICKNESS=',E12.4,' NANOMETRES',/)
20    FORMAT('DO YOU WANT A FIXED ANGLE OR FREQUENCY SCAN?'
     1   ,'---TYPE ANGLE OR FREQUENCY')
21    FORMAT(10X,'FIXED FREQUENCY CALCULATION',/,10X,27(1H-),/)
22    FORMAT('YOU HAVE CHOICE OF A FREE ELECTRON MODEL ',/
     1   ,'OR TO INPUT DATA FOR THE REFRACTIVE INDEX',/)
23    FORMAT('TYPE MODEL OR DATA')
24    FORMAT(10X,'FREE ELECTRON MODEL CALCULATION',/,10X,31(1H-),/)
```

```
25      FORMAT('MODEL TREATS MEDIUM AS A PLASMA WITH A NATURAL',/
       1 ,'RESONANCE FREQUENCY (PLASMA FREQUENCY).MODEL REQUIRES',/
       2 ,'KNOWLEDGE OF THE HIGH FREQUENCY DIELECTRIC CONSTANT',/
       3 ,'OF THE MEDIUM, WHICH FOR A METAL IS UNITY AND FOR A ',/
       4 ,'SEMICONDUCTOR IS USUALLY BETWEEN 10 AND 20.',/,
       5 2X,'MODEL ALSO REQUIRES THE PLASMA FREQUENCY AND THE ',/
       6 ,'DAMPING PARAMETER WHICH IS GENERALLY NOT A CRITICAL',/
       7 ,'FACTOR AND CAN OFTEN BE GIVEN APPROXIMATELY AS ONE',/
       8 ,'PER CENT OF THE PLASMA FREQUENCY.',/)
26      FORMAT('TYPE PLASMA FREQUENCY')
27      FORMAT('TYPE FREQUENCY FOR ATR ANGLE SCAN')
28      FORMAT('TYPE DAMPING PARAMETER')
29      FORMAT('TYPE HIGH FREQUENCY DIELECTRIC CONSTANT')
30      FORMAT('PLASMA FREQUENCY= ',E12.4,/
       1 ,'SCAN FREQUENCY= ',E12.4,/
       2 ,'DAMPING TERM   = ',E12.4,/
       3 ,'HIGH FREQ. DIELECTRIC CONSTANT= ',E12.4)
31      FORMAT('MODEL COMPLEX REFRACTIVE INDEX=',E12.4,' +J',E12.4,/)
32      FORMAT(10X,'CALCULATION USING REFRACTIVE INDEX DATA',
       1 /,10X,39(1H-),/)
33      FORMAT('TYPE REAL PART OF REFRACTIVE INDEX')
34      FORMAT('TYPE IMAGINARY PART OF REFRACTIVE INDEX')
35      FORMAT('SCAN FREQUENCY= ',E12.4,/
       1 ,'GIVEN COMPLEX REFRACTIVE INDEX= ',E12.4,' +J',E12.4,/)
36      FORMAT('TYPE STARTING ANGLE FOR ATR SCAN (DEGREES)')
37      FORMAT('TYPE ANGLE INCREMENT (DEGREES)')
38      FORMAT('TYPE NUMBER OF ANGLE VALUES IN SCAN(UP TO 200)')
39      FORMAT('   ANGLE(DEG)',7X,'R12',11X,'R23',12X,'R')
40      FORMAT(4(E12.4,2X))
41      FORMAT(10X,'FIXED ANGLE CALCULATION',/,10X,23(1H-),/)
42      FORMAT('PLASMA FREQUENCY= ',E12.4,/
       1 ,'DAMPING TERM= ',E12.4,/
       2 ,'HIGH FREQ. DIELECTRIC CONSTANT= ',E12.4)
43      FORMAT('TYPE STARTING FREQUENCY FOR ATR SCAN')
44      FORMAT('TYPE FREQUENCY INCREMENT')
45      FORMAT('TYPE NUMBER OF FREQUENCY VALUES IN SCAN(UP TO 200)')
46      FORMAT('THE ATR EFFECT IS OBSERVED FOR INCIDENT ANGLES',/
       1 ,'GREATER THAN ,OR NEAR TO THE CRITICAL ANGLE FOR THE',/
       2 ,'PRISM MATERIAL (ABOUT 41 DEG. FOR GLASS WITH N=1.5)',/
       3 ,'   SOME STRUCTURE WILL BE OBSERVED BELOW THIS ANGLE',/
       4 ,'HOWEVER DUE TO MULTIPLE REFLECTION EFFECTS.',/)
47      FORMAT('TYPE ANGLE FOR ATR SCAN (IN DEGREES)')
48      FORMAT('ANGLE FOR ATR SCAN= ',E12.4,' DEGREES',//,
       1 '   FREQ.',10X,'R12',11X,'R23',12X,'R')
49      FORMAT(60(1H-),/,60(1H-),//)
50      FORMAT('DO YOU WISH TO RUN ANOTHER CASE?-TYPE YES OR NO')
C
C       ------------------------------------------------------------
C
        STOP
        END
C
C       ------------------------------------------------------------
```

```
      C
            SUBROUTINE REF(TH1,RP,R2,R3,W,D,R12,R23,R)
            COMPLEX A12,A23,A,CTH2,CTH3,J,R2,R3
      C
      C     SUBROUTINE TO CALCULATE REFLECTED INTENSITY AT EACH INTERFACE
      C     AND FOR COUPLED SYSTEM.
      C
            J=CMPLX(0.0,1.0)
            THR=TH1*3.14159/180.0
            STH1=SIN(THR)
            CTH1=COS(THR)
            CTH2=CSQRT(1.0-(STH1*RP/R2)**2)
            CTH3=CSQRT(1.0-(STH1*RP/R3)**2)
            A12=(R2*CTH1-RP*CTH2)/(R2*CTH1+RP*CTH2)
            A23=(R3*CTH2-R2*CTH3)/(R3*CTH2+R2*CTH3)
            R12=REAL(A12)**2+AIMAG(A12)**2
            R23=REAL(A23)**2+AIMAG(A23)**2
            A=(A12+A23*CEXP(J*2.0*W*R2*D/2.998E8*CTH2))
          1  /(1.0+A12*A23*CEXP(J*2.0*W*R2*D/2.998E8*CTH2))
            R=REAL(A)**2+AIMAG(A)**2
            RETURN
            END
```

Physics Programs
Edited by A. D. Boardman
© 1980 John Wiley & Sons Ltd.

CHAPTER 3

Computer-Generated Holograms

A. D. BOARDMAN and M. E. S. CHAPMAN

1. INTRODUCTION

Since the advent of the laser considerable interest has been shown in the topic known as holography. However, this is basically an imaging process that takes place in two stages and can be done with optical, microwave, or acoustic waves. In the first stage a diffraction pattern is created by interference between waves scattered off an object and a powerful direct reference beam, but using the same coherent source (see Figure 1). The recorded version of this diffraction pattern is called the hologram and contains both phase and amplitude information. In the second stage, all that is necessary to reconstruct a three-dimensional image of the object is to illuminate the hologram with the original reference beam and employ an appropriate imaging system. In fact it is not difficult to show that the image is located in the plane conjugate to the plane of the object.

This form of holography, is, of course, important, and very exciting. On the other hand, digital holography, in which computers are used both to generate holograms, and to reconstruct objects from holograms, in the form of data arrays, is also very striking with many areas of application. Some of these include optical character recognition, optical surface testing, archival data storage, image intensifiers, automobile and aeroplane design, and other forms of three-dimensional computer displays.[1-3]

An advantage of computer-based holography is that the object does not have to be real and hence exist physically. This is a very useful feature because it is often convenient to specify an object in purely mathematical terms. In other words digital holograms can be used as optical elements, in a general sense, dealing with properties that do not have a physical analogue. It also almost goes without saying that computer-generated holograms are free of the complications normally encountered in making holograms experimentally.

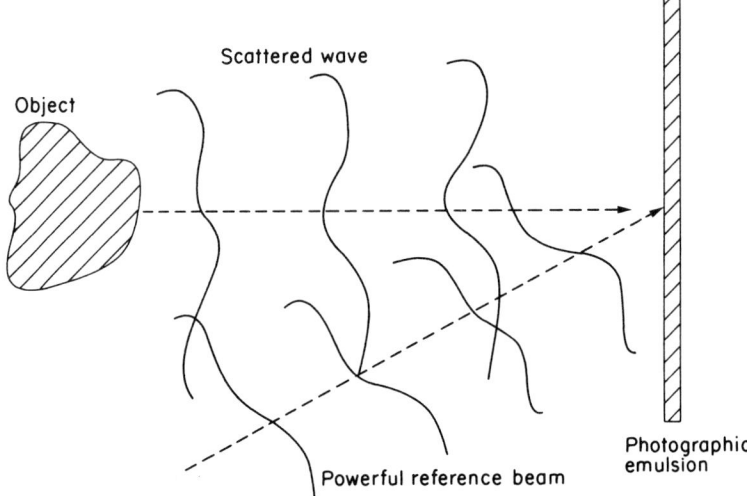

Figure 1. Schematic holographic arrangement. The hologram is the photographically recorded version of the resultant diffraction pattern

2. FRAUNHOFER DIFFRACTION PATTERNS AND FOURIER TRANSFORMS

Naturally there are many hologram-forming geometries, but this chapter focuses attention on what are known as Fourier transform holograms whose computer generation will be based on Lohmann's method. Attention is further restricted to two-dimensional objects but there is nothing, in principle, to prevent a generalization to three dimensions. The basic idea of Fourier transform holography is that the hologram is a record of the interference between waves that are the exact Fourier transforms of the object and the reference source. Since a Fraunhofer diffraction pattern is the Fourier transform of the aperture,[4] the Fourier transform holograms are sometimes referred to as Fraunhofer holograms.[5] Indeed, the Fourier transforms required in this chapter are approached most easily by considering the familiar Fraunhofer diffraction theory.

Scattering and diffraction are not synonymous terms, but they are aspects of a single physical process; one does not exist without the other. A diffraction field is the total field in the presence of the scattering object. In particular, Fraunhofer diffraction describes the interaction between a plane wave and an aperture when both the source and point of observation are at infinity. The observation of a Fraunhofer pattern is performed with a convex lens, designed, or assumed, not to introduce any spurious affects of its own. In this way the Fourier transform of the aperture is located in its back focal plane that can be called the Fourier plane.

Generally speaking, if X and Y are the coordinates of a point in the aperture and plane waves arrive at and emerge from the aperture with respective direction cosines l_i, m_i and l_e, m_e, then the complex Fraunhofer diffraction amplitude is [1,4] the Fourier transform

$$F(p, q) = \int\int G(X, Y)\exp[(2\pi i/\lambda)(pX + qY)]\,dX\,dY, \qquad (1)$$

where $p = l_e - l_i$, $q = m_e - m_i$, λ is the wavelength and $G(X, Y)$ is a function that is non-zero in the aperture space. The latter restriction means that although the integration extends to infinity in both X- and Y-directions, the contributions to the integral arise from the aperture space. The sign in the exponential can be chosen as positive[1] or negative[4] and the choice corresponds to defining a plane wave, propagating in the direction defined by wave number k, as $\exp\{i(\omega t - \mathbf{k}\cdot\mathbf{r})\}$ or $\exp\{i(\mathbf{k}\cdot\mathbf{r} - \omega t)\}$. The positive sign is selected here, but obviously it does not matter, physically, which one is used.

If the Fourier plane is the back focal plane of a converging lens, of focal length f, then as shown in Figure 2 the complex Fourier amplitude at the

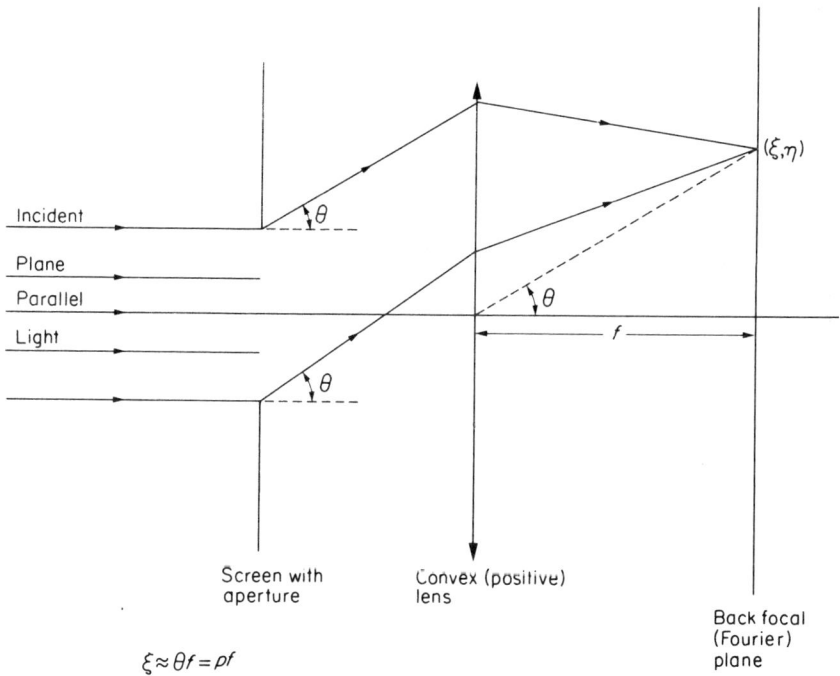

Figure 2. Fraunhofer diffraction by an aperture where the diffraction pattern is in the Fourier plane, or the back focal plane of a positive lens

point (ξ, η) is, since $\xi = pf$, $\eta = qf$,

$$F(\xi, \eta) = \int\int G(X, Y) \exp[(2\pi i/f\lambda)(\xi X + \eta Y)] \, dX \, dY, \qquad (2)$$

i.e. there is a spot $(\xi\eta)$ in the focal plane that is the focal point for a diffracted beam from the object. Furthermore, in the application discussed in this chapter, $G(X, Y)$ is going to represent the light distribution in some mathematically defined two-dimensional object where X, Y are spatial coordinates in the object plane. The computer problem amounts to devising a means of encoding $F(\xi, \eta)$, in graphical form, and thereby computer-generating a hologram.

It is not very elegant to use equation (2) in its present form, because the adoption of another coordinate system would effectively scale out f and λ. Indeed, if the coordinates x and y are introduced through the relationships

$$x = \frac{2\pi}{f\lambda}\xi, \qquad y = \frac{2\pi}{f\lambda}\eta, \qquad (3)$$

then equation (2) takes the tidier, universal, form

$$F(x, y) = \int\int G(X, Y) e^{i(xX+yY)} \, dX \, dY. \qquad (4)$$

The integrations in equation (4) are fairly easily performed for simple objects, like rectangular or circular apertures. In fact, as pointed out only recently,[6] the analytical evaluation of the integrals is not too difficult for quite a variety of polygonal apertures. Computer generation of holograms, however, is usually illustrated by using groups of letters or numerals as the object, and this arrangement is also used here.

An object that happens to be a letter, such as A or M is made up of straight lines but even curved letters, in the large N limit, can be constructed from N straight lines. It is not necessary, in the first instance, to do the latter since a decent approximation to letters like C and S, say, is obtained by forming them as ⌐ and ⌐. The device of using straight lines makes the mathematical specification of letters a lot easier and leads to analytical Fourier transforms, thus obviating any need for fast Fourier transform techniques to save computer time.

3. LETTER GROUP TRANSFORMS

It is probably not immediately obvious to a student how a letter such as A is to be specified as a function $G(X, Y)$. In order to get some ideas about it let us consider light passing through the rectangular aperture shown in Figure 3. $G(X, Y)$ is zero everywhere, except in the aperture region where it is some constant K, say. The integration limits are, therefore, the aperture limits so that equation (4), for this case, is

$$F(x, y) = K \int_{-a}^{a} \int_{-b}^{b} e^{i(xX+yY)} \, dX \, dY. \tag{5}$$

The integral in equation (5) is

$$I = \int_{-a}^{a} \int_{-b}^{b} e^{i(xX+yY)} \, dX \, dY = 4ab \, \text{sinc}(xa)\text{sinc}(yb), \tag{6}$$

where $\text{sinc}(xa) = [\sin(xa)]/xa$ etc.

The aperture area is $A = 4ab$ and Parseval's theorem for Fourier transforms gives[4] $K = (1/\lambda)\sqrt{E/A}$, where E is the total energy incident upon the aperture. Therefore

$$F(x, y) = \frac{\sqrt{EA}}{\lambda} \text{sinc}(xa)\text{sinc}(yb). \tag{7}$$

Now imagine that the aperture is gradually closed by letting $a \to 0$. In this limiting process the aperture approaches a slit lying on the interval $(-b, b)$. During the approach to the limit $\text{sinc}(xa) \to 1$ and $A \to 0$ so that in the limit the expected result that $F \to 0$ occurs. Suppose, however, that $E \to \infty$ as

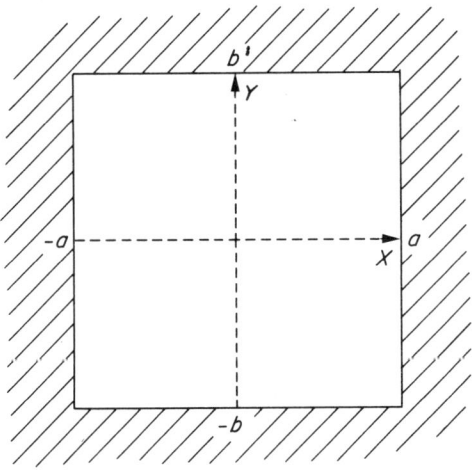

Figure 3. Rectangular aperture in an opaque screen

$A \to 0$ such that aE remains finite then F remains finite and retains the functional form sinc(yb). This is exactly what is obtained if $G(X, Y)$ is a delta function $\delta(X)$, representing a straight line parallel to the y-axis. This limiting process is the way in which a line in a letter object is specified, so for all the straight-line components of a complex letter object $G(X, Y)$ is defined as a Dirac delta function.

A hologram is a record of amplitude and phase variation over the (x, y) plane so it is not necessary to normalize $F(x, y)$ in anything but an arbitrary sense. Since this is the case equation (4) can be used, absolutely, by making $G(X, Y)$ exactly equal to a delta function such as[5] $\delta(X)$, for example. Thus for a line of length L drawn along the Y-axis and symmetrically placed at the origin

$$G(X, Y) = \delta(X), \quad |Y| \leq L/2 \qquad (8)$$
$$ 0 \quad\quad\quad\; |Y| > L/2,$$

so that equations (8) and (4) give

$$F(x, y) = L \frac{\sin(yL/2)}{(yL/2)} = L \operatorname{sinc}\left(\frac{yL}{2}\right). \qquad (9)$$

Of course, not all lines lie along a coordinate axis. For example, the letter M involves two lines set at some angles to the axes. Furthermore, lines need not be symmetrically disposed about the origin. Both of these features can be very easily incorporated by rotation or translation.

A line at some angle θ to the X-axis can be thought of as lying along a new axis denoted by X'. The rotated X' axis is then related to X and Y through the equation

$$X' = X \cos\theta + Y \sin\theta \qquad (10)$$

In the rotated coordinate system a line lying along the X'-axis has a Fourier transform $L \operatorname{sinc}(x'L/2)$. However, x and y, obviously transform from one coordinate system to another like X' and Y' so that the Fourier transform of a line orientated at an angle θ to the X-axis is simply $L \operatorname{sinc}(L/2[x \cos\theta + y \sin\theta])$.

If the midpoint of a line, instead of being at $X_0 = 0$, $Y_0 = 0$, is at some arbitrary point $X_0 \neq 0$, $Y_0 \neq 0$ then the Fourier transform of a line is the same as if it was centred on the origin except that it is multiplied by a factor $\exp i(xX_0 + yY_0)$. This is most easily seen for the example defined through equation (8). In this case a line whose centre is at X_0 must be represented by $\delta(X - X_0)$ so that it follows immediately that equation (9) contains a factor $\exp(ixX_0)$. The general case is just as easy to see, and the whole mathematical process is usually said to be an example of the shift theorem.[1]

An object such as the letter A, shown in Figure 4, has three straight lines, labelled 1, 2, and 3, with coordinates $(X_1, Y_1; X_2, Y_2)$ where the (X, Y)

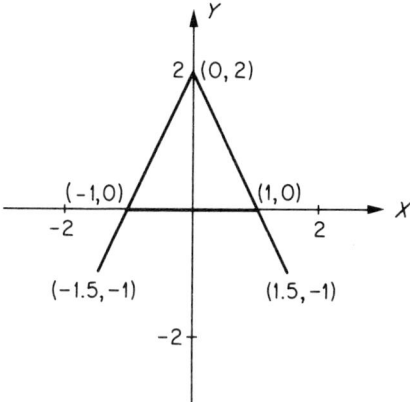

Figure 4. The letter A in the (X, Y) coordinate system

pairs define the ends of the line from left to right. In Figure 4 the lines have coordinates $(-1.5, -1; 0, 2)$, $(-1, 0; 1, 0)$ and $(0, 2; 1.5, -1)$. Already, this example shows that two sets of circumstances can arise in which $Y_2 > Y_1$ or $Y_2 < Y_1$ and these must be distinguished from each other.

General formulae for the length L of the line and its mid-point coordinates X_0, Y_0 are

$$L = [(X_2 - X_1)^2 + (Y_2 - Y_1)^2]^{\frac{1}{2}}, \quad (11)$$

and

$$X_0 = \frac{X_1 + X_2}{2} = |(X_1 - X_2)/2| + X_1, \quad (12)$$

$$Y_0 = \frac{Y_1 + Y_2}{2} = \begin{aligned} &|Y_1 - Y_2)/2| + Y_1, & Y_2 > Y_1 \\ &|Y_1 - Y_2)/2| + Y_2, & Y_2 < Y_1. \end{aligned} \quad (13)$$

The formulae for $\sin \theta$ and $\cos \theta$ are

$$\sin \theta = |(Y_2 - Y_1)|/L, \quad (14)$$

$$\cos \theta = \begin{aligned} &|(X_2 - X_1)|/L, & Y_2 > Y_1 \ (\theta \text{ acute}) \\ &-|(X_2 - X_1)|/L, & Y_2 \le Y_1 \ (\theta \text{ obtuse}). \end{aligned} \quad (15)$$

The Fourier transform of the mathematical object A, in Figure 4, is

$$F(x, y) = F_1 + F_2 + F_3 = \sum_{i=1}^{3} F_i(x, y), \quad (16)$$

where

$$F_1 = \frac{3}{2}\sqrt{5}\operatorname{sinc}\left(\frac{3}{4}x + \frac{3}{2}y\right)\exp\left(-3\mathrm{i}\frac{x}{4} + \mathrm{i}\frac{y}{2}\right), \quad (17)$$

$$F_2 = 2\operatorname{sinc}(x), \quad (18)$$

$$F_3 = \frac{3}{2}\sqrt{5}\operatorname{sinc}\left(-\frac{3}{4}x + \frac{3}{2}y\right)\exp\left(3\mathrm{i}\frac{x}{4} + \mathrm{i}\frac{y}{2}\right), \quad (19)$$

and a letter, or an object, constructed from N straight lines has a Fourier transform

$$F(x, y) = \sum_{i=1}^{N} F_i(x, y). \quad (20)$$

4. SAMPLING THE FOURIER TRANSFORM

The computer-generated Fourier transform hologram, in contradistinction to an experimentally generated hologram, only samples the Fourier transform $F(x, y)$. It cannot be calculated, at an array of points in the (x, y) plane, with an indefinitely fine mesh because of lack of storage space and the undesirability of using a vast amount of computer time. A simple answer to this problem is to limit the number of points in the (x, y) plane at which $F(x, y)$ is calculated. This could be done by arbitrarily limiting the density of calculation points (spatial frequency limiting).

However, as might be expected, limiting the number of sampling points, if not done carefully, can introduce serious defects into the eventual reconstructed image. Fortunately a way out of this problem exists, because it is possible to approach the whole question of sampling in a systematic manner that leads to a prescription for obtaining satisfactory reconstructed images.

Sampling simply means calculating the function $F(x, y)$ at a limited number of points. After this is done a hologram is constructed by a computer graphical technique (discussed later on) and an optical method of interrogating the hologram is ultimately used to produce an image field that should be a reproduction of this original object. That, at least, is the ideal aim. Now the hologram being a representation of $F_s(x, y)$, the sampled Fourier transform, rather than $F(x, y)$ the true transform, does not behave like this. Some features are present that are very interesting. The most fundamental feature is that, on reconstruction from a sampled Fourier transform hologram, several images of the object are formed. These are called higher order spectra. Another property of these images is that they can be well separated or overlap.

The appearance of spectra can be understood from an analysis of the sampling process. Suppose a function $f(x)$ is sampled at intervals Δx along x

at the points $x_m = m\,\Delta x$ then the act of sampling $f(x)$ is mathematically equivalent to combining $f(x)$ with the curious function $\text{comb}(x/\Delta x)$ in the manner[1,7]

$$f_s(x) = \frac{f(x)}{\Delta x}\,\text{comb}(x/\Delta x), \qquad (21)$$

where

$$\text{comb}\left(\frac{x}{\Delta x}\right) = \Delta x \sum_{m=-\infty}^{\infty} \delta(x - m\,\Delta x). \qquad (22)$$

Equation (21) is therefore the same as

$$f_s(x) = \sum_{m=-\infty}^{\infty} f(m\,\Delta x)\,\delta(x - m\,\Delta x). \qquad (23)$$

A simple application of the convolution theorem gives the Fourier transform of $f_s(x)$ as

$$G_s(X) = \frac{1}{\Delta x} \sum_{m=-\infty}^{\infty} G\!\left(X - \frac{m}{\Delta x}\right), \qquad (24)$$

an expression that shows, immediately, that a consequence of the sampling is the appearance of an infinite set of periodically shifted versions of the original function.

The hologram is two-dimensional, and represents the function $F_s(x, y)$. The same conclusions apply, however, that the reconstruction produces, in principle, an infinite set of periodically shifted versions of the object, centred now at points $(m\,\Delta x, n\,\Delta y)$. Normally $\Delta x \neq \Delta y$ and the sampling is performed at intervals Δx or Δy, along both x- and y-directions, according to whichever is the largest. Assuming that $\Delta x > \Delta y$, then the Fourier transform of the sampled function leads to the appearance of the object in square cells all over the (X, Y) plane.

This is illustrated in Figure 5, (a), (b), and (c), for the letter X, representing critical sampling, oversampling, and undersampling. These three categories refer to whether the object, i.e. the letter X, just fills a cell, falls well within the cell, overlaps adjacent cells. The critical sampling rate is Δx_c and over- and undersampling is performed at rates $\Delta x < \Delta x_c$ and $\Delta x > \Delta x_c$ respectively. Of the three pictures the oversampled case is the most aesthetic because the images are nicely separated, but obviously oversampling must not be taken too far. If Δx is smaller, as we will see later on, the brightness inhomogeneity across the object is not as great as in the Δx_c case. This, however, is difficult to detect; more important is the implication that a small $\Delta x < \Delta x_c$ indicates a fine sample so that, for a fixed (x, y) field, many more points have to be used or for a fixed number of cells the hologram is magnified which leads to poor image quality.

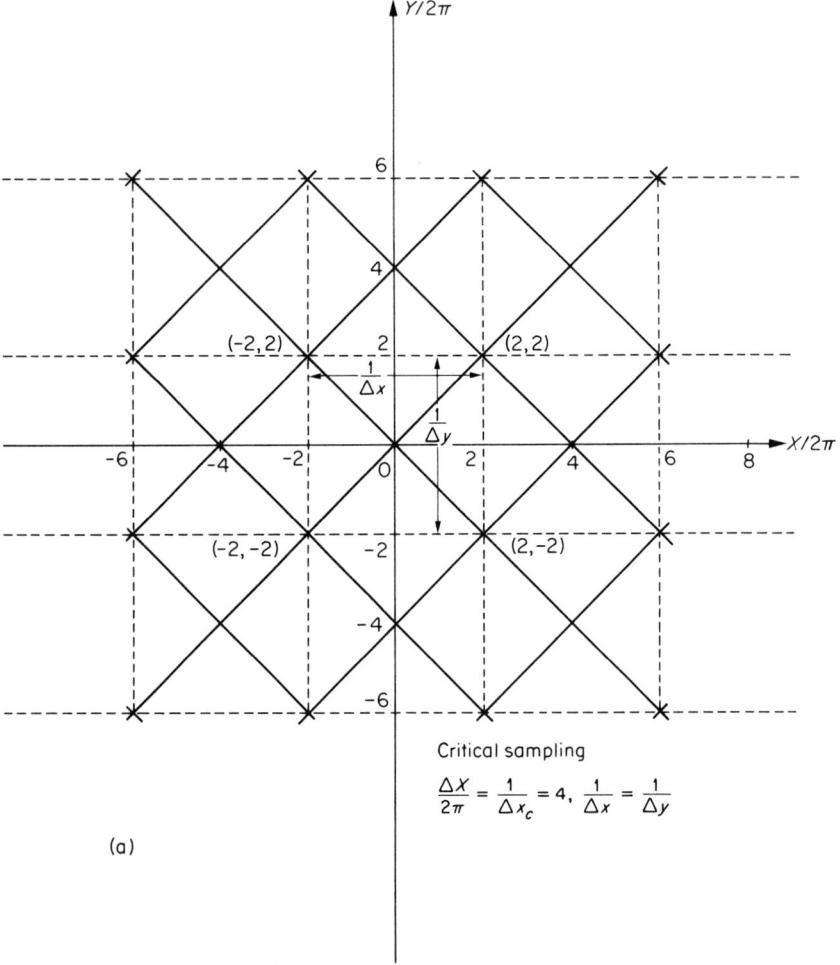

(a)

Figure 5. Illustration, using the letter X, of the consequences of using a sampled Fourier transform to reconstruct the original object with an inverse transform. (a) Critical sampling, (b) oversampling, (c) undersampling. Note that $\Delta X \Delta x = 2\pi$

The critical sampling rate Δx_c is 2π divided by the reciprocal of the maximum size of the object in the X-direction. This is more formally understandable by considering $F(x, y)$ that has, in general, for $y = 0$, a term in it of the form $\sin[(L/2)x \cos \theta] \exp(iX_0 x)$, where L is the length of a line in the object and X_0 is its mid-point. Such a function involves the products $\sin[(L/2)x \cos \theta]\cos(xX_0)$ and $\sin[(L/2)x \cos \theta]\sin(xX_0)$. This means

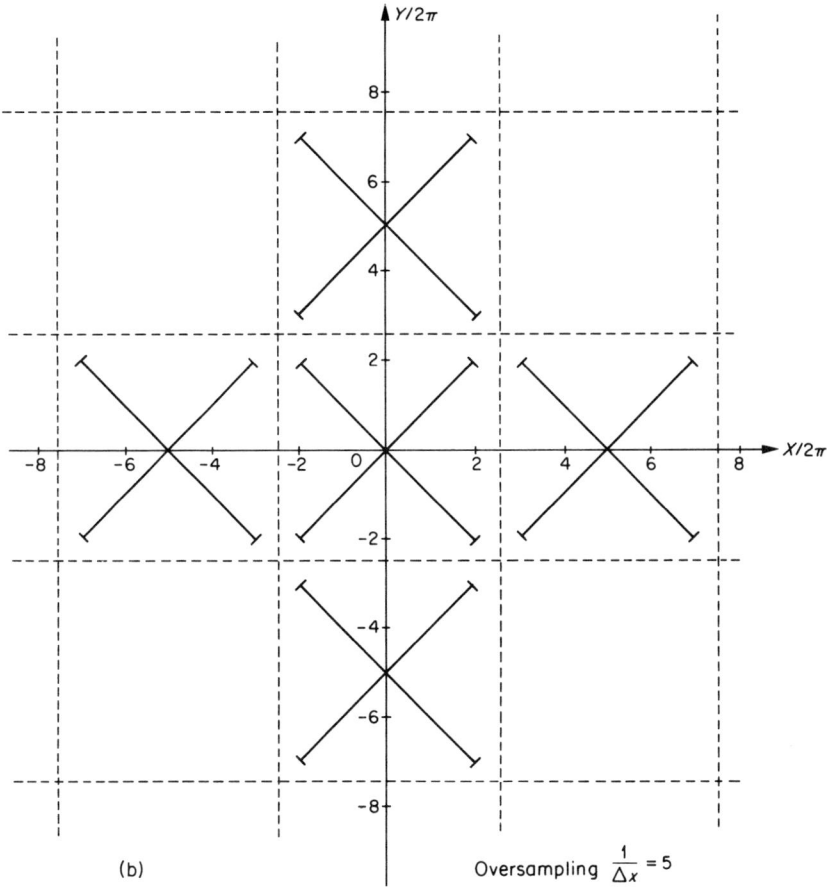

Figure 5b

that, in the x-direction, there are spatial frequencies

$$f_x = \left(X_0 \pm \frac{L}{2}\cos\theta\right)\bigg/2\pi. \qquad (25)$$

Similarly, for the y-direction, there are spatial frequencies

$$f_y = \left(Y_0 \pm \frac{L}{2}\sin\theta\right)\bigg/2\pi. \qquad (26)$$

The spatial frequency bandwidths are, therefore,

$$\Delta f_x = f_x^{(\max)} - f_x^{(\min)} = (L\cos\theta)/2\pi, \qquad (27)$$
$$\Delta f_y = f_y^{(\max)} - f_y^{(\min)} = (L\sin\theta)/2\pi. \qquad (28)$$

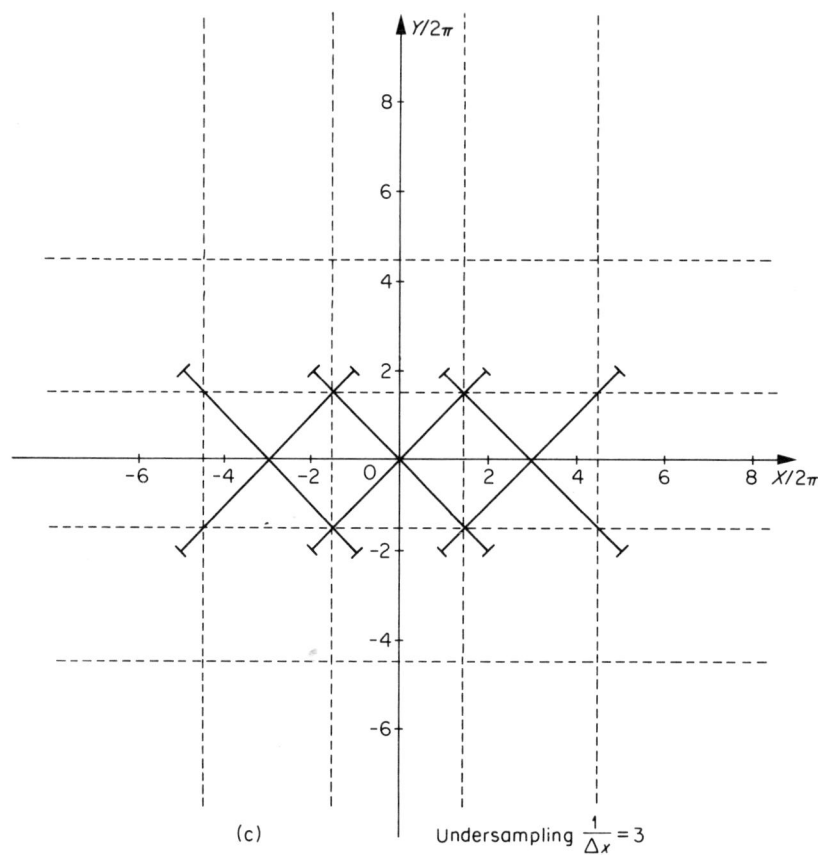

Figure 5c

$2\pi\Delta f_x$ and $2\pi\Delta f_y$ are the projections of the line on the X- or Y-axis and $2\pi f_x^{(\max)}$, $2\pi f_y^{(\max)}$ are the coordinates of the ends of the line. Δf_x and Δf_y are found for a complex object, i.e. a sequence of letters, by taking the difference between the maximum and minimum X-values and the maximum and minimum Y-values. Hence Δx_c is equal to Δf_x^{-1} or Δf_y^{-1}, whichever is the largest. It is evident that if the periodic array is spaced at intervals of $1/\Delta x_c$ then, along X say, $X_{\max} - X_{\min} = 2\pi/\Delta x_c$.

5. BINARY FOURIER TRANSFORM HOLOGRAMS

5.1 Representation of $F(x, y)$ at sampling points

The Fourier transform $F(x, y)$ is a complex quantity of the form

$$F(x, y) = A(x, y)e^{-i\alpha(x, y)}, \qquad (29)$$

that is computed at intervals Δx and Δy along the x- and y-directions. A representation of $F(x, y)$ at these sampling points must be devised before the hologram can be generated by the computer. If, initially, the (x, y) plane is imagined to be opaque, then the hologram can be constructed by making in it a set of rectangular apertures. Each aperture is completely transparent, while its surroundings are opaque. In this sense, the transmittance at all points of the hologram is 0 or 1. Hence the holograms are binary. This makes them like the familiar, halftone newspaper pictures. Continuous-tone pictures were investigated briefly[8] but, as pointed out by Brown and Lohmann,[9] binary holograms allow more light to reach the reconstructed images than the usual, thin emulsion, grey holograms. This and other features[9] make binary holograms an attractive proposition.

It should be recalled that the problem in hand is the representation of $F_s(x, y)$ and this is done, essentially, by constructing an aperture at each sampling point, whose height W_{nm} is proportional to the amplitude of $F_s(x, y)$ at that point and then shifting the aperture off-centre, by an amount P_{nm}, to represent the phase of $F_s(x, y)$. This aperture is located within the mnth sampling cell of the xy plane and centred upon the mnth sampling point $(n \Delta x, m \Delta x)$ (setting $\Delta x = \Delta y$). A variety of aperture shapes could be used, but the rectangular one is extensively discussed in the literature which is why it is used here.

An aperture, like the one shown in Figure 6, has a constant width $c \Delta x$, variable height $W_{nm} \Delta x$, and variable location along x, within the cell, $P_{nm} \Delta x$. The transfer function of the whole hologram is[10]

$$H(x, y) = \sum_n \sum_m \mathrm{rect}\left(\frac{x - [n + P_{nm}]\Delta x}{c \Delta x}\right) \mathrm{rect}\left(\frac{y - m \Delta x}{W_{nm} \Delta x}\right), \qquad (30)$$

where the standard rectangle function

$$\mathrm{rect}(x) = \begin{matrix} 1 & 0 \leq |x| \leq \tfrac{1}{2} \\ 0 & \tfrac{1}{2} < |x| < \infty, \end{matrix} \qquad (31)$$

implies that

$$\left|\frac{x - (n + P_{nm}) \Delta x}{c \Delta x}\right| \leq \tfrac{1}{2}, \qquad (32)$$

$$\left|\frac{y - m \Delta x}{W_{nm} \Delta x}\right| \leq \tfrac{1}{2}, \qquad (33)$$

and hence, explicitly,

$$-c\frac{\Delta x}{2} + (n + P_{nm}) \Delta x \leq x \leq c\frac{\Delta x}{2} + (n + P_{nm}) \Delta x, \qquad (34)$$

$$-\frac{W_{nm}}{2} \Delta x + m \Delta x \leq y \leq \frac{W_{nm}}{2} \Delta x + m \Delta x, \qquad (35)$$

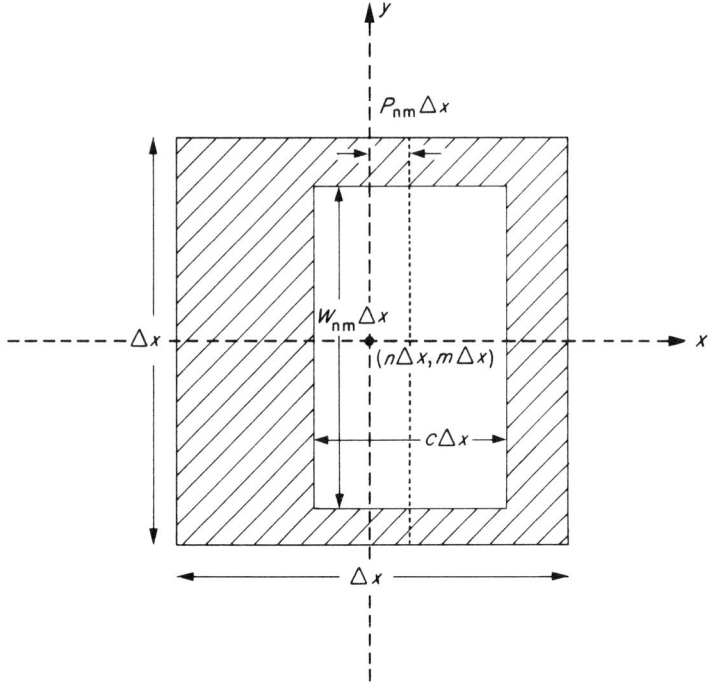

Figure 6. An aperture in a cell in (x, y) space. The area of the cell is $(\Delta x)^2$ and it is centred on the point $(n\,\Delta x, m\,\Delta x)$. The aperture is offset from this point by an amount $P_{nm}\,\Delta x$ but is symmetric about the x-axis

showing that measurements may just as well be made, with respect to the cell centre $m\,\Delta x$, $n\,\Delta x$, in the coordinate system x', y' where

$$\left(P_{nm} - \frac{c}{2}\right)\Delta x \leq x' \leq \left(P_{nm} + \frac{c}{2}\right)\Delta x, \tag{36}$$

$$-\frac{W_{nm}}{2}\Delta x \leq y' \leq \frac{W_{nm}}{2}\Delta x. \tag{37}$$

Suppose that a basic reconstruction system like the one shown in Figure 7 is used.[1,10] An off-axis plane wave $\exp(-ixS)$ from a point source S illuminates the hologram. This choice, incidentally, allows a phase variation along x, and this is consistent with encoding the phase in the hologram by means of shifts P_{mn} in the x-direction only. The amplitude at the hologram plane is $\exp(-ixS)\,H(x, y)$ so that $G(X, Y)$, the inverse Fourier transform, appears

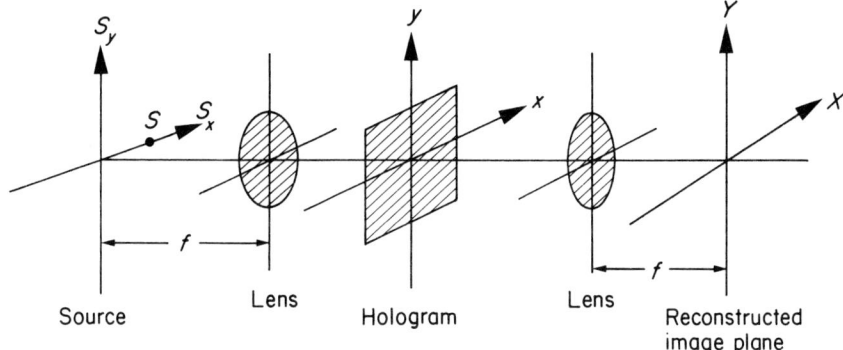

Figure 7. Optical reconstruction system used in the theory of section 5

at the image plane, i.e.

$$G(X, Y) = \sum_n \sum_m \int_{x_{min}}^{x_{max}} \int_{y_{min}}^{y_{max}} e^{-i(xX+yY)} e^{-ixS} \, dx \, dy, \qquad (38)$$

where use is made of equations (30) to (35). Here, x_{min}, x_{max} and y_{min}, y_{max} are, in fact, given by equations (34) and (35). Since $G(X, Y)$ is also required to be a reconstruction of the object, a prescription for identifying W_{nm} and P_{nm} therefore suggests itself.

The integrations in equation (38) are quite easy to perform and lead to

$$G(X, Y) = c(\Delta x)^2 \mathrm{sinc}\left(c\frac{\Delta x}{2}[X+S]\right) \sum_n \sum_m W_{nm} \mathrm{sinc}\left(\frac{W_{nm}}{2}\Delta x Y\right)$$
$$\times \exp\{-[i\Delta x(X+S)(n+P_{nm}) + imY]\}. \qquad (39)$$

This is not a regular series and in order to identify it term by term with a regular Fourier series representation of the object the following approximations[10] ought to be made:

$$R = \mathrm{sinc}\left(c\frac{\Delta x}{2}[X+S]\right) \simeq 1,$$

$$\mathrm{sinc}\left(\frac{W_{nm}}{2}\Delta x Y\right) \simeq 1, \qquad (40)$$

$$\exp(-iXP_{nm}\Delta x) \simeq 1,$$

so that

$$G(X, Y) \simeq c(\Delta x)^2 \sum_n \sum_m W_{nm} \exp[-i\Delta x S(n+P_{nm})] \exp[-i\Delta x(Xn + Ym)]. \qquad (41)$$

If the distribution in the image plane is a reconstruction of the object then

$$G(X, Y) = \int\int F(x, y)\exp[-i(xX+yY)]\,dx\,dy$$

$$\approx \sum_n \sum_m A_{nm}\exp(-i\alpha_{nm})\exp[-i(Xn+Ym)\Delta x]c(\Delta x)^2. \quad (42)$$

Making equations (41) and (42) for $G(X, Y)$ equivalent yields the simple results

$$W_{nm} = A_{nm} \quad (43)$$

$$(\Delta xS)(n+P_{nm}) = \alpha_{nm} \quad (44)$$

to within a multiple $2\pi L_{nm}$, where L_{nm} is an integer.[9] The neglect of L_{nm} can cause cell overlap,[9] but we neglect it here without serious consequences. Hence, in the construction of the hologram, the height W_{nm}, and position P_{nm}, of the aperture in the cell directly record the amplitude and the phase of $F_s(x, y)$ for the sample point $(n\,\Delta x, m\,\Delta x)$. Obviously the aperture should be fairly close to either side of the centre of the cell so that, on average, $\langle P_{nm}\rangle = 0$ and $\langle\alpha_{nm}\rangle = 0$. Equation (44) can be simplified by setting $\Delta xS = 2\pi M$, where M is an integer. This step ensures that

$$e^{i\Delta xSn} = e^{i2\pi Mn} = 1$$

so that it can, in effect, be forgotten. Hence the tidier version of equation (44) is

$$P_{nm} = \alpha_{nm}/2\pi M. \quad (45)$$

The brightness[10] of the image at $X=0$, the centre of the object, is proportional to $[\sin(\pi cM)/\pi M]^2$. The latter expression has maxima when

$$|cM| = \tfrac{1}{2}, \tfrac{3}{2}, \ldots, \quad (46)$$

thus making the brightness at $X=0$ proportional to $1/(\pi M)^2$. The brightness varies as M^{-2} so the lowest value $M=1$ must be selected. This choice of M, from (46), makes $c=\tfrac{1}{2}$. The next choice is $c=\tfrac{3}{2}$ which is rejected on the rather obvious grounds that it defines an aperture larger than the cell size. Even $c=\tfrac{1}{2}$ can cause cell overlap but it is not a serious problem.

Extreme values of R, in equation (40), correspond to $X=\pm(\Delta X/2)$ where $\pm(\Delta X/2)$ refers to a reconstructed object lying between $-(\Delta X/2)$ and $\Delta X/2$. Now $\Delta xS = 2\pi M$, $cM=\tfrac{1}{2}$, and $\Delta x\,\Delta X = 2\pi$ so that R becomes $c\,\text{sinc}[(\pi/2)(1\pm c)]$. Hence, if this factor is not suppressed there is a brightness ratio of $[(1+c)/(1-c)]^2$ from one side of the object to the other. The neglect of it is not very important, because the eye adjusts quite well to its absence, and it is quite easy, if necessary, to incorporate into any computer program. Furthermore, if oversampling is used the ratio is, in any case, reduced. For example, a sampling rate of $\sim\frac{2}{3}$ of the critical sampling rate makes Δx smaller by a factor $\frac{2}{3}$, implying the use of $\Delta xS = 3\pi M$. This gives a brightness ratio of $[(3+2c)/(3-2c)]^2$ which is 4:1 for $c = 1/2$. The other sinc function approximation in equations (40) can be analysed in the same way.

The last approximation in equations (40) amounts to making an error in phase equal to

$$\phi = P_{nm} X \Delta x = \alpha_{nm}/2, \qquad (47)$$

where $-\Delta X/2 \leq X \leq \Delta X/2$, $\Delta X \Delta x = 2\pi$ and $M = 1$. Therefore, since α lies between π and $-\pi$, the error in phase cannot amount to more than $\pi/2$ or a path difference of $\lambda/4$ where λ is the wavelength. This sort of error is within the Rayleigh criterion so it is probably not very serious.

5.2 Amplitude clipping and random phasing

(a) Amplitude clipping

The amplitude and phase of $F_S(x, y)$ are calculated for each cell in the hologram plane and a typical distribution is shown in Figures 8 and 9 in the form of histograms using 50 channels or levels. The amplitude plot is normalized to make all the amplitudes lie between 0 and 1. The interesting point about the amplitude plot is that nearly all the amplitudes are crowded

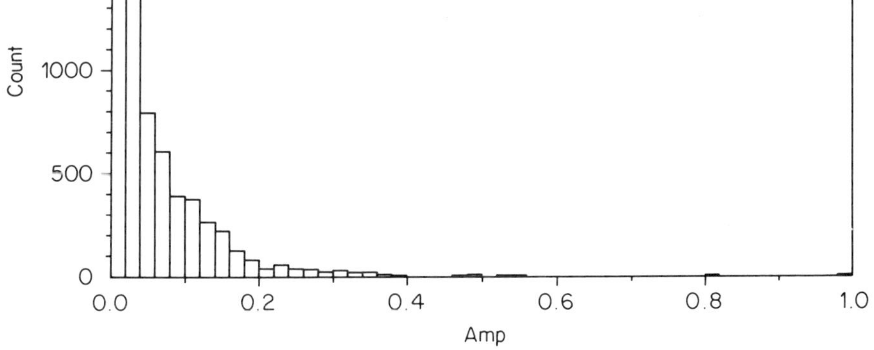

Figure 8. Amplitude histogram: no random phasing

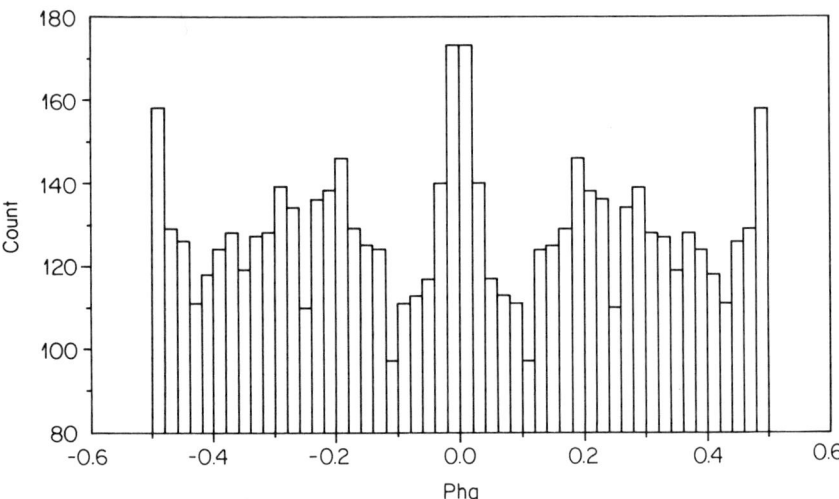

Figure 9. Phase histogram: no random phasing. Note that the phase distribution is symmetrical about the origin

into the low-amplitude range. This feature is a consequence of using letters as an object. It would seem prudent to introduce some form of clipping to chop off the top end of the amplitude range, where so few counts are recorded and then renormalize. This has the effect of making the apertures wider, for the other amplitudes, than they would otherwise be and hence the image field ought to be brighter.

The nature of the clipping used in the computer program is as follows. The original amplitudes are in the range $0 \leq A \leq 1$ and the number of histogram levels, or channels, is 50 (incidentally this number is easily located in the program and is easily changed). Suppose that by the time some channel number n_L is reached 99 per cent of the amplitudes are counted. A clipping factor $50/n_L$ can then be formed and all the original amplitudes converted to $A' = (50/n_L)A$. This scales up all the amplitudes and effectively makes the original amplitude $A = A_L$, at n_L, equal to unity but permits a very small number of amplitudes to become greater than unity. This is only an apparent difficulty because for any $A' > 1$ the aperture is completely open and so few cells are involved that it does not matter. An examination of Table 1, the contents of which come out on output channel 3, confirms the points discussed above. The critical sampling rate Δx_c and the modified sampling rate actually used is also given in Table 1 as is some information used later concerning the plotting of the hologram. Note from Table 1 and Figure 9 that the phase distribution is symmetrical about the origin.

Table 1. Typical output on channel 3 for the pen plotter with linewidth = 0.5 mm. 13 lines fill the cell. Units = 3.375

```
SAMPLE RATE =    0.967
MODIFIED SAMPLE RATE =   0.638
DISTRIBUTION OF AMPLITUDE AND PHASE IN HOLOGRAM

   AMP    NUMBER    PHASE    NUMBER
  0.010   1700.0    -0.49    158.0
  0.030   1566.0    -0.47    129.0
  0.050    792.0    -0.45    126.0
  0.070    604.0    -0.43    111.0
  0.090    390.0    -0.41    118.0
  0.110    374.0    -0.39    124.0
  0.130    260.0    -0.37    128.0
  0.150    220.0    -0.35    119.0
  0.170    124.0    -0.33    127.0
  0.190     78.0    -0.31    128.0
  0.210     36.0    -0.29    139.0
  0.230     56.0    -0.27    134.0
  0.250     36.0    -0.25    110.0
  0.270     34.0    -0.23    136.0
  0.290     22.0    -0.21    138.0
  0.310     30.0    -0.19    146.0
  0.330     18.0    -0.17    129.0
  0.350     20.0    -0.15    125.0
  0.370      8.0    -0.13    124.0
  0.390      4.0    -0.11     97.0
  0.410      0.0    -0.09    111.0
  0.430      0.0    -0.07    113.0
  0.450      0.0    -0.05    117.0
  0.470      4.0    -0.03    140.0
  0.490      8.0    -0.01    173.0
  0.510      0.0     0.01    173.0
  0.530      4.0     0.03    140.0
  0.550      4.0     0.05    117.0
  0.570      0.0     0.07    113.0
  0.590      0.0     0.09    111.0
  0.610      0.0     0.11     97.0
  0.630      0.0     0.13    124.0
  0.650      0.0     0.15    125.0
  0.670      0.0     0.17    129.0
  0.690      0.0     0.19    146.0
  0.710      0.0     0.21    138.0
  0.730      0.0     0.23    136.0
  0.750      0.0     0.25    110.0
  0.770      0.0     0.27    134.0
  0.790      0.0     0.29    139.0
  0.810      4.0     0.31    128.0
  0.830      0.0     0.33    127.0
  0.850      0.0     0.35    119.0
  0.870      0.0     0.37    128.0
  0.890      0.0     0.39    124.0
  0.910      0.0     0.41    118.0
  0.930      0.0     0.43    111.0
  0.950      0.0     0.45    126.0
  0.970      0.0     0.47    129.0
  0.990      4.0     0.49    158.0
SUGGESTED CLIP =  2.941 TO CHANGE TYPE 1 ELSE 0
```

(b) *Random phasing*

It will be seen later that the hologram that can be plotted using the theoretical work dealt with up to now has a star-like structure. Furthermore, each line of the object is associated with a high-density region of the hologram. In a typical letter-group object there is a considerable number of horizontal lines and vertical lines so that a significant overlap occurs. If the hologram is produced automatically as a film output from the computer or is graph-plotted and then photographically reduced, very poor image quality can be expected. This troublesome feature would be overcome if it were possible to spread out the hologram (Fourier) amplitudes, over the hologram plane, in some way so as to reduce the area of overlaps. A method of doing this that is very easy to include in a computer program is to multiply the object amplitudes by a random phase factor. After doing this the Fourier amplitude distribution becomes more uniform.

Suppose the object function $G(X, Y)$, is arbitrarily multiplied by a phase factor $\exp(-iaX - ibY)$ then, as we discussed earlier, its Fourier transform shifts to $F(x-a, x-b)$. The introduction of such a phase factor into the function $G(X, Y)$ does not make any physical difference because it is eliminated from the observable $|G(X, Y)|^2$. Since this is the case, then a and b can be quite arbitrarily chosen and an excellent choice would be to make them random numbers. This, in fact, is what is done in the program and they are selected by a random number routine that produces uniformly distributed numbers $0 \leq r \leq 1$. The random number r is then converted to a set $r' = 2r - 1$ that lies on the interval $-1 \leq r' \leq 1$ and a fraction of the maximum size of x and y is used to form

$$a = \frac{r'}{4} x_{\max}, \qquad (48)$$

$$b = \frac{r'}{4} y_{\max}. \qquad (49)$$

The effect of using a and b is to spread out or equalize the amplitudes. This means that, before clipping, there is a greater occupancy of the higher channel numbers as is seen in Table 2 and Figure 10. Note also that now the phase distribution is no longer symmetrical.

5.3 Drawing the hologram

The computer graphics package is used to generate the hologram in the form of a large-scale pen plotter plot that is later reduced photographically or in the form of a directly usable 35 mm microfilm slide.

In order to find W_{nm} and P_{nm} the computer program determines the complex Fourier transform as an amplitude $FXY(i, j)$ and a phase ANGLE

Table 2. Typical output on channel 3 for the microfilm unit with linewidth = 0.2 mm. 33 lines fill the cell. Units = 3.375

```
SAMPLE RATE =    0.967
MODIFIED SAMPLE RATE =   0.638
DISTRIBUTION OF AMPLITUDE AND PHASE IN HOLOGRAM

  AMP    NUMBER    PHASE    NUMBER
 0.010   280.0    -0.49    145.0
 0.030   636.0    -0.47    140.0
 0.050   734.0    -0.45    112.0
 0.070   554.0    -0.43    121.0
 0.090   455.0    -0.41    131.0
 0.110   399.0    -0.39    124.0
 0.130   311.0    -0.37    129.0
 0.150   278.0    -0.35    107.0
 0.170   310.0    -0.33    116.0
 0.190   260.0    -0.31    124.0
 0.210   223.0    -0.29    108.0
 0.230   244.0    -0.27    120.0
 0.250   213.0    -0.25    113.0
 0.270   211.0    -0.23    116.0
 0.290   188.0    -0.21    131.0
 0.310   135.0    -0.19    135.0
 0.330   164.0    -0.17    131.0
 0.350   124.0    -0.15    120.0
 0.370    95.0    -0.13    116.0
 0.390    78.0    -0.11    120.0
 0.410    70.0    -0.09    120.0
 0.430    52.0    -0.07    144.0
 0.450    49.0    -0.05    141.0
 0.470    59.0    -0.03    158.0
 0.490    43.0    -0.01    117.0
 0.510    30.0     0.01    162.0
 0.530    30.0     0.03    136.0
 0.550    28.0     0.05    148.0
 0.570    22.0     0.07    119.0
 0.590    16.0     0.09    117.0
 0.610    14.0     0.11    139.0
 0.630    11.0     0.13    106.0
 0.650    11.0     0.15    120.0
 0.670    14.0     0.17    114.0
 0.690    14.0     0.19    107.0
 0.710     4.0     0.21    131.0
 0.730    10.0     0.23    134.0
 0.750     3.0     0.25    147.0
 0.770     4.0     0.27    125.0
 0.790     4.0     0.29    140.0
 0.810     2.0     0.31    119.0
 0.830     3.0     0.33    110.0
 0.850     2.0     0.35    120.0
 0.870     1.0     0.37    116.0
 0.890     3.0     0.39    122.0
 0.910     4.0     0.41    164.0
 0.930     2.0     0.43    157.0
 0.950     0.0     0.45    148.0
 0.970     0.0     0.47    120.0
 0.990     3.0     0.49    140.0
SUGGESTED CLIP =  1.471 TO CHANGE TYPE 1 ELSE 0
```

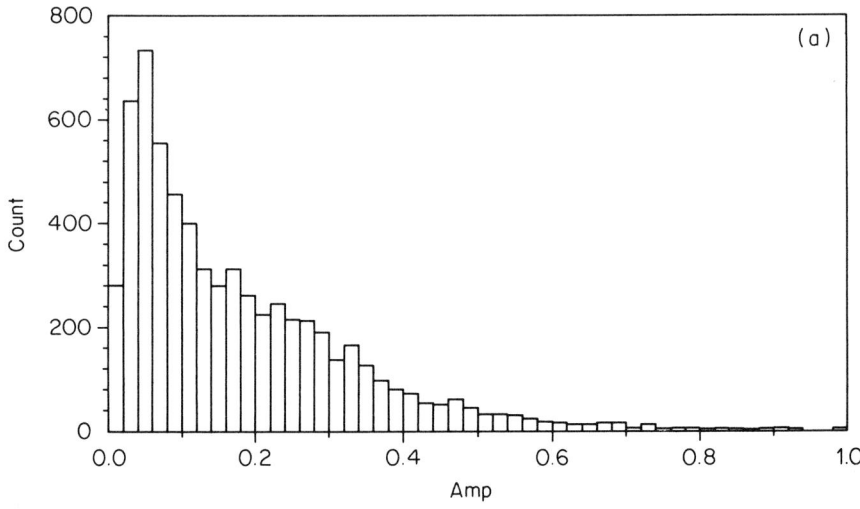

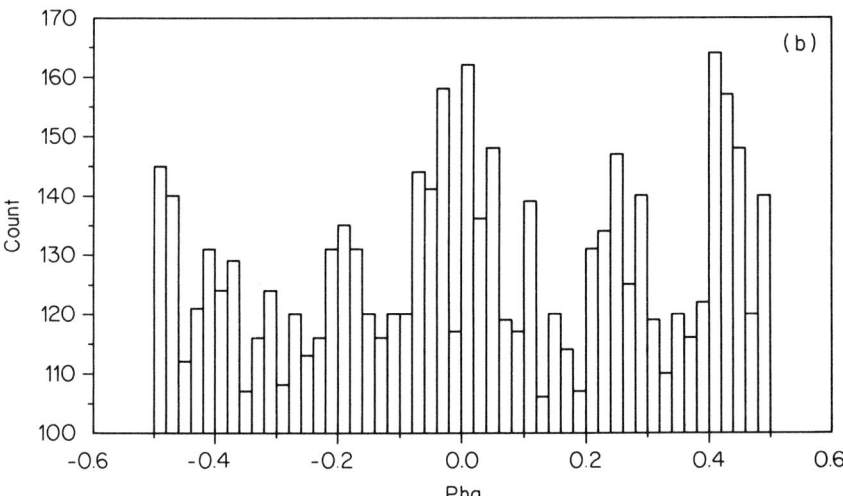

Figure 10. (a) Amplitude histogram: with random phasing. (b) Phase histogram: with random phasing

(i, j) that is normalized to 2π. The sampling points are the centres of cells covering the x–y plane whose dimensions are $(\Delta x)^2$ and are labelled (i, j). The origin of the coordinate system is in the centre of the array with a cell symmetrically disposed about the origin, as shown in Figure 11. Hence, if the number of sampling points in the x- and y-directions are i_m, j_m

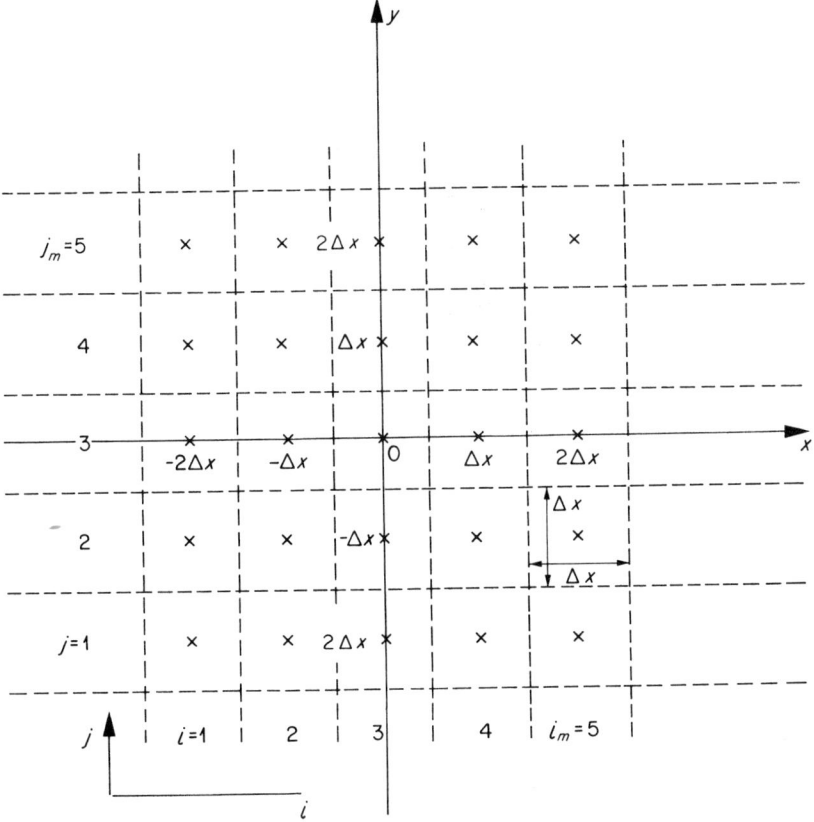

Figure 11. Distribution of sampling points in the (x, y) plane

(ISAMPS and JSAMPS in the program) then the coordinates of the sampling points are

$$x = [i - \tfrac{1}{2}(i_m + 1)] \Delta x, \qquad (50)$$
$$y = [j - \tfrac{1}{2}(j_m + 1)] \Delta x. \qquad (51)$$

An 80×80 cell array is used. The sampling point is $(n \Delta x, m \Delta x)$ and it is assumed that $F(x, y)$ is virtually constant over the cell area. Figure 6 shows the cell drawn, conventionally, as completely opaque, except for an aperture of height $W_{nm} \Delta x$ and width $c \Delta x$ that is displaced a distance $P_{nm} \Delta x$ from the centre. This is the aperture that is the basis of the preceding theoretical discussions. It is not, however, computer-generated in this way. It is, in fact, drawn, by the computer, in the manner of Figure 12 in which the aperture is made opaque by using a black ink pen plotter or by exposing a 35 mm film

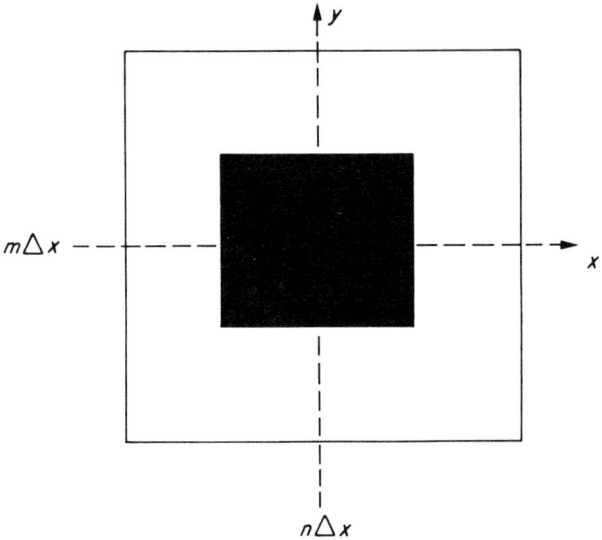

Figure 12. Appearance of opaque region of the unit cell of the hologram as drawn by the computer

unit controlled by the computer. A negative can be made of the holograms produced in this way, but it is not necessary to do this.

The plotting begins with the pen located at the extreme bottom left corner of the hologram and progresses from left to right, scanning each cell. It is then relocated at the extreme left, once again, to begin a scan of the next row of cells. The shading in of the apertures creates a black rectangular shape in each cell as shown in Figure 12.

The details of the way the program controls the behaviour of the plotter will now be discussed. The first important point is that the pen itself has a finite width, as also does the beam that is used in the film plotter device. A reasonable estimate of the pen width is 0.5 mm and for the beam 0.2 mm. This means that the minimum line thickness that can be made by the pen is $\sim 1/50$ in. Hence the minimum amplitude that can be represented graphically is limited by the linewidth of the plotter. The width of each aperture is a constant equal to half the width of the unit cell, and the unit cell size is found from a knowledge of the drawing area.

The area of the hologram is selected as either 3600×270 mm^2, 3600×840 mm^2, or 430×270 mm^2. Now if i_m and j_m (NXDIM and NYDIM in the subroutines) are the number of cells in the x- and y-directions then, for the 3600×270 area, say, the two quantities $x_s = 3600/i_m$ and $y_s = 840/j_m$ can be compared. The smallest of the two values x_s, y_s is the number of millimetres per unit cell that will just fill the drawing area, and for an 80×80 cell array $x_s = 45$ mm and $y_s = 3.375$ mm. The cell size is therefore 3.375×3.375 mm^2

and the drawing units are upgraded, by the program, from the default value of 1 mm to $S = 3.375$ mm. It is now possible to determine the number of lines drawn by the pen that will just fill a unit cell.

The pen begins at the bottom left-hand cell of the hologram, labelled $i = 1$, $j = 1$. It is then moved along what in the plotting space is now called the y-axis a distance equal to $\frac{1}{4}$ (all distances refer to fractions of a unit cell width). If the phase is zero this is the edge of the aperture. The program then decides if a line is to be drawn. If it is not, the pen is moved on without drawing a line through $\frac{1}{2}$. By this time the pen is $\frac{1}{4}$ from the edge of the (1, 1) unit cell and is then moved through $\frac{1}{4}$ to test if a line needs to be drawn in the next unit cell. Now if a line does need to be drawn the pen first moves on by an amount ANGLE (I, J) to incorporate the phase (P_{nm} in the theory given earlier) and then draws a line of length $\frac{1}{2}$. It is then moved back by an amount ANGLE (I, J) to cancel the phase shift before moving on $\frac{1}{4}$ to the next cell. This is done across the complete row of cells. After this the pen is relocated at the left-hand side of the hologram and is incremented up the cell by $\frac{1}{2}$ a linewidth. The $\frac{1}{2}$ linewidth increment arises because the minimum that can be drawn in a cell is one linewidth, but the lines must be symmetrically drawn about the centre of the aperture, i.e. if a single line only is drawn it must symmetrically occupy the centre of the aperture. Using an increment of $\frac{1}{2}$ linewidth causes all the lines to overlap in the manner of Figure 13.

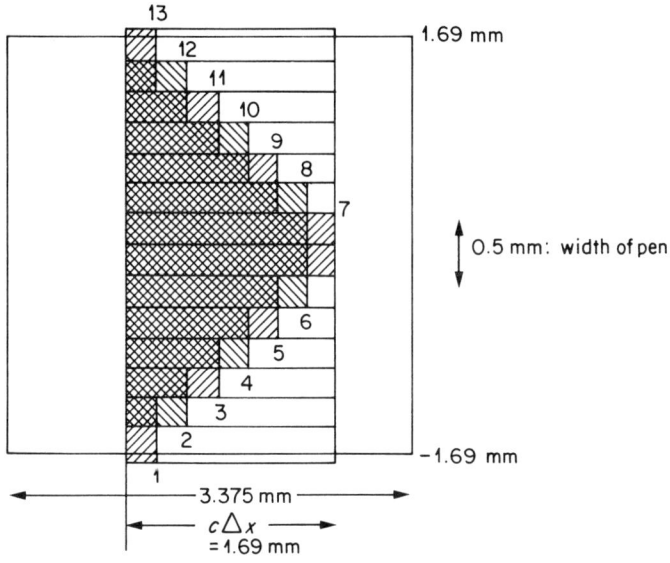

Figure 13. Scale cut-away magnification of the overlapping pen lines for the case $W_{nm} = 1$. 13 lines are used with a small overlap at the top and bottom of the cell

The number of lines that fill a unit cell of the hologram, using a $\frac{1}{2}$ linewidth increment, is

$$\text{NULINE} = 2\left[\!\left[\frac{S}{\text{ALINWD}} + 0.5\right]\!\right] - 1, \quad (52)$$

where the cell dimension is S, ALINWD is the linewidth of the plotter pen, or beam, and $[\![\ ]\!]$ denotes the truncated value of the number in integer form (equivalent to using IFIX in FORTRAN). For S = 3.375, and ALINWD = 0.5, NULINE is 13, as shown in Figure 13. Thinner lines are possible for the film output device so ALINWD = 0.2 is chosen in the program, giving NULINE = 33. The formula (52) ensures that, if the aperture is as wide as the cell, the cell is completely filled with lines. There is, with this method, a small overspill. This, however, is variable because the pen-width tends to vary. This possibility of variation in pen-width really makes overlapping advisable, even if there are other non-overlapping methods of filling the cell. The overspill does not prevent the hologram from behaving satisfactorily, since there are relatively few completely filled cells.

The height of the aperture is equal to the amplitude W = FXY(I, J) × CLIP associated with the cell, where $0 \leq W \leq 1$. The program forms two fractions X = N/NULINE and X1 = (N − 1)/NULINE where N labels a line somewhere between the bottom and the top of the cell and $1 \leq N \leq \text{NULINE}$. For an aperture of height W the upper end is positioned at (1 + W)/2 and the lower end at (1 − W)/2. So for X1 < $\frac{1}{2}$(1 − W) and X > $\frac{1}{2}$(1 + W) no lines are drawn.

6. THE COMPUTER PROGRAM

6.1 Description

The full computer program is listed at the end of the chapter and contains a list of comments sufficiently comprehensive for a user to make any necessary changes. Obviously a program like this must use a sophisticated graphics package. The one used here is called GINO-F that has practically universal use in the U.K. and is available in some other countries. Changing the program to fit an alternative system should present few problems.

The program can be run interactively and if a good graphics package is available this is an extremely satisfactory mode of operation. This is because many features can be demonstrated quickly and vividly in an interactive mode using video display units. It must be emphasized, though, that it is necessary to produce high-quality hard copies if the eventual reconstruction process is to be successful.

If interactive facilities are not available then it is submitted as a normal job to the computer, for batch processing, and the interactive mode requests, that are redundant in this mode, are simply filed via channel 2.

A control file, fed in through channel 1, contains the data that controls the whole program. The object can be any arrangement of 30 straight lines and the coordinates of these are read in through channel 4. The example chosen to illustrate this chapter is the letter group MESC that consists of 16 straight lines. The possibility of using 30 lines should allow most students to develop a hologram for their own initials or even their names. If 30 lines is not enough this can be easily increased through an elementary program modification.

The amplitude and phase distributions, together with the sample rate value, its modified value, the clipping factor, the drawing units of the hologram, and the number of lines needed to fill the aperture appear through channel 3.

Most of the rest of the program was discussed, in one form or another, in the earlier sections. The final stages, however, includes the possibility of producing a line printer output of the amplitude distribution. This is done by overprinting and, if required, appears through channel 3. We also thought that it may be interesting to place a window over the hologram and produce a set of holograms of varying size by simply changing the window. Finally, a facility for controlling the hologram development, whilst using the program in an interactive mode with a video display unit is included. However, the quality of the hard copies of such holograms needs careful monitoring.

6.2 Running the program

If it is assumed that a hologram is to be produced in a non-interactive manner then the following FORTRAN command is used:

FORTRAN PROGRAM = HOLO, NONFABS,   LIB = NAG, PLOT,
    DATA1 = CONTROL,   DATA4 = OBJECT,   WRITE2 = OUT1,
    WRITE3 = OUT2, JD(JT2000, MZ80K, URF).

The purpose in stating this here, and in this form, is that it literally shows what the program does.

The command shows that the program is in a file called HOLO and, since hologram production is likely to be a reasonably lengthy process, that the quick turn-round, limited batch service is not required (hence NONFABS). The comprehensive NAG library is requested and the desire to do plotting is expressed through PLOT. The control data are in a file called CONTROL, object data are in a file OBJECT, the interactive mode requests are dumped into a file OUT1, and the amplitude phase distribution, etc., is placed in file OUT2. The job description, JD(...), contains the job time JT that, for the 1904S computer, is $\simeq 2000$ s, the core requirement MZ of 80 K and an urgency code URF (fast).

The two data files are CONTROL and OBJECT. OBJECT contains:

	Format	Name in program
(1) Number of line segments in the object	I2	NULINE
(2) Coordinates of each end of a line (X1, Y1, X2, Y2)	4F(4.1, 1X)	APOINT

Note: Later in the program the definition of NULINE is changed to mean the number of lines that completely cover the aperture.

For the group of letters MESC, in the form MESC the file OBJECT contains 16:

−3.5	1.0	−3.5	−1.0	
−3.5	1.0	−2.5	−1.0	
−2.5	−1.0	−1.5	1.0	
−1.5	1.0	−1.5	−1.0	
−1.0	1.0	−1.0	−1.0	The units are set as centimetres for drawing
−1.0	1.0	0.0	1.0	the object as part of the output.
−1.0	0.2	−0.5	0.2	The object is restricted to the area
−1.0	−1.0	0.0	−1.0	$-10.0 \leq X \leq 10.0$
0.5	1.0	1.5	1.0	$-7.5 \leq Y \leq 7.5$
0.5	1.0	0.5	0.2	
0.5	0.2	1.5	0.2	
1.5	0.2	1.5	−1.0	
0.5	−1.0	1.5	−1.0	
2.0	1.0	3.0	1.0	
2.0	1.0	2.0	−1.0	
2.0	−1.0	3.0	−1.0	

The file CONTROL contains:

	Format	Name in program
(1) Output device 1 = lineprinter: for overprinting display of amplitude distribution. 2 = Narrow plotter: normal size 3 = Wide plotter: uses very much wider paper. 4 = Microfilm output.	I1	NDEV
(2) Critical sampling rate multiplier, i.e. sampling rate actually used = critical rate × AMULT	F4.1	AMULT
(3) Number of sampling points in x-direction	I3	ISAMPS (or NXDIM)

(4)	Number of sampling points in y-direction	I3	JSAMPS (or NYDIM)
(5)	1: With random phasing 0: Without random phasing	I1	NRAN
(6)	1: Modification of clipping factor 0: No modification of clipping factor	I1	NCLIP
(7)	New clipping factor if 1 is used for (6)	F4.2	CLIP
(8)	Number of extra (smaller holograms required by window variation	I1	NSM
(9)	In the final section of subroutine DISPLAY, a further opportunity to modify the clipping factor occurs. It is appropriate only to the interactive mode and is suppressed by using 0.0 as data	F3.1	CLIP
(10)	Related to the interactive opportunity, outlined in (9). Entering 0 as data suppresses this part of the program.	I1	NDEV

A typical example of this file is as follows:

 2   : narrow plotter selected
 0.66: AMULT
 80 : number of cells in x-direction
 80 : number of cells in y-direction
 0   : no random phasing required
 0   : no modification of clipping factor required
 0   : no extra holograms requested
 0.0 : suppression of interactive mode opportunity
 0   : suppression of interactive mode opportunity

In the interactive mode all program input and output takes place at the terminal so that in the FORTRAN command = CONTROL, = OBJECT, = OUT1 and = OUT2 are omitted. The interactive mode is extremely valuable for instantly monitoring the influence of parameter variation on the hologram development, but the non-interactive mode is generally satisfactory for production runs.

6.3 Output from the program

The hologram, of course, appears on the graphical output device but, as pointed out earlier, in the non-interactive mode the interactive requests are dumped on to channel 2 and the amplitude and phase tables appear through channel 3.

Before the hologram is drawn, the object is plotted out and, as the comment in the program states, although it is not actually necessary, this action does provide a check on the object data. The amplitude and phase distributions are plotted next and, finally, the hologram (or holograms) are drawn.

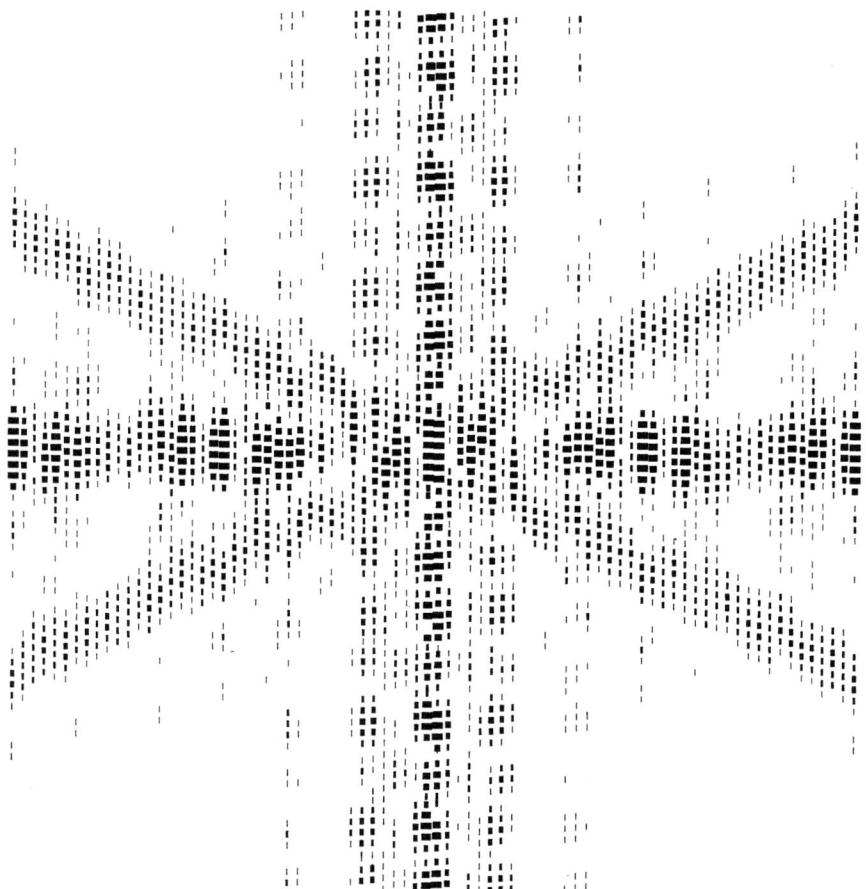

Figure 14. Hologram of the letter-group object MESC: without random phasing

For the object MESC holograms, such as those shown in Figures 14 and 15, are produced according to whether or not random phasing is included. These holograms cover an 80×80 unit cell array using AMULT = 0.66 and the unmodified clipping factor, set inside the program. Both holograms exhibit the star-like structure associated with the lines of the object. Figure 15 shows quite clearly the way in which the random phasing 'spreads' the picture out.

7. THE RECONSTRUCTION

The computer-generated holograms shown in Figures 14 and 15 are Fourier transform holograms.[1,7,11] The reconstruction process is, in principle, the

Figure 15. Hologram of the letter-group object MESC: with random phasing

optical process shown in Figure 16. An optically produced Fourier transform hologram uses the Fourier transforms of the object and reference source to produce an interference pattern. The reconstruction process produces the usual twin images, but in the special case of the Fourier transform hologram they are located at infinity and a lens is normally used to relocate them in its focal plane. The twin images are related to each other by an inversion operation, through the zero order position.

The computer-generated hologram behaves in just the same way, with respect to the twin images and the inversion, but sampling the true Fourier transform leads to spectra, i.e. many other images are present. Optically produced holograms do not have this feature.

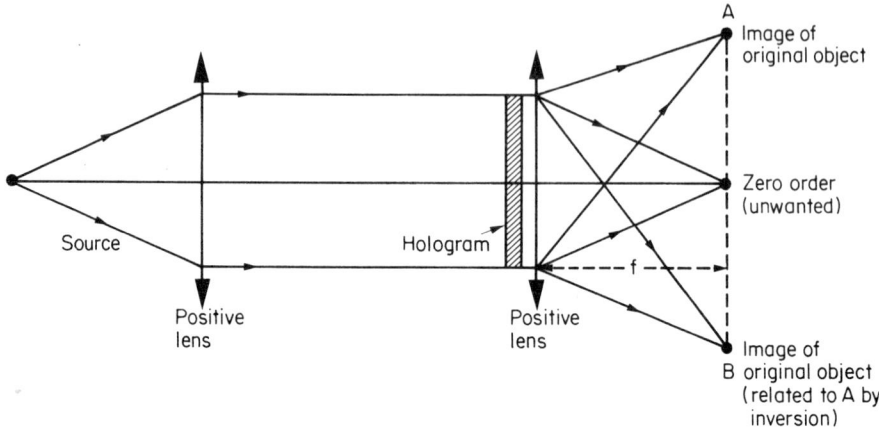

Figure 16. Reconstruction with a Fourier transform hologram

The results, obtained by reconstructing the random phase hologram, generated by the computer program given here, are shown in the photographs in Figures 17 and 18. These are obtained from holograms consisting of photographically reduced graph plotter output. Very good results are also obtained by using the 35 mm slide film directly produced by the computer, without any further photographic work. A He–Ne laser was used by focusing it to a point source with an ordinary microscope lens of 25 mm focal length. It was then collimated to a plane wave of sufficient cross-sectional area to illuminate the hologram. The reconstructions are generally small so it was found easier to observe the images on a screen in the far field, at several metres, rather than use a second lens. A positive or a negative of the hologram works equally well.

Figure 17 shows clearly the original MESC object and also that sampling has the effect of producing other images, even though the light intensity in them falls away quite rapidly. The effect of oversampling is shown by the fact that the images are nicely separated. Figure 18 is a novel photograph obtained by allowing the laser beam to enter the camera directly. Thus Figure 18 is an aerial photograph in which the objective lens is used to mask off the rest of the image field.

8. EXERCISES AND PROJECT SUGGESTIONS

The program can be used as a demonstration to produce holograms and reconstructions of the initials or name of the user. A deeper investigation will lead to an understanding of the effects of oversampling and undersampling, but it should be noted that certain straight-line objects, such as the

COMPUTER-GENERATED HOLOGRAMS 111

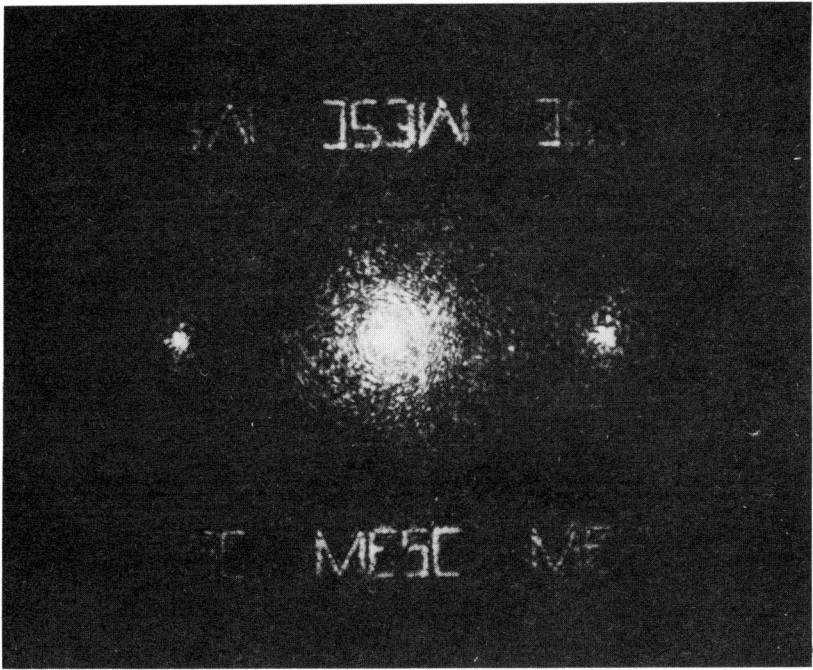

Figure 17. Photograph of the image field obtained using the hologram in Figure 15 produced directly on microfilm

letter H, for instance, have only vertical and horizontal lines. Without random phasing the hologram, for such a letter, has the appearance of a vertical line at right angles to a horizontal line. The introduction of random phasing will split and separate the vertical line into two lines, but the horizontal line is only shifted. A way out of this is to express each line as the sum of N lines where N is large enough to make a difference. If this is done then random phasing will produce many lines in the hologram. This procedure is obviously valuable for general objects and rounded letters that are the limit of an N-line object as N becomes large.

Other investigations could include varying the clipping, taking another look at the approximations, windowing the hologram and considering other apertures and shapes such as circular or elliptic. Finally a line printer was used recently[12] to produce binary holograms. This was done by using the real part of the Fourier transform to control the overprinting, rather than just the amplitude, as is used in the program here. Overprinting is allowed if the real part of the transform is greater than zero, if it is not, then a blank space is left. The computer program given here could be adapted for this purpose by altering the overprinting section and selecting device 1 = LP in the data.

Figure 18. Aerial photograph of reconstructed object

REFERENCES

1. R. J. Collier, C. B. Burckhardt, and L. H. Lin, *Optical Holography* (Academic Press, New York, 1971).
2. D. C. Chu and J. R. Fienup, *Opt. Eng.*, **13,** 189 (1974).
3. J. N. Butters, *Holography and its Technology* (Peter Peregrinus Ltd. on behalf of I.E.E., 1971).
4. M. Born and E. Wolf, *Principles of Optics* (Pergamon Press, Oxford, New York, 1970).
5. J. S. Marsh and R. C. Smith, *Am. J. Phys.*, **44,** 774 (1974).
6. R. C. Smith and J. S. Marsh, *J. Opt. Soc.*, **64,** 798 (1974).
7. J. W. Goodman, *Introduction to Fourier Optics* (McGraw-Hill, New York, 1968).
8. W. H. Lee, *App. Opt.*, **9,** 639 (1970).
9. B. R. Brown and A. W. Lohmann, *IBM J. Res. Dev.*, **13,** 160 (1969).
10. A. W. Lohmann and D. P. Paris, *App. Opt.*, **6,** 1739 (1967).
11. T. S. Huang, *I.E.E.E.*, **59,** 1335 (1971).
12. K. Nagashima, *Opt. Comm.*, **17,** 273 (1976).

DIGITAL HOLOGRAMS

```
C     THIS PROGRAM GENERATES DIGITAL HOLOGRAMS FOR OBJECTS
C     MADE UP OF STRAIGHT LINES. IT IS WRITTEN
C     TO RUN INTERACTIVELY(OR  OTHERWISE) IN STANDARD FORTRAN. THE
C     GRAPHICAL PROCEDURES USE ELEMENTS OF GINO-F(GRAPHICAL INPUT/OUTPUT
C     -FORTRAN)DEVELOPED BY THE COMPUTER AIDED DESIGN CENTRE ,
C     CAMBRIDGE UNIVERSITY.IT IS READILY AVAILABLE ON DIFFERENT
C     MACHINES IN THE U.K. AND ELSEWHERE.UPTO 30 LINES MAYBE
C     SPECIFIED BY THEIR CO-ORDINATES IN AN AREA DEFINED BY
C     -10.0<X<10.0 , -7.5<Y<7.5 IN ARBITRARY UNITS.THE HOLOGRAM CAN
C     BE AS LARGE AS 80 BY 80 CELLS-THIS IS SET TO CONTAIN
C     PROGRAM WITHIN A CORE REQUIREMENT OF 80K

C     DEFINITIONS OF CHANNELS USED IN
C     FORMAT STATEMENTS.
C        CHANNEL 1=CONTROLFILE-INPUT
C        CHANNEL 2=OUTPUTS REQUESTS FOR DATA IN AN INTERACTIVE
C        MODE-FILED IF NOT REQUIRED
C        CHANNEL 3=RESULTS-OUTPUTS AMPLITUDE AND PHASE DISTRIBUTION.
C        CHANNEL 4=OBJECT-INPUTS

C     DEFINITION OF VARIABLES
C        FXY(X,Y)    = AMPLITUDE OF FOURIER TRANSFORM
C        ANGLE(X,Y)  = PHASE OF FOURIER TRANSFORM
C        F           = COMPLEX FOURIER TRANSFORM
C        APOINT      = ARRAY CONTAINING CO-ORDINATES
C                      OF LINE ENDS
C        COSTH,SINTH = COSINE AND SINE OF ANGLE LINE
C                      MAKES WITH X-AXIS
C        L           = ARRAY CONTAINING LINE LENGTHS
C        A,B         = ARRAYS USED FOR RANDOM PHASE
C        XM,YM       = ARRAYS CONTAINING CO-ORDINATES
C                      OF MID-POINTS OF LINE SEGMENTS
C     FREE FORMATS ARE NOT USED BECAUSE THEY ARE NOT AVAILABLE
C     IN THE DESIRED FORM ON ALL MACHINES

      REAL L
      DIMENSION FXY(80,80),ANGLE(80,80),APOINT(30,4),
     1COSTH(30),SINTH(30),L(30),A(30),B(30)
      DIMENSION XM(30),YM(30)
      COMMON FXY,ANGLE,APOINT
      COMPLEX F
C     SET ARRAYS AND CONSTANTS
      DO 9 I=1,30
      A(I)=0.0
      B(I)=0.0
    9 CONTINUE
C     STARTING VALUES
      XMAX=-100.0
      XMIN=100.0
      YMAX=-100.0
      YMIN=100.0
      PI=4.0*ATAN(1.0)
      WRITE(2,300)
  300 FORMAT(1X,'TYPE OUTPUT DEVICE 1=LP,2=NARROW,3=WIDE,4=FILM'/)
      READ(1,301)NDEV
  301 FORMAT(I1)
```

DIGITAL HOLOGRAMS

```
C       THE NUMBER OF LINES THEN THEIR COORDINATES ARE READ IN

        WRITE(2,99)
 99     FORMAT(1X,'TYPE NUMBER OF LINE SEGMENTS --MAX 30 I2 FORM'/)
        READ(4,21)NULINE
 21     FORMAT(I2)
        WRITE(2,100)
 100    FORMAT(1X,'SPECIFY ALL LINE SEGMENTS FROM LEFT TO RIGHT'/)
        DO 3 I=1,NULINE
        WRITE(2,101)
 101    FORMAT(1X,'TYPE -- X1,Y1,X2,Y2 IN 4(F4.1,1X) FORM'/)
        READ(4,200)(APOINT(I,II),II=1,4)
 200    FORMAT(4(F4.1,1X))

C       MAXIMUM AND MINIMUM CO-ORDINATES
C       IN X AND Y ARE CALCULATED
C       CALCULATE THE LINE LENGTHS AND ALSO THE COSINE AND SINE
C       OF THE ANGLE THE LINE MAKES WITH THE X- AXIS

        IF(APOINT(I,3).GT.XMAX)XMAX=APOINT(I,3)
        IF(APOINT(I,1).LT.XMIN)XMIN=APOINT(I,1)
        IF(APOINT(I,2).GE.APOINT(I,4))GOTO 5
        IF(APOINT(I,4).GT.YMAX)YMAX=APOINT(I,4)
        IF(APOINT(I,2).LT.YMIN)YMIN=APOINT(I,2)
        GOTO 6
 5      IF(APOINT(I,2).GT.YMAX)YMAX=APOINT(I,2)
        IF(APOINT(I,4).LT.YMIN)YMIN=APOINT(I,4)
 6      L(I)=SQRT((APOINT(I,3)-APOINT(I,1))**2+(APOINT(I,4)-APOINT
       1(I,2))**2)
        COSTH(I)=ABS(APOINT(I,3)-APOINT(I,1))/L(I)
        SINTH(I)=ABS(APOINT(I,4)-APOINT(I,2))/L(I)
        IF(APOINT(I,4).LT.APOINT(I,2))COSTH(I)=-COSTH(I)
        XM(I)=ABS((APOINT(I,1)-APOINT(I,3))/2.0+APOINT(I,1)
        YM(I)=ABS((APOINT(I,2)-APOINT(I,4))/2.0)
        IF(APOINT(I,2).LE.APOINT(I,4))YM(I)=YM(I)+APOINT(I,2)
        IF(APOINT(I,4).LT.APOINT(I,2))YM(I)=YM(I)+APOINT(I,4)
 3      CONTINUE

C       HERE THE PROGRAM CALCULATES THE BANDWIDTH
C       IN BOTH DIRECTIONS(X AND Y).IT THEN SETS THE CRITICAL
C       SAMPLING RATE AS 2.0*PI/(LARGEST VALUE)

        IF(XMAX-XMIN.GE.YMAX-YMIN)SAMPLE=2.0*PI/(XMAX-XMIN)
        IF(YMAX-YMIN.GT.XMAX-XMIN)SAMPLE=2.0*PI/(YMAX-YMIN)
        WRITE(3,202)SAMPLE
 202    FORMAT(1X,'SAMPLE RATE = ',F7.3)

C       OVERSAMPLING LEADS TO
C       A RECONSTRUCTION WITH IMAGES NICELY SEPARATED
C       BUT HERE THE SAMPLING RATE CAN BE MODIFIED
C       TO ALLOW BOTH UNDER AND OVER
C       SAMPLING TO BC INVESTIGATED

        WRITE(2,203)
```

DIGITAL HOLOGRAMS

```
  203 FORMAT(1X,'DO YOU WISH TO MODIFY THE SAMPLING RATE,'/
     1' TYPE MULTIPLICATION FACTOR IN F4.2 FORM'/)
      READ(1,204)AMULT
  204 FORMAT(F4.2)
      SAMPLE=SAMPLE*AMULT
      WRITE(3,205)SAMPLE
  205 FORMAT(1X,'MODIFIED SAMPLE RATE = ',F7.3)
      WRITE(2,103)
  103 FORMAT(1X,'HOW MANY SAMPLES IN X -- MAX 80 I3 FORM'/)
      READ(1,201)ISAMPS
      WRITE(2,104)
  104 FORMAT(1X,'HOW MANY SAMPLES IN Y -- MAX 80 I3 FORM'/)
      READ(1,201)JSAMPS
  201 FORMAT(I3)

C     THE PROGRAM CALCULATES THE ANALYTICAL FOURIER TRANSFORM
C     FOR EACH CELL

      WRITE(2,105)
  105 FORMAT(1X,'TO EQUALISE AMPLITUDE DISTRIBUTION',
     1' TYPE 1 ELSE 0')
      READ(1,106)NRAN
  106 FORMAT(I1)

C     TO EVEN OUT AMPLITUDES IN HOLOGRAM THE PHASE OF EACH
C     ELEMENT IN THE OBJECT IS ALLOWED TO VARY LINEARLY
C     ALONG ITS LENGTH -- IF NRAN IS SET TO 1

      IF(NRAN.EQ.0)GOTO 10
      X=(FLOAT(ISAMPS)/2.0-0.5)*SAMPLE
      Y=(FLOAT(JSAMPS)/2.0-0.5)*SAMPLE

      DO 8 I=1,NULINE

C     G05AAF(X) IS A ROUTINE THAT PRODUCES
C     RANDOM NUMBERS BETWEEN 0 AND 1

      A(I)=(G05AAF(XXX)*2.0-1.0)*X/4.0
      B(I)=(G05AAF(YYY)*2.0-1.0)*Y/4.0
    8 CONTINUE

C     SUM THE FOURIER TRANSFORM OF EACH LINE
C     AT EACH SAMPLING POINT

   10 DO 1 J=1,JSAMPS
      Y=(-(FLOAT(JSAMPS)/2.0+0.5)+FLOAT(J))*SAMPLE
      DO 2 I=1,ISAMPS
      X=(-(FLOAT(ISAMPS)/2.0+0.5)+FLOAT(I))*SAMPLE
    7 F=CMPLX(0.0,0.0)
      DO 20 NFOUR=1,NULINE
      SINC=L(NFOUR)/2.0
      SINC=SINC*((X-A(NFOUR))*COSTH(NFOUR)+(Y-B(NFOUR))*SINTH(NFOUR))

C     A CHECK TO STOP OVERFLOW I.E. 0.0/0.0

      IF(ABS(SINC).GT.10.0**(-20))GOTO 50
```

DIGITAL HOLOGRAMS

```
      SINC=1.0
      GOTO 51
   50 SINC=SIN(SINC)/SINC
   51 SINC=L(NFOUR)*SINC
      F=F+CEXP(CMPLX(0.0,(X-A(NFOUR))*XM(NFOUR)+(Y-B(NFOUR))*YM(NFOUR)))
     1*CMPLX(SINC,0.0)
   20 CONTINUE

C     THE FOURIER TRANSFORM IS OF COMPLEX FORM AND IS NOW SPLIT
C     INTO AN AMPLITUDE TERM AND A PHASE TERM. THE PHASE TERM
C     IS NORMALISED OVER THE INTERVAL 0 TO 2*PI.

      FXY(I,J)=CABS(F)
      BB=AIMAG(F)
      AA=REAL(F)
      ANGLE(I,J)=ATAN2(BB,AA)/(2.0*PI)
    2 CONTINUE
    1 CONTINUE
      WRITE(2,108)
  108 FORMAT(1X,'AMPLITUDE AND NORMALISED PHASE CALCULATED')
      CALL AOBJCT(NULINE,NDEV)
      WRITE(2,109)
  109 FORMAT(1X,'INPUT PLOTTED')
      CALL RELACS(ISAMPS,JSAMPS,NDEV,CLIP)
      CALL DISPLAY(ISAMPS,JSAMPS,NDEV,CLIP)

C     GINO-F SUBROUTINE - DEVFIN TERMINATES
C     THE GRAPHIC DEVICE USED-SWITCHES OFF GLOBAL CALLS

      CALL DEVFIN
      STOP
      END



      SUBROUTINE AOBJCT(NULINE,NDEV)

C     THIS SUBROUTINE PLOTS THE INPUTTED OBJECT
C     AND PROVIDES A CHECK OF THE DATA.IT COULD BE OMITTED.

      DIMENSION APOINT(30,4),FXY(80,80),ANGLE(80,80)
      COMMON FXY,ANGLE,APOINT

C     ANGLE IS REDUNDANT HERE.IT IS KEPT FOR CONVENIENCE TO
C     PRESERVE THE FORM OF THE COMMON STATEMENT.
C     GINO-F SUBROUTINES
C        NARROW   - CALLS NARROW PAPER CALCOMP
C        PRINTR   - CALLS LINE PRINTER
C        WIDE     - CALLS WIDE PAPER CALCOMP
C        FILM     - CALLS 35MM SLIDE PLOT-PRODUCES HOLOGRAMS DIRECTLY.
C        PENSEL   - SELECTS LINE THICKNESS ON FILM
C        UNITS(S) - S=NUMBER OF MM'S IN CURRENT DRAWING UNITS
C        WINDO2   - SETS UP 2-D WINDOW
C        SHIFT2   - SHIFTS REFERENCE AXIS BY VECTOR INCREMENT (X,Y)
C        MOVTO2   - POSITION PEN-BEAM AT A POINT X,Y
C        LINTO2   - DRAW A 2-D LINE FROM CURRENT POSITION TO X,Y
C        MOVBY2   - POSITION THE PEN -BEAM (2-D) (A,B)
```

DIGITAL HOLOGRAMS

```
C                       INCREMENTAL CO-ORDINATE DISTANCE
C          LINBY2    -  DRAW A 2-D LINE FROM THE CURRENT POSITION
C                       POSITION (A,B) INCREMENTAL CO-ORDINATE DISTANCE
C          PICCLE    - CLEARS DRAWING AREA OF ALL PREVIOUS PICTURES
C          TRANSF(0)- SWITCHES OFF SHIFT TRANSFORMATION
C      PAPENQ(X,Y,C)-DRAWING AREA.X,Y:- PAPER SIZE.C:- PAPER TYPE.
C      DEVPAP(A,B,C)-SPEC. OF  PLOTTER PAPER,DIMENSIONS A*B,TYPE:- C.
C      C IS DUMMY IF NO CHOICE PAPER AVAILABLE

       IF(NDEV.EQ.1)CALL PRINTR
       IF(NDEV.EQ.2)CALL NARROW
       IF(NDEV.EQ.3)CALL WIDE
       IF(NDEV.EQ.4) CALL FILM
C      NOTE:CALLS TO OUTPUT DEVICES ARE GLOBAL.
       CALL UNITS(10.0)
C      DRAWING UNITS ARE NOW CM
       IF(NDEV.EQ.4)CALL PENSEL(1,0.15,1)
C      COLOUR,LINEWIDTH,TYPE-FIRST AND LAST ARE IRRELEVANT FOR FILM
       CALL PAPENQ(XPAP,YPAP,IP)
       CALL WINDO2(0.0,20.0,0.0,15.0)
C      WINDOW DEFINED XLEFT,XRIGHT,YBOTTOM,YTOP
       CALL SHIFT2(10.0,7.5)
       CALL MOVTO2(0.0,7.5)
       CALL LINBY2(0.0,-15.0)
       CALL MOVTO2(-10.0,0.0)
       CALL LINBY2(20.0,0.0)
       DO 3 I=1,NULINE
       CALL MOVTO2(APOINT(I,1),APOINT(I,2))
       CALL LINTO2(APOINT(I,3),APOINT(I,4))
     3 CONTINUE
       CALL WINDO2(0.,XPAP,0.,YPAP)
C      RESET WINDOW TO FULL PAPER SIZE
       CALL TRANSF(0)
       CALL PICCLE
       RETURN
       END

       SUBROUTINE RELACS(NXDIM,NYDIM,NDEV,CLIP)

C      THIS SUBROUTINE NORMALISES THE AMPLITUDE TO 1
C      IT ALSO GENERATES A HISTOGRAM OF BOTH THE NORMALISED
C      AMPLITUDE AND PHASE TO AID CHOOSING A CLIPPING
C      FACTOR IN THE DISPLAYING OF THE HOLOGRAM.THE AMPLITUDE
C      DISTRIBUTION IS  CLIPPED TO 99 PER CENT,
C      UNLESS OTHERWISE ALTERED.

C      VARIABLE DEFINITION
C         HISTA    - ARRAY CONTAINING NUMBER AT COUNTA
C         COUNTA   - ARRAY CONTAINING AMPLITUDE LEVELS
C         HISTP    - ARRAY CONTAINING NUMBER AT COUNTP
C         COUNTP   - ARRAY CONTAINING PHASE LEVELS
C         ITA      - ARRAY CONTAINING TITLES FOR AMPLITUDE HISTOGRAM
C         ITP      - ARRAY CONTAINING TITLES FOR PHASE HISTOGRAM

       DIMENSION FXY(80,80),HISTA(50),COUNTA(50),ITA(15),ANGLE(80,80)
       DIMENSION APOINT(30,4),HISTP(50),COUNTP(50),ITP(14)
```

DIGITAL HOLOGRAMS

```
C       SCDTA IS REQUIRED BY GRAPH PLOTTING SUBROUTINE FGPLT

        COMMON/SCDTA/RV(1)
        COMMON FXY,ANGLE,APOINT
        DATA ITA/34,'AMPLITUDE DISTRIBUTION IN HOLOGRAM',3,'AMP',
       15,'COUNT'/
        DATA ITP/30,'PHASE DISTRIBUTION IN HOLOGRAM',3,'PHA',5,'COUNT'/

C       RESET UNITS

        CALL UNITS(1.)
        LEVELS=50
C       NUMBER OF HISTOGRAM CHANNELS

C       SETS UP INITIAL VALUES OF HISTA,HISTP,COUNTA,COUNTP


C       SEARCH FXY FOR MAXIMUM VALUE THEN NORMALISE FXY TO 1.
C       CONSTRUCT AMPLITUDE AND PHASE HISTOGRAMS

        DO 10 K=1,LEVELS
        HISTA(K)=0.0
        HISTP(K)=0.0
        COUNTA(K)=FLOAT(K)/FLOAT(LEVELS)
        COUNTP(K)=COUNTA(K)-0.5
     10 CONTINUE
        AMAX=0.0
        DO 1 J=1,NYDIM
        DO 1 I=1,NXDIM
        IF(FXY(I,J).GT.AMAX)AMAX=FXY(I,J)
      1 CONTINUE
        DO 2 J=1,NYDIM
        DO 2 I=1,NXDIM
        FXY(I,J)=FXY(I,J)/AMAX
        DO 3 K=1,LEVELS
        IF(FXY(I,J).GT.COUNTA(K))GOTO 3
        HISTA(K)=HISTA(K)+1.0
        GOTO 6
      3 CONTINUE
      6 DO 5 K=1,LEVELS
        IF(ANGLE(I,J).GT.COUNTP(K))GOTO 5
        HISTP(K)=HISTP(K)+1
        GOTO 2
      5 CONTINUE
      2 CONTINUE
        WRITE(3,102)
        WRITE(3,103)


C       MAKE HISTOGRAM MARKERS CENTRAL FOR USE WITH FGPLT

        DO 8 K=1,LEVELS
        X=1.0/FLOAT(LEVELS)
        COUNTA(K)=COUNTA(K)-X/2.0
        COUNTP(K)=COUNTP(K)-X/2.0
      8 CONTINUE
```

DIGITAL HOLOGRAMS

```
C       WRITE OUT VALUES IN TABLE FORM

        DO 4 K=1,LEVELS
        WRITE(3,101)COUNTA(K),HISTA(K),COUNTP(K),HISTP(K)
 101    FORMAT(1X,F5.3,4X,F6.1,5X,F5.2,7X,F5.1)
   4    CONTINUE
 102    FORMAT(1X,'DISTRIBUTION OF AMPLITUDE AND PHASE IN HOLOGRAM'/)
 103    FORMAT(2X,'AMP',4X,'NUMBER',5X,'PHASE',9X,'NUMBER')
        WRITE(2,100)
 100    FORMAT(1X,'AMPLITUDE NORMALISED'/)
        DENOM=FLOAT(NXDIM*NYDIM)
        TOTAL=0.0
C       DEFINE A CLIPPING FACTOR TO EFFECTIVELY


C       CLIP ONE PER CENT OF POINTS OFF TOP END OF AMPLITUDE DISTRIBUTION
C             IF THIS IS NOT THE BEST CLIP THEN THE USER IS ALLOWED TO
C       ADJUST IT.


        DO 9 I=1,LEVELS
        TOTAL=HISTA(I)+TOTAL
        IF(TOTAL/DENOM.LT.0.99)GOTO 9
        CLIP=FLOAT(LEVELS)/FLOAT(I)
        WRITE(3,105)CLIP
        GOTO 11
   9    CONTINUE
 105    FORMAT(1X,'SUGGESTED CLIP = ',F6.3,'TO CHANGE TYPE 1 ELSE 0')
   7    FORMAT(I1)
  11    READ(1,106)NCLIP
 106    FORMAT(I1)
        IF(NCLIP.EQ.0)GOTO 12
        WRITE(3,107)
 107    FORMAT(1X,'TYPE NEW CLIP VALUE F4.2 FORM'/)
        READ(1,108)CLIP
 108    FORMAT(F4.2)


C       GINO-F SUBROUTINES
C           A4      - SETS PLOT TO A4 SIZE (29 CM*21 CM)
C           FGPLT   - PLOTS A HISTOGRAM WHERE
C                     HISTA IS THE NUMBER OF POINTS
C                     AT COUNTA - THESE ARE BOTH REAL ARRAYS
C                     OF SIZE LEVELS
C           NEWPAG  - CALLS A NEW PAGE

  12    CALL A4

C       HISTOGRAMS OF AMPLITUDE AND PHASE DISTRIBUTIONS
C       FGPLT IS A STANDARD PLOTTING ROUTINE

        CALL FGPLT(COUNTA,HISTA,LEVELS,6,0,1,0,ITA)
        CALL FGPLT(COUNTP,HISTP,LEVELS,6,0,1,0,ITP)
        CALL NEWPAG
        RETURN
        END
```

DIGITAL HOLOGRAMS

```
      SUBROUTINE DISPLAY(NXDIM,NYDIM,NDEV,CLIP)

C     THIS SUBROUTINE DRIVES THE O/P  DEVICE
C     DRAWING THE HOLOGRAM
C     WHEN USING THE FILM OUTPUT IT IS DIRECTLY
C     USABLE AS A HOLOGRAM IN AN OPTICAL RECONSTRUCTION.
C     BOTH THE NARROW AND WIDE CALCOMP PLOTTER OUTPUT
C      WILL REQUIRE PHOTOGRAPHIC REDUCTION-THE PHOTOGRAPHIC
C     FILM THEN BEING USED AS THE HOLOGRAM.
C     THE LINEPRINTER OUTPUT IS JUST AN AMPLITUDE
C     PLOT AND IS INCLUDED TO PROVIDE IMMEDIATE OUTPUT
C     SO THAT THE USER CAN CHECK THE CLIPPING CHOSEN.


C     VARIABLE DEFINITION
C        CHAR   - INTEGER ARRAY CONTAINING CODE
C                 FOR SYMBOL OUTPUT ON QUICK LINEPRINTER
C                 OVERPRINTING AMPLITUDE PLOT

      INTEGER CHAR,E
      DIMENSION FXY(80,80),ANGLE(80,80),CHAR(10),E(130),APOINT(30,4)
      COMMON FXY,ANGLE,APOINT
      DATA CHAR/' ',':','-','I','/','+','L','=','U','0'/
      IXX=0
   25 IF(NDEV.EQ.1)GOTO 6
   51 FORMAT(I1)
      WRITE(2,995)
  995 FORMAT(1X,'HOW MANY SMALLER HOLOGRAMS - TYPE NUMBER')
      READ(1,51)NSM

C     NEXT SECTION CALLS O/P DEVICE AND CHOOSES SCALING

C        NUMERATOR OF XS AND YS ARE THE MAX SIZE OF
C        DRAWING AREA IN MM. MINIMUM XS OR YS GIVES
C        MM PER UNIT CELL,IN ORDER TO FILL THE
C        FULL DRAWING AREA.

C     VARIABLE DEFINITION
C        ALINWD - GIVES A VALUE OF THE PEN LINEWIDTH
C        GINO-F SUBROUTINES
C        DEVPAP - DEFINES THE PAPER SIZE


      CALL PICCLE
      IF(NDEV.NE.2)GOTO 20
      XS=3600.0/FLOAT(NXDIM)
      YS=270.0/FLOAT(NYDIM)
      ALINWD=0.5
C     BASED UPON THE ACTUAL PEN WIDTH IN MM
      CALL DEVPAP(3600.,270.,III)
      GOTO 30
   20 IF(NDEV.NE.3)GOTO 21
      XS=3600.0/FLOAT(NXDIM)
      YS=840.0/FLOAT(NYDIM)
      ALINWD=0.5
      CALL DEVPAP(3600.,840.,III)
```

DIGITAL HOLOGRAMS

```
      GOTO 30
   21 XS=430.0/FLOAT(NXDIM)
      YS=270.0/FLOAT(NYDIM)
      ALINWD=0.2
C     THINNER LINES  ALLOWED ON FILM
      CALL DEVPAP(430.,270.,III)

C     AMIN1 IS A STANDARD FORTRAN FUNCTION TO FIND
C     A MINIMUM VALUE

C     VARIABLE DEFINITION
C         NULINE IS THE NUMBER OF PEN LINES THAT WILL FILL
C         A SAMPLING CELL IN THE OUTPUT HOLOGRAM

   30 S=AMIN1(XS,YS)
      CALL UNITS(S)
      SS=1.0/S
      IF(NDEV.EQ.4)CALL PENSEL(1,SS,1)
C     DIRECT HOLOGRAM PRODUCTION AS 35MM FILM SLIDE
      NULINE=2*IFIX(S/ALINWD+0.5)-1
      WRITE(3,200)S,NULINE
  200 FORMAT(1X,'UNITS = ',F7.3,' NUMBER OF LINES = ',I2/)
      CALL WINDO2(0.0,FLOAT(NXDIM),0.0,FLOAT(NYDIM))

C     THE CENTRE OF THE SQUARE SAMPLING CELL IS THE DATUM
C     THE PEN IS POSITIONED IN THE FIRST CELL IN THE FIRST
C     ROW AT (-0.5,-0.5) WITH RESPECT TO DATUM.


C     THE PEN IS STEPPED FROM SAMPLING CELL TO SAMPLING CELL IN ONE
C     ROW AND * IN EACH CELL COMPARE THE POSITION OF THE PEN WITH THE
C     AMPLITUDE FXY AND DECIDE WHETHER OR NOT TO DRAW
C     A LINE. THEN INCREMENT THE PEN BY HALF A LINE WIDTH(DRAWN BY PEN)
C     AND REPEAT FROM *. WHEN ONE ROW IS COMPLETE STEP TO
C     THE NEXT AND REPEAT. THE LINE IS HALF THE LENGTH OF
      A CELL WIDTH AND ITS CENTRE IS MOVED FROM THE CENTRE
      OF THE CELL BY THE VALUE IN ANGLE.

   57 DO 1 I=1,NXDIM
      DO 2 N=1,NULINE
      DO 3 J=1,NYDIM
      CALL MOVBY2(0.0,0.25)
      X=FLOAT(N)/FLOAT(NULINE)
      X1=FLOAT(N-1)/FLOAT(NULINE)
      IF(X1.LT.0.5*(1.0-(FXY(I,J)*CLIP)))GOTO 4
      IF(X.GT.0.5*(1.0+(FXY(I,J)*CLIP)))GOTO 4
      IF(FXY(I,J)*CLIP.LT.ALINWD/S)GOTO 4
      CALL MOVBY2(0.0,ANGLE(I,J))
      CALL LINBY2(0.0,0.5)
      CALL MOVBY2(0.0,-ANGLE(I,J))
      GOTO 5
    4 CALL MOVBY2(0.0,0.5)
    5 CALL MOVBY2(0.0,0.25)
    3 CONTINUE
      CALL MOVBY2(1.0/FLOAT(NULINE),-FLOAT(NYDIM))
    2 CONTINUE
```

DIGITAL HOLOGRAMS

```
     1   CONTINUE

C        IF MORE(SMALLER) HOLOGRAMS ARE REQUIRED,
C        THIS NEXT SECTION RESETS
C        THE WINDO2 AND DRAWS A SMALLER AREA OF THE PREVIOUS HOLOGRAM-
C        IT SUBTRACTS 10 CELLS  FROM EACH AXIS.THIS IS REPEATED UNTIL
C        THE NUMBER OF SMALLER HOLOGRAMS REQUIRED IS PLOTTED

    53   IF(IXX.EQ.NSM)GOTO 56
         CALL PICCLE
         XLEFT=5.0*FLOAT(NSM-IXX)
         XRIGHT=FLOAT(NXDIM)-XLEFT
         YLEFT=XLEFT
         YRIGHT=FLOAT(NYDIM)-YLEFT
         CALL WINDO2(XLEFT,XRIGHT,YLEFT,YRIGHT)
         IXX=IXX+1
         GOTO 57
    56   GOTO 7

C        OVERPRINTING QUICK AMPLITUDE CHECK ON LINEPRINTER

     6   CONTINUE
         DO 8 I=1,NYDIM
         WRITE(3,101)
   101   FORMAT(1H )
         DO 9 L=1,10
         FLAG=0.0
         DO 10 M=1,NXDIM
         IF(FXY(M,I)*CLIP.LT.FLOAT(L)*0.1)GOTO 11
         E(M)=CHAR(L)
         FLAG=1.0
         GOTO 10
    11   E(M)=CHAR(1)
    10   CONTINUE
         IF(FLAG)12,12,9
     9   WRITE(3,102)(E(K),K=1,NXDIM)
    12   CONTINUE
     8   CONTINUE
         WRITE(3,103)
   102   FORMAT(1H+,130A1)
   103   FORMAT(1X,'FINISHED')

C        THE FINAL SECTION ALLOWS THE CLIPPING OR
C        O/P DEVICE TO BE CHANGED WHILST MONITORING THE HOLOGRAM PRODUCTION

C        IT IS INCLUDED TO ALLOW THE USER TO EXPT WITH
C        DATA ON A VIDEO DISPLAY AND THEN OBTAIN THE
C        HARD COPY OF REQUIRED  HOLOGRAM


     7   WRITE(2,105)
   105   FORMAT(1X,'TO VARY CLIPPING TYPE NEW VALUE ELSE <CR>')
         READ(1,152)CLIP
   152   FORMAT(F3.1)
         IF (CLIP.LT.0.1)GOTO 50
         GOTO 25
```

DIGITAL HOLOGRAMS

```
 50  WRITE(2,106)
106  FORMAT(1X,'TO CHANGE O/P DEV TYPE 1=LP,2=NARROW,3=WIDE,',
    1'4=FILM ELSE <CR>'/)
     READ(1,150)NDEV
150  FORMAT(I1)
     IF(NDEV.EQ.0)RETURN
     GOTO 25
     END
```

PART 2

Magnetism

Physics Programs
Edited by A.D. Boardman
© 1980 John Wiley & Sons Ltd.

CHAPTER 4

Calculation of the Fields Near Permanent Magnets

M. I. DARBY

1. INTRODUCTION

Partial differential equations are important in almost all branches of physics, and often they can only be solved numerically. Owing to the diversity of boundary conditions and other factors that may apply, it is impracticable to produce computer library routines capable of solving more than one specific type of problem. For this reason it is valuable to have some practical experience of the difficulties involved in applying one of the common numerical techniques in a relatively simple situation.

The problem considered here is the calculation of the magnetic field in the vicinity of a uniformly magnetized rectangular permanent magnet. The magnet is assumed to be infinite in one direction, so that the problem reduces to two dimensions. The basic magnetostatic equations are given in sections 2 and 3. There it is shown that the fields are conveniently written in terms of a scalar magnetostatic potential, which satisfies Laplace's equation, and which is completely determined by the boundary conditions. Sometimes it is possible to obtain an analytical solution for the potential, but usually Laplace's equation must be solved numerically. One method of doing so, and that adopted here, is to replace the partial differential equation with a set of (linear) finite difference equations. These can then be solved by standard methods, either directly by elimination or by iteration. The latter method is employed below.

2. THE MAGNETOSTATIC POTENTIAL

The magnetic induction vector **B** produced by a steady electric current I satisfies Ampère's law,[1]

$$\oint_C \mathbf{B} \cdot \mathbf{dl} = \mu_0 I, \tag{1}$$

where C is a contour enclosing the conductor carrying I. Employing Stokes's integral theorem, this equation can be written as

$$\text{curl } \mathbf{B} = \mu_0 \mathbf{J}, \tag{2}$$

where $\mathbf{J}$ is the current density (Am^{-2}). The other basic property of $\mathbf{B}$ is that it forms closed loops, i.e. it satisfies

$$\text{div } \mathbf{B} = 0. \tag{3}$$

A small current loop produces a field $\mathbf{B}$ which resembles the electric field near an electric dipole, and consequently a magnetic dipole moment can be identified with the loop. A magnetic material may be thought of as containing a large number of elementary loops, giving rise to a dipole moment per unit volume, $\mathbf{M}$, known as the magnetization. The magnetization contributes to $\mathbf{B}$ and it can be shown[1] quite generally that equation (2) is replaced by

$$\text{curl } \mathbf{B} = \mu_0 \mathbf{J} + \mu_0 \text{ curl } \mathbf{M}, \tag{4}$$

where $\mathbf{J}$ is the real current density. It is convenient to define a magnetic field $\mathbf{H}$ by

$$\mathbf{B} = \mu_0 (\mathbf{H} + \mathbf{M}), \tag{5}$$

and from equation (4) $\mathbf{H}$ satisfies

$$\text{curl } \mathbf{H} = \mathbf{J}, \tag{6}$$

There are usually no true currents in a permanent magnet so that (6) reduces to

$$\text{curl } \mathbf{H} = 0, \tag{7}$$

and therefore it is possible to define a scalar magnetic potential ϕ by

$$\mathbf{H} = -\nabla \phi. \tag{8}$$

From equations (3) and (5),

$$\text{div } \mathbf{H} = -\text{div } \mathbf{M}, \tag{9}$$

or in terms of ϕ,

$$\nabla^2 \phi = \text{div } \mathbf{M}. \tag{10}$$

This is Poisson's equation for the potential and, by analogy with electrostatics, the term div $\mathbf{M}$ plays the role of a volume magnetic charge density, and is frequently referred to as the pole density.

3. BOUNDARY CONDITIONS ON INTERFACES

The boundary conditions on the magnetostatic potential at the interface between two media in which true currents are absent can be derived from

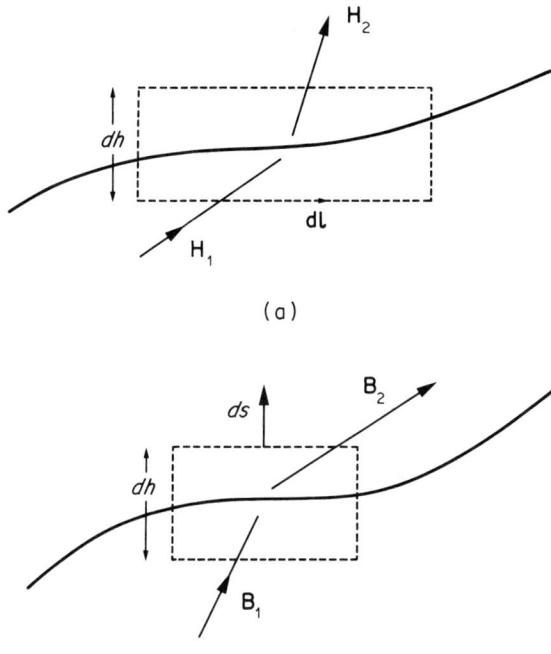

Figure 1. Fields near the boundary between two media illustrating the geometry and notation employed in the text to discuss the boundary conditions on (a) **H**, and (b) **B**

equations (7) and (3). Integration of (7) yields

$$\oint \mathbf{H} \cdot d\mathbf{l} = 0, \tag{11}$$

and evaluating the line integral around a rectangular contour intersecting the boundary between two regions, in which the magnetic fields are $\mathbf{H}_1$ and $\mathbf{H}_2$, as shown in Figure 1(a),

$$\mathbf{H}_1 \cdot d\mathbf{l} - \mathbf{H}_2 \cdot d\mathbf{l} + 0(dh) = 0, \tag{12}$$

(assuming that dh can be made arbitrarily small). Equation (12) implies that the tangential components of **H** are continuous across the boundary, and can also be expressed as

$$\hat{\mathbf{n}} \times (\mathbf{H}_1 - \mathbf{H}_2) = 0, \tag{13}$$

where $\hat{\mathbf{n}}$ is a unit vector normal to the surface. In terms of the scalar potentials,

$$\hat{\mathbf{n}} \times (\nabla \phi_1 - \nabla \phi_2) = 0 \tag{14}$$

and, integrating along the boundary yields, in many circumstances,

$$\phi_1 = \phi_2. \tag{15}$$

Thus the potential is continuous across the boundary.

The second boundary condition is derived from $\text{div}\,\mathbf{B} = 0$ by applying Gauss's theorem to yield

$$\int_S \mathbf{B} \cdot d\mathbf{S} = 0. \tag{16}$$

For a small cylindrical volume intersecting the boundary, indicated by the dotted lines in Figure 1(b), the surface integral yields

$$\mathbf{B}_2 \cdot \hat{\mathbf{n}}\, dS - \mathbf{B}_1 \cdot \hat{\mathbf{n}}\, dS + 0(dh) = 0, \tag{17}$$

indicating that the normal component of $\mathbf{B}$ is continuous. If the two regions have magnetizations $\mathbf{M}_1$ and $\mathbf{M}_2$, substitution of $\mathbf{B} = \mu_0(-\nabla\phi + \mathbf{M})$ yields

$$(-\nabla\phi_1 + \mathbf{M}_1) \cdot \hat{\mathbf{n}} = (-\nabla\phi_2 + \mathbf{M}_2) \cdot \hat{\mathbf{n}}. \tag{18}$$

This is the required boundary condition on the gradient of ϕ. It can be seen from equation (18) that, by analogy with electrostatics, the term $\mathbf{M} \cdot \hat{\mathbf{n}}$ plays the role of a surface magnetic charge (pole) density.

The normal component of the magnetic field $\mathbf{H}$ has a discontinuity equal to the difference of the components of the magnetizations. Consequently the magnetic field inside the magnet opposes the magnetization and is known as the demagnetizing field.

4. THE MODEL PROBLEM

The computational problem is to determine the magnetic field in the regions inside and outside a two-dimensional rectangular magnet by solving Poisson's equation (10) for the scalar potential. It is assumed that the magnet is uniformly magnetized, so that $\mathbf{M}(\mathbf{r})$ is a constant vector, in which case equation (10) reduces to Laplace's equation,

$$\frac{\partial^2 \phi}{\partial x^2} + \frac{\partial^2 \phi}{\partial y^2} = 0, \tag{19}$$

everywhere except on the boundary of the magnet. On the latter, the potential is continuous, but the components of its gradient normal to the surface change by the normal component of the magnetization, in accordance with equation (18).

At large distances from the magnet the potential will resemble that of a small magnetic dipole of moment $\mathbf{m} = \mathbf{M}V$, where V is the volume of the

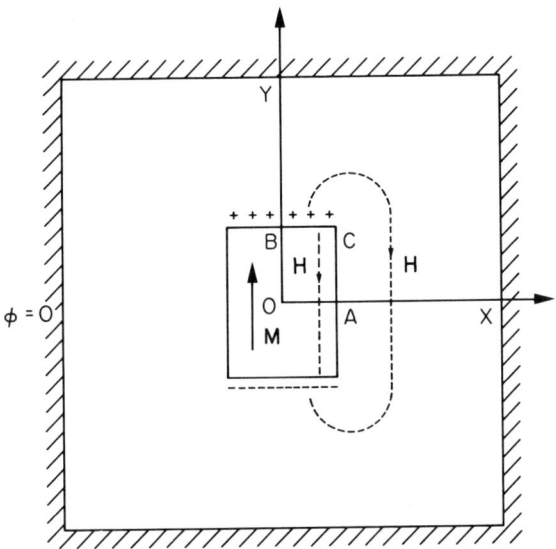

Figure 2. Schematic diagram showing the positions of the magnet and the outer boundary, and indicating the directions of the magnetic field. Because of symmetry it is necessary to consider only the region in the positive quadrant

magnet, i.e.

$$\phi(\mathbf{r}) = -\frac{1}{4\pi} \mathbf{m} \cdot \text{grad}\left(\frac{1}{r}\right). \tag{20}$$

In principle it is possible to match the numerical solutions of (19) to this expression on a distant rectangular boundary, but to avoid this added complication it will be assumed here that ϕ is essentially zero on that boundary, as shown in Figure 2. The effect of this approximation on the final solutions can be investigated by increasing the size of the large rectangle.

5. THE FINITE DIFFERENCE EQUATIONS

Laplace's equation (19) can be solved by approximating the derivatives of ϕ by finite difference formulae. The second derivative of a function of a single variable $f(x)$ tabulated at equal intervals of x can be approximated,[2] using Taylor's expansion, by

$$\frac{d^2 f}{dx^2} \approx \frac{1}{h^2} \{f(x+h) + f(x-h) - 2f(x)\}, \tag{21}$$

where h is the interval. For a function of two variables, $f(x, y)$ can be specified at points on a square mesh labelled by integers i and j, so that $x = ih$; $y = jh$ ($i, j = 1, 2, 3, \ldots$).

Equation (21) enables Laplace's equation to be replaced by a set of finite element equations:

$$\frac{1}{h^2}\{\phi_{i+1,j} + \phi_{i-1,j} + \phi_{i,j+1} + \phi_{i,j-1} - 4\phi_{i,j}\} = 0, \tag{22}$$

for each point (i, j). Near the boundaries of the region this equation must be modified in an appropriate way, described in detail below, to take into account the physical boundary conditions. The resulting set of equations, one equation for each mesh point, can be solved for the $\phi_{i,j}$ either by direct matrix methods or by an iterative process (see, for example, ref. 3). For large matrices the second method has the advantage that the zeros are preserved throughout, and consequently less computer storage is required. An iterative approach is employed here.

The five values of ϕ in equation (22) are said to form a star (Figure 3). If four of the values are known approximately, equation (22) can be employed to determine an improved value for the fifth. In an iterative process initial values of ϕ, $\phi_{i,j}^{(0)}$ say, must be assigned to each mesh point. Generally, it is not essential that these initial values are a close approximation to the final solution. On the boundaries the ϕ-values may be known exactly from the outset, but often the $\phi_{i,j}^{(0)}$-values inside the region can only be chosen somewhat arbitrarily. Frequently the $\phi_{i,j}^{(0)}$ are set equal to a constant value

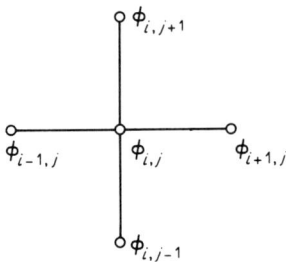

Figure 3. The star of function values required to approximate $\nabla^2 \phi$ at a general mesh point (i, j). The mesh is assumed to be square and the mesh size is h

(0.5 in the program presented here). The initial $\phi_{i,j}^{(0)}$ can be improved by applying equation (22) to each mesh point in turn, giving quantities $\phi_{i,j}^{(1)}$. Explicitly,

$$\phi_{i,j}^{(1)} = \tfrac{1}{4}\{\phi_{i+1,j}^{(0)} + \phi_{i-1,j}^{(0)} + \phi_{i,j-1}^{(0)} + \phi_{i,j+1}^{(0)}\}, \quad (23)$$

If the new $\phi_{i,j}^{(1)}$ are substituted into equation (22) the bracket on the left-hand side will not be exactly zero, but will have a residual value, R_{ij} say, which is some measure of the discrepancy between $\phi_{i,j}^{(1)}$ and the true solution ϕ. Repeating the procedure, new values $\phi_{i,j}^{(2)}$ can be computed from the $\phi_{i,j}^{(1)}$ using a formula similar to (23). This iterative process is continued until the ϕ values do not alter, within a specified accuracy, between one scan of the mesh points and the next.

Iterative formulae like (23) employing only the old $\phi^{(0)}$ values on the right-hand side are said to be of Jacobi type. When performing the calculations with a computer it is more natural to use the newly calculated ϕ values in the right-hand side of (23) as soon as possible. Thus, for example, if the mesh is scanned column by column the quantities $\phi_{i-1,j}^{(1)}$ and $\phi_{i,j-1}^{(1)}$ will have been computed before $\phi_{i,j}^{(1)}$ is evaluated, and equation (23) can be replaced by

$$\phi_{i,j}^{(1)} = \tfrac{1}{4}\{\phi_{i+1,j}^{(0)} + \phi_{i-1,j}^{(1)} + \phi_{i,j-1}^{(1)} + \phi_{i,j+1}^{(0)}\}. \quad (24)$$

This expression gives rise to a Gauss–Seidel scheme. It can be shown[3] that this iterative process converges more rapidly than the simpler Jacobi method.

More complicated iteration formulae than (24) have been devised which give even more rapid convergence than the Gauss–Seidel method. One such procedure,[3] which changes the old field value by adding to it a small fraction of the old residual R_{ij} is known (amongst other names) as successive over-relaxation (SOR). For the nth iteration, assuming column-by-column scanning of the mesh, the appropriate formula is

$$\phi_{i,j}^{(n)} = \phi_{i,j}^{(n-1)} + \frac{\alpha}{4}\{\phi_{i+1,j}^{(n-1)} + \phi_{i-1,j}^{(n)} + \phi_{i,j-1}^{(n)} + \phi_{i,j+1}^{(n-1)} - 4\phi_{i,j}^{(n-1)}\}, \quad (25)$$

where α is a parameter which usually lies between 1 and 2 in practice.[3] This method has been employed in the present work.

Because of the symmetry of the problem it is necessary to consider only one-quarter of the total region, for example the first quadrant shown in Figure 4. The iteration formulae (23)–(25) cannot be employed for points on the surface of the magnet because Laplace's equation is not valid there. Nor can these formulae be used as they stand for points on the boundaries of the region, because some of the ϕ-values in the star formulae for the boundary points will be outside the region. The latter situation can be illustrated by

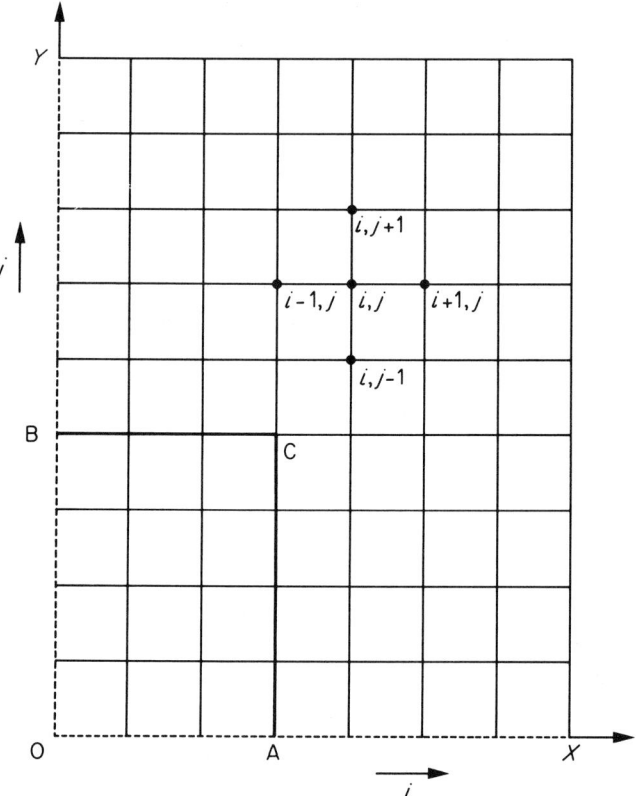

Figure 4. A typical mesh covering the region of interest. The area $OACB$ represents the portion of the magnet in this quadrant

considering the boundary OY, for which $i=1$, giving for the residuals R_{1j}

$$R_{1j} = \phi_{2,j} + \phi_{0,j} + \phi_{1,j+1} + \phi_{1,j-1} - 4\phi_{1,j}. \tag{26}$$

The values $\phi_{0,j}$ lying outside the field region are sometimes called 'fictitious' values, and are often labelled by an asterisk added as a superscript, e.g. $\phi_{0,j}^*$. Because of the special conditions applying at the boundaries the fictitious values required to calculate the residuals can usually be expressed as functions of the $\phi_{i,j}$ values inside the region. For the present problem the appropriate special finite difference formulae required for the region boundaries and the magnet surface are now derived in detail.

5.1 The outer boundary

All $\phi_{i,j} = 0$, and this boundary is excluded from the iterative process.

5.2 The symmetry axis OY

The line OY lies along the y-coordinate axis (Figure 4). The field in the negative x region ($y > 0$) is the mirror image of that in the first quadrant, i.e.

$$\phi(-x, y) = \phi(x, y).$$

Hence the fictitious field values $\phi^*_{0,j}$ (see Figure 5a) are given by

$$\phi^*_{0,j} = \phi_{2,j}. \tag{27}$$

Employing this in the star formula for R_{ij} gives a SOR formula:

$$\phi^{(n)}_{1,j} = \phi^{(n-1)}_{1,j} + \frac{\alpha}{4}\{2\phi^{(n')}_{2,j} + \phi^{(n')}_{1,j-1} + \phi^{(n')}_{1,j+1} - 4\phi^{(n-1)}_{1,j}\}, \tag{28}$$

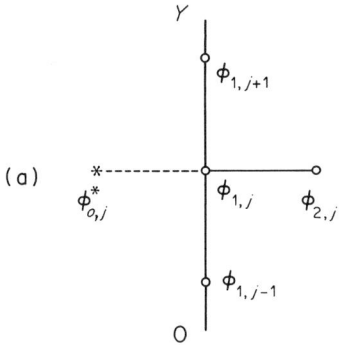

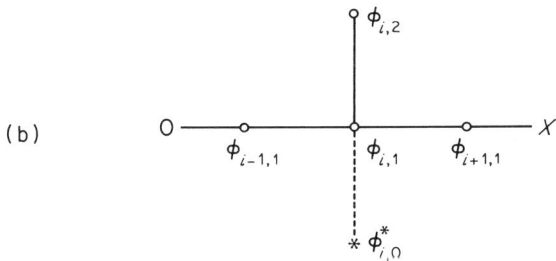

Figure 5. Function values involved in treating typical points on the boundaries: (a) symmetry line OY showing the position of the fictitious point $\phi^*_{0,j}$; (b) symmetry line OX showing the position of the fictitious point $\phi^*_{i,0}$; (c) magnet boundary AC; (d) magnet boundary BC

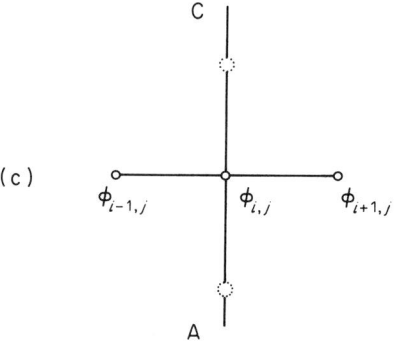

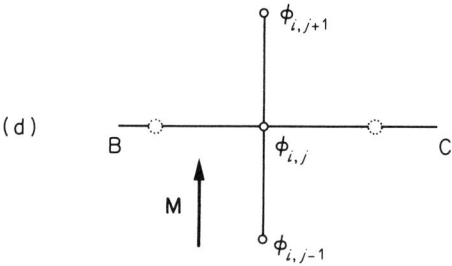

Figure 5c and d

where n' is n or $(n-1)$ and indicates that the most recently calculated ϕ-values are to be inserted. This formula applies for $j > 1$, and the origin O and point B are excluded as they also lie on other symmetry axes.

5.3 The symmetry axis OX

Line OX lies along the x-axis. From the symmetry, the magnetic field $\mathbf{H}$ is everywhere parallel to OY, so that

$$H^y = -\frac{\partial \phi}{\partial y} = H, \tag{29}$$

$$H^x = -\frac{\partial \phi}{\partial x} = 0, \tag{30}$$

The second equation means that all the ϕ-values on OX are equal. Furthermore, the continuity of $\mathbf{H}$ across OX requires that the gradient $\partial \phi / \partial y$ should be continuous. Defining fictitious values $\phi_{i,0}^*$ as in Figure 5(b), and employing a simple forward difference formula for the gradient, this condition gives

$$\frac{1}{h}\{\phi_{i,2}-\phi_{i,1}\}=\frac{1}{h}\{\phi_{i,1}-\phi_{i,0}^*\},$$

or

$$\phi_{i,0}^*=2\phi_{i,1}-\phi_{i,2}. \tag{31}$$

Substitution into the star formula for the residual, the SOR iteration formula becomes

$$\phi_{i,1}^{(n)}=\phi_{i,1}^{(n-1)}+\frac{\alpha}{4}\{\phi_{i+1,1}^{(n')}+\phi_{i-1,1}^{(n')}-2\phi_{i,1}^{(n-1)}\}, \tag{32}$$

where n' is n or $(n-1)$ depending on how the boundary mesh points are scanned. In fact this equation is redundant because the ϕ are all equal by equation (30). Hence, since ϕ is zero on the outer boundary, it is necessary to set

$$\phi=0,$$

at all points on the line, including the origin O.

5.4 The magnet boundary AC

The condition on tangential **H** requires that ϕ is continuous on the boundary, and this is automatically ensured in the iteration. Since the magnetization has no component perpendicular to AC, the continuity of the normal component of **B** implies that $\partial\phi/\partial x$ is continuous. Employing a forward difference formula, see Figure 5c,

$$\frac{1}{h}\{\phi_{i+1,j}-\phi_{i,j}\}=\frac{1}{h}\{\phi_{i,j}-\phi_{i-1,j}\},$$

or

$$\phi_{i,j}=\tfrac{1}{2}\{\phi_{i+1,j}+\phi_{i-1,j}\}. \tag{33}$$

Hence on this boundary, excluding point C, which requires a different treatment because it also lies on boundary BC,

$$\phi_{i,j}^{(n)}=\phi_{i,j}^{(n-1)}+\frac{\alpha}{2}\{\phi_{i+1,j}^{(n')}+\phi_{i-1,j}^{(n')}-2\phi_{i,j}^{(n-1)}\}, \tag{34}$$

where again n' is either n or $(n-1)$ as appropriate.

5.5 The magnet boundary BC

Again ϕ is continuous, but now, from equation (18), the gradient in y has a discontinuity equal to $|\mathbf{M}|$, i.e.

$$-\left.\frac{\partial \phi}{\partial y}\right|_{\text{in}} + M = -\left.\frac{\partial \phi}{\partial y}\right|_{\text{out}}. \tag{35}$$

In terms of finite differences, this becomes (see Figure 5d):

$$-\frac{1}{h}\{\phi_{i,j+1} - \phi_{i,j}\} = -\frac{1}{h}\{\phi_{i,j} - \phi_{i,j-1}\} + M,$$

or

$$\phi_{i,j} = \tfrac{1}{2}\{\phi_{i,j+1} + \phi_{i,j-1} + Mh\}. \tag{36}$$

Therefore, excluding point C, but including point B,

$$\phi_{i,j}^{(n)} = \phi_{i,j}^{(n-1)} + \frac{\alpha}{2}\{\phi_{i,j+1}^{(n')} + \phi_{i,j-1}^{(n')} + Mh - 2\phi_{i,j}^{(n-1)}\}. \tag{37}$$

5.6 The point C

This point is difficult to treat satisfactorily. Only a very approximate expression for ϕ will be used here, it being assumed that a simple average of the expressions for AC and BC, given by (33) and (36), is appropriate. Hence, adding,

$$2\phi_{i,j} = \tfrac{1}{2}\{\phi_{i+1,j} + \phi_{i-1,j}\} + \tfrac{1}{2}\{\phi_{i,j+1} + \phi_{i,j-1} + Mh\}, \tag{38}$$

and

$$\phi_{i,j}^{(n)} = \phi_{i,j}^{(n-1)} + \frac{\alpha}{4}\{\phi_{i+1,j}^{(n')} + \phi_{i,j+1}^{(n')} + \phi_{i,j+1)}^{(n')} + \phi_{i,j-1}^{(n')} + Mh - 4\phi_{i,j}^{(n-1)}\}. \tag{39}$$

6. THE COMPUTER PROGRAM

The iteration scheme is readily programmed for a computer, and a suitable FORTRAN coding is given at the end of the chapter. The flow diagram Figure 6 indicates a few practical details. The iteration loop for the $\phi_{i,j}$ is included in the main routine. Subroutines calculate the magnetic field and handle the output of data.

It is necessary to keep a running count of the number of iterations performed, so that the process can be terminated if it shows no sign of converging to the required accuracy. There are several criteria for satisfactory convergence that can be employed, and in the appended program two different criteria are used in series. In the first, the residual with the largest

CALCULATION OF THE FIELDS NEAR PERMANENT MAGNETS

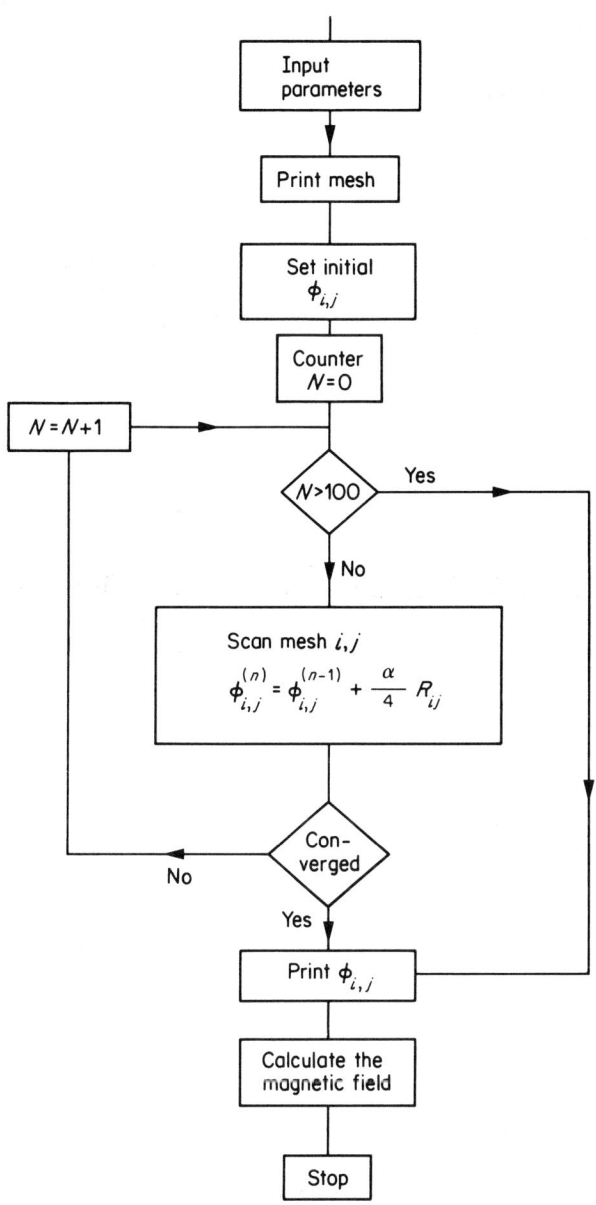

Figure 6. Flow diagram for the computer program

magnitude is determined for the nth iteration, and if this is less than some prescribed value, R_m say, the process is terminated. In some cases this criterion might be too strong, for it is conceivable that all but a few residuals are much smaller than R_m, and the solution for the $\phi_{i,j}$ might be acceptable at that stage. Therefore the second criterion used is that the root-mean-square average residual over the total number of scanned mesh points N is less than some small value R_{ms}, i.e.

$$\left(\frac{1}{N}\sum_{i,j} R_{i,j}^2\right)^{\frac{1}{2}} < R_{ms}. \tag{40}$$

The iteration is terminated when one of these two criteria is first satisfied.

The program produces, on a line printer or a VDU, a mesh pattern showing the position and size of the magnet. It also outputs the potential ϕ_{ij} and magnetic field magnitudes and directions in array form.

The components of the magnetic field **H** are calculated by subroutine THEMA using the simple finite difference formulae for the gradients:

$$H_{i,j}^x = -\frac{1}{h}(\phi_{i,j} - \phi_{i-1,j}); \qquad H_{i,j}^y = -\frac{1}{h}(\phi_{i,j} - \phi_{i,j-1}). \tag{41}$$

The lines OX and OY are not included, but from symmetry $H^x = 0$ in these directions. It may be possible, depending on the facilities available, to use graph plotting facilities to display the vector field **H** directly. However, the present program outputs arrays containing the directions $\theta_{i,j}$ (in degrees with line OX corresponding to $\theta = 0°$) and magnitudes $H_{i,j}$ of the field. These are calculated from

$$\theta_{i,j} = \tan^{-1}(H^y/H^x); \qquad H_{i,j} = [(H^x)^2 + (H^y)^2]^{\frac{1}{2}}, \tag{42}$$

care being taken to adjust the arc tangent as appropriate, depending on the signs of H^x and H^y. To illustrate the use of these formulae, consider the calculation of **H** at the point $i = 5$, $j = 5$ from some typical values of the potential, noting that $h = 0.1$, i.e.

$$H^x = -\frac{1}{h}\{\phi_{5,5} - \phi_{4,5}\} = -10.0\{0.38 - 0.65\} = 2.7,$$

$$H^y = -\frac{1}{h}\{\phi_{5,5} - \phi_{5,4}\} = -10.0\{0.38 - 0.30\} = -0.8. \tag{43}$$

Hence expression (42) gives

$$H_{5,5} = \{(2.7)^2 + (0.8)^2\}^{\frac{1}{2}} = 2.82,$$
$$\theta_{5,5} = \tan^{-1}(-0.8/2.7) = -16° \equiv +344°. \tag{44}$$

These answers can be located in the typical output given in this chapter.

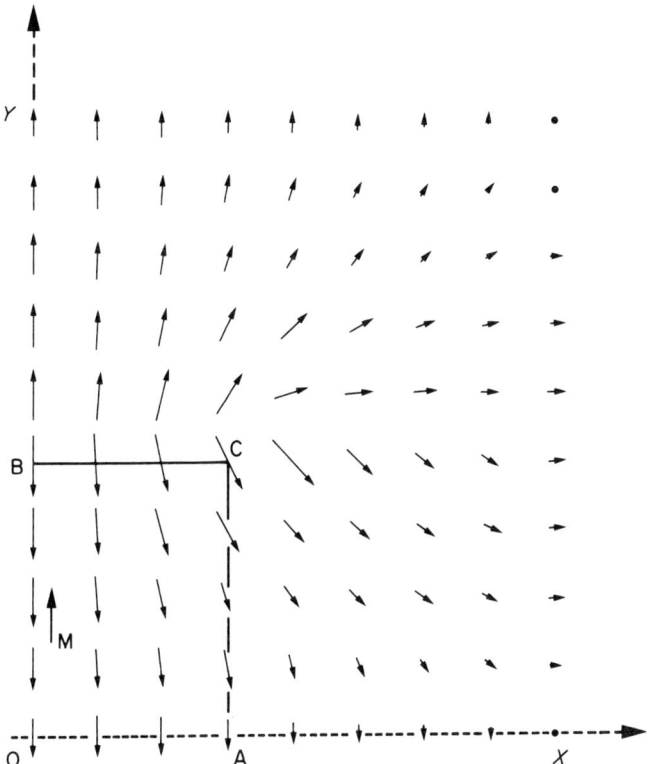

Figure 7. A typical magnetic field distribution in the positive quadrant plotted from results obtained from the computer program

The values of $H_{i,j}$ and $\theta_{i,j}$ enable the vector magnetic field to be plotted at each mesh point. For the mesh point (5, 5) for example, the vector **H** can be represented by a line of length 2.82 units (on some suitable scale) drawn to make an angle of +344° with the OX-axis. A plot of all the field vectors corresponding to the typical output is shown in Figure 7.

7. RUNNING THE PROGRAM

The small list of input parameters is as follows.

IN, JN   Number of mesh points in the x- and y-directions, respectively.
IM, JM   Number of mesh points in the x- and y-directions occupied by the magnet.
H   Mesh size (metres) (written as h in the text).
ALPHA   Convergence parameter $(1 < \alpha < 2)$.
AMAG   The magnetization (Am^{-1}).

The basic equations of section 2 are appropriate when the physical quantities are measured in SI units. In this system the magnetic induction vector **B** has units of tesla (T), and both the magnetization **M** and magnetic field **H** have units of ampere metre^{-1} (Am^{-1}). The values of $|\mathbf{M}|$ for typical permanent magnets are in the range 10^5–10^6 Am^{-1}. Spatial dimensions are measured in metres. Since M and h enter the potential equations only through the product Mh, the solution for a given value of Mh is applicable to many problems with different combinations of these parameters. However, it should be noted that the magnitude of the magnetic field **H** depends on the value of h in each case, through the gradient, and further, the magnitude of **H** scales linearly with M.

A typical set of input parameters is as follows:

9	10	4	5	0.1	1.5	5.0
(IN)	(JN)	(IM)	(JM)	(H)	(ALPHA)	(AMAG)

The first six parameters are read in on one line using FORMAT (4I4, 2F5.2) and AMAG is read separately using FORMAT (F5.2). It should be noted that the same device number has been employed for both input and output. The values of the input data are printed out immediately followed by a mesh pattern showing the position of the magnet.

The next three numbers printed give information on the convergence of the iteration process. The parameters determining the conditions for termination of the iteration have been preset in the program. The largest residual, REMAX, and the root mean square of the residuals, RESUM, are determined after each scan of the mesh. These are compared with two parameters, EPMAX and EPSUM respectively, which have both been given values of 10^{-4}. If REMAX is less than EPMAX, or if RESUM is less than EPSUM, it is assumed that the solution is sufficiently accurate for the present purpose. From the typical results given here it is seen that iteration has terminated when RESUM ($=0.95 \times 10^{-4}$) became smaller than 10^{-4}. The largest residual at that time was reasonably close to 10^{-4} (REMAX $= 2.4 \times 10^{-4}$), giving confidence in the solution obtained. The number of iterations needed is also printed out, being 28 in the example given. A control in the program stops the calculations if the integer NCOUN, counting the number of iterations, exceeds 100. It is a simple matter to change the preset parameters by making minor modifications to the program coding.

Finally, the arrays containing $\phi_{i,j}$, $\theta_{i,j}$ and $H_{i,j}$ are printed out. To understand the solution for the vector field **H**, the latter should be plotted in a form like Figure 7. It can be seen from that figure that inside the magnet the field is almost parallel to the direction of magnetization, but is in the opposite direction. This is the demagnetizing field. Outside the magnet the field is strongest near the magnet's face BC, where it is roughly parallel to the direction of **M**. This behaviour is that expected from the simplest model of a bar magnet in which (fictitious) magnetic poles are placed on the faces. The field falls off rapidly with distance in the x-direction, and smoothly changes direction, except near the corner of the magnet C, until on the symmetry axis OX it is in the opposite direction to **M**.

There are several exercises that can be performed with the computer program given here, simply by changing the input parameters. For instance it is instructive to investigate how the magnetic field distribution depends on the various constants, and how the convergence of the iterative process depends on α. Four typical exercises of this type are:

(1) Determination of the dependence of the magnetic field distribution on the size, shape, and magnetization of the magnet (vary input parameters IN, IM, JN, JM and AMAG).
(2) Investigation of the effects of changing the mesh size h and the determination of the optimum size for h (vary H in conjunction with IN, IM, JN and JM to retain the same region and magnet sizes).
(3) Investigation of the sensitivity of the magnetic field distribution near the magnet to the distance of the outer boundary from the magnet (vary IN and IM). Ideally the far boundary should be at infinity.
(4) Determination of the dependence of the rate of convergence of the iterative process for the $\phi_{i,j}$ on the value of the successive over-relaxation parameter (vary ALPHA) (cf. ref. 3 for the theory).

More insight into the method of solution of the problem can be obtained by making modifications to the program. Five typical exercises, involving only minor changes to the coding, could be based on using the following:
(a) other convergence parameters EPMAX and EPSUM;
(b) changing the coding to calculate the magnetic induction field **B** (from equation (5));
(c) more complicated difference formulae;[4]
(d) better formulae for special points,[4] e.g. for point C;
(e) other convergence criteria.[3,4]

The next stage of investigation would involve considering different models requiring considerable changes to the basic formulae, and effectively requiring completely new programs to be written. Exercises of this type are more advanced and could form the basis of project work. Typical examples include the consideration of the following:
(i) magnets of different shapes (but with plane boundaries);
(ii) composite rectangular magnets of materials with different magnetizations;
(iii) the effect of using a different mesh size over some part of the region of the field;
(iv) two or more separated rectangular magnets;
(v) cylindrical magnets.

Having moved away from finite difference formulae in Cartesian coordinates, as in (v) above, many other projects are possible. There are also problems in many other areas of physics that require solutions to elliptical partial differential equations obtainable by a similar numerical treatment to that considered here.

REFERENCES

1. W. K. H. Panofsky and M. Philips, *Classical Electricity and Magnetism* (Addison-Wesley, Reading, Mass., 1962).
2. C. D. Smith, *Numerical Solution of Partial Differential Equations* (Oxford University Press, London, 1965).
3. *Modern Computing Methods*, 2nd edn. (H.M.S.O., London, 1961).
4. K. J. Binns and P. J. Lawrenson, *Analysis and Computation of Electric and Magnetic Field Problems* (Pergamon Press, London, 1963).

PERMANENT MAGNETS

```
      DIMENSION PHI(50,50),TH(50,50),FE(50,50)
C   INPUT AND PRINT OUT PARAMETERS
      READ(1,100)IN,JN,IM,JM,H,ALPHA
      READ(1,101)AMAG
  100 FORMAT(4I4,2F5.2)
  101 FORMAT(F5.2)
      WRITE(1,105)IN,JN
      WRITE(1,106)IM,JM
      WRITE(1,107)H
      WRITE(1,108)AMAG
      WRITE(1,109)ALPHA
  105 FORMAT('REGION SIZE=',I2,'X',I2)
  106 FORMAT('MAGNET SIZE=',I2,'X',I2)
  107 FORMAT('MESH INTERVAL=',F5.2)
  108 FORMAT('MAGNETIZATION=',F5.2)
  109 FORMAT('CONVERGENCE FACTOR=',F5.2)
      WRITE(1,221)
      WRITE(1,221)
      WRITE(1,222)
  222 FORMAT(5X,'MESH USED')
      WRITE(1,221)
      EPMAX=1.0E-04
      EPSUM=1.0E-04
      HAM=H*AMAG
C   PRINT MESH PATTERN
      CALL MESH(IN,JN,IM,JM)
      INL=IN-1
      JNL=JN-1
C   SET OUTER BOUNDARIES
      DO 1 J=1,JN
      PHI(1,J)=0.0
    1 PHI(IN,J)=0.0
      DO 2 I=1,IN
      PHI(I,1)=0.0
    2 PHI(I,JN)=0.0
C   SET TRIAL INITIAL VALUES
      DO 5 I=2,INL
      DO 6 J=2,JNL
    6 PHI(I,J)=0.5
    5 CONTINUE
      ALPF=ALPHA/4.0
C   BEGIN ITERATION LOOP
      NCOUN=0
   10 NCOUN=NCOUN+1
      IF(NCOUN.GT.100) GOTO 11
      REMAX=0.0
      RESUM=0.0
      DO 3 J=1,JNL
      DO 4 I=1,INL
      IF((I.EQ.1).AND.(J.EQ.1)) GOTO 26
      IF((I.EQ.1).AND.(J.EQ.JM)) GOTO 23
      IF(I.EQ.1) GOTO 20
      IF(J.EQ.1) GOTO 21
      IF((I.EQ.IM).AND.(J.LT.JM)) GOTO 22
      IF((J.EQ.JM).AND.(I.LT.IM)) GOTO 23
      IF((J.EQ.JM).AND.(I.EQ.IM)) GOTO 24
```

PERMANENT MAGNETS

```
      R=PHI(I+1,J)+PHI(I-1,J)+PHI(I,J-1)+PHI(I,J+1)-4.0*PHI(I,J)
      GOTO 25
   20 R=2.0*PHI(2,J)+PHI(1,J-1)+PHI(1,J+1)-4.0*PHI(1,J)
      GOTO 25
   21 R=PHI(I+1,1)+PHI(I-1,1)-2.0*PHI(I,1)
      GOTO 25
   22 R=2.0*(PHI(IM+1,J)+PHI(IM-1,J))-4.0*PHI(IM,J)
      GOTO 25
   23 R=2.0*(PHI(I,JM+1)+PHI(I,JM-1)+HAM)-4.0*PHI(I,JM)
      GOTO 25
   24 R=PHI(IM+1,JM)+PHI(IM-1,JM)+PHI(IM,JM+1)+PHI(IM,JM-1)+
     1HAM-4.0*PHI(IM,JM)
      GOTO 25
   26 R=0.0
   25 CONTINUE
      PHI(I,J)=PHI(I,J)+ALPF*R
      AR=ABS(R)
      IF(AR.GT.REMAX) REMAX=AR
      RESUM=RESUM+R*R
    4 CONTINUE
    3 CONTINUE
C     TEST CONVERGENCE
      RESUM=SQRT(RESUM/FLOAT(INL*JNL))
      IF(REMAX.LT.EPMAX) GOTO 12
      IF(RESUM.GT.EPSUM) GOTO 10
   12 CONTINUE
   11 CONTINUE
      WRITE(1,120)NCOUN,REMAX,RESUM
  120 FORMAT(1H ,' NO OF ITERATIONS =',I3,'   REMAX= ',
     1E10.4,'      RESUM=',E10.4)
      WRITE(1,221)
      WRITE(1,221)
      WRITE(1,121)
  121 FORMAT(1H ,15X,'MAGNETOSTATIC POTENTIAL',)
      WRITE(1,221)
  221 FORMAT(1H )
C     CALCULATE MAGNETIC FIELD AND OUTPUT ARRAYS
      CALL OUT(IN,JN,PHI,9,3)
      CALL THEMA(IN,JN,H,PHI,TH,FE)
      WRITE(1,122)
  122 FORMAT(15X,'MAGNETIC FIELD DIRECTIONS',)
      WRITE(1,221)
      CALL OUT(IN,JN,TH,9,2)
      WRITE(1,123)
  123 FORMAT(15X,'MAGNETIC FIELD MAGNITUDES',)
      WRITE(1,221)
      CALL OUT(IN,JN,FE,9,2)
  111 FORMAT(1H ,5E10.2)
      STOP
      END
      SUBROUTINE OUT(IN,JN,F,NPRIN,NCON)
      DIMENSION F(50,50)
C     REORDERS ARRAY F AND PRINTS NPRIN COLUMNS AT A TIME
C     IF NCON=1 OUTPUTS CHARACTERS FOR MESH
      IP=0
    1 IP=IP+NPRIN
```

CALCULATION OF THE FIELDS NEAR PERMANENT MAGNETS

PERMANENT MAGNETS

```
      INP=IN
      IF(IP.LT.IN) INP=IP
      IL=IP-NPRIN+1
      JLN=1
      DO 2 JT=JLN,JN
      J=JN-JT+1
      IF(NCON.EQ.1) GOTO 3
      WRITE(1,111)(F(IK,J),IK=IL,INP)
      GOTO 2
    3 WRITE(1,113)(F(IK,J),IK=IL,INP)
    2 CONTINUE
      WRITE(1,112)
      WRITE(1,112)
      WRITE(1,112)
      IF(IP.LT.IN) GOTO 1
  111 FORMAT(1H ,12F7.2)
  112 FORMAT(1H )
  113 FORMAT(1H ,20A4)
      RETURN
      END
      SUBROUTINE MESH(IN,JN,IM,JM)
      DIMENSION Q(50,50)
      INTEGER T1,T2
C     SETS ARRAY OF CHARACTERS FOR MESH REGIONS
      DATA T1,T2/1H.,1H1/
      DO 1 I=1,IN
      DO 2 J=1,JN
      Q(I,J)=T1
    2 IF((I.LE.IM).AND.(J.LE.JM)) Q(I,J)=T2
    1 CONTINUE
      CALL OUT(IN,JN,Q,20,1)
      RETURN
      END
      SUBROUTINE THEMA(IN,JN,H,F,T,HM)
      DIMENSION F(50,50),HM(50,50),T(50,50)
C     INPUTS POTENTIAL AS F AND OUTPUTS MAGNETIC FIELD IN
C     ARRAYS T(DIRECTIONS) AND HM(MAGNITUDES)
      DO 1 I=2,IN
      DO 2 J=2,JN
      HX=-(F(I,J)-F(I-1,J))/H
      HY=-(F(I,J)-F(I,J-1))/H
      IF(HX.EQ.0.0) HX=0.01
      HXY=HY/HX
      TT=ATAN(ABS(HXY))*180.0/3.14159
C     ADJUST ARCTAN FOR DIFFERENT QUADRANTS
      IF((HX.GE.0.0).AND.(HY.GE.0.0)) P=TT
      IF((HX.LT.0.0).AND.(HY.GE.0.0)) P=180.0-TT
      IF((HX.LE.0.0).AND.(HY.LT.0.0)) P=180.0+TT
      IF((HX.GE.0.0).AND.(HY.LT.0.0)) P=360.0-TT
      T(I,J)=P
      HM(I,J)=SQRT(HX*HX+HY*HY)
    2 CONTINUE
    1 CONTINUE
      DO 3 J=2,JN
      HY=-(F(1,J)-F(1,J-1))/H
      T(1,J)=90.0
```

PERMANENT MAGNETS

```
      IF(HY.LT.0.0) T(1,J)=270.0
   3  HM(1,J)=ABS(HY)
      DO 4 I=1,IN
      T(I,1)=270.0
   4  HM(I,1)=HM(I,2)
      T(IN,1)=0.0
      RETURN
      END
```

PERMANENT MAGNETS **RESULTS**

REGION SIZE= 9X10
MAGNET SIZE= 4X 5
MESH INTERVAL= 0.10
MAGNETIZATION= 5.00
CONVERGENCE FACTOR= 1.50

    MESH USED

```
. . . . . . . . .
. . . . . . . . .
. . . . . . . . .
. . . . . . . . .
. . . . . . . . .
1 1 1 1 . . . . .
1 1 1 1 . . . . .
1 1 1 1 . . . . .
1 1 1 1 . . . . .
1 1 1 1 . . . . .
```

NO OF ITERATIONS = 28   REMAX= 0.2361E-03   RESUM=0.9454E-04

MAGNETOSTATIC POTENTIAL

```
     0.00  0.00  0.00  0.00  0.00  0.00  0.00  0.00  0.00
     0.14  0.14  0.13  0.11  0.08  0.06  0.04  0.02  0.00
     0.30  0.29  0.26  0.22  0.17  0.12  0.08  0.04  0.00
     0.47  0.45  0.41  0.34  0.26  0.18  0.11  0.05  0.00
     0.66  0.65  0.59  0.48  0.33  0.22  0.13  0.06  0.00
  j  0.90  0.88  0.82  0.65  0.38  0.23  0.14  0.06  0.00
  ↑  0.63  0.61  0.55  0.43  0.30  0.20  0.12  0.06  0.00
     0.40  0.38  0.34  0.27  0.21  0.14  0.09  0.04  0.00
     0.19  0.19  0.17  0.13  0.10  0.07  0.05  0.02  0.00
     0.00  0.00  0.00  0.00  0.00  0.00  0.00  0.00  0.00
  └──→ i
```

MAGNETIC FIELD DIRECTIONS

```
 90.00  89.59  89.55  89.47  89.33  89.07  88.54  86.99    0.00
 90.00  88.36  84.74  80.29  74.81  68.41  59.39  41.49    0.00
 90.00  86.96  80.14  71.18  59.51  48.09  36.15  20.81    0.00
 90.00  85.96  76.82  63.07  42.18  28.34  18.73   9.87    0.00
 90.00  85.63  76.12  57.47  17.95   7.57   3.70   1.62    0.00
270.00 273.85 282.27 306.79 344.04 346.72 349.77 354.07    0.00
270.00 274.63 286.28 309.11 321.70 330.51 337.35 346.67    0.00
270.00 273.77 282.85 296.28 304.04 313.13 322.43 336.79    0.00
270.00 272.02 276.75 283.16 286.97 293.02 300.97 317.54    0.00
270.00 270.00 270.00 270.00 270.00 270.00 270.00 270.00    0.00
```

MAGNETIC FIELD MAGNITUDES

```
1.44  1.40  1.27  1.08  0.85  0.61  0.39  0.19  0.01
1.52  1.48  1.35  1.14  0.89  0.64  0.43  0.27  0.19
1.70  1.65  1.52  1.28  0.98  0.74  0.55  0.42  0.37
1.97  1.93  1.82  1.54  1.16  0.90  0.71  0.58  0.52
2.32  2.33  2.38  2.05  1.53  1.15  0.87  0.70  0.62
2.68  2.68  2.76  2.81  2.82  1.50  0.98  0.74  0.64
2.31  2.27  2.15  1.96  1.57  1.19  0.86  0.65  0.56
2.05  1.99  1.82  1.56  1.23  0.94  0.68  0.50  0.41
1.92  1.86  1.67  1.38  1.08  0.79  0.53  0.32  0.22
1.92  1.86  1.67  1.38  1.08  0.79  0.53  0.32  0.22
```

Physics Programs
Edited by A. D. Boardman
© 1980 John Wiley & Sons Ltd.

CHAPTER 5

Particle Capture in High Gradient Magnetic Separation

R. GERBER

1. INTRODUCTION

High gradient magnetic separation (HGMS) is a technique for the removal of weakly magnetic particles from suspensions that has many practical applications such as sewage and water treatments and the processing of industrial slurries. The method is based on the utilization of a magnetic traction force which extracts the particles from the fluid when the suspension passes through the separator system. The magnetic traction force, $\mathbf{F}_m$, is proportional to the difference, $\chi_p - \chi_f$, between the particle and fluid susceptibilities and, since curl $\mathbf{H} = 0$, to the product, H grad H, of the magnitude of the magnetic field and its gradient at the position of the particle. The difference $\chi_p - \chi_f$ is usually very small for weakly magnetic particles, and also the magnetic field magnitude H cannot be increased above a certain upper limit for technical reasons. Thus an efficient extraction, which results from a large value of $\mathbf{F}_m$, requires that the value of grad H be high. Hence the name of the method.

An example of a HGMS system is shown schematically in Figure 1. It consists of a canister, filled with an ordered matrix of very thin magnetic wires, which is placed in a magnetic field, large enough to saturate the wires. The canister, using an appropriate plumbing, conducts either the suspension of mixed magnetic and non-magnetic particles or flush water through the system. The thin wires of the matrix dehomogenize the background magnetic field and give rise to a high value of its gradient in their immediate surroundings. When the magnetic field is on, the suspension of particles to be separated flows through the matrix, the magnetic particles are captured on to the wires, and the purified suspension leaves the system. At intervals, when the retention capacity of the matrix is reached, the feed is halted, the magnetic field is switched off, and the magnetic particles are flushed out of the separator. Then the cycle is repeated.

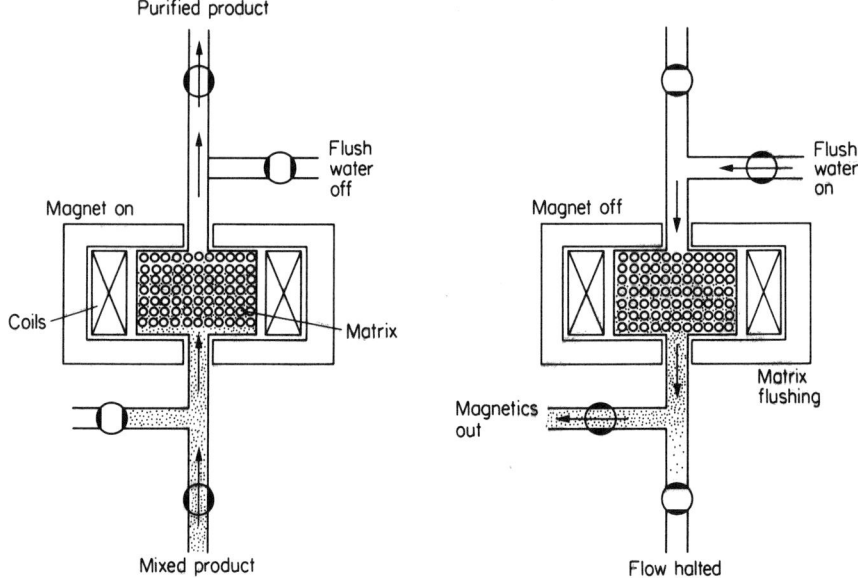

Figure 1. Schematic view of a high gradient magnetic separator

The packing fraction, F (which is defined as the volume occupied by the wires to that of the canister), of the matrix is typically between 0.05 and 0.1. Thus most of the space in the canister is actually free. This has two important consequences. First, the fluid impedance of the system is low. Hence a large amount of material can be passed through and treated by the separator in a short period of time. Second, the capture process can be considered in a single-wire approximation.[1] In this approximation, since the wires in the canister are relatively far apart, it is assumed that the particles interact with only one individual wire at a time.

An important quantity characterizing any separation system is the filter performance. It is given as the ratio N_{out}/N_{in}, where N_{out} and N_{in} are the number of magnetic particles leaving and entering the separator, respectively. The smaller the ratio the better the performance.

In order to obtain some idea of filter performance let us consider an ordered matrix comprised of a large number of stacked-up sheets of gauze which are woven in a regular rectangular pattern from a thin ferromagnetic wire of radius a. If the magnetic field is normal to the surface of each sheet and if certain limiting conditions are fulfilled, it can be shown[2] that the filter performance is given by the formula

$$N_{out} = \exp\left(\frac{-2R_{ca}FL}{\pi a}\right) N_{in}, \qquad (1)$$

where L is the total length of the matrix stack and R_{ca} is the capture radius (in units of a) associated with an individual wire.

It is clear that particle trajectories and capture radii are of fundamental importance for HGMS. A useful starting point for an investigation of this technique is a computational study dealing with particle capture in the single-wire approximation, and this is presented in this chapter.

2. SINGLE-WIRE APPROXIMATION

2.1 Configuration

The background magnetic field in the separator becomes inhomogeneous due to the appearance of magnetic charges on the ferromagnetic wires of the matrix. This effect is a maximum if the ordered matrix is arranged so that the direction of the field is always perpendicular to the axes of the wires. Therefore, in the single-wire approximation, we consider a configuration as shown in Figure 2. A ferromagnetic wire of radius a, saturation magnetization M_s, is placed axially along the z-axis of an orthogonal coordinate system. A magnetic field, H_0, sufficient to saturate the wire, is applied in the x-direction and a fluid of viscosity η flows with a background velocity V_0 along a direction which lies in the xy plane making an angle α with the x-axis.

A completely arbitrary direction of fluid flow is perfectly possible in

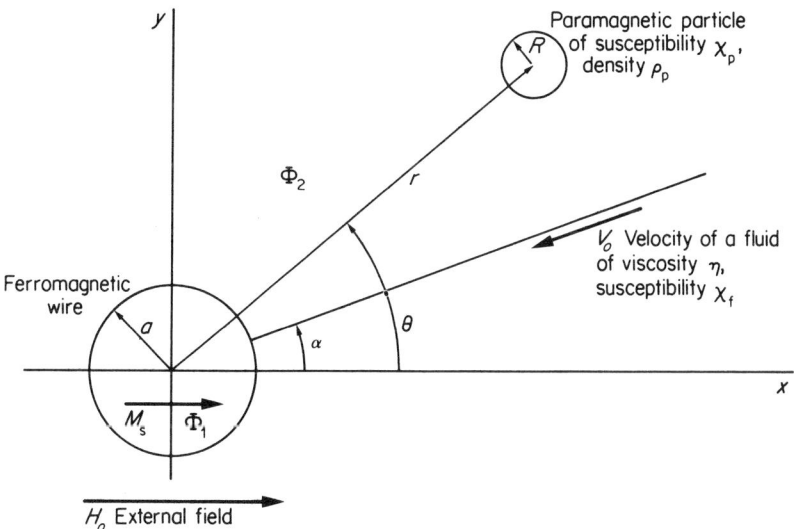

Figure 2. Configuration of the particle capture problem in the single-wire approximation

HGMS. However, such direction can be resolved in one lying in the xy plane and in one parallel with the z-axis. The latter case, the so-called axial configuration, can be solved analytically[3] and is therefore outside the scope of this chapter.

Paramagnetic particles of spherical shape with radius $R(R \ll a)$, volume V_p, susceptibility χ_p and density ρ_p, are carried by the fluid. The flow of the fluid around the wire is considered to be streamlined and frictionless, but the particles are assumed to experience a viscous drag force $\mathbf{F}_v$ given by the Stokes equation $\mathbf{F}_v = -6\pi\eta R(\mathbf{v}-\mathbf{V})$, where $(\mathbf{v}-\mathbf{V})$ is the relative velocity of the particle with respect to the fluid.

Physical quantities used throughout this chapter are in SI units (Sommerfeld system). The metric part of their dimensions may be sometimes expressed in terms of the wire radius a. Then, to indicate this renormalization, the symbols are provided with the subscript a e.g. $R_a = R/a$, $V_{pa} = V_p/a^3$, etc.).

2.2 Magnetic field

The magnetic field $\mathbf{H}$ of a wire magnetized by a uniform background field $\mathbf{H}_0$ can be expressed as $\mathbf{H} = -\text{grad}\,\Phi$, where Φ is the magnetic scalar potential. Since $\text{div}\,\mathbf{B} = 0$, we can write for a linear medium $\text{div}\,\mathbf{B} = \text{div}\,(\mu\mathbf{H}) = \mu\,\text{div}\,\mathbf{H} = \mu\,\text{div}\,\text{grad}\,\Phi = 0$. Thus the potential Φ must satisfy Laplace's equation $\nabla^2\Phi = 0$ and also the boundary conditions requiring the tangential components of $\mathbf{H}$ and the normal components of $\mathbf{B}$ to be continuous at the surface of the wire.

In plane polar coordinates, since $\partial^2\Phi/\partial z^2 = 0$, Laplace's equation becomes

$$r\frac{\partial}{\partial r}\left(r\frac{\partial \Phi}{\partial r}\right) + \frac{\partial^2 \Phi}{\partial \theta^2} = 0 \tag{2a}$$

and the boundary conditions at the surface of the wire ($r = a$) can be written as

$$\frac{\partial \Phi_1}{\partial \theta} = \frac{\partial \Phi_2}{\partial \theta}, \tag{2b}$$

$$\mu_0\left(-\frac{\partial \Phi_1}{\partial r} + M_s \cos\theta\right) = -\mu_f \frac{\partial \Phi_2}{\partial r}, \tag{2c}$$

where μ_0 and μ_f are the permeabilities of free space and the fluid and Φ_1 and Φ_2 are the magnetic scalar potentials inside and outside the wire, respectively (Figure 2).

Equations (2a), (2b), and (2c) can be simultaneously satisfied by a linear combination

$$\Phi_1 = -C_1 r \cos\theta + A_1 r^{-1} \cos\theta, \tag{3a}$$

$$\Phi_2 = -C_2 r \cos\theta + A_2 r^{-1} \cos\theta \tag{3b}$$

of cylindrical harmonics.[4]

The field in the origin (on the wire axis) is finite, hence $A_1 = 0$. At large distances, the field tends to H_0, consequently $C_2 = H_0$. The remaining two constants, C_1 and A_2, are obtained by substitution of equations (3a, b) into (2b, c). We obtain

$$C_1 = H_0 - A_2 a^{-2} \tag{4}$$

and

$$A_2 = \frac{\mu_0 - \mu_f}{\mu_0 + \mu_f} H_0 a^2 + \frac{\mu_0}{\mu_0 + \mu_f} M_s a^2. \tag{5}$$

Since for most practical cases $|\mu_0 - \mu_f| \leq 10^{-3}$, the first term in equation (5) can be neglected in comparison with the second. Thus we obtain from (4) and (5)

$$A_2 = \tfrac{1}{2} M_s a^2, \tag{6}$$
$$C_1 = H_0 - \tfrac{1}{2} M_s. \tag{7}$$

Substituting $A_1 = 0$, $C_2 = H_0$, equations (6) and (7) into (3a, b) we obtain the scalar potentials, the negative gradient of which gives the respective magnetic fields, $\mathbf{H}_1$ and $\mathbf{H}_2$, inside and outside the wire. Their components are

$$H_{1r} = (H_0 - \tfrac{1}{2} M_s)\cos\theta, \tag{8a}$$
$$H_{1\theta} = -(H_0 - \tfrac{1}{2} M_s)\sin\theta, \tag{8b}$$

and

$$H_{2r} = (\tfrac{1}{2} M_s a^2 r^{-2} + H_0)\cos\theta, \tag{9a}$$
$$H_{2\theta} = (\tfrac{1}{2} M_s a^2 r^{-2} - H_0)\sin\theta. \tag{9b}$$

2.3 Fluid velocity distribution

The fluid is assumed to be viscous when interacting with the particles, but when interacting with the wires of the filter it is considered to be ideal. Consequently, if irrotational flow is assumed, the velocity of the fluid motion is given as $\mathbf{V} = -\operatorname{grad} \psi$, where ψ is the velocity potential. Again, the potential ψ must satisfy Laplace's equation $\nabla^2 \psi = 0$ and also the boundary condition that the normal component of the velocity must vanish at the surface of the wire.

Therefore, in plane polar coordinates, ψ has to satisfy simultaneously equation (2a), when substituted instead of Φ, and the condition $\partial \psi / \partial r = 0$ for $r = a$ and any θ. Bearing in mind that at large distances the flow takes place at an angle α with a uniform velocity V_0, we can see immediately that the function

$$\psi = -V_0 r \cos(\theta - \alpha) + A r^{-1} \cos(\theta - \alpha), \tag{10}$$

which is analogous to (3a, b), will satisfy both equation (2a) and the boundary condition. Substitution of (10) into $(\partial \psi/\partial r)_{r=a} = 0$ gives $A = -V_0 a^2$. This value can now be used in (10), when the negative gradient is taken of this expression. Thus we obtain the components of the fluid velocity:

$$V_r = V_0(1 - a^2 r^{-2})\cos(\theta - \alpha), \tag{11a}$$

$$V_\theta = -V_0(1 + a^2 r^{-2})\sin(\theta - \alpha). \tag{11b}$$

2.4 Equations of particle motion

The equation of motion can be written in a vector form as

$$m\mathbf{a} = \mathbf{F}_v + \mathbf{F}_m, \tag{12}$$

where $m = \frac{4}{3}\pi R^3 \rho_p$ is the mass of the particle, $\mathbf{a}$ is the acceleration, $\mathbf{F}_v$ and $\mathbf{F}_m$ are the viscous drag and magnetic traction force, respectively. In order to be able to solve equation (12) we need to establish the components of all its terms in plane polar coordinates r and θ.

Consider two mutually perpendicular unit vectors $\hat{\mathbf{r}}$ and $\hat{\boldsymbol{\theta}}$ which have directions of increasing $\mathbf{r}$ and increasing θ. Their directions change with time and since the derivative of a unit vector is perpendicular to the vector, we have

$$\frac{d\hat{\mathbf{r}}}{dt} = \frac{d\theta}{dt}\hat{\boldsymbol{\theta}}, \tag{13a}$$

$$\frac{d\hat{\boldsymbol{\theta}}}{dt} = -\frac{d\theta}{dt}\hat{\mathbf{r}}. \tag{13b}$$

Now, the position of the particle in plane polar coordinates is given by $\mathbf{r} = r\hat{\mathbf{r}}$. Differentiating this equation and using (13a, b), we obtain the particle velocity $\mathbf{v}$, the components of which are $v_r = dr/dt$ and $v_\theta = r(d\theta/dt)$. By second differentiation, we obtain the particle acceleration $\mathbf{a}$ and its components

$$a_r = \frac{d^2 r}{dt^2} - r\left(\frac{d\theta}{dt}\right)^2, \tag{14a}$$

$$a_\theta = r\frac{d^2\theta}{dt^2} + 2\frac{dr}{dt}\frac{d\theta}{dt}. \tag{14b}$$

The components of the viscous drag force $\mathbf{F}_v$ are

$$F_{vr} = -6\pi\eta R\left(\frac{dr}{dt} - V_r\right) \tag{15a}$$

and

$$F_{v\theta} = -6\pi\eta R\left(r\frac{d\theta}{dt} - V_\theta\right), \tag{15b}$$

where V_r and V_θ are given by equations (11a, b).

To establish the components of the magnetic traction force $\mathbf{F}_m$ we have to consider first the magnetic energy of a system comprising a fluid and a particle.

The magnetic energy density is expressed in general as $\frac{1}{2}HB$. Imagine a volume V_p demarcated inside the fluid. The magnetic energy of the fluid enclosed in this volume is $\frac{1}{2}HB_fV_p = \frac{1}{2}\mu_f V_p H^2$. Let us now remove the fluid from the volume V_p and replace it by the particle. The energy associated with the particle itself is $\frac{1}{2}\mu_p V_p H^2$. The energy increment U of the system (particle + fluid) is given as the difference between these two energies, i.e. $U = \frac{1}{2}(\mu_p - \mu_f)V_p H^2$. Taking ∇U (positive gradient, since the magnetic energy is of the same nature as the kinetic energy) we get

$$\mathbf{F}_m = \frac{1}{2}\mu_0 \chi V_p \nabla(H^2), \tag{16}$$

where $\chi = \chi_p - \chi_f$ is the difference between susceptibilities of the particle and the fluid.

Considering a spherical particle, $V_p = \frac{4}{3}\pi R^3$, and substituting $H^2 = H_{2r}^2 + H_{2\theta}^2$ from (9a, b) into (16), we obtain, after some manipulation, the components of the magnetic traction force

$$F_{mr} = -\frac{4\pi\mu_0\chi M_s a^2 R^3}{3r^3}\left(\frac{M_s a^2}{2r^2} + H_0\cos 2\theta\right), \tag{17a}$$

$$F_{m\theta} = -\frac{4\pi\mu_0\chi M_s a^2 R^3}{3r^3}H_0\sin 2\theta. \tag{17b}$$

Combining equations (14a, b), (15a, b), and (17a, b) with (12) we obtain the complete equations of motion: in radial direction

$$\frac{2\rho_p R^2}{9\eta}\left(\frac{d^2 r_a}{dt^2} - r_a\left(\frac{d\theta}{dt}\right)^2\right) + \frac{dr_a}{dt}$$

$$= V_{0a}\left(1 - \frac{1}{r_a^2}\right)\cos(\theta - \alpha) - V_{ma}\left(\frac{M_s}{2H_0 r_a^2} + \cos 2\theta\right)\frac{1}{r_a^3} \tag{18a}$$

and in azimuthal direction

$$\frac{2\rho_p R^2}{9\eta}\left(r_a\frac{d^2\theta}{dt^2} + 2\frac{dr_a}{dt}\frac{d\theta}{dt}\right) + r_a\frac{d\theta}{dt}$$

$$= -V_{0a}\left(1 + \frac{1}{r_a^2}\right)\sin(\theta - \alpha) - V_{ma}\frac{\sin 2\theta}{r_a^3}, \tag{18b}$$

where $r_a = r/a$, $V_{0a} = V_0/a$, $V_{ma} = V_m/a = \frac{2}{9}(\chi\mu_0 M_s H_0 R_a^2/\eta)$ and $R_a = R/a$. The quantity V_m, the dimensions of which are m s^{-1}, is called the magnetic velocity.

The first (inertial) terms in equations (18a, b) can usually be neglected in comparison with others in case of HGMS of small particles in liquids. Then R^2 is very small and η moderately high and (18a, b) reduce to a pair of first-order differential equations:

$$\frac{dr_a}{dt} = V_{0a}\left(1 - \frac{1}{r_a^2}\right)\cos(\theta - \alpha) - V_{ma}\left(\frac{K}{r_a^2} + \cos 2\theta\right)\frac{1}{r_a^3}, \tag{19a}$$

$$\frac{d\theta}{dt} = -V_{0a}\left(1 + \frac{1}{r_a^2}\right)\frac{\sin(\theta - \alpha)}{r_a} - V_{ma}\frac{\sin 2\theta}{r_a^4}, \tag{19b}$$

where $K = M_s/2H_0$. Equations (19a, b) will be analysed and numerically solved in the main text of this chapter.

If the particle bearing fluid is a gas, η is rather small and the inertial term cannot be neglected. Complete equations of motion are therefore required for a numerical solution of this problem.[5] An appropriate extension of our computer programs to perform this task also is included in the suggested exercises.

3. NUMERICAL SOLUTION OF ORDINARY DIFFERENTIAL EQUATIONS

3.1 Predictor–corrector method

Consider a first-order initial-value problem

$$y' = f(x, y), \tag{20a}$$

$$y(x_0) = y_0, \tag{20b}$$

where x_0, y_0, and the function $f(x, y)$ are given, y' denotes dy/dx. Under quite general conditions this problem has a unique solution $y(x)$ in the interval[6] $\langle x_0, \bar{x} \rangle$, $\bar{x} \neq x_0$.

We would like to find a numerical approximation to the solution $y(x)$. Let us divide the interval $\langle x_0, \bar{x} \rangle$ into m subintervals, each of length h. Thus, there are $m + 1$ points

$$x_n = x_0 + nh, \quad n = 0, 1, 2, \ldots, m,$$

where we are seeking the values $y_1, y_2, \ldots, y_m$ of the solution $y(x)$.

The numerical method of solving differential equations such as (20a, b) involves the procedure which yields y_{n+1}, given the sequence $y_n, y_{n-1}, y_{n-2} \ldots$ up to y_{n-s}. The value of s may be 0, 1, 2, etc. When y_{n+1} is found, the procedure is repeated until the whole interval $\langle x_0, \bar{x} \rangle$ is covered.

The Euler method, which has $s = 0$, is the simplest numerical method for solving initial-value problems of the type (20a, b). To derive this method, assume that the value y_n is known and integrate (20a) from x_n to x_{n+1}. We get

$$y_{n+1} - y_n = \int_{x_n}^{x_{n+1}} f(x, y(x))\, dx. \tag{21}$$

The integral in (21) can be approximated by the rectangle rule as $hf(x_n, y(x_n))$. Thus we obtain the formula (procedure)

$$y_{n+1} - y_n = hf(x_n, y_n). \tag{22}$$

Starting with $n = 0$, i.e. substituting first the known value y_0, we find by repetition all required values.

Evidently, the accuracy of determination of y_{n+1} depends on h. It can be shown that the single step (quadrature) error is approximately proportional to h^2 for the Euler method. We can, therefore, increase the accuracy by decreasing the size of h. This will, however, increase the number of steps and hence the computing time. Moreover, with each step an accumulated error will enter into the process. This error will partly offset the improvement in accuracy which was gained by the reduction of h. Clearly, one would like to increase the accuracy without the need to reduce h. This can be achieved by using numerical quadrature formulae, based on Lagrangian interpolation, for the approximation of the integral in equation (21). These quadrature formulae take into account the behaviour of the function $f(x, y(x))$ in a wider surroundings of the interval of integration. Thus the integral in (21) can be expressed to a high degree of accuracy for a relatively large h.

We have two types of formulae available. The open or predictor formula

$$y_{n+1} - y_{n+1-q} = h \sum_{i=1}^{M} \beta_i f(x_{n+1-i}, y_{n+1-i}), \tag{23a}$$

and the closed or corrector formula

$$y_{n+1} - y_{n+1-q} = h \sum_{i=0}^{M'} \beta'_i f(x_{n+1-i}, y_{n+1-i}), \tag{23b}$$

where q is a fixed integer, and β_i and β'_i are the numerical coefficients. A full analysis and values of q, β_i, and β'_i of various methods can be found in Young and Gregory.[6]

Note that equation (23a) contains y_{n+1} explicitly, whereas (23b) contains y_{n+1} implicitly, hence the names 'open' and 'closed' formula, respectively.

Formulae (23a, b) together form the predictor–corrector method. It works as follows: let us assume that we are at the nth point and that we already know, by some means, the values $y_n, y_{n-1}, \ldots, y_{n-M+1}$. Substituting these known values in (23a) we obtain the predicted value $y^{(P)}$. Substituting $y_{n+1}^{(P)}$ together with $y_n, y_{n-1}, \ldots$ etc. in (23b) we obtain the corrected value $y_{n+1}^{(C)}$. If $|(y_{n+1}^{(P)} - y_{n+1}^{(C)})/y_{n+1}^{(C)}| < E$, where E is the stipulated error, we put $y_{n+1} = y_{n+1}^{(C)}$ and repeat the cycle for the next point, i.e. for $n+1$. If the relative error between the predicted and corrected value is larger than E we have to reduce h and calculate the next $M-1$ values by some other means (see section 3.2) before the predictor–corrector method can again be employed.

There is a large family of predictor–corrector methods available. The Adams–Moulton method, which is employed in our program, is particularly attractive, since it is of a good stability and high precision (quadrature error $\simeq 0(h^5)$). The numerical quantities characterizing this method are $q = 1$, $M = 4$, $\beta_i = 55/24$, $-59/24$, $37/24$, $-9/24$, $M' = 3$, $\beta'_i = 9/24$, $19/24$, $-5/24$, $1/24$.

3.2 Runge–Kutta method

To get a predictor–corrector method into action we need to know the values of y at several starting points $x_0 + h$, $x_0 + 2h, \ldots, x_0 + sh$, in addition to the given value of $y(x_0)$. Hence the desire for an accurate self-starting method, i.e. a high order method which would require a knowledge of y at only one initial point. The Runge–Kutta method is of this type and its derivation will be briefly described here. Consider again equation (21). We can write

$$y_{n+1}^{(P)} = y_n + hf(x_n, y_n), \tag{24a}$$

$$y_{n+1} = y_n + \frac{h}{2}[f(x_n, y_n) + f(x_{n+1}, y_{n+1}^{(P)})], \tag{24b}$$

where the predicted value $y_{n+1}^{(P)}$ and the corrected value y_{n+1} were obtained by applying the rectangle and the trapezoidal rule, respectively. Formulae (24a, b), which constitute the Heun predictor–corrector method, can be rewritten in the following form:

$$y_{n+1} = y_n + \alpha_1 \Delta_1 + \alpha_2 \Delta_2, \tag{25}$$

where

$$\Delta_1 = hf(x_n, y_n), \tag{26a}$$

$$\Delta_2 = hf(x_n + \rho h, y_n + \rho \Delta_1), \tag{26b}$$

and where the numerical coefficients have the values of $\rho = 1$, $\alpha_1 = \frac{1}{2}$, $\alpha_2 = \frac{1}{2}$.

Thus instead of two formulae, (24a, b), we have one formula, (25), with two 'differential' increments Δ_1 and Δ_2. The first increment Δ_1 serves not only in (25) but also as a 'predictor' for the second increment Δ_2 in (26b).

Generalizing formulae (25) and (26a, b) to involve four increments and determining the appropriate numerical coefficients by comparison of $y_{n+1} = \sum_{j=1}^{4} \alpha_j \Delta_j$ with y_{n+1} given as a Taylor's series expansion of $y(x)$ at x_n, we obtain the (fourth-order) Runge–Kutta method as

$$y_{n+1} = y_n + \tfrac{1}{6}(\Delta_1 + 2\Delta_2 + 2\Delta_3 + \Delta_4), \tag{27}$$

where

$$\Delta_1 = hf(x_n, y_n), \tag{28a}$$
$$\Delta_2 = hf(x_n + \tfrac{1}{2}h, y_n + \tfrac{1}{2}\Delta_1), \tag{28b}$$
$$\Delta_3 = hf(x_n + \tfrac{1}{2}h, y_n + \tfrac{1}{2}\Delta_2), \tag{28c}$$
$$\Delta_4 = hf(x_n + h, y_n + \Delta_3). \tag{28d}$$

The Runge–Kutta method, the single step error of which is $0(h^5)$, is used in our programs either in its own right or in connection with the Adams–Moulton method as a starter. This choice satisfies the rule that the starter should have approximately the same accuracy as the main execution method.

3.3 Systems of equations and higher order equations

Many physical problems involve coupled differential equations such as (18a, b) and (19a, b). Consequently, it is necessary to develop a method of solving them. Consider a first-order system

$$\begin{aligned} y_1'(x) &= f_1(x, y_1, y_2, \ldots, y_n), \\ y_2'(x) &= f_2(x, y_1, y_2, \ldots, y_n), \\ &\vdots \\ y_n'(x) &= f_n(x, y_1, y_2, \ldots, y_n), \end{aligned} \tag{29a}$$

with initial conditions

$$y_1(x_0), y_2(x_0), \ldots, y_n(x_0), \tag{29b}$$

where x_0, initial conditions (29b), and the functions

$$f_j(x, y_1, y_2, \ldots, y_n), j = 1, 2, \ldots, n,$$

are given.

Let us introduce the vector notation:[6]

$$\mathbf{y}(x) = \begin{pmatrix} y_1(x) \\ y_2(x) \\ \vdots \\ y_n(x) \end{pmatrix}, \quad \mathbf{y}'(x) = \begin{pmatrix} y'_1(x) \\ y'_2(x) \\ \vdots \\ y'_n(x) \end{pmatrix},$$

$$\mathbf{f}(x, \mathbf{y}) = \begin{pmatrix} f_1(x, y_1, y_2, \ldots, y_n) \\ f_2(x, y_1, y_2, \ldots, y_n) \\ \vdots \\ f_n(x, y_1, y_2, \ldots, y_n) \end{pmatrix} = \begin{pmatrix} f_1(x, \mathbf{y}) \\ f_2(x, \mathbf{y}) \\ \vdots \\ f_n(x, \mathbf{y}) \end{pmatrix},$$

etc. The system (29a) and the initial conditions (29b) can then be rewritten as follows:

$$\mathbf{y}' = \mathbf{f}(x, \mathbf{y}), \tag{30a}$$

$$\mathbf{y}(x_0) = \mathbf{y}_0. \tag{30b}$$

Equations (30a, b) are of the same form as (20a, b). Consequently, the Adams–Moulton method for solution of a system of first-order differential equations can be obtained from (23a, b) if scalar quantities are replaced by vectors and the numerical coefficients β_i, β'_i are left unchanged. Thus we get

$$\mathbf{y}^{(P)}_{n+1} = \mathbf{y}_n + \frac{h}{24}(55\mathbf{f}(x_n, \mathbf{y}_n) - 59\mathbf{f}(x_{n-1}, \mathbf{y}_{n-1})$$
$$+ 37\mathbf{f}(x_{n-2}, \mathbf{y}_{n-2}) - 9\mathbf{f}(x_{n-3}, \mathbf{y}_{n-3})), \tag{31a}$$

$$\mathbf{y}_{n+1} = \mathbf{y}_n + \frac{h}{24}(9\mathbf{f}(x_{n+1}, \mathbf{y}^{(P)}_{n+1}) + 19\mathbf{f}(x_n, \mathbf{y}_n)$$
$$- 5\mathbf{f}(x_{n-1}, \mathbf{y}_{n-1}) + \mathbf{f}(x_{n-2}, \mathbf{y}_{n-2})). \tag{31b}$$

Similarly, the Runge–Kutta method is obtained from (27) and (28a, b, c, d) as

$$\mathbf{y}_{n+1} = \mathbf{y}_n + \tfrac{1}{6}(\mathbf{\Delta}^{(1)}_n + 2\mathbf{\Delta}^{(2)}_n + 2\mathbf{\Delta}^{(3)}_n + \mathbf{\Delta}^{(4)}_n), \tag{32}$$

where

$$\mathbf{\Delta}^{(1)}_n = h\mathbf{f}(x_n, \mathbf{y}_n), \tag{33a}$$

$$\mathbf{\Delta}^{(2)}_n = h\mathbf{f}(x_n + \tfrac{1}{2}h, \mathbf{y}_n + \tfrac{1}{2}\mathbf{\Delta}^{(1)}_n), \tag{33b}$$

$$\mathbf{\Delta}^{(3)}_n = h\mathbf{f}(x_n + \tfrac{1}{2}h, \mathbf{y}_n + \tfrac{1}{2}\mathbf{\Delta}^{(2)}_n), \tag{33c}$$

$$\mathbf{\Delta}^{(4)}_n = h\mathbf{f}(x_n + h, \mathbf{y}_n + \mathbf{\Delta}^{(3)}_n). \tag{33d}$$

It is easy to show that the problem of solving a single differential equation of higher order can be reduced to the problem of solving a system of first-order differential equations.
Consider
$$y^{(n)} = f(x, y, y', \ldots, y^{(n-1)}), \tag{34a}$$
where
$$y(x_0) = y_0, \, y'(x_0) = y'_0, \ldots, y^{(n-1)}(x_0) = y_0^{(n-1)} \tag{34b}$$
are the given initial conditions at the point x_0.

We can define a set of new functions as
$$y_1(x) = y(x), \, y_2(x) = y'(x), \ldots, y_n(x) = y^{(n-1)}(x).$$

It is obvious that the functions $y_1(x), y_2(x), \ldots, y_n(x)$ satisfy the system

$$\begin{aligned} y'_1 &= y_2 \\ y'_2 &= y_3 \\ &\vdots \\ y'_{n-1} &= y_n \\ y'_n &= f(x, y_1, y_2, \ldots, y_n), \end{aligned} \tag{35a}$$

and the initial conditions
$$y_1(x_0) = y_0, \quad y_2(x_0) = y'_0, \ldots, \quad y_n(x_0) = y_0^{(n-1)}. \tag{35b}$$

Thus by solving (35a, b) we find the function $y_1(x) = y(x)$ and its n derivative which are the solution to (34a, b).

Consequently, formulae (31a, b) or (32) and (33a, b, c, d) can also be used for finding a numerical solution of an equation of higher order if the indicated transformation to a system of first-order equations is used.

4. COMPUTER PROGRAMS

4.1 Main program

The objective of the main program is to calculate particle trajectories and capture radii R_{ca} with a prescribed accuracy for a wide variety of input parameters and initial conditions. The program consists of the MASTER section and the subroutines FUNCTN, RK, AM and XSECT.

The MASTER section handles the input–output operations and calls the relevant subroutines to perform the computation.

The input data are:

    J—the control integer which selects the iterating subroutine;
    K—the control integer which selects whether the output is directed to a lineprinter or/and to a file;
    IO—the number of trajectories or capture radii R_{ca} to be calculated;
    ALPHA—the angle α which the fluid flow makes with the external field H_0. These directions also enable us to define two convenient frames of reference, the primary has an x-axis parallel to H_0 and the secondary has an x-axis anti-parallel to the flow (Figure 3).

And

(a) if either the RK or AM subroutine has been chosen:
XAO and YAO—the starting point in the secondary frame of reference. These values are transformed in primary coordinates and

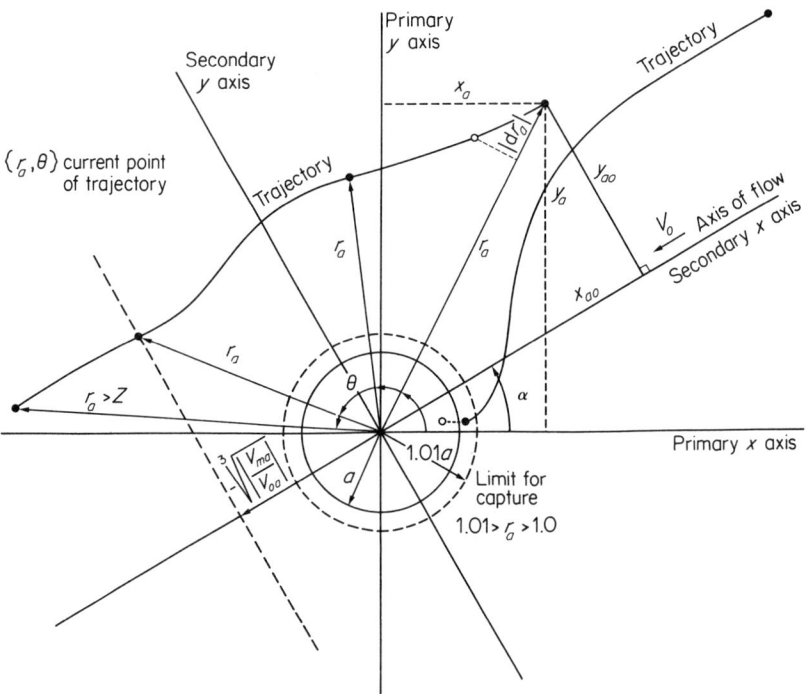

Figure 3. Definition of the primary and secondary frame of coordinates and the limiting conditions of capture and escape

later (by subroutines) in plane polar coordinates, where the iteration takes place;

VMA—the magnetic velocity V_{ma};

VOA—the fluid velocity V_{0a};

AK—the constant $K = M_s/2H_0$;

DTO—the time step length (a good empirical choice is DTO = $1/(10\, V_{0a})$), DTO is converted to the variable DT which can be adjusted during the iteration in subroutines;

E—the stipulated error (e.g. $E = 0.001$) of the Admas–Moulton predictor–corrector method.

Or

(b) if the XSECT subroutine has been chosen:
VMA, VOA, AK, DTO, ALPHA, XAO;

YLA and YHA—the lower and upper estimate of the y-coordinate (in the secondary frame) of a point lying on the critical curve.

The output data are:

(a) if either the RK or AM subroutine has been chosen:
AK, VMA, VOA, VMA/VOA, ALPHA, XAO, YAO;

X and Y—the x–y coordinates of the trajectory points in the primary frame of reference;

V—the velocity of the particle at each trajectory point;

DAR and DIT—the radial and angular components, dr_a/dt and $d\theta/dt$, of the velocity at each point;

DEET—the time step length dt between the points;

(b) if the XSECT subroutine has been chosen:
ITERATIONS and then AK, VMA, VOA, VMA/VOA, ALPHA;

XC—the x-coordinate (in the secondary frame) corresponding to the capture radius R_{ca} on the critical curve;

YC—the capture radius R_{ca}.

The flow chart of the MASTER section is shown in Figure 4.

The subroutine FUNCTN contains the equations of motion in the form of (19a, b).

The subroutines RK and AM each transfer the input parameters and the starting point from the MASTER section, compute one trajectory, return it to the MASTER section (to be printed out), and receive a new set of initial conditions to calculate the next trajectory.

The RK and AM subroutines use the following limiting conditions:

(i) the particle is captured $\leftrightarrow$ $1.0 < r_a < 1.01$;
(ii) the particle is beyond the range of interest $\leftrightarrow$ $r_a > Z = \max(14.2, r_{a0} + 0.2)$;
(iii) the impenetrability of the wire is violated $\leftrightarrow$ $r_a < 1.0$;
(iv) the overall step limit is violated $\leftrightarrow$ $|dr_a| > 0.1$.

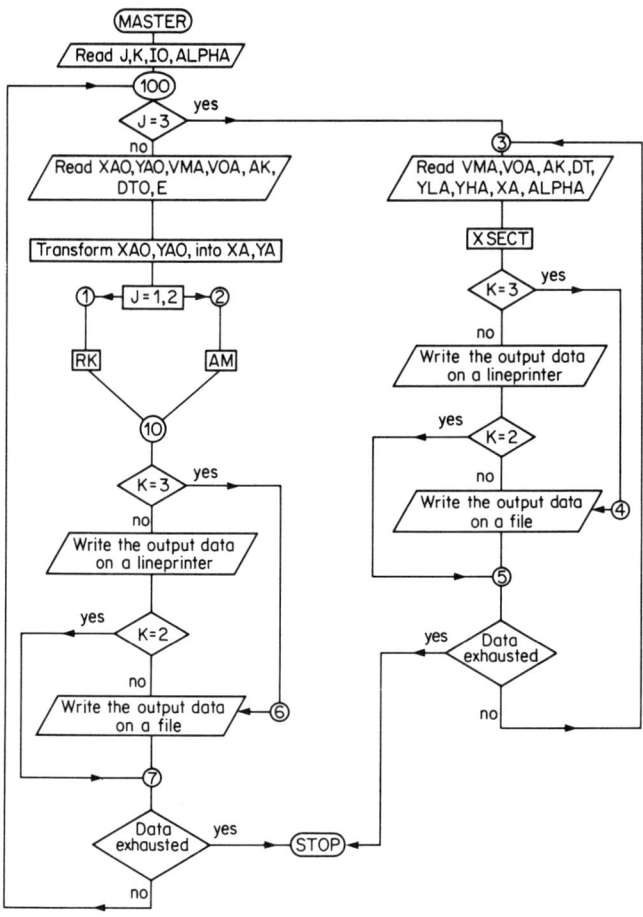

Figure 4. Flow chart of the MASTER section

If the condition (i) or (ii) is satisfied the control is returned to the MASTER section; if (iii) or (iv) takes place the step length DT is cut down.

The RK subroutine is based on the Runge–Kutta method as expressed by (32) and (33a, b, c, d). The AM subroutine uses the Runger–Kutta method as a starter and the Adams–Moulton method, given by (31a, b) as the main procedure. The role of the starter is not limited only to the initial point. After every cut in the step length DT the execution process is reversed to the Runge–Kutta method to provide four equally spaced points which are necessary for the action of the main predictor–corrector procedure. The process of execution in both subroutines, RK and AM, is apparent from the flow charts, which are shown in Figures 5 and 6, respectively.

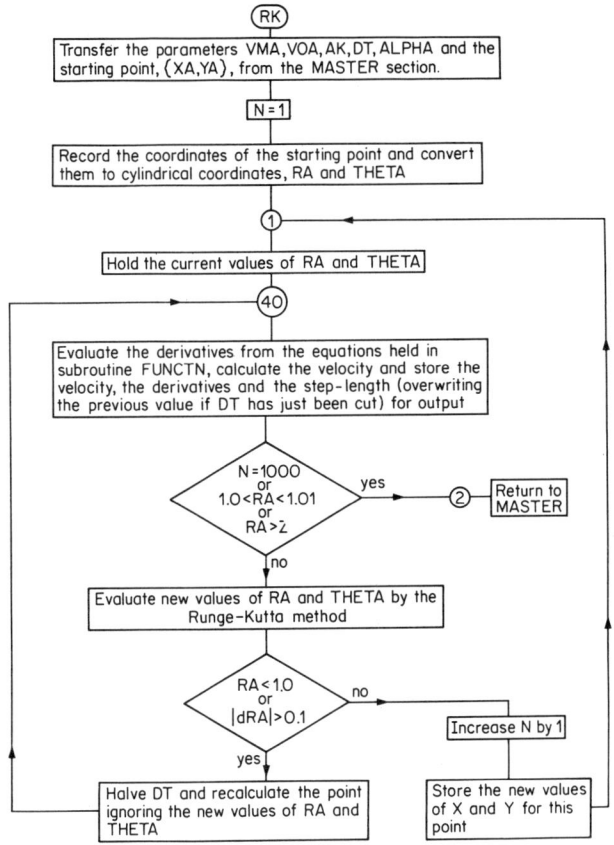

Figure 5. Flow chart of the RK subroutine

The XSECT subroutine calculates the capture radius R_{ca}. Again, the input data are transferred from the MASTER section, value of R_{ca} is computed and returned with iterations to be printed out. A new set of input data is transferred and the cycle repeated.

Before describing the flow of control in XSECT we have to consider a few points of a general nature.

It is apparent from equations (11a, b), (15a, b) and (17a, b) that the influence which the wire has upon the motion of the particle falls off with the distance r_a. Thus, beyond a certain region, the dimensions of which increases with $|V_{ma}/V_{0a}|$, the effect of the wire can be neglected and the trajectories are essentially parallel to the direction of the fluid flow (i.e. to

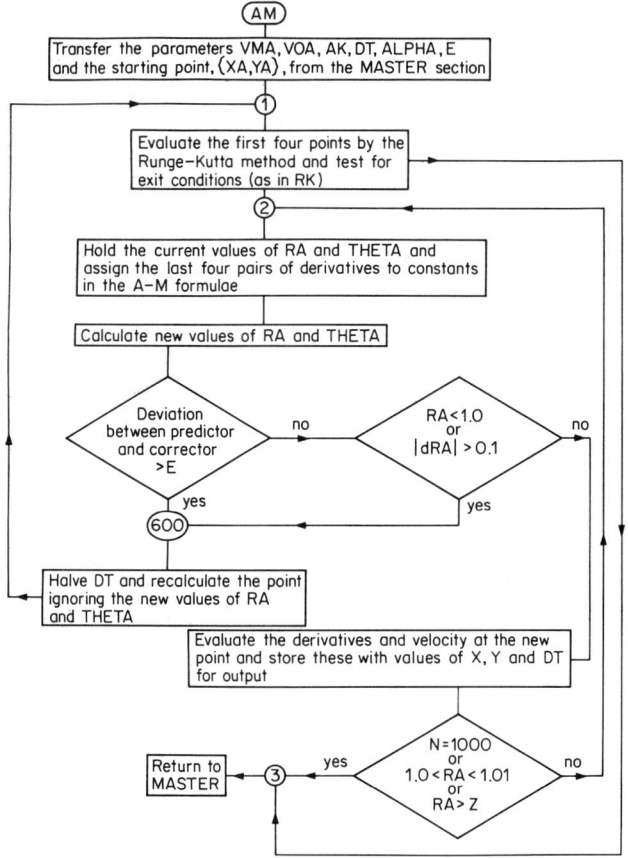

Figure 6. Flow chart of the AM subroutine

the secondary x-axis). The trajectories can be divided into two kinds, differing whether the particle is captured or not. The borderline case of a trajectory for which the particle is just captured is called the critical trajectory. The distance between the parallel part of the critical trajectory and the x-axis defines the capture radius (see Figure 7).

There are two critical trajectories, one above and one below the axis of flow. Thus we have two capture radii, R_{ca1} and R_{ca2}. The average capture radius $\frac{1}{2}(R_{ca1}+R_{ca2})$ is exactly the quantity R_{ca} which appears in equation (1). The proof that this single-wire analysis applies to the multiwire regular matrix can be found in Birss, Gerber, and Parker.[2]

The XSECT subroutine always computes in principle the capture radius, R_{ca1}, above the axis of flow. To obtain R_{ca2}, the computation has to be

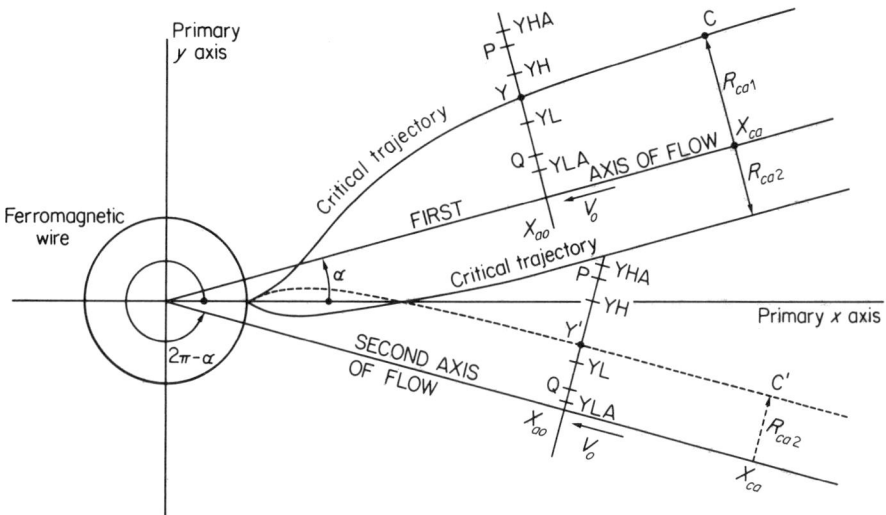

Figure 7. Arrangement of the axes of flow and various parameters used in the XSECT subroutine

performed in the mirror image of the original flow with respect to the primary x-axis. This is achieved by the same input data as for R_{cal}, only instead of α an angle $\beta = 2\pi - \alpha$ is used.

The XSECT subroutine uses the same limiting conditions (i), (iii), and (iv) as the previous subroutines. However, instead of (ii) a more rigorous condition, which would be linked to the ratio $|V_{ma}/V_{0a}|$ and would define the escape of the particle, is required. Bearing in mind that the escape occurs when the fluid drag ultimately exceeds the magnetic traction force, we can find, by analysing (19a), the relation

(ii')
$$r_a \cos(\theta - \alpha) < -\sqrt[3]{\left|\frac{V_{ma}}{V_{0a}}\right|}$$

which is the condition used in XSECT in place of (ii).

The action of the XSECT subroutine can be described in terms of values YLA, YL, YHA, YH, which are respectively the initial and current estimates of the lower and upper value of Y, the ordinate of a point on the critical trajectory, and in terms of auxiliary parameters P and Q (see Figure 7).

At the beginning YL = YLA and YH = YHA. The particle is started at the point {XAO, YAO = $\frac{1}{2}$(YLA + YHA)} and the trajectory is produced. If the particle is captured, YL is raised to YAO; if it is not captured, YH is lowered to YAO. The new starting point is set, in either case, as $\frac{1}{2}(YL+YH)$. This process is repeated. If the initial estimates have been chosen correctly, i.e. if the unknown Y lies indeed between YLA and YHA, the sequential values YL and YH will converge towards Y. When YL and YH have come sufficiently close to give Y with a required accuracy, the process is halted.

If Y does not lie between YLA and YHA, the starting point would converge upon one of these values. This is, however, avoided by defining the parameters P and Q which lie below YHA and above YLA. Every time the starting point enters the region either between YHA and P or YLA and Q, the values YHA or YLA are redefined until the point Y lies between them. Then Y is found as mentioned previously.

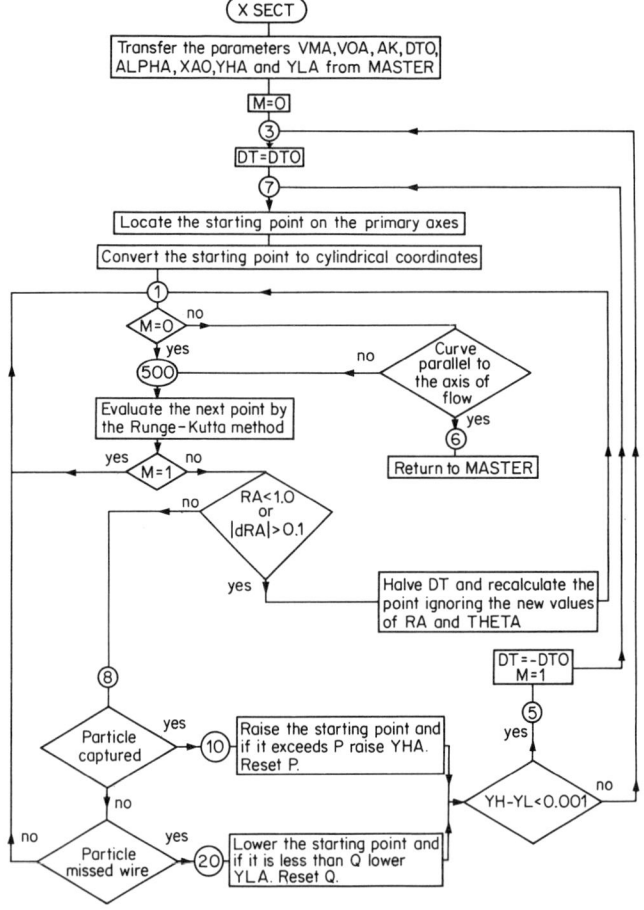

Figure 8. Flow chart of the XSECT subroutine

Having found Y, the time step length is made negative and the trajectory is iterated backwards up to a point C, where the difference between its ordinate YCA and that of the previous iteration is less than a stipulated error. The coordinate YCA is taken as the capture radius, the value XCA gives the distance from the origin to the place where the capture radius has been gauged.

The action of the XSECT subroutine, which has been just described, can be followed from its flow chart shown in Figure 8.

4.2 Graph-plotting programs

The curves of the trajectories and velocities produced by the main program require a pictorial representation. Any graph-plotting package can be used for this purpose. As an example we present here two simple graph-plotting programs based on the GINO-F package. Details of the specific commands can be found in the manual *Graph Plotting from Fortran Programs* available from the Computing Laboratory, University of Salford.

The program TRAJECTORIES reads the number of trajectories, the points of the wire contour, and the points of individual trajectories via channels 1, 5, and 6, respectively. The program VELOCITIES reads the number of velocity curves, the velocity range, the angle α via channel 1, and the points of the individual velocity curves via channel 6.

4.3 Example

As an example a brief investigation of FeO spherical particles carried by water and captured in the single-wire approximation is presented. The following data, relevant to the problem, have been used in the computation:

saturation magnetization of the wire	$M_s = 1.6 \times 10^6 \text{ Am}^{-1}$
external magnetic field	$H_0 = 1.0 \times 10^6 \text{ Am}^{-1}$
radius of the wire	$a = 10 \ \mu\text{m}$
radius of the paramagnetic particle	$R = 1 \ \mu\text{m}$
difference between the susceptibilities of the particle and the fluid	$\chi = 7.178 \times 10^{-3}$
viscosity of the fluid	$\eta = 1.0 \times 10^{-3} \text{ Nm}^{-2} \text{ s}$
density of the particle	$\rho_p = 5.7 \times 10^3 \text{ kgm}^{-3}$
magnetic velocity	$V_{ma} = V_m/a = \frac{2}{9}(\chi\mu_0 M_s H_0 R_a^2/\eta)$ $= 3.207 \times 10^4 \text{ s}^{-1}$
fluid velocity	$V_{0a} = -6.288 \times 10^2 \text{ s}^{-1}$

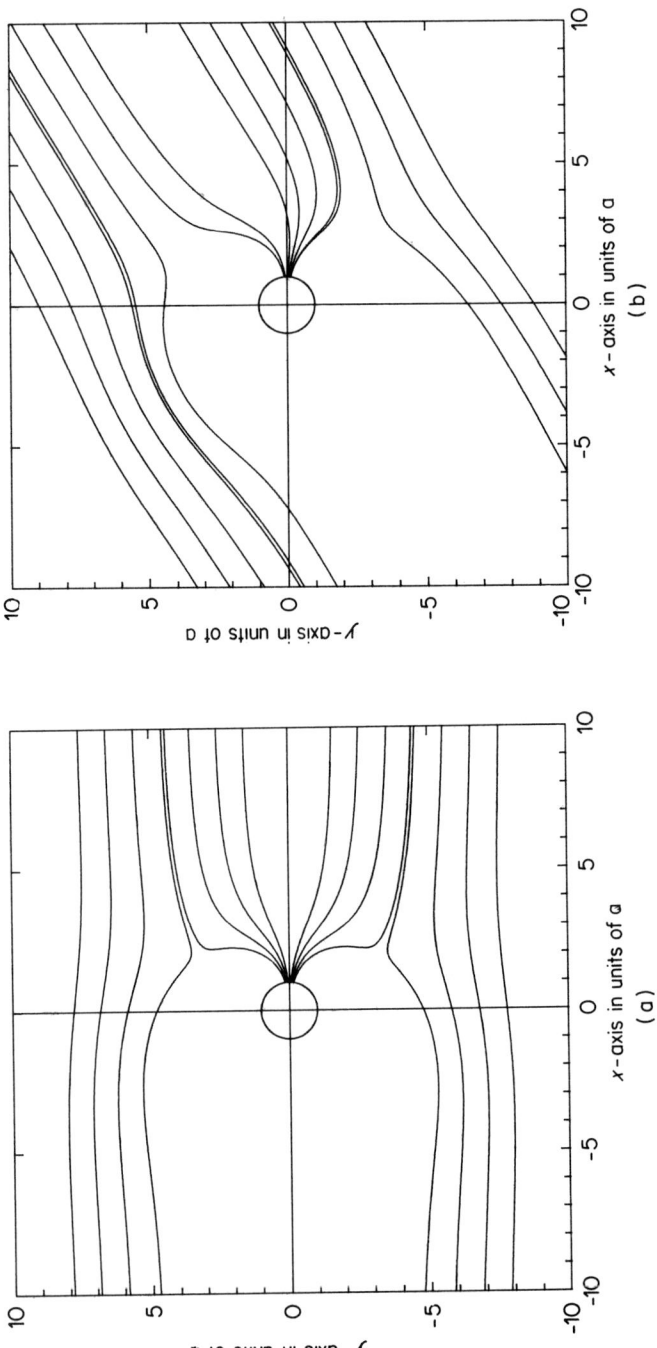

PARTICLE CAPTURE IN HIGH GRADIENT MAGNETIC SEPARATION 171

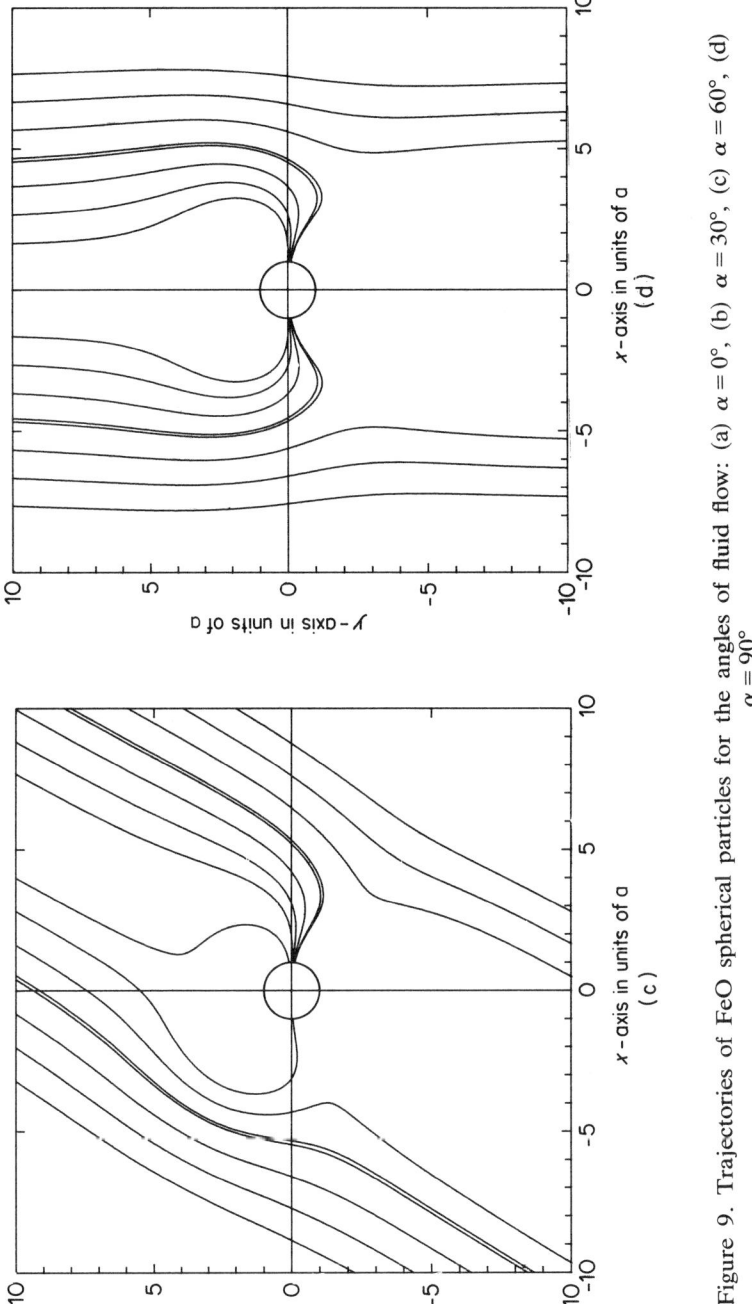

Figure 9. Trajectories of FeO spherical particles for the angles of fluid flow: (a) $\alpha = 0°$, (b) $\alpha = 30°$, (c) $\alpha = 60°$, (d) $\alpha = 90°$

Figures 9(a), (b), (c) and (d) show the trajectories computed for the angle $\alpha = 0°$, 30°, 60°, and 90°, respectively. The respective values of R_{ca} resulting from the computation are 4.613, 4.367, 4.410, and 4.697.

Note that the trajectories which are just super- and subcritical for, say, $\alpha = 0°$ need not remain so for different values of α.

It is obvious that by varying the numerical values of the quantities mentioned in this example one can obtain particle trajectories and capture radii for a wide choice of possible physical situations. Thus, as an exercise, one can vary the input data (which reflect the physical properties of the wire, the fluid, and the particles) and observe what will happen to the trajectories and capture radii.

However, to obtain meaningful results, it is necessary to adopt values which would correspond to physical reality. For instance, suppose that the wire is made from an ordinary ferromagnetic material which exhibits only small hysteresis and is characterized by a value M_s of the saturation magnetization. If H_s is a value of the internal magnetic field which is large enough to saturate the wire (this value can be read off from the magnetization curve of the material from which the wire is made) then, due to demagnetizing effects, the value of the external field H_0, which is perpendicular to the wire axis, has to satisfy the relation

$$H_0 \geq H_s + \tfrac{1}{2}M_s. \tag{36}$$

Consequently, the constant $K = M_s/2H_0$ must lie within the limits

$$0 \leq K \leq 1. \tag{37}$$

The values of K, depending on M_s and H_0, are usually in the range 0.4 to 0.9.

If the material of the wire is paramagnetic, the value H_0 can be chosen without any constraint and M_s is given by

$$M_s = \chi_m H_0, \tag{38}$$

where χ_m is the susceptibility of the material of the wire. Note that in this case M_s is not the saturation magnetization but only a uniform magnetization. Nevertheless the theory is still valid; it is the uniformity not the saturation of the magnetization which is the necessary condition.

The quantities R and a are restricted as far as their ratio R/a is concerned, namely $R/a \leq 0.1$, since the theory is formulated for a small particle limit. In practice a is larger than 1 μm.

The quantity $\chi = \chi_p - \chi_f$ can be either positive or negative depending on the value and signs of the χ_p and χ_f which are respectively the susceptibilities of the particles and the fluid involved in the separation. The magnitudes of χ are typically between 10^{-5} and 10^{-2}.

The velocity of the fluid V_0 (which enters in the calculation as the normalized velocity $V_{0a} = V_0/a$) can vary over a very broad range, say, from 1 mm s^{-1} to 50 mm s^{-1}.

To facilitate the legibility of the data and results, the input and output in our programs (sections 4.1 and 4.2) are handled simply in the F-format. The numerical field widths of the format stipulated in the programs should be sufficient to accomodate the results for the data within the indicated range of values. If, however, some results were to require an extension of the field widths, it is easy to change them accordingly or to use the E- instead of the F- format.

5. FURTHER EXERCISES

Apart from varying the numerical data, in the manner described in section 4.3 (for instance calculating the trajectories for $\chi < 0$, i.e. for $V_{ma} < 0$), the following, more substantial, exercises are suggested.

(1) Apply the method of solving higher order differential equations (see section 3.3) to equations of particle motion (18a, b), which contain the second-order inertial term. Generalize the main separation program to produce the numerical solutions of these equations.

(2) Consider a capture process, instead of in the single wire, in the single-sphere approximation. Find the magnetic field and the fluid velocity distribution around a ferromagnetic sphere magnetized to saturation by an external field H_0 and immersed in fluid flow of a background velocity V_0. In analogy with section 2 derive the appropriate equations of particle motion. Modify the main separation program to solve these equations numerically. Compare the capture efficiency of the single wire and the single-sphere approximations.

(3) Analyse a capture process in the single-wire approximation, where the wire has an elliptic cross-section. Similarly, as suggested in section 5.2, find the magnetic field, the fluid velocity distribution, and derive the appropriate equations of particle motion. Modify the FUNCTN subroutine accordingly and use the main separation program for finding numerical solutions of these equations. Investigate the capture process for various values of eccentricity and various orientations of the elliptic cross-sections in respect to the directions of fluid flow and magnetic field.

REFERENCES

1. J. H. P. Watson, *J. Appl. Phys.*, **44,** 4209 (1973).
2. R. R. Birss, R. Gerber, and M. R. Parker, *Filtration & Separation*, **July/August,** 1 (1977).

3. R. R. Birss, R. Gerber, and M. R. Parker, *I.E.E.E. Trans. Magn.*, **MAG-12,** 892 (1976).
4. B. I. Bleaney and B. Bleaney, *Electricity and Magnetism*, 2nd edn. (Clarendon Press, Oxford) 47 (1965).
5. W. F. Lawson Jr., W. H. Simons, and R. P. Treat, *J. Appl. Phys.*, **48,** 3213 (1977).
6. D. M. Young and R. T. Gregory, *A Survey of Numerical Mathematics* (Addison-Wesley, Reading, Mass.) 422 (1972).

PARTICLE CAPTURE IN HIGH GRADIENT MAGNETIC SEPARATION 175

TYPICAL INPUT DATA FOR THE MAIN PROGRAM (FOR ALPHA = 30 DEGR.)

```
   2    1   16    0.5236
  15.0                1.6
32071.0    -628.8    0.8    0.0002    0.01
  15.0                2.6
32071.0    -628.8    0.8    0.0002    0.01
  15.0                3.6
32071.0    -628.8    0.8    0.0002    0.01
  15.0                4.48
32071.0    -628.8    0.8    0.0002    0.01
  15.0                4.6
32071.0    -628.8    0.8    0.0002    0.01
  15.0                5.6
32071.0    -628.8    0.8    0.0002    0.01
  15.0                6.6
32071.0    -628.8    0.8    0.0002    0.01
  15.0                7.6
32071.0    -628.8    0.8    0.0002    0.01
  15.0               -1.6
32071.0    -628.8    0.8    0.0002    0.01
  15.0               -2.6
32071.0    -628.8    0.8    0.0002    0.01
  15.0               -3.6
32071.0    -628.8    0.8    0.0002    0.01
  15.0               -4.48
32071.0    -628.8    0.8    0.0002    0.01
  15.0               -4.6
32071.0    -628.8    0.8    0.0002    0.01
  15.0               -5.6
32071.0    -628.8    0.8    0.0002    0.01
  15.0               -6.6
32071.0    -628.8    0.8    0.0002    0.01
  15.0               -7.6
32071.0    -628.8    0.8    0.0002    0.01
```

PHYSICS PROGRAMS

TYPICAL OUTPUT FROM THE MAIN PROGRAM (FOR ALPHA = 30 DEGR.)

```
  K=.80    VMA=  32071.00    VOA=   -628.80
 VMA/VOA= -51.00    ALPHA=0.523600
  XAO=   15.0000
  YAO=    1.6000
  NUMBER OF COMPUTED POINTS   Q= 219
```

X	Y	V	DRA/DT	DTH/DT
12.19037	8.88566	628.09008	-625.39780	3.85099
12.13641	8.85351	628.08329	-625.37877	3.87576
12.08246	8.82135	628.07635	-625.35959	3.90071
12.02852	8.78918	628.06925	-625.34026	3.92585
11.97458	8.75701	628.06199	-625.32078	3.95118
11.92065	8.72482	628.05457	-625.30115	3.97669
11.86672	8.69263	628.04698	-625.28136	4.00238
11.81280	8.66042	628.03921	-625.26142	4.02827
11.75889	8.62820	628.03127	-625.24132	4.05433
11.70499	8.59598	628.02315	-625.22107	4.08058
11.65109	8.56374	628.01484	-625.20066	4.10702
11.59720	8.53149	628.00634	-625.18009	4.13364
11.54332	8.49923	627.99764	-625.15935	4.16045
11.48944	8.46697	627.98873	-625.13846	4.18744
11.43558	8.43468	627.97962	-625.11740	4.21462
11.38172	8.40239	627.97030	-625.09618	4.24198
11.32787	8.37009	627.96075	-625.07479	4.26952
11.27403	8.33777	627.95098	-625.05324	4.29724
11.22019	8.30545	627.94097	-625.03151	4.32515
11.16637	8.27311	627.93073	-625.00962	4.35323
11.11255	8.24075	627.92024	-624.98757	4.38149
11.05874	8.20839	627.90950	-624.96533	4.40993
11.00495	8.17601	627.89850	-624.94293	4.43854
10.95116	8.14362	627.88723	-624.92036	4.46733
10.89738	8.11121	627.87569	-624.89761	4.49628
10.84361	8.07879	627.86387	-624.87469	4.52541
10.78985	8.04636	627.85176	-624.85159	4.55470
10.73610	8.01391	627.83935	-624.82832	4.58416
10.68236	7.98145	627.82664	-624.80487	4.61378
10.62863	7.94898	627.81362	-624.78124	4.64356
10.57491	7.91649	627.80028	-624.75744	4.67349
10.52120	7.88398	627.78660	-624.73345	4.70358
10.46750	7.85146	627.77259	-624.70929	4.73381
10.41382	7.81892	627.75822	-624.68495	4.76418
10.36014	7.78636	627.74350	-624.66043	4.79470
10.30648	7.75379	627.72842	-624.63573	4.82535
10.25283	7.72121	627.71295	-624.61085	4.85613
10.19919	7.68860	627.69710	-624.58579	4.88703
10.14556	7.65598	627.68085	-624.56055	4.91806
10.09195	7.62334	627.66418	-624.53513	4.94919
10.03835	7.59068	627.64710	-624.50953	4.98043
9.98476	7.55800	627.62959	-624.48375	5.01177
9.93119	7.52531	627.61164	-624.45779	5.04320
9.87763	7.49259	627.59323	-624.43166	5.07472

PARTICLE CAPTURE IN HIGH GRADIENT MAGNETIC SEPARATION 177

TYPICAL OUTPUT FROM THE MAIN PROGRAM (FOR ALPHA = 30 DEGR.)

9.82409	7.45986	627.57436	-624.40535	5.10631
9.77056	7.42710	627.55500	-624.37886	5.13796
9.71704	7.39433	627.53516	-624.35220	5.16968
9.66354	7.36153	627.51481	-624.32537	5.20144
9.61006	7.32872	627.49394	-624.29837	5.23323
9.55659	7.29588	627.47255	-624.27120	5.26506
9.50313	7.26302	627.45061	-624.24386	5.29689
9.44970	7.23013	627.42811	-624.21636	5.32872
9.39628	7.19722	627.40504	-624.18870	5.36055
9.34288	7.16429	627.38138	-624.16088	5.39234
9.28949	7.13134	627.35713	-624.13291	5.42409
9.23612	7.09836	627.33225	-624.10479	5.45578
9.18278	7.06536	627.30675	-624.07652	5.48740
9.12945	7.03233	627.28060	-624.04811	5.51893
9.07614	6.99927	627.25379	-624.01956	5.55034

AND SO ON UNTIL THE END OF THE CURVE:

1.99124	0.41735	5001.67286	-4677.03717	-871.26219
1.93562	0.38296	5480.27835	-5160.53217	-934.83109
1.87304	0.34697	6097.52330	-5784.10413	-1013.04001
1.80132	0.30904	6929.73002	-6624.48599	-1112.93695
1.71689	0.26863	8125.14141	-7830.53890	-1247.62033
1.66812	0.24723	8948.72743	-8660.67437	-1335.55755
1.61342	0.22484	10016.44810	-9736.00489	-1444.80580
1.55086	0.20120	11465.39899	-11193.90579	-1586.01225
1.47731	0.17593	13564.56137	-13303.78436	-1779.20508
1.38701	0.14839	16933.11739	-16685.52037	-2068.28761
1.33225	0.13340	19551.81429	-19312.12550	-2279.52085
1.26754	0.11726	23409.29863	-23178.74971	-2574.57313
1.18736	0.09943	29766.09748	-29546.41012	-3029.53802
1.13808	0.08959	34867.61021	-34654.26075	-3373.57563
1.07897	0.07882	42661.96776	-42455.80182	-3872.16617
1.00376	0.06665	56335.52544	-56137.70355	-4688.95543

THEN THE SAME OUTPUT FORMAT FOR CURVES NO.:2,3,4,.....,16.

178   PHYSICS PROGRAMS

MAIN MAGNETIC SEPARATION PROGRAM

```
C      THE MAIN PROGRAM SERVES AS A CARRIAGE TO INPUT DATA AND OUTPUT RESULTS
C   AFTER THE EQUATIONS HAVE BEEN SOLVED.IT ALSO DIRECTS THE DATA TO THE
C   APPROPRIATE SUBROUTINE THAT IS TO PERFORM THE ITERATIONS ON THE EQUATIONS
C   WHICH ARE HELD IN THE SUBROUTINE 'FUNCTN'.THERE ARE TWO METHODS OF
C   ITERATION,THE RUNGE-KUTTA AND THE ADAMS-MOULTON METHOD.
C   IN ADDITION TO THIS THE SUBROUTINE 'XSECT' WILL FIND THE CAPTURE CROSS-
C   SECTION OF THE WIRE FOR DIFFERENT FLUID VELOCITIES.
C
C      VMA,VOA,AK AND DTO ARE THE MAGNETIC AND FLUID VELOCITIES,
C      THE SHORT RANGE CONSTANT AND THE TIME STEP,RESPECTIVELY.
C      E IS THE MAXIMUM DEVIATION BETWEEN PREDICTOR AND CORRECTOR IN THE
C      PREDICTOR-CORRECTOR METHODS.
C      YLA AND YHA ARE PARAMETERS IN 'XSECT' SUBROUTINE.
C      J IS THE CONTROL INTEGER WHICH SELECTS THE METHOD OF ITERATION.
C      J=1,2,3.
C      K IS THE CONTROL INTEGER WHICH SELECTS THE MODE OF OUTPUT.
C      K=1,2,3.
C      IO IS THE NUMBER OF INITIAL VALUES OF X AND Y,READ IN AS XA AND YA,
C      FOR WHICH CURVES WILL BE PRODUCED.
C      Z IS THE RATIO OF VMA TO VOA.
C      YC IS THE CAPTURE CROSS-SECTION.
C      V IS THE VELOCITY OF THE PARTICLE AT A GIVEN POINT.
C      THE SUBROUTINES WILL PRODUCE Q POINTS(I.E. Q X-Y PAIRS)FOR
C      EACH CURVE.THIS PARAMETER IS IMPORTANT IN THE PLOTTING OF THE
C      CURVES.
C      DAR,DIT AND DEET ARE ARRAYS OF THE GRADIENTS, I.E. THE RADIAL
C      AND ANGULAR VELOCITIES, AND THE TIME STEP,RESPECTIVELY.
C      ALPHA IS THE INCLINATION OF THE FLUID VELOCITY TO THE X-AXIS.
C
       INTEGER Q
       DIMENSION X(1000),Y(1000),G(1000),F(1000),V(1000)
       DIMENSION DAR(1000),DIT(1000),DEET(1000)
       READ(1,1001)J,K,IO,ALPHA
       I=0
  100  I=I+1
       IF(J.EQ.3)GOTO 3
       READ(1,1002)XAO,YAO
       READ(1,1000)VMA,VOA,AK,DTO,E
       XA=XAO*COS(ALPHA)-YAO*SIN(ALPHA)
       YA=YAO*COS(ALPHA)+XAO*SIN(ALPHA)
       ZZ=VMA/VOA
       DT=DTO
       GOTO(1,2),J
    1  CALL RK(XA,YA,VMA,VOA,AK,DT,NN,X,Y,V,ALPHA,DAR,DIT,DEET)
       GOTO 10
    2  CALL AM(XA,YA,VMA,VOA,AK,DT,NN,X,Y,V,E,ALPHA,DAR,DIT,DEET)
       GOTO 10
    3  READ(1,1004)VMA,VOA,AK,DT,YLA,YHA,XAO,ALPHA
       CALL XSECT(YLA,YHA,VMA,VOA,AK,DT,Z,YC,XC,XAO,ALPHA)
       IF(K.EQ.3)GOTO4
       WRITE(2,1006)AK,VMA,VOA,Z,ALPHA,XC,YC
       IF(K.EQ.2)GOTO5
    4  WRITE(3,1005)Z,YC,XC,ALPHA
    5  IF(I.EQ.IO)GOTO 20
       I=I+1
       GOTO 3
```

PARTICLE CAPTURE IN HIGH GRADIENT MAGNETIC SEPARATION 179

MAIN MAGNETIC SEPARATION PROGRAM

```
   10 Q=NN
      IF(K.EQ.3)GOTO6
      WRITE(2,1009) AK,VMA,VOA,ZZ,ALPHA,XAO,YAO
      WRITE(2,1007)Q
      WRITE(2,1008)
      WRITE(2,1002)(X(N),Y(N),V(N),DAR(N),DIT(N),DEET(N),N=1,Q)
      WRITE(2,1010)
      IF(K.EQ.2)GOTO7
    6 WRITE(3,1003)Q
      WRITE(3,1002)(X(N),Y(N),V(N),DAR(N),DIT(N),DEET(N),N=1,Q)
    7 IF(I.EQ.IO)GOTO 20
      GOTO 100
   20 STOP
 1000 FORMAT(2F10.2,F10.5,2F10.8,2F10.5)
 1001 FORMAT(3I5,F10.5)
 1002 FORMAT(1X,F10.5,2F20.5,2F15.5,F15.10)
 1003 FORMAT(1X,I10)
 1004 FORMAT(2F10.2,F10.5,F10.8,5F10.5)
 1005 FORMAT(1X,F10.5,3F20.5)
 1006 FORMAT(5H0   K=,F3.2,7H    VMA=,F10.2,
     C7H    VOA=,F10.2/10H  VMA/VOA=,
     CF7.2,9H   ALPHA=,F8.6/7H    XCA=,
     CF9.4/7H    RCA=,F9.4)
 1007 FORMAT(1X,31H  NUMBER OF COMPUTED POINTS   Q=,I4///)
 1008 FORMAT(1X,88H       X                   Y                    V
     C      DRA/DT             DTH/DT       DT///)
 1009 FORMAT(5H0   K=,F3.2,7H    VMA=,F10.2,
     C7H    VOA=,F10.2/10H  VMA/VOA=,
     CF7.2,9H   ALPHA=,F8.6/7H    XAO=,
     CF9.4/7H    YAO=,F9.4)
 1010 FORMAT(////)
      END
C
C
C   THE FUNCTN SUBROUTINE.
C
C   THIS SUBROUTINE CONTAINS THE EQUATIONS TO BE ITERATED.
C
      SUBROUTINE FUNCTN(RA,THETA,VMA,VOA,AK,DRADT,DTHDT,ALPHA)
      DRADT=VOA*(1.0-(RA**(-2)))*COS(THETA-ALPHA)-VMA*AK/(RA**5)-VMA*
     1COS(2.0*THETA)/(RA**3)
      DTHDT=-VOA*(1.0+(RA**(-2)))*SIN(THETA-ALPHA)/RA-VMA*SIN(2.0*THETA
     1)/(RA**4)
      RETURN
      END
C
C
C   THE RUNGE-KUTTA METHOD.
C
C   THIS IS A SELF-STARTING ,FOURTH-ORDER METHOD OF HIGH ACCURACY.
C   IT INCORPORATES THE STEP LENGTH ADJUSTMENT.
C
      SUBROUTINE RK(XA,YA,VMA,VOA,AK,DT,N,X,Y,U,ALPHA,DAR,DIT,DEET)
      REAL L,KR1,KT1,KR2,KT2,KR3,KT3,KR4,KT4
      DIMENSION X(1000),Y(1000),V(1000)
      DIMENSION DAR(1000),DIT(1000),DEET(1000)
```

MAIN MAGNETIC SEPARATION PROGRAM

```
      N=1
      X(N)=XA
      Y(N)=YA
      RA=SQRT(XA*XA+YA*YA)
      THETA=ATAN2(YA,XA)
      Z=RA+0.2
      IF(Z.LE.14.2)Z=14.2
    1 R=RA
      TH=THETA
   40 CALL FUNCTN(RA,THETA,VMA,VOA,AK,DRADT,DTHDT,ALPHA)
      V(N)=SQRT(DRADT*DRADT+RA*RA*DTHDT*DTHDT)
      DAR(N)=DRADT
      DIT(N)=DTHDT
      DEET(N)=DT
      IF(N.EQ.1000)GOTO 2
      IF(RA.LT.1.01.AND.RA.GT.1.00)GOTO 2
      IF(RA.GT.Z)GOTO 2
      KR1=DRADT*DT
      KT1=DTHDT*DT
      RA=R+0.5*KR1
      THETA=TH+0.5*KT1
      CALL FUNCTN(RA,THETA,VMA,VOA,AK,DRADT,DTHDT,ALPHA)
      KR2=DRADT*DT
      KT2=DTHDT*DT
      RA=R+0.5*KR2
      THETA=TH+0.5*KT2
      CALL FUNCTN(RA,THETA,VMA,VOA,AK,DRADT,DTHDT,ALPHA)
      KR3=DRADT*DT
      KT3=DTHDT*DT
      RA=R+KR3
      THETA=TH+KT3
      CALL FUNCTN(RA,THETA,VMA,VOA,AK,DRADT,DTHDT,ALPHA)
      KR4=DRADT*DT
      KT4=DTHDT*DT
      DRA=(KR1+2.0*(KR2+KR3)+KR4)/6.0
      DTH=(KT1+2.0*(KT2+KT3)+KT4)/6.0
      RA=R+DRA
      THETA=TH+DTH
      IF(RA.LT.1.00.OR.ABS(DRA).GT.0.1)GOTO 4
      N=N+1
      X(N)=RA*COS(THETA)
      Y(N)=RA*SIN(THETA)
      GOTO 1
    4 DT=DT/2.0
      RA=R
      THETA=TH
      GOTO 40
    2 RETURN
      END
C
C
C   THE ADAMS-MOULTON METHOD.
C
C   THIS METHOD USES THE FIFTH-ORDER ADAMS-MOULTON PREDICTOR-
C   CORRECTOR METHOD UTILISING THE RUNGE-KUTTA METHOD AS ITS STARTER
C   TO FIND THE FIRST FOUR POINTS.
```

PARTICLE CAPTURE IN HIGH GRADIENT MAGNETIC SEPARATION

MAIN MAGNETIC SEPARATION PROGRAM

```
C   THE ERROR IN THIS TECHNIQUE IS GOVERNED BY E.
C   IT ALSO INCORPORATES THE STEP LENGTH ADJUSTMENT
C
      SUBROUTINE AM(XA,YA,VMA,VOA,AK,DT,N,X,Y,V,E,ALPHA,F,G,DEET)
      REAL L,KR1,KT1,KR2,KT2,KR3,KT3,KR4,KT4
      DIMENSION X(1000),Y(1000),F(1000),G(1000),V(1000)
      DIMENSION DEET(1000)
      NO=4
      N=1
      X(N)=XA
      Y(N)=YA
      RA=SQRT(XA*XA+YA*YA)
      THETA=ATAN2(YA,XA)
      Z=RA+0.2
      IF(Z.LE.14.2)Z=14.2
C   START OF RUNGE-KUTTA
    1 R=RA
      TH=THETA
   40 CALL FUNCTN(RA,THETA,VMA,VOA,AK,DRADT,DTHDT,ALPHA)
      V(N)=SQRT(DRADT*DRADT+RA*RA*DTHDT*DTHDT)
      DEET(N)=DT
      F(N)=DRADT
      G(N)=DTHDT
      IF(N.EQ.1000)GOTO 3
      IF(RA.LE.1.01.AND.RA.GE.1.00)GOTO 3
      IF(RA.GE.Z)GOTO 3
      IF(N.GE.NO)GOTO 2
      KR1=DRADT*DT
      KT1=DTHDT*DT
      RA=R+0.5*KR1
      THETA=TH+0.5*KT1
      CALL FUNCTN(RA,THETA,VMA,VOA,AK,DRADT,DTHDT,ALPHA)
      KR2=DRADT*DT
      KT2=DTHDT*DT
      RA=R+0.5*KR2
      THETA=TH+0.5*KT2
      CALL FUNCTN(RA,THETA,VMA,VOA,AK,DRADT,DTHDT,ALPHA)
      KR3=DRADT*DT
      KT3=DTHDT*DT
      RA=R+KR3
      THETA=TH+KT3
      CALL FUNCTN(RA,THETA,VMA,VOA,AK,DRADT,DTHDT,ALPHA)
      KR4=DRADT*DT
      KT4=DTHDT*DT
      DRA=(KR1+2.0*(KR2+KR3)+KR4)/6.0
      DTH=(KT1+2.0*(KT2+KT3)+KT4)/6.0
      RA=R+DRA
      THETA=TH+DTH
      IF(RA.LT.1.00.OR.ABS(DRA).GT.0.1)GOTO 4
      N=N+1
      X(N)=RA*COS(THETA)
      Y(N)=RA*SIN(THETA)
      GOTO 1
    4 DT=DT/2.0
      NO=N+4
      RA=R
```

MAIN MAGNETIC SEPARATION PROGRAM

```
      THETA=TH
      GOTO 40
C   START OF ADAMS-MOULTON
    2 R=RA
      TH=THETA
      F1=F(N)
      G1=G(N)
      N=N-1
      F2=F(N)
      G2=G(N)
      N=N-1
      F3=F(N)
      G3=G(N)
      N=N-1
      F4=F(N)
      G4=G(N)
      N=N+3
      AMR1=R+DT*(55.0*F1-59.0*F2+37.0*F3-9.0*F4)/24.0
      AMT1=TH+DT*(55.0*G1-59.0*G2+37.0*G3-9.0*G4)/24.0
      RA=AMR1
      THETA=AMT1
      CALL FUNCTN(RA,THETA,VMA,VOA,AK,DRADT,DTHDT,ALPHA)
      AMR2=R+DT*(9.0*DRADT+19.0*F1-5.0*F2+F3)/24.0
      AMT2=TH+DT*(9.0*DTHDT+19.0*G1-5.0*G2+G3)/24.0
      ETA=ABS(AMR1-AMR2)/AMR2
      IF(ETA.GE.E)GOTO 600
      IF(AMT2.LT.1.0E-10)GOTO 700
      ZETA=ABS(AMT1-AMT2)/AMT2
      IF(ZETA.GE.E)GOTO 600
      GOTO 700
  600 DT=DT/2.0
      NO=N+4
      RA=R
      THETA=TH
      GOTO 1
  700 RA=AMR2
      THETA=AMT2
      IF(RA.LT.1.00.OR.ABS(RA-R).GT.0.1)GOTO 600
      CALL FUNCTN(RA,THETA,VMA,VOA,AK,DRADT,DTHDT,ALPHA)
      N=N+1
      F(N)=DRADT
      G(N)=DTHDT
      V(N)=SQRT(DRADT*DRADT+RA*RA*DTHDT*DTHDT)
      DEET(N)=DT
      X(N)=RA*COS(THETA)
      Y(N)=RA*SIN(THETA)
      IF(N.EQ.1000)GOTO 3
      IF(RA.LE.1.01.AND.RA.GE.1.00)GOTO 3
      IF(RA.GT.2)GOTO 3
      GOTO 2
    3 RETURN
      END
C
C
C   THE XSECT SUBROUTINE.
C
```

MAIN MAGNETIC SEPARATION PROGRAM

```
C   THIS SUBROUTINE CALCULATES THE CAPTURE CROSS SECTION OF THE WIRE.
C   IT DOES SO BY ITERATING A CURVE AND ASSESING WHETHER
C   CAPTURE HAS TAKEN PLACE OR NOT.IF IT HAS, THE STARTING POINT WAS
C   TOO LOW AND IT IS THEN RAISED TO HALF-WAY BETWEEN ITS PRESENT
C   POSITION AND THE STARTING VALUE OF THE CURVE WHEN THE PARTICLE LAST
C   MISSED THE WIRE WHICH IS A LITTLE TOO HIGH.SIMILARLY THE HIGH VALUE
C   IS LOWERED IF THE PARTICLE MISSES THE WIRE.BY REPEATING THIS
C   PROCESS UNTIL THE HIGH AND LOW STARTING POINTS CONVERGE THE CAPTURE
C   CROSS-SECTION IS FOUND.THIS IS REPEATED AS  A FUNCTION OF THE
C   RATIO VMA:VOA.
C
      SUBROUTINE XSECT(YLA,YHA,VMA,VOA,AK,DTO,Z,YC,XC,XAO,ALPHA)
      REAL L,KR1,KT1,KR2,KT2,KR3,KT3,KR4,KT4
      WRITE(2,1040)
 1040 FORMAT(13H0  ITERATIONS//)
      M=0
      D=0.0
      Z=VMA/VOA
      YAO=(YLA+YHA)/2.0
      YH=YHA
      YL=YLA
    3 DT=DTO
    7 YA=XAO*SIN(ALPHA)+YAO*COS(ALPHA)
      XA=XAO*COS(ALPHA)-YAO*SIN(ALPHA)
      RA=SQRT(XA*XA+YA*YA)
      THETA=ATAN2(YA,XA)
    1 R=RA
      TH=THETA
      CALL FUNCTN(RA,THETA,VMA,VOA,AK,DRADT,DTHDT,ALPHA)
      IF(M.EQ.0)GOTO 500
      YC=RA*SIN(THETA-ALPHA)
      XC=RA*COS(THETA-ALPHA)
      DRC=ABS(YC-D)
      IF(DRC.LT.0.0005)GOTO 6
      D=YC
  500 KR1=DRADT*DT
      KT1=DTHDT*DT
      RA=R+0.5*KR1
      THETA=TH+0.5*KT1
      CALL FUNCTN(RA,THETA,VMA,VOA,AK,DRADT,DTHDT,ALPHA)
      KR2=DRADT*DT
      KT2=DTHDT*DT
      RA=R+0.5*KR2
      THETA=TH+0.5*KT2
      CALL FUNCTN(RA,THETA,VMA,VOA,AK,DRADT,DTHDT,ALPHA)
      KR3=DRADT*DT
      KT3=DTHDT*DT
      RA=R+KR3
      THETA=TH+KT3
      CALL FUNCTN(RA,THETA,VMA,VOA,AK,DRADT,DTHDT,ALPHA)
      KR4=DRADT*DT
      KT4=DTHDT*DT
      DRA=(KR1+2.0*(KR2+KR3)+KR4)/6.0
      DTH=(KT1+2.0*(KT2+KT3)+KT4)/6.0
      RA=R+DRA
      THETA=TH+DTH
```

MAIN MAGNETIC SEPARATION PROGRAM

```
      IF(M.EQ.1)GOTO 1
      IF(RA.GT.1.00.AND.ABS(DRA).LT.0.1)GOTO 8
      RA=R
      THETA=TH
      DT=DT/2.0
      GOTO 1
    8 IF(RA.GT.1.00.AND.RA.LT.1.01)GOTO 10
      RL=-(ABS(Z)**0.3333)
      IF(RA*COS(THETA-ALPHA).LT.RL)GOTO 20
      GOTO 1
   10 YL=YAO
      P=YHA-0.1
      WRITE(2,1020)YAO
 1020 FORMAT(1X,F20.5)
      IF(YL.LT.P)GOTO 100
      YHA=YHA+0.2
      YH=YHA
  100 YAO=(YH+YL)/2.0
      W=YH-YL
      IF(W.LT.0.001)GOTO 5
      GOTO 3
   20 YH=YAO
      Q=YLA+0.1
      WRITE(2,1030)YAO
 1030 FORMAT(1X,F30.5)
      IF(YH.GT.Q)GOTO 200
      YLA=YLA/2.0
      IF(YLA.LT.0.05)YLA=0.0
      YL=YLA
  200 YAO=(YH+YL)/2.0
      W=YH-YL
      IF(W.LT.0.001)GOTO 5
      GOTO 3
    5 DT=-DTO
      M=1
      WRITE(2,1010)YAO
 1010 FORMAT(1X,F10.5//)
      GOTO 7
    6 RETURN
      END
```

GRAPH PLOTTING PROGRAM: TRAJECTORIES

```
C   THIS PROGRAM WILL PLOT OUT THE WIRE AND THE PATHS OF THE PARTICLES.
C   THE POINTS FOR THE WIRE SHOULD BE READ IN VIA CHANNEL NO.5.
C   THE DATA FOR THE PATHS SHOULD BE READ IN VIA CHANNEL NO.6.
C
      INTEGER Q
      DIMENSION X(1000),Y(1000),IPLOT(19)
      DATA IPLOT/22,22H PARTICLE TRAJECTORIES,20,20HX-AXIS IN UNITS OF A
     *,20,20HY-AXIS IN UNITS OF A/
      I=0
      READ(1,1001)IO
      READ(5,1000)(X(N),Y(N),N=1,120)
      CALL NARROW
      CALL RSIZE(150.0,150.0)
      CALL FIXAXS(-9.99,10.0,-9.99,10.0)
      CALL FGPLT(X,Y,120,5,0,1,0,IPLOT)
    1 I=I+1
      READ(6,999)Q
      READ(6,1000)(X(N),Y(N),N=1,Q)
      CALL FGPLT(X,Y,Q,15,0,1,1,IPLOT)
      IF(I.EQ.IO)GOTO 100
      GOTO 1
  100 CALL DEVFIN
      STOP
  999 FORMAT(I10)
 1000 FORMAT(1X,F10.5,F20.5)
 1001 FORMAT(I5)
      END
```

GRAPH PLOTTING PROGRAM: VELOCITIES

```
C   THIS PROGRAM WILL PLOT OUT THE LINE REPRESENTING THE
C   VELOCITIES OF THE PARTICLES AS A FUNCTION OF X-DISTANCE FROM
C   THE WIRE,BUT NOT THE WIRE ITSELF AS THE SCALE IS TOO LARGE.
C   THE POINTS FOR THE CURVE SHOULD BE READ IN VIA CHANNEL NO.6.
C
      INTEGER Q
      DIMENSION X(1000),Y(1000),R(1000),V(1000),IPLOT(22)
      DATA IPLOT/19,19HPARTICLE VELOCITIES,22,22HPOSITION IN UNITS OF A,
     *31,31HVELOCITY IN UNITS OF A PER SEC./
      I=0
      READ(1,1001)IO,Z,ALPHA
      READ(6,999)Q
      N=1
    2 READ(6,1000)X(N),Y(N),V(N)
      THETA=ATAN2(Y(N),X(N))
      R(N)=SQRT(X(N)*X(N)+Y(N)*Y(N))*COS(THETA-ALPHA)
      N=N+1
      IF(N.LE.Q)GOTO 2
      CALL NARROW
      CALL RSIZE(200.0,100.0)
      CALL FIXAXS(-9.99,10.0,0.0,Z)
      CALL CHAPEN(2)
      CALL FGPLT(R,V,Q,15,0,1,0,IPLOT)
      I=I+1
      IF(IO.EQ.I)GOTO 1
      DO 1 I=2,IO
      READ(6,999)Q
      N=1
    3 READ(6,1000)X(N),Y(N),V(N)
      THETA=ATAN2(Y(N),X(N))
      R(N)=SQRT(X(N)*X(N)+Y(N)*Y(N))*COS(THETA-ALPHA)
      N=N+1
      IF(N.LE.Q)GOTO 3
      CALL CHAPEN(2)
      CALL FGPLT(R,V,Q,15,0,1,1,IPLOT)
    1 CONTINUE
      CALL DEVFIN
      STOP
  999 FORMAT(I10)
 1000 FORMAT(1X,F10.5,2F20.5)
 1001 FORMAT(I5,2F15.5)
      END
```

Physics Programs
Edited by A. D. Boardman
© 1980 John Wiley & Sons Ltd.

CHAPTER 6

Magnetization in the Crystal Field System Praseodymium

J. A. G. TEMPLE

1. INTRODUCTION

The rare earth metals, atomic numbers 58–71, have unfilled 4f shells lying inside closed 5s 5p shells. The outermost electrons, that is $5d^1$ $6s^2$, are lost to the conduction band and become a part of the nearly free electron sea. The small spatial extent of the 4f wavefunctions makes overlap of these wave functions on neighbouring ions very unlikely in the solid. This implies a magnetic moment which is well localized on the ionic site. This situation is very different from the more conventional magnetic systems (Fe, Co, Ni, etc.), in which the magnetism is due entirely to the free electrons, and cannot be attributed to well-localized magnetic moments. For this reason the rare earths are intrinsically interesting, especially so when the range and type of magnetic orderings (ferromagnetic, antiferromagnetic, helical, fan, and spiral types often with magnetic unit cells many times larger than the crystal unit cell) are noted. They are becoming increasingly important in technology too, with uses in high-power permanent magnets and in lasers.

This localized magnetic moment is, to some extent, screened by the 5s 5p electrons and does not experience the full effect of the Coulomb interaction due to the periodic array of ions. Another way of describing this is to say that the crystalline electric field is weaker. Each of the localized magnetic moments experiences the effect of many others in the crystal through a double interaction with the conduction electrons. A local moment will scatter a conduction electron with a spin-dependent interaction so that the conduction electron carries away with it a 'memory' of the orientation of the first moment. This electron, on a second scattering, with a, presumably, different local moment, conveys this memory to the second moment. Thus it appears as though there is a direct interaction between local moments i and j of the form $F(\mathbf{R}_{ij})(\mathbf{J}_i \cdot \mathbf{J}_j)$. $F(\mathbf{R}_{ij})$ oscillates rapidly with the distance $\mathbf{R}_{ij}$

between total ionic angular momenta $\mathbf{J}_i$ and $\mathbf{J}_j$. This is the well-known RKKY interaction (Rudermann and Kittel,[1] Kasuya,[2] Yosida[3]) which dominates the magnetic behaviour of the 14 rare earth elements. In its simplest form, this interaction may be represented by an internal local magnetic field at each site.

Hund's rules give the ground state of the 4f configuration. This is separated from the other multiplets of total angular momentum by an energy which is typically of the order of 1000 K. At low temperatures (i.e. a few kelvin) therefore, only the lowest J-multiplet is occupied; J being the total angular momentum of the ion. This multiplet is $(2J+1)$-fold degenerate in the absence of an internal magnetic field and a crystalline electric field. In the presence of a crystal field, the multiplet is split into a series of singlets, doublets and triplets depending on the symmetry of the charge distributions of neighbouring ions.

At low temperatures (i.e. tens of kelvin) when only a few of these singlets, doublets, and triplets are occupied the system will, in general, be magnetically ordered with all the localized moments arranged in some well-defined pattern. The free electrons, that is those responsible for the electrical conductivity, will sense this ordering and will be scattered differently in the ordered state to the disordered state. There will thus be, possibly, interesting anomalies in the conductivity, or its reciprocal the resistivity, at low temperatures, as well as the magnetization itself. These anomalies reflect the competition between the internal magnetic fields at each lattice site and the local electrical fields due to the Coulomb interaction with neighbouring ions. For praseodymium this competition is finely balanced and, as explained in section 2, no other metal is known to sit so close to a threshold between the ability to order spontaneously as the temperature is reduced and the ability to remain paramagnetic even towards the absolute zero of temperature. The magnetization anomalies may thus be expected to be rather interesting in the presence of an externally applied magnetic field.

2. THE MODEL

Praseodymium (Pr) is element number 59 and has therefore a $4f^2$ configuration for which Hund's rules give the quantum numbers $S=1$, $L=5$ and, since the shell is less than half full, $J=L-S=4$. The ground state is denoted $^3H_4$ and is ninefold degenerate in the absence of a crystalline electric field or internal magnetic field. Other ways of adding together the total orbital angular momentum L and the total spin S would give the configurations $^3H_5^3H_6$. These lie at much greater energies; in praseodymium it requires at least 1000 K to excite the ions into these higher energy states. Because of this large energy requirement it is safe to assume only the lowest multiplet, namely $^3H_4$, is populated at room temperature or below.

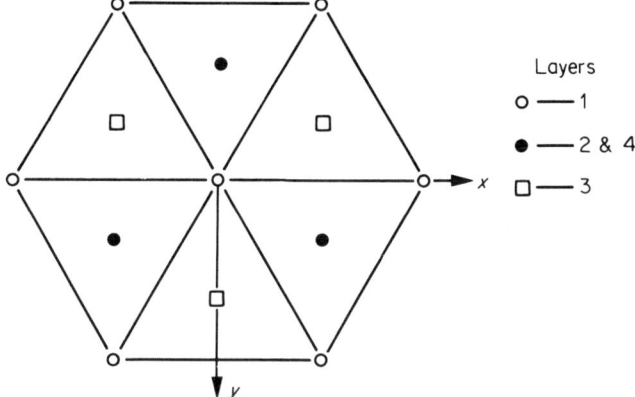

Figure 1. Plan view of dhcp crystal structure showing layers of ions in the planes perpendicular to the crystal c-axis. Ions in successive layers are denoted by 1, 2, 3, and 4

The solid-state form of Pr has the double hexagonal close-packed structure (dhcp) as shown in Figure 1. This is usually denoted as a stacking sequence ABCB of hexagonal layers. In a material with cubic symmetry the layer sequence is ABC and in a hexagonal material the sequence is ABAB. It is clear that, based on the nearest-neighbour environments, the dhcp structure has layers of alternate designation, cubic or hexagonal. Two types of crystallographic symmetry imply that there are two different magnetic behaviours; in this material two interpenetrating sublattices each with its own molecular field.

Suppose that there is an externally applied magnetic field. This acts at each lattice site, together with an additional magnetic field due to an extra alignment of all the magnetic moments on the neighbouring lattice sites, since each ion responds to the applied field by a tendency to align with it. This additional internal field is called a 'molecular field' after the original concept of Weiss.[4] In this approximation, the magnetic field due to the partial alignment of neighbouring moments is linearly related to this alignment or magnetization. Although the applied field will be the same at all lattice sites, the internal field will vary according to the physical arrangement of the neighbours, that is to the symmetry and number of neighbouring ions and their magnetization, i.e.

$$\mathbf{H}_c = \frac{\lambda}{g^2 \mu_B^2} \mathbf{M}_c + \frac{\mu}{g^2 \mu_B^2} \mathbf{M}_h + \mathbf{H}_{applied},$$

$$\mathbf{H}_h = \frac{\lambda}{g^2 \mu_B^2} \mathbf{M}_h + \frac{\mu}{g^2 \mu_B^2} \mathbf{M}_c + \mathbf{H}_{applied},$$

(1)

where $\mathbf{H}_c$ and $\mathbf{H}_h$ are the total internal magnetic fields at the cubic and hexagonal sites, $\mathbf{M}_c$ and $\mathbf{M}_h$ their associated magnetizations, λ and μ are the two Weiss molecular field parameters, g is the Landé factor, and μ_B is the Bohr magneton. It has been assumed in equation (1) that the cubic–cubic and hexagonal–hexagonal interactions are equivalent. This is a reasonable assumption since the interaction between two moments of a pure rare earth metal depends only on the conduction electrons and the distance between the ions.

As the temperatures used in an experimental study of the magnetization will be only a few tens of Kelvin at most, only the lowest J-multiplet will enter the calculations. This ninefold degenerate set of levels will split up into singlets, doublets, and triplets under the influence of the electric field of all the neighbouring ions. This is the crystal field and its effect is determined principally by the symmetry of the neighbouring ions. As we are concerned only with a set of levels, all of which possess the same total angular momentum J, it is possible to calculate the effect of the Coulomb interaction of all the ions on each other; a kind of Stark effect, by using the method of Stevens operators. These make easier the task of determining the splitting of the levels with the same J but differing J_z through the first-order perturbation caused by the electric field of neighbouring ions. Using these Stevens operator equivalents (Stevens[5]), the ionic Hamiltonian, or total energy operator, becomes

$$\mathcal{H}_\alpha = V_\alpha - g\mu_B \mathbf{H} \cdot \mathbf{J}, \tag{2}$$

where α = cubic or hexagonal, V_α is the energy in the crystal field, and $\mathbf{H}$ is the total internal magnetic field.

$$V_{\text{cubic}} = B_2 O_2^0 + B_4(O_4^0 + 20\sqrt{2} O_4^3) + B_6 \left(O_6^0 - \frac{35}{\sqrt{8}} O_6^3 + \frac{77}{8} O_6^6 \right), \tag{3}$$

$$V_{\text{hexagonal}} = B_2 O_2^0 + B_4 O_4^0 + B_6(O_6^0 + \tfrac{77}{8} O_6^6), \tag{4}$$

where B_l are constants with dimensions of energy and the O_l^m are the Stevens operators that are polynomials of degree l in the quantum mechanical operators J_z, $J+$, $J-$. Some typical Stevens operators are:

$$\begin{aligned} O_2^0 &= 3J_z^2 - J(J+1), \\ O_6^6 &= \tfrac{1}{2}\{(J+)^6 + (J-)^6\}, \end{aligned} \tag{5}$$

and a full list of them may be found in Hutchings.[6] The resulting operator $\mathcal{H}_\alpha$ may be expressed as a matrix whose elements are the numbers obtained from $\langle i | \mathcal{H}_\alpha | j \rangle$, where $|i\rangle$ and $|j\rangle$ are states of total angular momentum J and z-component J_z. That is, a state of the system is written $|J; J_z\rangle$ where L and

S have been taken as fixed ($L = 5$, $S = 1$ for two 4f electrons). The magnitude of $J = \sqrt{J(J+1)}\hbar$ and the basis set of states is taken to be

$$|J; J_z = J\rangle, \quad |J; J_z = J-1\rangle, \ldots, \quad |J; J_z = -J\rangle.$$

As we are dealing only with the lowest J multiplet, this notation is further abbreviated to $|i\rangle$, $|j\rangle$, etc. where only the J_z component of the total angular momentum is shown within the Bra-Ket notation of Dirac. If the resulting matrix is diagonalized then the eigenvalues displayed down the diagonal are the energies of the different crystal field states in the total internal magnetic field. The corresponding eigenvectors are the wave functions of the state produced and will, in general, be linear combinations of the basis states. In the absence of any internal magnetic field, the crystal field lifts the nine-fold degeneracy of the $^3H_4$ ground state. However, there still remains a partial degeneracy, for example $|+1\rangle$ will have the same energy as $|-1\rangle$ and this degeneracy will be removed by an internal magnetic field. Also, the crystal field reorders the energy levels so that, for example, $|0\rangle$ lies below both $|+1\rangle$ and $|-1\rangle$. When a magnetic field develops due to either an applied magnetic field or spontaneous ordering, this level scheme will change and in the limit of very large internal magnetic fields the levels will be, in ascending order: $|-4\rangle$, $|-3\rangle$, $|-2\rangle$, $|-1\rangle$, $|0\rangle$, $|1\rangle$, $|2\rangle$, $|3\rangle$, $|4\rangle$.

The ground state, which is the state of lowest energy, may be a pure $|J_z\rangle = |0\rangle$ state and is then an example of a 'singlet ground state system'. A characteristic of such a system is that it will not order even at the lowest obtainable temperature because the magnetic moment operator vanishes identically (i.e. $\langle 0|J_z|0\rangle = 0$). Such a system may exist if the ratio of crystal field energy to internal magnetic energy exceeds a certain threshold value. Praseodymium is a unique metal because the ratio of exchange energy (the magnetic interaction between two local moments), to the energy of the ions in the crystal field is found to be about 0.96, whereas the threshold value is 1.0. For a ratio greater than 1.0 Pr would order spontaneously at some small non-zero temperature, but for ratios less than 1.0 it will remain paramagnetic until temperatures of a few millikelvin at which the hyperfine interaction becomes important. No other metal is known to sit so close to this threshold value. The level schemes, from neutron-scattering experiments, in dhcp Pr are shown in Figure 2. From this figure it will be observed that the cubic sites may be expected to play a rather small role in the magnetization at low temperatures since no significant population of levels with resultant moments can occur until temperatures of about 80 K are reached. It is, therefore, instructive to look at the bottom three levels on the hexagonal sites and investigate their behaviour in magnetic fields applied parallel to the z- or x-axes. This will give physical insight into the behaviour of the system as a whole.

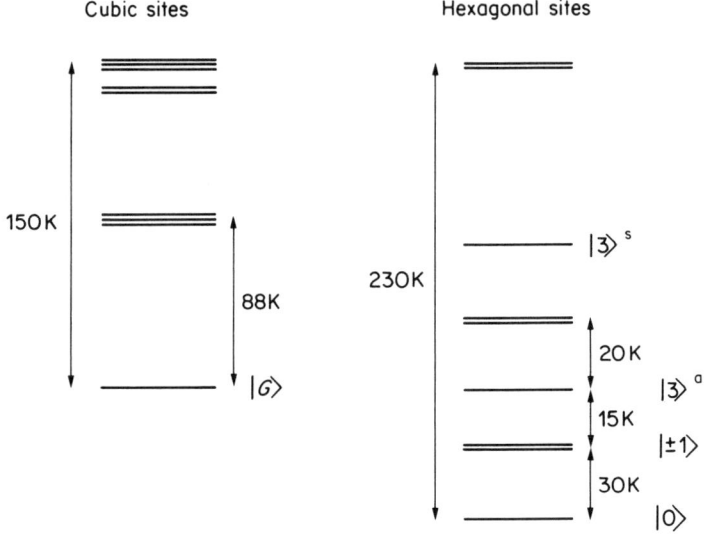

Figure 2. Praseodymium: energy levels in zero applied magnetic field

2.1 Example of way model is used

Ignoring, for the present, any molecular field, we may consider the artificial system of a $J = 1$ spin sitting in a crystal field of axial symmetry. If instead of V_α we write $\mathcal{H}_{cf}$ for the effect of the crystalline electric field, then equation (2) becomes

$$\mathcal{H} = \mathcal{H}_{cf} - g\mu_B \mathbf{H} \cdot \mathbf{J}, \qquad (6)$$

where $\mathbf{H}$ is the applied magnetic field only. Considering first the field applied parallel to the z-axis, then $\mathbf{H} \cdot \mathbf{J} = H_z J_z$. Hence, if we write $\alpha = g\mu_B$ and the familiar result for $\mathcal{H}_{cf}$ we may express (6) as

$$\mathcal{H} = D J_z^2 - \alpha J_z, \qquad (7)$$

where D and α are numbers with dimensions of energy and D represents the crystal field splitting in a system of uniaxial symmetry. For a $J = 1$ system the corresponding matrix Hamiltonian becomes (see for example Dicke and Wittke[7])

$$\mathcal{H} = \begin{pmatrix} D-\alpha & 0 & 0 \\ 0 & 0 & 0 \\ 0 & 0 & D+\alpha \end{pmatrix},$$

where we have used a quantization scheme along the z-axis with eigenvalues 0, $D-\alpha$, $D+\alpha$, so that the energy level diagram has the form shown in Figure 3.

The magnetization is calculated from

$$M = -g\mu_B \langle\langle J_z \rangle\rangle, \qquad (8)$$

where $\langle\langle A \rangle\rangle$ denotes the thermal average of the expectation value of the operator A. It is necessary to use the thermal average as the energy splitting of the levels caused by either crystal field or magnetic field is of the same order as the temperature and therefore there will be a statistical population of the levels throughout the 10^{23} sites in a macroscopic crystal.

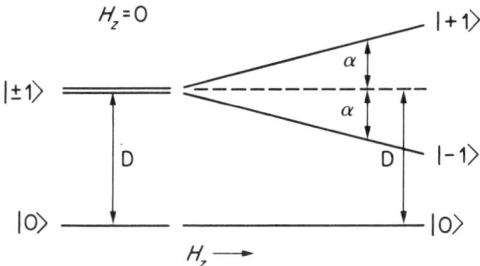

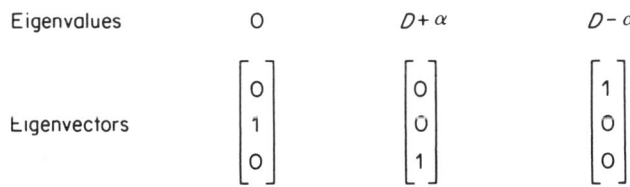

Figure 3. Energy levels of a $J=1$ system with uniaxial crystal field showing how the levels split with a magnetic field applied along the z-axis.

With the eigenvalues and eigenvectors given in Figure 3 we have

$$\langle\langle J_z\rangle\rangle = \left[\langle 0|J_z|0\rangle + \langle -1|J_z|-1\rangle \exp\left\{-\frac{(D-\alpha)}{kT}\right\}\right.$$
$$\left. + \langle +1|J_z|+1\rangle \exp\{-(D+\alpha)/kT\}\right]Z^{-1}, \quad (9)$$

where Z, the partition function, is given by

$$Z = 1 + \exp\{-(D-\alpha)/kT\} + \exp\{-(D+\alpha)/kT\}, \quad (10)$$

k is Boltzmann's constant and T is the temperature in kelvin. Clearly, when $\alpha = 0$ ($H_z = 0$) then $\langle\langle J_z\rangle\rangle = 0$ and the system can never order no matter how low the temperature to which it is subjected. This is characteristic of a singlet ground state system. However, when $\alpha \neq 0$ (since $\langle 0|J_z|0\rangle = 0$) the magnetization looks like

$$M = -\frac{g\mu_B}{Z}[-\exp\{-(D-\alpha)/kT\} + \exp\{-(D+\alpha)/kT\}], \quad (11)$$

This is very small for $\alpha \ll D$, (i.e. energy of the spin in the applied magnetic field is very much less than the crystal field splitting), showing that the initial susceptibility is small. If the applied field is increased, then, eventually, the $|-1\rangle$ level will fall below the $|0\rangle$ level. This occurs at a field for which $D - \alpha = 0$ or $H_z = D/g\mu_B$. In a more realistic model, there will also be a molecular field which tends to amplify the applied field so that the condition for the $|-1\rangle$ level to fall below the $|0\rangle$ level would be $H_z + \beta M_z = D/g\mu_B$, where β represents the strength of the molecular field. At a field infinitesimally greater than this critical field the magnetization will become

$$M = \frac{g\mu_B[1 - \exp(-2D/kT)]}{2 + \exp(-2D/kT)}, \quad (12)$$

which for very small temperatures tends to $M = (\frac{1}{2})g\mu_B$. This appears as a jump discontinuity in the magnetization, i.e. for α just smaller than D, as the temperature tends to zero, the magnetization remains zero; but for α just larger than D, under the same circumstances, the magnetization is $(\frac{1}{2})g\mu_B$.

Now consider what happens if the field is applied along the x-direction, in the simple model. If we put $\alpha = g\mu_B H_x$ then

$$\mathcal{H} = DJ_z^2 - \alpha J_x, \quad (13)$$

or in matrix form, still using quantization along the z-axis,

$$\mathcal{H} = \begin{pmatrix} D & 0 & 0 \\ 0 & 0 & 0 \\ 0 & 0 & D \end{pmatrix} - \frac{\alpha}{\sqrt{2}}\begin{pmatrix} 0 & 1 & 0 \\ 1 & 0 & 1 \\ 0 & 1 & 0 \end{pmatrix}.$$

MAGNETIZATION IN THE CRYSTAL FIELD SYSTEM PRASEODYMIUM   195

The resultant matrix has eigenvalues, found in the usual way from the characteristic equation,

$$D, \tfrac{1}{2}\{D \pm \sqrt{D^2 + 4\alpha^2}\}.$$

This gives the form of energy level diagram shown in Figure 4 together with the corresponding eigenvectors. The magnetic behaviour in this case is rather different from applying a field parallel to the z-axis. Applying the field along the x-axis polarizes the ground state and induces a moment on it.

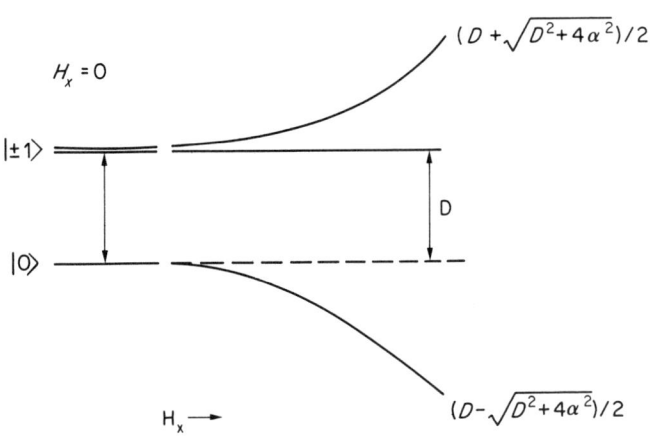

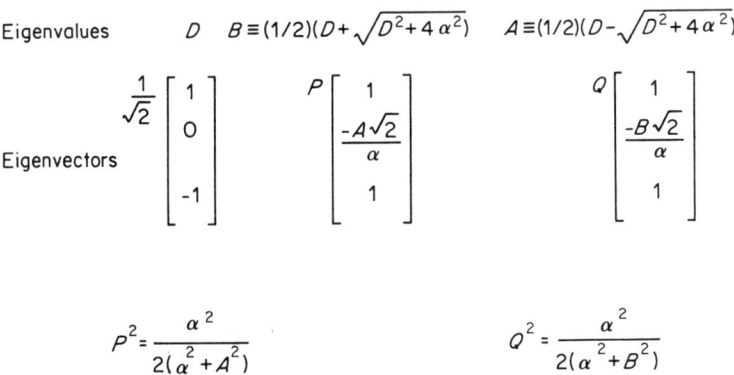

Figure 4. Energy levels of a $J = 1$ system with uniaxial crystal field showing how the levels split with a magnetic field applied along the x-axis

Since we have chosen the crystallographic c-axis as the axis of quantization, and the particular form for the crystal field energy given in equation (13), the required magnetic moment operator is J_x. Hence the magnetization is

$$M = -g\mu_B \langle\langle J_x \rangle\rangle. \tag{14}$$

From the eigenvectors and eigenvalues given in Figure 4 this expression is, for small non-zero fields $(0 < \alpha \ll D)$,

$$M = -\frac{g\mu_B}{Z}\left\{0 \cdot \exp\left[-\frac{D}{kT}\right] - \frac{4AP^2}{\alpha} \cdot \exp\left[-\frac{(D+\sqrt{D^2+4\alpha^2})}{2kT}\right]\right.$$
$$\left. - \frac{4BQ^2}{\alpha} \cdot \exp\left[-\frac{(D-\sqrt{D^2+4\alpha^2})}{2kT}\right]\right\}, \tag{15}$$

where A and B are two eigenvalues, and P and Q are wave-function normalization constants. Their values are

$$A = (\tfrac{1}{2})\{D - \sqrt{D^2+4\alpha^2}\}, \qquad B = (\tfrac{1}{2})\{D + \sqrt{D^2+4\alpha^2}\},$$

and

$$P = \left[\frac{\alpha^2}{2(\alpha^2+A^2)}\right]^{\frac{1}{2}}, \qquad Q = \left[\frac{2}{2(\alpha^2+B^2)}\right]^{\frac{1}{2}}.$$

The initial susceptibility is found by taking $\alpha \ll D$ and expanding the square root. It is given approximately by

$$M \sim 2g\mu_B \alpha / DZ, \tag{16}$$

where

$$Z = \exp[-D/kT] + \exp[-(D+\sqrt{D^2+4\alpha^2})/2kT]$$
$$+ \exp[-(D-\sqrt{D^2+4\alpha^2})/2kT].$$

The magnetization, although small, for small α, is in fact much larger than the case when the field is applied along the z-axis.

2.2 Application to the real system

The behaviour of the simple $J=1$ system gives the clue to the observed magnetization of Pr at low temperatures and high fields. The experimental magnetization, for fields along the [001] axis and the [110] axis is shown in Figure 5 and is taken from McEwan et al.[8] The z-axis of the model is the [001] axis, whereas the x-axis of Figure 1 corresponds to the [110] axis of the experimental work. For Pr the Landé g-factor is (4/5), so therefore the saturated magnetic moment is 3.2 Bohr magnetons per atom. Applying the field along the x-axis causes the magnetization to rise rapidly, although,

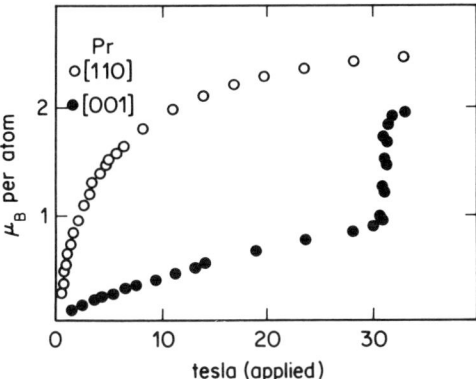

Figure 5. Experimentally observed magnetization of Pr single crystals with applied magnetic field. From McEwen, Cock, Roeland, and Mackintosh (1973)

even in 30 T, the saturation value is not achieved. However, when the field is applied along the z-axis there is only a small increase in magnetization until an applied field of ~ 32 T is reached, whereupon a large jump occurs in the magnetization. This is due to a level crossing the $|0\rangle$ ground state. Unlike our simple model, it is in fact the $|-3\rangle$ level crossing the $|0\rangle$ level that gives this jump.

The computer program that is given at the end of this chapter calculates the magnetization on each of the two crystallographic sites (i.e. the cubic and hexagonal sites of the dhcp structure) for a given applied field and temperature. This is achieved by setting up the matrix Hamiltonian of equation (2) for the given applied field and diagonalizing it. The eigenvalues and eigenvectors are then used in equations (14) or (8) to calculate a magnetization at each site. This magnetization is converted to an internal field by equation (1) and this new total internal field is compared with that obtained on the previous iteration for the same applied field. If they are not close enough the program iterates these steps until the self-consistent solution is found. This is stored and a new field is chosen by adding on a certain amount, decided from the input data, and the whole process repeated for a particular temperature. When all the field steps have been calculated a table of results is printed on the line printer, then simple graphs are drawn on the line printer of the magnetization of the cubic sites, the hexagonal sites, and their average against the applied field. More temperatures may be requested and the entire process is repeated until a negative temperature is encountered, whereupon the program terminates.

The experimental magnetization can be reproduced reasonably well by choosing the values of the molecular field parameters $\lambda = -1.20$, $\mu = 4.14$ K, together with the values of B_2, B_4, and B_6 that are set within the program. If these values of the parameters are used it will be found that the jump discontinuity is 'smoothed out' over some range ΔH where $\Delta H \sim (kT/g\mu_B)$. This is due to a hidden sophistication in that the free energy curve against magnetization has two minima separated by an energy barrier. At the critical field there are two allowed magnetizations, the smaller and the larger, and only if the temperature is sufficiently large can the system sample both and go smoothly from one to the other. For small temperatures the second state is not 'sensed' until sufficient energy, in the form of the applied field, has been delivered to drive the system over the potential barrier. In certain cases the program may fail to converge at this point as the allowed number of iterations, set within the program, is exceeded. The program then lists the results obtained up to that point and proceeds with the next temperature.

3. THE PROGRAM DATA

The necessary data to be presented to the program take the following form where each FORMAT refers to a new line of data. An integer number of field steps (FORMAT I3) is used. If this number is negative the graphs produced on the line printer will be approximately 7 in wide. This is a useful feature if the graphs are to be returned to a teletype, or are to form part of a report. The next data are the lowest field in tesla and the increment (FORMAT 2F6.2), two molecular field parameters λ, μ in kelvin (FORMAT 2F7.2) and an integer (FORMAT I1) that must be either a 0 or 1. If it is desired to apply the field along the z-axis use zero and if the field is to be applied along the x-axis use unity. Next there is an integer (FORMAT I1) which, if zero, inhibits the printing of all the eigenvalues and eigenvectors; it is this form that should generally be used. If the eigenvectors are desired, inputting unity here will print out nine eigenvalues across the page, followed underneath by the eigenvectors. These latter will each be columns of nine numbers and are to be interpreted as the coefficients of successively smaller $|J_z\rangle$ values. Thus, for example, in zero applied field, on the cubic sites the lowest eigenvalue would have below it (see Figure 2):

0.000
0.609
0.000
0.000   which is interpreted as
0.000 $0.609\,|+3\rangle - 0.509\,|0\rangle - 0.609\,|-3\rangle$
−0.509
0.000
0.000
−0.609
0.000

The next numbers in the data are real and represent the required temperatures in kelvin (FORMAT F7.2). There can be any (reasonable) number of these, but the last one must be negative in order to stop the execution of the program.

4. A NOTE ON THE EIGENVECTORS

The eigenvectors are normalized so that, if V_i represents a nine-tuple eigenvector, V_i^T is its transpose, and $V_i^T V_i = 1$ for all i. They are also orthogonal in the sense that $V_i^T V_j = \delta_{ij}$. These properties are readily demonstrated in simple cases by examination of the eigenvectors produced by the program (to an accuracy of three decimal places). In a very large applied magnetic field the energy level scheme on both sites will tend towards nine levels, equally spaced, the lowest being the $|-4\rangle$ state, then $|-3\rangle$, and so on up to the $|+4\rangle$ state. If one was interested in a neutron-scattering experiment then the intensity of neutrons scattered with momentum change along the z-axis is proportional to the Boltzmann population of the initial energy level $|i\rangle$ and to the square of matrix elements connecting initial and final states $|f\rangle$, that is

$$I(E_{if}) \propto \frac{\exp(-E_i/kT)}{Z} \cdot |\langle i| J_z |f\rangle|^2,$$

where E_{if} is the energy lost or gained by the neutron beam. With the subroutines provided with the program it is fairly simple to produce intensity spectra for different level schemes and different temperatures.

5. RUNNING THE PROGRAM

The data required to run the program and the input formats required were discussed in section 3. A useful starting point will be to use the program in an attempt to reproduce some recent experimental data (McEwan et al.[8]). The experiments were done at the temperature of liquid helium, 4.2 K, and with applied magnetic fields up to about 34 T. These fields were applied to the x- and z-axes in turn. This range of applied magnetic field represents the limit of high field experimental equipment currently available. Fields in excess of this may be applied in the program. It may be interesting to determine at what field you would expect to obtain magnetic saturation of a pure crystal of praseodymium. Choosing molecular field parameters $\lambda = -1.20$, $\mu = 4.14$ K and a temperature of 4.2 K gives reasonable representation of the experimental results. For convenience, only the regions where the curves begin to become asymtotic are drawn at the end of the chapter, although the computer program will always give the complete curves, starting at 0 tesla. The deviations from experiment are discussed briefly in section 2.2.

Keeping λ and μ fixed, for the present, the temperature should be varied. By pumping on the vapour above the liquid helium, temperatures down to 1 K are possible experimentally so temperatures down to this should be tried in the simulation. Lower temperatures may also be used, the effect being to sharpen up the transition that occurs for magnetic fields applied parallel to the z-axis. However, even though a temperature as low as a few millikelvin (~ 50 mK) may be produced by using the technique of magnetic cooling (a good description may be found in Mendelssohn[9]), temperatures as low as this are not recommended for the program for two reasons. Firstly, at these very low temperatures some new physics occurs; the magnetic moments of the nuclei may become aligned through the hyperfine interaction and would, therefore, add an extra contribution to the molecular field. The approximations made in setting up the model used here are inadequate to represent this effect correctly. Secondly, overflows may occur when calculating the exponential factors, thus causing the program to end abnormally, the exact effect will depend on the computer operating system under which the job is being run.

Higher temperatures are also well worth investigating. The next convenient experimental temperature would be 77 K, the temperature of liquid nitrogen. Increasing the temperature allows more of the energy levels to be populated in proportion to the usual Boltzmann factor. This changes the averaged magnetic moment of each ion and smears out the magnetic ordering over a greater range of applied magnetic fields. For higher temperatures, stronger magnetic fields will be required to induce the ordering. Note that praseodymium melts at about 1300 K so temperatures larger than this would not be sensible.

Having investigated the role of temperature, the other parameters easily varied are the two molecular field constants λ and μ. Equation (1) of section 2 shows how these parameters are related to the strengths of the internal magnetic fields. By applying a large hydrostatic pressure to the crystal one might expect to change the lattice spacings and hence alter λ and μ. Because the interaction between two local magnetic moments has the oscillatory RKKY form, changing the lattice spacing by small amounts may change both the magnitude and the sign of either λ or μ, or both. Unfortunately, the way in which this would happen is rather complex, and lies on the frontiers of current research, so it is difficult to give an easy physical picture. Nevertheless it is instructive to vary both λ and μ. Try varying them one at a time, in small steps, to build up a picture of the behaviour of the system you have created within a parameter space limited by, say, $-5 \leq \lambda, \mu \leq 5$ K. This will represent a lot of computing and may well need more than one session to do it properly.

As an extension to the work, there is the possibility of building up the sort of pictures you would get from a neutron diffraction experiment. This was discussed briefly in section 4.

Not only are there anomalies in the magnetization curves, there are also associated anomalies in the resistivity curves or magneto-resistance, if it is measured as a function of applied magnetic field. For those who would like to study this aspect, and who are willing to write their own programs it is an easy matter. The equations necessary may be found in either of the two references[10] at the end of this chapter, the only extra ones required are those to calculate two matrices that contain the scattering probability information.

Another extension of the work, which requires no extra programming, is to keep $\lambda = -1.20$, $\mu = 4.14$ and to reset the constants BH2, BH4, BH6, BC2, BC4, BC6 used in the program. These are the constants multiplying the various Stevens operators and represent the strengths of each of the terms of different symmetry in the electric crystal field. Altering their values will have the effect of altering the energy levels and the order in which they occur in the absence of an applied magnetic field. Thus a hypothetical praseodymium may be 'produced' in which a completely different arrangement of the levels is investigated. This extension will be most suitable for those who wish to delve into the current research literature to obtain the most up-to-date values for the crystal field constants in a point charge model.

REFERENCES

1. M. A. Rudermann and C. Kittel, *Phys. Rev.*, **96,** 99 (1954).
2. T. Kasuya, *Prog. Theor. Phys.*, **16,** 45 (1956).
3. K. Yosida, *Phys. Rev.*, **106,** 893 (1957).
4. P. Weiss, *J. de Phys.*, **6,** 661 (1907).
5. K. W. H. Stevens, *Proc. Phys. Soc. (Lond.)*, **A65,** 209 (1952).
6. M. T. Hutchings, *Solid State Phys.*, **16,** 227 (1964).
7. R. H. Dicke, and J. P. Wittke, *Introduction to Quantum Mechanics* (Addison-Wesley, Reading, Mass., 1961).
8. K. A. McEwan, G. J. Cock, L. W. Roeland, and A. R. Mackintosh, *Phys. Rev. Lett.*, **30,** 287 (1973).
9. K. Mendelssohn, *The Quest for Absolute Zero* (Weidenfeld and Nicolson, London, 1966).
10. J. A. G. Temple, K. A. McEwan, 'Magneto-resistance of praseodymium' in *Crystal Field Effects in Metals*, Ed. A. Furrer (Plenum, New York, 1977); K. A. McEwan, J. A. G. Temple, and G. D. Webber, *Physica*, **86–88**(B), 533 (1977).

PRASEODYMIUM

```
      DIMENSION RHH(9,9),RHC(9,9),EVALC(9),EVALH(9)
      DIMENSION RZO(9,9),RMO(9,9),RPO(9,9),RES(50,5)
      DIMENSION RXO(9,9),USE(9)
      LOGICAL A4
      A4=.FALSE.
CCCCCCCCCCCCCCCCCCCCCCCCCCCCCCCCCCCCCCCCCCCCCCCCCCCCCCCCCCCC
C
C     THE LOGICAL VARIABLE A4 GOVERNS THE SIZE OF THE GRAPHS PRODUCED ON
C 120 COLUMNS THAT IS 12 INCHES. A4=TRUE GIVES GRAPHS OCCUPYING
C  ONLY 71 COLUMNS WIDTH AND IS THEREFORE SUITABLE FOR OUTPUTTING
C RESULTS TO NORMAL TELETYPES
C  A4 IS SET TRUE IF THE NUMBER OF FIELD STEPS (THE FIRST DATA ITEM)
C  IS NEGATIVE, OTHERWISE IT REMAINS FALSE.
C
CCCCCCCCCCCCCCCCCCCCCCCCCCCCCCCCCCCCCCCCCCCCCCCCCCCCCCCCCCCC
C
C SET THE OPERATOR MATRICES TO ZERO BEFORE SETTING THEIR
C  NON ZERO ELEMENTS
C
      DO 100 I=1,9
      DO 100 J=1,9
      RZO(I,J)=0.0
      RMO(I,J)=0.0
      RXO(I,J)=0.0
      RPO(I,J)=0.0
  100 CONTINUE
C
C SET THE NON ZERO ELEMENTS OF THE FOUR OPERATOR MATRICES
C
      DO 200 I=1,9
      FI=FLOAT(5-I)
      RZO(I,I)=FI
  200 CONTINUE
      DO 205 I=2,9
      J=I-1
      FJ=FLOAT(J)
      FI=FLOAT(10-I)
      XX=FI*FJ
      XX=SQRT(XX)
      RPO(J,I)=XX
      RMO(I,J)=XX
  205 CONTINUE
      DO 210 I=1,9
      DO 210 J=1,9
      RXO(I,J)=0.50*(RPO(I,J)+RMO(I,J))
  210 CONTINUE
CCCCCCCCCCCCCCCCCCCCCCCCCCCCCCCCCCCCCCCCCCCCCCCCCCCCCCCCCCCC
C
C DATA IS READ IN WITH THE FOLLOWING FORMATS
C
C NUMBER OF FIELD STEPS FOR EACH TEMPERATURE   FORMAT   I3
C  IF NUMBER OF FIELD STEPS IS NEGATIVE
C   THEN THE LINEPRINTER GRAPHS ARE PRODUCED A4 SIZE.
C
C LOWEST FIELD AND FIELD INCREMENT IN TESLA    FORMAT 2F6.2
C  TWO MOLECULAR FIELD PARAMETERS IN KELVIN    FORMAT 2F7.2
```

PRASEODYMIUM

```
C  (FOR DETAILS OF MOLECULAR FIELD CONSTANTS SEE EQN.(1) AND ASSOCIATED
C      TEXT).
C
C INTEGER=0  FOR FIELD PARALLEL TO THE Z-AXIS FORMAT  I1
C        =1  FOR FIELD PARALLEL TO THE X-AXIS
C INTEGER=0  FOR NO PRINTING OF EIGENVECTORS FORMAT   I1
C        =1  FOR PRINTING OF ALL EIGENVECTORS
C SERIES OF TEMPERATURES THE LAST OF WHICH MUST
C  BE NEGATIVE.                         FORMAT  F7.2
C
C
CCCCCCCCCCCCCCCCCCCCCCCCCCCCCCCCCCCCCCCCCCCCCCCCCCCCCCCCCCCCC
C
      READ(1,1008) NHS
      IF(NHS.LT.0) A4=.TRUE.
      NHS=IABS(NHS)
      IF(NHS.GT.50) NHS=50
      READ(1,1010) PHL,RHS
      READ(1,1011) RM1,RM2
      READ(1,1009) ID
      READ(1,1009) IP
      WRITE(2,1200)
      WRITE(2,1201) RHL
      WRITE(2,1202) NHS,RHS
      IF(ID.EQ.0) GOTO 220
      WRITE(2,1203)
      GOTO 230
 220  WRITE(2,1204)
 230  CONTINUE
      WRITE(2,1205) RM1,RM2
CRYSTAL FIELD PAPAMETER ARE SET
      BH2=4.920
      BH4=1.390
      BH6=1.240
      BC2=0.0
      BC4=BH4
      BC6=BH6
 300  READ(1,1012) T
      IF(T.LT.0.0) GOTO 940
      IM=30
      DO 120 I=1,5
      DO 120 J=1,50
      RES(J,I)=0.0
 120  CONTINUE
      IH=0
      RMAGC=0.0
      RMAOH=0.0
CU IS CONVERSION FACTOR FROM TESLA TO KELVIN
      CU=(0.80*0.92730)/1.38060
      IF(RHL.GT.0.0) IM=100
 310  IT=0
      IH=IH+1
 320  IT=IT+1
      IF(IT.LT.IM  GOTO 325
      WRITE(2,2900)
      WRITE(2,2901) IH,II,IM
```

PRASEODYMIUM

```
      GOTO 900
  325 IF(IH.GT.NHS) GOTO 900
      IF(IH.GT.50) GOTO 900
      FIH=FLOAT(IH-1)
CALCULATE APPLIED MAGNETIC FIELD RHA
      RHA=RHL+FIH*RHS
      IF(ID.EQ.0) GOTO 330
C
CONSTPUCT THE HAMILTONIAN MATRICES FOR THE ENERGY OF THE IONS
CONSIDERING THE INTERNAL MAGNETIC FIELD AND THE
CRYSTAL FIELD ENERGY LEVELS. UNITS ARE KELVIN.
C
      HZC=0.0
      HZH=0.0
      HXC=(RHA+PM1*RMAGC+RM2*RMAGH)*CU*0.50
      HXH=(RHA+RM1*RMAGH+RM2*RMAGC)*CU*0.50
      GOTO 340
  330 HXC=0.0
      HXH=0.0
      HZC=(RHA+RM1*RMAGC+RM2*RMAGH)*CU
      HZH=(RHA+RM1*RMAGH+RM2*RMAGC)*CU
  340 CALL PRH(RHH,0,HXH,HZH,BH2,BH4,BH6)
      CALL PRH(RHC,1,HXC,HZC,BC2,BC4,BC6)
CCCCCCCCCCCCCCCCCCCCCCCCCCCCCCCCCCCCCCCCCCCCCCCCCCCCCCCCCCCCCCCC
C
CALCULATE THE EIGENVALUES AND EIGENVECTORS OF THE TWO MATRICES
C
C     ROUTINES F01AJF+F02AMF ARE STANDARD LIBRARY ROUTINES(NAG)
C     WHICH CALCULATE EIGENVALUES AND EIGENVECTORS OF REAL
C     SYMMETRIC MATRICES BY HOUSEHOLDER REDUCTION TO TRI-
C     DIAGONAL FORM FOLLOWED BY THE QR ALGORITHM TO COMPLETE
C     THE DIAGONALISATION.
C     FOR DETAILS OF HOUSEHOLDER AND QP ALGORITHMS SEE BOOKS ON
C     NUMERICAL METHODS, FOR EXAMPLE ACTON R.S. "NUMERICAL METHODS
C     THAT WORK" PUBLISHED BY HARPER,N.Y. (1970) PAGE 347
C
C
C  THE PARAMETER LIST OF F01AJF(N,TOL,A,IA,D,E,Z,IZ) IS
C     N=INTEGER,THE OPDER OF MATRIX A
C     TOL=REAL,MACHINE DEPENDENT CONSTANT, FOR ICL 1900 TOL=2.0**(-218)
C     A=REAL ARRAY,OF DIMENSION AT LEAST(N,N) CONTAINING THE SYMMETRIC
C          MATRIX,THE LOWER TRIANGLE ONLY IS REQUIRED. THE ARRAY IS
C          NOT OVERWRITTEN BY THE ROUTINE.
C     IA=INTEGER,THE FIRST DIMENSION OF A AS DEFINED IN THE CALLING.
C     D=REAL ARRAY,OF DIMENSION AT LEAST (N), ON EXIT IT CONTAINS THE
C          DIAGONAL ELEMENTS OF TRIDIAGONAL MATRIX.
C     E=REAL ARRAY,OF DIMENSION AT LEAST (N), ON EXIT IT CONTAINS THE
C          N-1 OFF DIAGONAL ELEMENTS OF TPIDIAGONAL MATRIX.
C     Z=REAL ARRAY,OF DIMENSION AT LEAST (N,N). ON EXIT IT CONTAINS
C          THE ORTHOGONAL MATRIX Q THE PRODUCT OF THE HOUSEHOLDEP
C          TRANSFORMATIONS.
C     IZ=INTEGER,THE FIRST DIMENSION OF Z AS DEFINED IN CALLING SEG.
C
C
C  THE PARAMETER LIST OF F02AMF(N,ACC,D,E,Z,IZ,IFAIL) IS
C     N=INTEGER,THE ORDER OF TRIDIAGONAL MATRIX T.
```

PRASEODYMIUM

```
C       ACC=REAL,SMALLEST NUMBER ON THE COMPUTER SUCH THAT 1+ACC=1
C            (ON ICL 1900 ACC=2.0**(-37))
C       D=REAL ARRAY,OF DIMENSION AT LEAST (N),CONTAINING DIAGONAL
C            ELEMENTS OF T.
C       E=REAL ARRAY,OF DIMENSION AT LEAST (N) CONTAINING THE SUB-
C            DIAGONAL ELEMENTS OF T STORED IN E(2)-=E(N). IT IS
C            OVERWRITTEN BY THE ROUTINE.
C       Z=REAL ARRAY, DIMENSION AT LEAST (N,N), IF EIGENVECTORS OF THE
C            FULL SYMMETRIC MATRIX ARE REQUIRED IT SHOULD CONTAIN THE
C            Q (SEE F01AJF). ON EXIT IT CONTAINS THE NORMALIZED
C            EIGENVECTORS SUCH THAT Z(I,J),I=1,N CORRESPONDS TO
C            EIGENVALUE J.
C       IZ=INTEGER,THE FIRST DIMENSION OF Z AS DECLARED.
C       IFAIL=INTEGER FLAG, GOVERNS ON ENTRY THE TYPES OF FAILURE
C            THAT WILL BE DETECTED,ON EXIT THE TYPE OF FAILURE.
C            IFAIL=0 ON EXIT FOR SUCCESSFUL COMPLETION.
C
C************************************************************
C
C   NOTE THAT IF NAG LIBRARY ROUTINES ARE NOT AVAILABLE THEN THESE
C      ROUTINES HAVE THEIR EXACT EQUIVALENTS IN MOST SCIENTIFIC
C      LIBRARIES( UNDER DIFFERENT NAMES AND WITH, POSSIBLY, DIFFERENT
C      ARGUMENT LISTS)
C************************************************************
C THOSE INTERESTED IN NUMERICAL EIGENVALUE PROBLEMS SHOULD CONSULT
C MARTIN R.S.,REINSCH C.,WILKINSON J.H.,NUM.MATH.BAND(11)181-95(1968)
C BOWDLER H,MARTIN R.S.,REINSCH C.,WILKINSON J.H,
C                    NUM MATH BAND(11) 293-306 (1968)
C************************************************************
       IFAIL=0
       CALL F01AJF(9,2.0**(-218),RHH,9,EVALH,USE,PHH,9)
       CALL F02AMF(9,2.0**(-37),EVALH,USE,PHH,9,IFAIL)
       IF(IFAIL.EQ.0) GOTO 345
       WRITE(2,2000)
       IFAIL=0
  345  CONTINUE
       CALL F01AJF(9,2.0**(-218),RHC,9,EVALC,USE,RHC,9)
       CALL F02AMF(9,2.0**(-37),EVALC,USE,RHC,9,IFAIL)
       IF(IFAIL.EQ.0) GOTO 346
       WRITE(2,2000)
       IFAIL=0
  346  CONTINUE
       IF(IP.EQ.0) GOTO 349
       IF(IT.GT.1) GOTO 349
       WRITE(2,2005)
       IF(ID.EQ.0) GOTO 347
       WRITE(2,2000) RHA
       GOTO 348
  347  WRITE(2,2007)RHA
  348  CONTINUE
       WRITE(2,2001)
C
C   PRINT EIGENVALUES AND EIGENVECTORS IF REQUIRED
C
       IE=1
       IV=9
```

PRASEODYMIUM

```
      CALL PRINTM(EVALH,1,9,IE)
      WRITE(2,2002)
      CALL PRINTM(RHH,9,9,IV)
      WRITE(2,2003)
      CALL PRINTM(EVALC,1,9,IE)
      WRITE(2,2002)
      CALL PRINTM(RHC,9,9,IV)
  349 CONTINUE
C
CALCULATE THE MAGNETIZATION ON BOTH SITES
C
      IF(ID.EQ.0) GOTO 350
      CALL PRT(9,RHC,EVALC,T,RXO,USE1)
      CALL PRI(9,RHH,EVALH,T,RXO,USE2)
      GOIO 360
  350 CALL PRT(9,RHC,EVALC,T,RZO,USE1)
      CALL PRT(9,RHH,EVALH,T,RZO,USE2)
  360 USE1=-0.80*USE1
      USE2=-0.80*USE2
C
CHECK FOR CONVERGENCE OF MAGNETIZATION
C
      XX=ABS(USE1-RMAGC)
      YY=ABS(USE2-RMAGH)
      IF(XX.GT.0.20) GOTO 390
      IF(YY.LE.0.20) GOTO 400
  390 RMAGC=USE1
      RMAGH=USE2
      GOTO 320
  400 CONTINUE
C
C STORE RESULTS IN ARRAY RES
C
      RES(IH,1)=T
      RES(IH,2)=RHA
      RES(IH,3)=USE1
      RES(IH,4)=USE2
      RES(IH,5)=0.50*(USE1+USE2)
C
CHECK WHETHER REQUISITE NUMBER OF FIELD STEPS HAVE BEEN
COMPLETED AND IF SO OUTPUT THE RESULTS TO LINEPRINTER
C
      IF(IH.LT.NHS) GOTO 310
  900 CONTINUE
C
CONSTRUCT A TABLE OF RESULTS
C
      WRITE(2,2004)
      WRITE(2,2008)
      CALL PRINTM(RES,50,5,NHS)
C
CONSTRUCT A GRAPHICAL DISPLAY ON THE LINEPRINTER
C
      CALL PRG(RES,50,5,IH,2,3,T,A4)
      CALL PRG(RES,50,5,IH,2,4,T,A4)
      CALL PRG(PES,50,5,IH,2,5,T,A4)
```

PRASEODYMIUM

```
      GOTO 300
 940 STOP
1009 FORMAT(I1)
1010 FORMAT(2F6.2)
1012 FORMAT(F7.2)
1008 FORMAT(I3)
1011 FORMAT(2F7.2)
2008 FORMAT(1H0,3X,6HTEMP K,5X,5HFIELD,5X,5HMAG C,5X,5HMAG H,5X
    A,7HAVERAGE)
2000 FORMAT(1H0,22HFAILURE IN NAG F02 AMF)
2001 FORMAT(1H0,37HEIGENVALUES HEXAGONAL SITES IN KELVIN)
2002 FORMAT(1H0,26HCORRESPONDING EIGENVECTORS)
2003 FORMAT(1H0,33HEIGENVALUES CUBIC SITES IN KELVIN)
2004 FORMAT(1H1,11HRESULTS ARE)
2005 FORMAT(1H0,25HAPPLIED FIELD IN TESLA IS)
2006 FORMAT(1H+,28X,F8.2,14HIN X DIRECTION)
2007 FORMAT(1H+,28X,F8.2,14HIN Z DIRECTION)
2900 FORMAT(1H0,31HMAGNETIZATION IS NOT CONVERGING)
2901 FORMAT(1H0,3HIH=,I3,3HIT=,I3,3HIM=,I3)
1200 FORMAT(1H1,27HINPUT DATA FOR PRASEODYMIUM)
1201 FORMAT(1H0,15HLOWEST FIELD IS,F8.2,6H TESLA)
1202 FORMAT(1H0,9HTHERE ARE,I3,9H STEPS OF,F8.2,6H TESLA)
1203 FORMAT(1H+,40X,22HAPPLIED IN X DIRECTION)
1204 FORMAT(1H+,40X,22HAPPLIED IN Z DIRECTION)
1205 FORMAT(1H0,32HMOLECULAR FIELD PARAMETERS USED=,2(F7.2,3X))
     END
     SUBROUTINE PRG(ROUT,NN,MM,NP,IX,IY,TT,A4)
     DIMENSION ROUT(NN,MM),ILINE(120),ICHAR(3)
     LOGICAL A4
     DATA ICHAR(1),ICHAR(2),ICHAR(3)/1H ,1H.,1H*/
     IF(NP.GT.NN) GOTO 900
CCCCCCCCCCCCCCCCCCCCCCCCCCCCCCCCCCCCCCCCCCCCCCCCCCCCCCCCCCC
C
CODE OUTPUTS A SIMPLE GRAPH OF ROUT ONTO LINEPRINTER
C
CCCCCCCCCCCCCCCCCCCCCCCCCCCCCCCCCCCCCCCCCCCCCCCCCCCCCCCCCCC
     ILIM=120
     IF(A4) ILIM=71
     SMALL=10.0**(-10)
     RYM=ROUT(1,IY)
     RYS=0.0
     DO 100 II=1,NP
     IF(ROUT(II,IY).GT.RYM) RYM=ROUT(II,IY)
     IF(ROUT(II,IY).LT.RYS) RYS=ROUT(II,IY)
 100 CONTINUE
     YB=RYS
     AYB=ABS(YB)
     IF(AYB.LT.SMALL) YB=0.0
     AYB=ABS(RYM-YB)
     IF(AYB.LT.SMALL) GOTO 900
     IF(.NOT.A4) RYS=109.0/(RYM-YB)
     IF(A4) RYS=69.0/(RYM-YB)
     IF(IY.EQ.3) GOTO 110
     IF(IY.EQ.4) GOTO 112
     WRITE(2,1902)
     GOTO 116
```

PRASEODYMIUM

```
  110 WRITE(2,1900)
      GOTO 116
  112 WRITE(2,1901)
  116 CONTINUE
      WRITE(2,1903) TT
      IF(IX.EQ.2) GOTO 120
      IF(IX.EQ.1) GOTO 122
      GOTO 130
  120 WRITE(2,2000)
      GOTO 130
  122 WRITE(2,2002)
  130 CONTINUE
  140 WRITE(2,2003)
      DO 320 II=1,ILIM
  320 ILINE(II)=ICHAR(2)
      IF(.NOT.A4) WRITE(2,2500) YB,RYM
      IF(A4) WRITE(2,2503) YB,RYM
      WRITE(2,2501)(ILINE(II),II=1,ILIM)
      RX=55.0/FLOAT(NP)
      ILX=IFIX(RX-0.50)
      DO 400 II=1,NP
      DO 360 JJ=1,ILIM
      ILINE(JJ)=ICHAR(1)
  360 CONTINUE
      ILINE(10)=ICHAR(2)
      Y=(ROUT(II,IY)-YB)*RYS
      Y=Y+0.50
      ILY=IFI/(Y)+10
      IF(ILY.GT.ILIM) ILY=ILIM
      ILINE(ILY)=ICHAR(3)
      WRITE(2,2501)(ILINE(JJ),JJ=1,ILIM)
      WRITE(2,2502) ROUT(II,IX)
      ILINE(ILY)=ICHAR(1)
      ILINE(10)=ICHAR(2)
      IF(ILX.LE.1) GOTO 390
      DO 380 JJ=1,ILX
      WRITE(2,2501)(ILINE(KK),KK=1,ILIM)
  380 CONTINUE
  390 CONTINUE
  400 CONTINUE
 1900 FORMAT(1H1,20X,11HCUBIC SITES)
 1901 FORMAT(1H1,20X,15HHEXAGONAL SITES)
 1902 FORMAT(1H1,20X,23HAVERAGE FROM BOTH SITES)
 1903 FORMAT(1H ,40X,14HTEMPERATURE IS,F8.2,7H KELVIN)
 2000 FORMAT(1H0,12H FIELD TESLA)
 2002 FORMAT(1H0,11H TEM KELVIN)
 2003 FORMAT(1H+,50X,17HMAGNETISATION(BM))
C BM=BOHR MAGNETONS
 2500 FORMAT(1H ,10X,E10.2,90X,E10.2)
 2501 FORMAT(1H ,120A1)
 2502 FORMAT(1H+,F9.3)
 2503 FORMAT(1H ,10X,E10.2,28X,E10.2)

  900 RETURN
      END
      SUBROUTINE PRH(RH,CH,HX,HZ,B2,B4,B6)
```

MAGNETIZATION IN THE CRYSTAL FIELD SYSTEM PRASEODYMIUM

PRASEODYMIUM

```
      DIMENSION RH(9,9)
      REAL B2,B4,B6,HX,HZ,H,G,X,XS,MU,SQP
      INTEGER CH
C
CCCCCCCCCCCCCCCCCCCCCCCCCCCCCCCCCCCCCCCCCCCCCCCCCCCCCCCCCCCC
C
CONSTRUCTS THE HAMILTONIAN MATRICES FOR PRASEODYMIUM AND STORES IN RH
CH IS=0 FOR HEXAGONAL SITES AND=1 FOR CUBIC SITES
CONSTANTS B2 B4 B6 ARE THE THREE CRYSTAL FIELD PARAMETERS
CODE USES HX AND HZ FOR THE TOTAL MAGNETIC FIELDS IN X AND Z DIRE
CTIONS RESPECTIVELY.
C
      DO 100 I=1,9
      DO 100 J=1,9
  100 RH(I,J)=0.0
      RH(1,1)=28.0*B2+14.0*B4+4.0*B6+4.0*HZ
      RH(1,2)=SQRT(8.0)*HX
      RH(2,1)=RH(1,2)
      RH(8,9)=RH(1,2)
      RH(9,8)=RH(8,9)
      RH(1,7)=SQRT(7.0)*5.50*B6
      RH(7,1)=RH(1,7)
      RH(3,9)=RH(1,7)
      RH(9,3)=RH(3,9)
      RH(2,2)=7.0*B2-21.0*B4-17.0*B6+3.0*HZ
      RH(3,2)=SQRT(14.0)*HX
      RH(2,3)=RH(3,2)
      RH(7,8)=RH(2,3)
      RH(8,7)=RH(7,8)
      RH(2,8)=(77.0*B6)/4.0
      RH(8,2)=RH(2,8)
      RH(3,3)=-8.0*B2-11.0*B4+22.00*B6+2.0*HZ
      RH(3,4)=SQRT(18.0)*HX
      RH(4,3)=RH(3,4)
      RH(6,7)=RH(3,4)
      RH(7,6)=RH(6,7)
      RH(4,4)=-17.0*B2+9.0*B4+B6+HZ
      RH(4,5)=SQRT(20.0)*HX
      RH(5,4)=RH(4,5)
      RH(5,6)=RH(4,5)
      RH(6,5)=RH(5,6)
      RH(5,5)=-20.0*B2+18.0*B4-20.0*B6
      RH(6,6)=RH(4,4)-2.0*HZ
      RH(7,7)=RH(3,3)-4.0*HZ
      RH(8,8)=RH(2,2)-6.0*HZ
      RH(9,9)=RH(1,1)-8.0*HZ
      IF(CH.EQ.0) GOTO 900
C
CUBIC SYMMETRY TERMS ADDED NOW
C
      RH(1,4)=SQRT(7.0)*(10.0*B4-5.0*B6)
      RH(4,1)=RH(1,4)
      RH(2,5)=SQRT(70.0)*(3.0*B4+1.250*B6)
      RH(5,2)=RH(2,5)
      RH(3,6)=10.0*B4+8.750*B6
      RH(6,3)=RH(3,6)
```

PRASEODYMIUM

```
      RH(6,9)=-RH(1,4)
      RH(9,6)=RH(6,9)
      RH(5,8)=-RH(2,5)
      RH(8,5)=RH(5,8)
      RH(4,7)=-RH(3,6)
      RH(7,4)=RH(4,7)
  900 RETURN
      END
      SUBROUTINE PRINTM(RMAT,I1,I2,IPOW)
      DIMENSION PMAI(I1,I2)
C
COLUMNS OF MATRIX RMAT(I1,I2) PRINTED BY SUBROUTINE (MAX 12 COLS)
C
      LIM=I2
      IF(I2.GT.LIM) LIM=12
      IF(IROW.GT.I1) GOTO 910
      WRITE(2,2001)
      DO 100 II=1,IROW
      WRITE(2,2000)(RMAT(II,JJ),JJ=1,LIM)
  100 CONTINUE
      GOTO 900
  910 WRITE(2,2003) IROW,I1
 2000 FORMAT(1H ,12F10.4)
 2001 FORMAT(1H0)
 2003 FORMAT(1H0,24HPRINTM CALLED WITH IROW=,I6,5H I1=,I6)
  900 RETURN
      END
      SUBROUTINE PRT(NN,EVEC,EVAL,TEM,ROP,RES)
      DIMENSION EVEC(NN,NN),EVAL(NN),ROP(NN,NN)
      DIMENSION RMOM(20),RU(20)
      REAL NUM,DEN
      INTEGER RR
      IF(NN.GT.20) GOTO 900
CCCCCCCCCCCCCCCCCCCCCCCCCCCCCCCCCCCCCCCCCCCCCCCCCCCCCCCC
C
CALCULATES THE THERMAL AVERAGE OF AN OPERATOR ROP  GIVEN THE
CORECT EIGENVALUES STORED IN ASCENDING ORDER IN EVAL AND THE
CORRESPONDING EIGENVECTORS STORED IN EVEC. TEM IS THE TEMPERATURE.
C
CCCCCCCCCCCCCCCCCCCCCCCCCCCCCCCCCCCCCCCCCCCCCCCCCCCCCCCC
      NUM=0.0
      DEN=0.0
      IF(TEM.LT.0.000000001) GOTO 800
      DO 100 II=1,20
      RMOM(II)=0.0
      RU(II)=0.0
  100 CONTINUE
      DO 200 KK=1,NN
      DO 110 II=1,NN
  110 RU(II)=0.0
      DO 160 II=1,NN
      DO 140 JJ=1,NN
      RU(II)=RU(II)+ROP(II,JJ)*EVEC(JJ,KK)
  140 CONTINUE
  160 CONTINUE
      DO 170 II=1,NN
```

PRASEODYMIUM

```
      RMOM(KK)=RMOM(KK)+EVEC(II,KK)*RU(II)
170 CONTINUE
200 CONTINUE
    NUM=EVAL(1)/TEM
    IF(NUM.LE.-60.0) GOTO 400
    NUM=0.0
    DO 210 II=1,NN
    XX=-EVAL(II)/TEM
    DEN=DEN+EXP(XX)
    NUM=NUM+RMOM(II)*EXP(XX)
210 CONTINUE
    RES=NUM/DEN
    GOTO 900
400 RR=0
    NUM=0.0
    DO 420 II=1,NN
420 IF(EVAL(II).EQ.EVAL(1)) RR=RR+1
    DO 430 II=1,RR
430 NUM=NUM+RMOM(II)
    JJ=RR+1
    XX=EVAL(RR)-EVAL(JJ)
    XX=EXP(XX/TEM)
    NUM=NUM+XX
    DEN=1.0+XX
    PES=NUM/DEN
    GOTO 900
800 DO 820 II=1,20
    RMOM(II)=0.0
820 RU(II)=0.0
    DO 830 II=1,NN
    DO 825 JJ=1,NN
825 RU(II)=RU(II)+ROP(II,JJ)*EVEC(JJ,1)
830 CONTINUE
    DO 840 II=1,NN
    RMOM(1)=RMOM(1)+EVEC(II,1)*RU(II)
840 CONTINUE
    RES=RMOM(1)

    END
```

X DIRECTION RESULTS

```
INPUT DATA FOR PRASEODYMIUM
LOWEST FIELD IS     0.00 TESLA
THERE ARE 40 STEPS OF    2.00 TESLA
                                          APPLIED IN X DIRECTION
MOLECULAR FIELD PARAMETERS USED=  -1.20      4.14
RESULTS ARE
    TEMP K     FIELD      MAG C      MAG H      AVERAGE

    4.2000     0.0000     0.0000    -0.0000     0.0000
    4.2000     2.0000     0.2789     0.5438     0.4113
    4.2000     4.0000     0.5092     1.2042     0.8567
    4.2000     6.0000     0.7421     1.6494     1.1957
    4.2000     8.0000     0.9151     1.9961     1.4556
    4.2000    10.0000     1.0634     2.2234     1.6434
    4.2000    12.0000     1.1603     2.3644     1.7624
    4.2000    14.0000     1.2907     2.5025     1.8966
    4.2000    16.0000     1.3748     2.5829     1.9788
    4.2000    18.0000     1.4722     2.6660     2.0691
    4.2000    20.0000     1.5455     2.7167     2.1311
    4.2000    22.0000     1.6143     2.7593     2.1868
    4.2000    24.0000     1.6897     2.8080     2.2488
    4.2000    26.0000     1.7500     2.8376     2.2938
    4.2000    28.0000     1.8067     2.8635     2.3351
    4.2000    30.0000     1.8601     2.8862     2.3732
    4.2000    32.0000     1.9149     2.9141     2.4145
    4.2000    34.0000     1.9622     2.9313     2.4467
    4.2000    36.0000     2.0069     2.9467     2.4768
    4.2000    38.0000     2.0493     2.9607     2.5050
    4.2000    40.0000     2.0895     2.9734     2.5315
    4.2000    42.0000     2.1289     2.9894     2.5591
    4.2000    44.0000     2.1651     2.9996     2.5824
    4.2000    46.0000     2.1997     3.0090     2.6044
    4.2000    48.0000     2.2327     3.0177     2.6252
    4.2000    50.0000     2.2642     3.0257     2.6449
    4.2000    52.0000     2.2943     3.0332     2.6637
    4.2000    54.0000     2.3232     3.0401     2.6816
    4.2000    56.0000     2.3507     3.0492     2.6999
    4.2000    58.0000     2.3773     3.0550     2.7162
    4.2000    60.0000     2.4028     3.0605     2.7317
    4.2000    62.0000     2.4275     3.0657     2.7466
    4.2000    64.0000     2.4511     3.0705     2.7608
    4.2000    66.0000     2.4740     3.0751     2.7745
    4.2000    68.0000     2.4960     3.0794     2.7877
    4.2000    70.0000     2.5173     3.0834     2.8004
    4.2000    72.0000     2.5378     3.0873     2.8125
    4.2000    74.0000     2.5570     3.0923     2.8247
    4.2000    76.0000     2.5762     3.0957     2.8360
    4.2000    78.0000     2.5948     3.0989     2.8468
```

X DIRECTION RESULTS

```
                 CUBIC SITES
                                TEMPERATURE IS    4.20 KELVIN
    FIELD TESLA
                                       MAGNETISATION(BM)
              0.00E 00                 0.26E 01
    ..............................................................
      8.000
         .                     *
     10.000
         .                      *
     12.000
         .                        *
     14.000
         .                          *
     16.000
         .                           *
     18.000
         .                             *
     20.000
         .                              *
     22.000
         .                                *
     24.000
         .                                 *
     26.000
         .                                   *
     28.000
         .                                    *
     30.000
         .                                     *
     32.000
         .                                       *
     34.000
         .                                        *
     36.000
         .                                          *
     38.000
         .                                           *
     40.000
         .                                            *
     42.000
         .                                             *
     44.000
         .                                              *
     46.000
         .                                               *
     48.000
         .                                                *
     50.000
         .                                                 *
     52.000
         .                                                  *
     54.000
         .                                                   *
     56.000
         .                                                    *
```

X DIRECTION RESULTS

HEXAGONAL SITES

TEMPERATURE IS   4.20 KELVIN

FIELD TESLA

MAGNETISATION(BM)

0.00E 00   0.31E 01

```
..........................................................
            *
   0.000
                  *
   2.000
                        *
   4.000
                           *
   6.000
                              *
   8.000
                                 *
  10.000
                                    *
  12.000
                                       *
  14.000
                                         *
  16.000
                                          *
  18.000
                                           *
  20.000
                                            *
  22.000
                                            *
  24.000
                                            *
  26.000
                                            *
  28.000
                                            *
  30.000
                                            *
  32.000
                                            *
  34.000
                                            *
  36.000
                                            *
  38.000
                                            *
  40.000
                                            *
  42.000
                                            *
  44.000
                                            *
  46.000
                                            *
  48.000
```

MAGNETIZATION IN THE CRYSTAL FIELD SYSTEM PRASEODYMIUM 215

X DIRECTION RESULTS

```
                    AVERAGE FROM BOTH SITES
                              TEMPERATURE IS    4.20 KELVIN
    FIELD TESLA
                                         MAGNETISATION(BM)
               0.00E 00                  0.28E 01
    ............................................................
                  *
       0.000
                     .
                       *
       2.000
                     .
                          *
       4.000
                     .
                            *
       6.000
                     .
                              *
       8.000
                     .
                                *
      10.000
                     .
                                 *
      12.000
                     .
                                   *
      14.000
                     .
                                    *
      16.000
                     .
                                     *
      18.000
                     .
                                      *
      20.000
                     .
                                       *
      22.000
                     .
                                        *
      24.000
                     .
                                         *
      26.000
                     .
                                          *
      28.000
                     .
                                           *
      30.000
                     .
                                            *
      32.000
                     .
                                             *
      34.000
                     .
                                              *
      36.000
                     .
                                               *
      38.000
                     .
                                               *
      40.000
                     .
                                                *
      42.000
                     .
                                                *
      44.000
                     .
                                                *
      46.000
                     .
                                                *
      48.000
```

Z DIRECTION RESULTS

```
INPUT DATA FOR PRASEODYMIUM
LOWEST FIELD IS      0.00 TESLA
THERE ARE 40 STEPS OF    2.00 TESLA
                                        APPLIED IN Z DIRECTION
MOLECULAR FIELD PARAMETERS USED=   -1.20      4.14
RESULTS ARE
  TEMP K      FIELD       MAG C      MAG H      AVERAGE

  4.2000     0.0000      0.0000    -0.0000     0.0000
  4.2000     2.0000      0.1304     0.0005     0.0654
  4.2000     4.0000      0.2398     0.0013     0.1206
  4.2000     6.0000      0.3670     0.0020     0.1845
  4.2000     8.0000      0.4743     0.0032     0.2387
  4.2000    10.0000      0.5945     0.0042     0.2994
  4.2000    12.0000      0.6956     0.0062     0.3509
  4.2000    14.0000      0.8060     0.0081     0.4071
  4.2000    16.0000      0.8994     0.0117     0.4556
  4.2000    18.0000      0.9985     0.0152     0.5068
  4.2000    20.0000      1.0914     0.0196     0.5555
  4.2000    22.0000      1.1671     0.0293     0.5982
  4.2000    24.0000      1.2485     0.0380     0.6433
  4.2000    26.0000      1.3240     0.0496     0.6868
  4.2000    28.0000      1.3907     0.0736     0.7322
  4.2000    30.0000      1.4555     0.0988     0.7772
  4.2000    32.0000      1.5154     0.1355     0.8254
  4.2000    34.0000      1.5706     0.1907     0.8807
  4.2000    36.0000      1.6360     0.3154     0.9757
  4.2000    38.0000      1.6818     0.4683     1.0751
  4.2000    40.0000      1.7556     0.6820     1.2188
  4.2000    42.0000      1.8109     0.9649     1.3879
  4.2000    44.0000      1.8628     1.2676     1.5652
  4.2000    46.0000      1.9070     1.5664     1.7367
  4.2000    48.0000      1.9440     1.8201     1.8821
  4.2000    50.0000      1.9669     2.0170     1.9919
  4.2000    52.0000      2.0003     2.1381     2.0692
  4.2000    54.0000      2.0192     2.2203     2.1198
  4.2000    56.0000      2.0369     2.2700     2.1535
  4.2000    58.0000      2.0535     2.3003     2.1769
  4.2000    60.0000      2.0689     2.3191     2.1940
  4.2000    62.0000      2.0834     2.3312     2.2073
  4.2000    64.0000      2.0970     2.3395     2.2182
  4.2000    66.0000      2.1098     2.3454     2.2276
  4.2000    68.0000      2.1257     2.3504     2.2380
  4.2000    70.0000      2.1368     2.3538     2.2453
  4.2000    72.0000      2.1472     2.3566     2.2519
  4.2000    74.0000      2.1570     2.3591     2.2580
  4.2000    76.0000      2.1663     2.3612     2.2638
  4.2000    78.0000      2.1751     2.3632     2.2691
```

MAGNETIZATION IN THE CRYSTAL FIELD SYSTEM PRASEODYMIUM

Z DIRECTION RESULTS

CUBIC SITES

TEMPERATURE IS 4.20 KELVIN

FIELD TESLA

MAGNETISATION(BM)

0.00E 00   0.22E 01

```
...............................................................
   8.000
                         *
  10.000
                           *
  12.000
                             *
  14.000
                              *
  16.000
                               *
  18.000
                                *
  20.000
                                 *
  22.000
                                  *
  24.000
                                   *
  26.000
                                    *
  28.000
                                     *
  30.000
                                      *
  32.000
                                       *
  34.000
                                        *
  36.000
                                         *
  38.000
                                          *
  40.000
                                           *
  42.000
                                            *
  44.000
                                             *
  46.000
                                              *
  48.000
                                              *
  50.000
                                               *
```

Z DIRECTION RESULTS

HEXAGONAL SITES

TEMPERATURE IS 4.20 KELVIN

FIELD TESLA

MAGNETISATION(BM)

0.00E 00   0.24E 01

....

```
   8.000
           *
  10.000
           *
  12.000
           *
  14.000
           *
  16.000
           *
  18.000
          .*
  20.000
          .*
  22.000
          .*
  24.000
          .*
  26.000
          . *
  28.000
          .  *
  30.000
          .  *
  32.000
          .   *
  34.000
          .    *
  36.000
          .     *
  38.000
          .      *
  40.000
          .       *
  42.000
          .        *
  44.000
          .         *
  46.000
          .          *
  48.000
          .           *
  50.000
          .            *
  52.000
          .             *
```

MAGNETIZATION IN THE CRYSTAL FIELD SYSTEM PRASEODYMIUM 219

Z DIRECTION RESULTS

AVERAGE FROM BOTH SITES
TEMPERATURE IS   4.20 KELVIN
FIELD TESLA

                    0.00E 00   MAGNETISATION(BM)
                                        0.23E 01

```
........................................ ..........................
    8.000
      .         *
   10.000
      .         *
   12.000
      .          *
   14.000
      .           *
   16.000
      .           *
   18.000
      .            *
   20.000
      .            *
   22.000
      .             *
   24.000
      .              *
   26.000
      .               *
   28.000
      .                *
   30.000
      .                 *
   32.000
      .                  *
   34.000
      .                   *
   36.000
      .                    *
   38.000
      .                     *
   40.000
      .                      *
   42.000
      .                       *
   44.000
      .                        *
   46.000
      .                         *
   48.000
      .                          *
   50.000
      .                          *
   52.000
      .                           *
   54.000
```

PART 3

Solid State and Quantum Physics

Physics Programs
Edited by A. D. Boardman
© 1980 John Wiley & Sons Ltd.

CHAPTER 7

Elastic Waves in Crystalline Solids

B. W. JAMES

1. INTRODUCTION

Several fundamental physical properties are related to the propagation of sound waves in solids and an understanding of these processes has led to the development of a number of devices. For example ultrasonic delay lines are widely used in colour television receivers, and diffraction grating dispersive filters are used in radar systems for pulse compression.

The behaviour of sound waves in gases, liquids, amorphous solids, and crystalline solids has been widely investigated. The study of the propagation of sound waves in crystalline solids is the most complex and most interesting of the forms of sound wave to investigate since account must be taken of the anisotropic elastic properties of crystalline solids (Love,[1] Musgrave,[2] and Pollard[3]). In each direction in a crystalline solid there will be three modes of propagation which will, in general, all have different phase velocities, particle motion directions, and energy flow directions. The largest velocity is associated with a longitudinal or nearly longitudinal wave and the other two velocities with transverse or nearly transverse waves. The particle motion directions of the three waves always form an orthogonal set. Pure longitudinal and transverse modes occur in high symmetry directions and in some accidental directions which may be obtained from the known physical properties.

2. TENSOR FORMULATION

2.1 Hooke's law

In order to discuss the elastic properties of anisotropic materials and hence the propagation of sound waves it is necessary to use a tensor formulation of Hooke's law as the stress and strain at a point are given by two second-rank tensors. Hence if σ_{ij} is the stress tensor and e_{kl} is the strain tensor then Hooke's law is written as

$$\sigma_{ij} = c_{ijkl} e_{kl} \quad (i, j, k, l = 1, 2, 3), \tag{1}$$

where c_{ijkl} is a fourth-rank tensor of 81 elements relating 9 stress components to 9 strain components. (Note that summation is assumed for repeated suffices, see Nye[4].)

2.2 The strain tensor

The particle displacements $\mathbf{U}$ of the strained material determine the nine elements of the general strain tensor E_{kl} and

$$E_{kl} = \frac{\partial U_k}{\partial x_l}, \tag{2}$$

where $\mathbf{U} = \mathbf{i}_1 U_1 + \mathbf{i}_2 U_2 + \mathbf{i}_3 U_3 = \mathbf{i}_i U_i$ and $\mathbf{i}_1, \mathbf{i}_2$, and $\mathbf{i}_3$ are unit vectors along the Cartesian axes x_1, x_2, and x_3 respectively.

The general tensor E_{kl} consists of a symmetrical part e_{kl} and an antisymmetrical part w_{kl}, where

$$e_{kl} = \frac{1}{2} \left(\frac{\partial U_k}{\partial x_l} + \frac{\partial U_l}{\partial x_k} \right), \tag{3}$$

and

$$w_{kl} = -\frac{1}{2} \left(\frac{\partial U_k}{\partial x_l} - \frac{\partial U_l}{\partial x_k} \right) = -w_{lk}. \tag{4}$$

Now if we first consider a rotation of the material about the origin of the axes without any deformation of the material, then in this rotation the displacement of any point is perpendicular to its radius vector so that

$$U_i x_i = 0 \quad \text{(scalar product)}, \tag{5}$$

or

$$E_{ij} x_i x_j = 0. \tag{6}$$

Since this is true for all x_i the coefficients on the left-hand side must all be zero. Hence

$$E_{ij} = 0 \quad \text{if } i = j; \quad E_{ij} = -E_{ji} \quad \text{if } i \neq j, \tag{7}$$

which is just the condition for E_{ij} to be antisymmetrical. So that in the special case of a rotation of the material about the origin of the axes without deformation, the general strain tensor E_{kl} becomes antisymmetrical, with the rotation of the material about the origin of the axes given by the antisymmetrical part of the tensor, w_{kl}. Thus, in the second case of a combined deformation of the material and of a rotation of it about the origin of the axes, the general strain tensor E_{kl} has a symmetrical part, e_{kl}, which gives the deformation of the material and an antisymmetrical part, w_{kl}, which gives the rotation of the material about the axes. The symmetrical part of the strain tensor (equation (3)) that does not contain rotation about the origin of the axes, and has only six independent elements, is used in the generalization of Hooke's law. This law relates the state of stress of the material to its state of deformation irrespective of the orientation of the material to the axes. Note that the strains $\partial U_i/\partial x_j$, $i=j$, correspond to longitudinal strains and those of the form $\partial U_i/\partial x_j$, $i \neq j$, correspond to shear strains.

2.3 The stress tensor

The stress tensor at a point can also be reduced to six independent components, and is defined in terms of the force per unit area acting across the faces of a unit cube as the volume of the cube tends to zero. The stresses σ_{ii} arise from forces acting normally to the faces and the stresses σ_{ij}, $i \neq j$, are shear stresses in which the force acts parallel to the plane of the face in the direction of the first subscript and across a plane perpendicular to the direction given by the second subscript. Clearly the forces acting across opposite pairs of faces of the cube of material must be equal and opposite in the case of equilibrium and this also applies in the limit under dynamic conditions. Furthermore, in equilibrium there is no rotation of the material, so by considering moments about each axis in turn,

$$\sigma_{ij} = \sigma_{ji}, \tag{8}$$

and again, this also applies in the limit under dynamic conditions. Thus there are just six independent components of stress.

2.4 The matrix notation

Returning to Hooke's law (equation (1)) it is apparent that in place of a possible 81 components of c_{ijkl} the above symmetry of σ_{ij} and e_{kl} lead to at most 36 independent elastic constants. At this point it is appropriate to introduce the more compact two-suffix notation for the elastic constants, that is used in the literature, in which the tensor for the c_{ijkl} ($i, j, k, l = 1, 2, 3$) is replaced by the matrix c_{ij}, $i, j = 1, 2, \ldots 6$ according to the following scheme of subscript equality:

Tensor notation	11	22	33	23,32	31,13	12,21
Matrix notation	1	2	3	4	5	6.

For example c_{1112} would be replaced by c_{16}. The stress σ_{ij} and the strain e_{kl} are also converted to the matrix notation according to this scheme, except that in the case of the strains a factor of $\frac{1}{2}$ is introduced in the matrix notation when $k \neq l$, so that

$$\sigma_i = c_{ij} e_j, \tag{9}$$

is the matrix form of Hooke's law.

3. ENERGY OF DEFORMATION

The potential energy Φ of the crystal may be expressed as a Taylor series in terms of the strain, and hence

$$\Delta\Phi = \Phi - \Phi(0) = \left(\frac{\partial \Phi}{\partial e_{ij}}\right)_0 e_{ij} + \frac{1}{2}\left(\frac{\partial^2 \Phi}{\partial e_{ij} \partial e_{kl}}\right)_0 e_{ij} e_{kl}$$
$$+ \frac{1}{6}\left(\frac{\partial^3 \Phi}{\partial e_{ij} \partial e_{kl} \partial e_{mn}}\right)_0 e_{ij} e_{kl} e_{mn} + \ldots, \tag{10}$$

where $\Phi(0)$ is the potential energy of the crystal in the equilibrium unstrained state and $\Delta\Phi$ is the change in potential energy from the equilibrium state to the strained state. Since the deformation is about the equilibrium position, $(\partial \Phi/\partial e_{ij})_0$, and, for small strains, terms in $e_{ij} e_{kl} e_{mn}$ and higher order terms may be neglected. For deformation about the equilibrium position $\Delta\Phi$ must be positive, which means that the right-hand side of equation (10) must be a positive quadratic form with the consequence that

$$\left(\frac{\partial^2 \Phi}{\partial e_{ij} \partial e_{kl}}\right)_0 = \left(\frac{\partial^2 \Phi}{\partial e_{kl} \partial e_{ij}}\right)_0. \tag{11}$$

Now during the application of the stress σ_{ij} a certain amount of work ΔW will be done to produce the strain e_{ij} and

$$\Delta W = \tfrac{1}{2} \sigma_{ij} e_{ij}. \tag{12}$$

The work ΔW may be equated to the strain energy $\Delta \Phi$ so that

$$\sigma_{ij} = \left(\frac{\partial^2 \Phi}{\partial e_{ij} \partial e_{kl}}\right)_0 e_{kl}, \tag{13}$$

and hence

$$c_{ijkl} = \left(\frac{\partial^2 \Phi}{\partial e_{ij} \partial e_{kl}}\right)_0 = c_{klij}. \tag{14}$$

which establishes another 15 equalities in the elastic constants and reduces the maximum number of independent constants to 21. It may be noted that the elastic constant tensor c_{ijkl} is equated to the second-order differential of the crystal potential energy function and for this reason they are sometimes referred to as second-order elastic constants. The next term in the Taylor series expansion gives rise to the third-order elastic constants, which give a measure of the anharmonic form of the interatomic forces or their deviation from the harmonic form of an ideal Hooke's law solid.

The number of independent elastic constants for each crystal class and their suffices are given by Nye[4] in graphical form and this is summarized in Table 1.

Table 1. The non-zero elastic constants for the various crystal systems and point groups. (Remember that $c_{ij} = c_{ji}$).

System	Number of point groups	Point group (Schoenflies)	Number of c_{ij}	c_{ij}
Triclinic	2	C_i and C_1	21	
Monoclinic	3	C_{2h}, C_s and C_2	13	$c_{11}, c_{12}, c_{13}, c_{15}, c_{22}, c_{23}, c_{25}, c_{33}, c_{35}, c_{44}, c_{46}, c_{55}$ and c_{66}
Orthorhombic	3	D_{2h}, C_{2v} and D_2	9	$c_{11}, c_{12}, c_{13}, c_{22}, c_{23}, c_{33}, c_{44}, c_{55}$ and c_{66}
Tetragonal	3	C_{4h}, S_4 and C_4	7	$c_{11} = c_{22}, c_{12}, c_{13} = c_{23}, c_{16} = -c_{26}, c_{33}, c_{44} = c_{55}$ and c_{66}
Tetragonal	4	D_{4h}, D_{2d}, C_{4v} and D_4	6	$c_{11} = c_{22}, c_{12}, c_{13} = c_{23}, c_{33}, c_{44} = c_{55}$ and c_{66}
Trigonal	2	C_{3i} and C_3	7	$c_{11} = c_{22}, c_{12}, c_{13} = c_{23}, c_{14} = -c_{24}, -c_{15} = c_{25}, c_{33}$ and $c_{44} = c_{55}$ whilst $c_{46} = 2c_{25}, c_{56} = 2c_{14}$ and $c_{66} = \frac{1}{2}(c_{11} - c_{12})$
Trigonal	3	D_{3d}, C_{3v} and D_3	6	$c_{11} = c_{22}, c_{12}, c_{13} = c_{23}, c_{14} = -c_{24}, c_{33}$ and $c_{44} = c_{55}$ whilst $c_{56} = 2c_{14}$ and $c_{66} = \frac{1}{2}(c_{11} - c_{12})$
Hexagonal	7	$D_{6h}, D_{3h}, C_{6v}, D_6, C_{6h}, C_{3h}$ and C_6	5	$c_{11} = c_{22}, c_{12}, c_{13} = c_{23}, c_{33}$ and $c_{44} = c_{55}$ whilst $c_{66} = \frac{1}{2}(c_{11} - c_{12})$
Cubic	5	O_h, T_d, O, T_h, and T	3	$c_{11} = c_{22} = c_{33}, c_{12} = c_{13} = c_{23}$ and $c_{44} = c_{55} = c_{66}$

4. THE WAVE EQUATION

The initial step in the derivation of the wave equation and hence of an expression for the velocity of the waves involves, as usual, the equations of motion for an element of the material, but now these must be expressed in the tensor format so that

$$\frac{\partial \sigma_{ij}}{\partial x_i} = \rho u_j, \tag{15}$$

where ρ is the density of the material. Now substituting equations (1) and (3) into (15) eliminates the stress components so that

$$\frac{1}{2} c_{ijkl} \frac{\partial}{\partial x_i} \left(\frac{\partial U_k}{\partial x_l} + \frac{\partial U_l}{\partial x_k} \right) = \rho u_j. \tag{16}$$

If it is now assumed that for an infinite anisotropic elastic solid the solution of equation (16) is a plane wave with constant amplitude $\mathbf{A}$, propagation wave number k, with wavefronts normal to a vector with direction cosines l_1, l_2, and l_3, then:

where v is the phase velocity ω/k in the direction given by l_i. The components of the amplitude may be denoted by α_i so that:

$$\mathbf{A} = \mathbf{i}_i \alpha_i. \tag{18}$$

Then when the assumed solution is substituted into equation (16) the following homogeneous equations result:

$$(\Gamma_{jk} - \delta_{jk} \rho v^2) \alpha_j = 0, \tag{19}$$

where

$$\Gamma_{jk} = \tfrac{1}{2} l_i l_l (c_{ijkl} + c_{ijlk}), \tag{20}$$

and

$$\delta_{jk} = 0, \quad \text{for } k \neq j; \qquad \delta_{jk} = 1 \quad \text{for } k = j. \tag{21}$$

These equations have a non-trivial solution only if the secular equation:

$$|\Gamma_{ij} - \delta_{ij} \rho v^2| = 0, \tag{22}$$

is satisfied. This equation is cubic in ρv^2 with three real roots, and from them the velocities v_1, v_2, and v_3 are obtained. The largest velocity corresponds to a longitudinal, or nearly longitudinal, wave and the other two velocities are to transverse or nearly transverse waves. The three values of ρv^2 are, of course, the eigenvalues of the matrix Γ_{jk}, and the corresponding solutions for the displacement vector $\mathbf{A}$ are the eigenvectors. For each value of v there is a corresponding solution, for which the relative values of the displacement components can be determined from equation (19).

The three displacement vectors or particle motion vectors are:

$$\mathbf{A}_1 = d_1 \mathbf{i}_i a_i; \quad \mathbf{A}_2 = d_2 \mathbf{i}_i b_i; \quad \mathbf{A}_3 = d_3 \mathbf{i}_i c_i, \qquad (23)$$

corresponding to three roots ρv_1^2, ρv_2^2, and ρv_3^2 respectively; d_1, d_2, and d_3 are constants dependent on the excitation, and

$$\frac{a_2}{a_1} = \frac{\Gamma_{22}(\Gamma_{11} - \rho v_1^2) - \Gamma_{12}\Gamma_{13}}{\Gamma_{13}(\Gamma_{22} - \rho v_1^2) - \Gamma_{12}\Gamma_{23}}$$

$$\frac{a_3}{a_1} = \frac{\Gamma_{23}(\Gamma_{11} - \rho v_1^2) - \Gamma_{12}\Gamma_{13}}{\Gamma_{12}(\Gamma_{33} - \rho v_1^2) - \Gamma_{23}\Gamma_{13}}, \qquad (24)$$

and

$$a_1^2 + a_2^2 + a_3^2 = 1.$$

There are similar expressions associated with v_2 and v_3 for the b_i and the c_i. The energy flow vector, as defined by Love,[1] is

$$E_i = \sigma_{ij} U_j, \qquad (25)$$

It represents the energy flow per unit time across a surface of unit area normal to this vector. An analogue of the Poynting's vector for the direction of energy flow is obtained when equations (1), (3), and (23) are substituted into (25) and the result is averaged over a cycle, namely:

$$S_i = \frac{1}{2} d_1^2 \frac{\omega^2}{v_1} c_{ijkl} a_j (l_l a_k + l_k a_l), \qquad (26)$$

with similar expressions for the other two waves.

5. THE COMPUTER PROGRAM

5.1 Program description

The program given here permits the calculation of the velocity, the particle motion direction and the energy flow direction of each mode of propagation for any crystal class in any direction, from the appropriate set of elastic constants and the density of the material. This is done by solving equations (22), (24), and (26). To specify a direction in the crystal two angular coordinates are used rather than the direction cosines used in the calculations, as the former are more easily visualized. The angles used are the zenith angle θ and the azimuth angle ϕ, where θ is the angle between the direction and the positive z-axis and ϕ is the angle between the projection of the direction on to the x–y plane and the positive x-axis. The calculations are performed by the program in a region specified by lower and upper angular limits and divided into a number of steps or intervals.

The master segment controls the stepping of the directions within the range given and the subroutine VELOCT is used to calculate the velocities from the elastic constants and density for the direction specified. Within the subroutine VELOCT, after setting up the appropriate equations, two library subroutines FO1AJF and FO2AMF are used to obtain the roots of equation (22), namely the eigenvalues. Both library subroutines are from the NAG library, but any equivalent subroutines could be used. The subroutine FO1AJF gives the Householder reduction of a real symmetric matrix to tridiagonal form for use by FO2AMF. The subroutine FO2AMF calculates all the eigenvalues and eigenvectors of the real symmetric tridiagonal matrix that has been produced by FO1AJF. The velocities v_i are then calculated from the square root of the eigenvalue divided by the density, after which control returns to the master segment. The particle motion direction is obtained from the eigenvectors produced by FO2AMF in VELOCT, instead of from equation (24), and the energy flow direction is obtained by evaluation of equation (26). The subroutine DIRECT is used to obtain the spherical polar directions from the direction cosine data.

The output from the program consists of the input data and, for each direction in the region specified, the three velocities and the associated particle motion directions and energy flow directions. In the output all directions are given both in terms of angles θ and ϕ, and in terms of the direction cosines l_1, l_2, and l_3. If the elastic constants are given in GN m^{-2} and the density in 1000 kg m^{-3}, then the velocity values will appear in km s^{-1} since $v = (F/\rho)^{\frac{1}{2}}$ where v is the velocity, F is the effective elastic constant for the mode, and ρ is the density. The effective elastic constant F has the same units as the elastic constants c_{ij} and is a function of them dependent on the mode considered.

5.2 Running the program

The elastic constants and densities of a number of materials are given in Table 2. Following the listing of the program a sample set of data based upon Table 2 is given, together with the associated output produced by a run of the program. The crystallographic axes for a tetragonal material, such as $CaMoO_4$, are an orthogonal set, with the z-axis parallel to the axis of fourfold rotational symmetry. The x- and y-axes are equivalent and they are chosen to be in the direction of the two equal dimensions of the smallest unit cell. A number of general observations may be made for sound wave propagation in tetragonal materials. For propagation along the z-axis $l_1 = 0$, $l_2 = 0$, and $l_3 = 1$, and after appropriate substitution into equations (19) and (22) it is clear that the three waves are all pure modes since the matrix is diagonal, and simple expressions can be obtained for the velocities. For propagation in the x–y plane one of the transverse waves is a pure mode

Table 2. The elastic constants and densities of some crystalline solids. All values of elastic constants are given, in GN m^{-2}, and only zero values have been omitted from the table. Note that the last row gives the source of reference

Material Point group	CaWO$_4$ C$_{4h}$	CaMoO$_4$ C$_{4h}$	SrMoO$_4$ C$_{4h}$	MgF$_2$ D$_{4h}$	CaSO$_4$, 2H$_2$O C$_{2h}$	NaBrO$_3$ T	Cu O$_h$
c_{11}	143.87	143.92	115.46	124.0	78.6	55.60	169.0
c_{12}	63.501	68.61	59.85	73.0	41.0	16.76	122.0
c_{13}	56.170	48.43	44.36	54.0	26.8	16.76	122.0
c_{14}							
c_{15}					−7.0		
c_{16}	16.355	12.72	12.09				
c_{22}	143.87	143.92	115.46	124.0	62.7	55.60	169.0
c_{23}	56.170	48.43	44.36	54.0		16.76	122.0
c_{24}					24.2		
c_{25}					3.1		
c_{26}	−16.355	−12.72	−12.09				
c_{33}	130.18	125.86	104.19	177.0	72.6	55.60	169.0
c_{34}							
c_{35}					−17.4		
c_{36}							
c_{44}	33.609	36.91	34.99	55.4	9.10	15.08	75.5
c_{45}							
c_{46}					−1.6		
c_{55}	33.609	36.91	34.99	55.4	26.4	15.08	75.5
c_{56}							
c_{66}	45.073	46.07	47.55	97.8	10.44	15.08	75.5
Density kg m^{-3}	6120	4255	4540	3177	2280	3339	8930
Reference	5	6	7	8	9	10	11

with particle motion parallel to the z-axis. According to Neighbours and Schacher,[12] the other two modes of propagation in this plane are both semi-pure. That is to say the direction of energy flow lies within the x–y plane, although it is not parallel to the direction of propagation. They also show that there are two pure mode directions in the x–y plane at 45° to each other, and the amount by which these are rotated from the x-axis depends on the magnitude of c_{16}. In planes defined by these directions and the z-axis, propagation is again by semi-pure modes, and accidental pure modes may occur in either of these planes at certain angles θ from the z-axis. Because of the crystal symmetry these planes are repeated at 90° intervals. By use of the computer program it is possible to investigate these and other features of wave propagation in crystalline materials.

Neglect of the features of energy flow in tetragonal materials can lead to incorrect identification of the wave modes in experimental measurements and consequent incorrect values of elastic constants, as the direction of propagation and the direction of energy flow may differ by up to 60°,

James.[7] These problems can be particularly acute when measurements are carried out on small specimens and unintended reflections can occur which give rise to misleading values for velocity. An extensive discussion of the features of the propagation of sound in solids is given by Musgrave.[2]

6. FURTHER DEVELOPMENT

Apart from the intrinsic interest in the results of this program there are other uses for the VELOCT subroutine of which two examples are given below:

(1) The Debye characteristic temperature θ_0 of a crystalline solid at 0 K may be calculated from the elastic constants and the density at 0 K using the expression:

$$\theta_0 = \frac{h}{k}\left(\frac{9N}{4\pi V}\right)^{\frac{1}{3}} \left(\int_0^{4\pi} \sum_{j=1}^{3} \frac{1}{v_j^3} \frac{d\Omega}{4\pi} \right)^{-\frac{1}{3}}, \qquad (27)$$

where N is Avogadro's number, h is Planck's constant, k is Boltzmann's constant, V is the molar volume at 0 K, and v_j are the velocities of the two transverse and longitudinal waves. The integration is carried out numerically with the aid of the VELOCT subroutine from this program. A program to evaluate θ_0 has been written by Gluyas, Hughes, and James[5] and from their elastic constant data for $CaWO_4$ extrapolated to 0 K they found $\theta_0 = 246.5$ K. In this case, in place of Avogradro's number N, $2N$ is used since $CaWO_4$ is taken to be 'diatomic' and consists of Ca^{2+} and WO_4^{2-} ions.

(2) Another application of the VELOCT subroutine of this program has been in the evaluation of the optimum values of the elastic constants of $CaMoO_4$ from experimental measurements of the velocities of ultrasonic waves (James[6]). The optimum set of elastic constants has been obtained from an over-determined set of measurements by use of the subroutine VELOCT with an iterative minimization subroutine (EO4CCF from the NAG library). The minimization routine was used to adjust the calculated c_{ij} to obtain the lowest value of

$$\text{SUMSQ} = \sum_{i=1}^{n} \left| \frac{v_i^2(\text{calc.})}{v_i^2(\text{meas.})} - 1 \right|, \qquad (28)$$

where v_i (calc.) is the value of the velocity calculated by the subroutine VELOCT from the current values of the c_{ij} and the density, v_i (meas.) is the measured value of that velocity, and n is the number of measurements that have been made. It can be shown that, for an appropriate set of velocity measurements taken from suitably orientated specimens, only as many velocities as the number of independent elastic constants have to be measured. However, once a particular orientation of specimen has been produced it is relatively straightforward to measure the velocity of all three modes and then an over-determined set of measurements becomes available for which this method is ideal.

REFERENCES

1. A. E. R. Love, *The Mathematical Theory of Elasticity* (Cambridge University Press, London 1927).
2. M. J. P. Musgrave, *Crystal Acoustics* (Holden Day, London, 1970).
3. H. F. Pollard, *Sound Waves in Solids* (Pion, London, 1977).
4. J. F. Nye, *Physical Properties of Crystals* (Oxford, London, 1957).
5. M. Gluyas, F. D. Hughes, and B. W. James, *J. Phys. D. Appl. Phys.*, **6,** 2025 (1973).
6. B. W. James, *J. Appl. Phys.*, **45,** 3201 (1974).
7. B. W. James, *Phys. Stat. Sol.* (a), **13,** 89 (1972).
8. H. R. Cutler, J. J. Gibson and K. A. McCarthy, *Solid State Commun.*, **6,** 431 (1968).
9. S. Haussuhl, *Z. Krist.*, **122,** 311 (1965).
10. M. Gluyas, R. Hunter and B. W. James, *J. Phys. D. Appl. Phys.*, **8,** 1 (1975).
11. Y. Hiki and A. V. Granato, *Phys. Rev.*, **144,** 411 (1966).
12. J. R. Neighbours and G. E. Schacher, *J. Appl. Phys.*, **38,** 5366 (1967).

VELOCITY PROGRAM

```
C     THE VELOCITY OF ULTRASONIC WAVES IN ANY DIRECTION IN A SOLID
C     ARE CALCULATED. THE DIRECTIONS IN WHICH THE VELOCITIES ARE
C     CALCULATED ARE SPECIFIED BY THE POLAR DIRECTIONS THETA AND PHI.
C     THE CALCULATIONS ARE DONE FOR A SERIES OF THETA AND PHI VALUES
C     GIVEN BY LOWER ANGULAR LIMITS TL AND PL, BY UPPER ANGULAR LIMITS
C     TU AND PU, AND BY THE NUMBER OF INTERMEDIATE DIRECTIONS NT AND NP.
      REAL LL
      DIMENSION C(6,6),LL(3),V(3),A(3,3),D(3),G(3,3),IA(20),
     1NOT(3,3),SS(3,3)
      DATA NOT(1,1),NOT(1,2),NOT(1,3),NOT(2,1),NOT(2,2),
     1NOT(2,3),NOT(3,1),NOT(3,2),NOT(3,3)/1,6,5,6,2,4,5,4,3/
      PI=3.141592654
      Z=PI/180.0
C     READ INPUT DATA
   23 READ(5,104)IC
      IF (IC-1) 1000,33,1
C     IC IS USED AS A CONTROL INTEGER AS FOLLOWS.
C     FOR IC=2 NEW TITLE,RHO,CIJ,TL,TH,NT,PL,PH,AND NP ARE SOUGHT,
C     IC MUST BE SET EQUAL TO 2 AT THE BEGINNING OF THE DATA.
C     FOR IC=1 NEW TL,TH,NT,PL,PH,AND NP ARE SOUGHT FOR USE WITH
C     THE PREVIOUS RHO AND CIJ.
C     FOR IC=0 RUN OF PROGRAM IS ENDED.
    1 READ(5,100)IA
      READ(5,101)RHO
      WRITE(6,200)IA
      WRITE(6,201)RHO
      WRITE(6,208)
      DO 2 I=1,6
      READ(5,103)(C(I,J),J=1,6)
    2 WRITE(6,202)(C(I,J),J=1,6)
   33 READ(5,102)TL,TH,NT
      READ(5,102)PL,PH,NP
      WRITE(6,203)TL,TH,NT
      WRITE(6,204)PL,PH,NP
      WRITE(6,206)
  100 FORMAT(20A4)
  101 FORMAT(1F0.0)
  102 FORMAT(2F0.0,I0)
  103 FORMAT(6F0.0)
  104 FORMAT(1I0)
  200 FORMAT(1H1,20A4)
  201 FORMAT(1H ,/,' DENSITY = ',F10.4,1H ,'*1000 KGM-3',/)
  202 FORMAT(1H ,'CIJ MATRIX',6F10.4)
  203 FORMAT(1H ,/,' THETA RANGE',F8.4,1H ,'TO',F8.4,1H ,'IN',I6,1H ,
     1'STEPS')
  204 FORMAT(1H ,'PHI RANGE  ',F8.4,1H ,'TO',F8.4,1H ,'IN',I6,1H ,
     1'STEPS')
  205 FORMAT(1H ,/,' MODE VELOCITY',F8.3,1H ,'KM/S')
  206 FORMAT(1H ,///,36X,'THETA           PHI      L       L2      L3',/)
  207 FORMAT(1H ,/,' ONE EIGENVALUE NEEDS MORE THAN 30 ITERATIONS',/)
  208 FORMAT(1H ,'ELASTIC CONSTANTS GNM-2')
      FNT=FLOAT(NT)
      FNP=FLOAT(NP)
      TS=(TH-TL)/FNT*Z
      PS=(PH-PL)/FNP*Z
      T=TL*Z
```

VELOCITY PROGRAM

```
      DO 3 I=1,NT
      P=PL*Z
      DO 4 J=1,NP
C     CACULATE VELOCITIES V,AND DISPLACEMENT VECTORS A
      IFAIL=1
      CALL VELOCT(RHO,T,P,C,V,LL,A,NOT,IFAIL)
      THETA=T/Z
      PHI=P/Z
      WRITE(6,209)THETA,PHI,LL(1),LL(2),LL(3)
      IF (IFAIL) 6,6,5
    5 WRITE(6,207)
      GOTO 4
    6 CONTINUE
  209 FORMAT(1H ,//,' PROPAGATION DIRECTION ',9X,5F10.4)
      DO 10 NR=1,3
C     CALCULATE ENERGY VECTORS S
      DO 11 IT=1,3
      SUM=0.0
      DO 12 JT=1,3
      DO 13 KT=1,3
      DO 14 LT=1,3
      M=NOT(IT,JT)
      N=NOT(KT,LT)
      SUM=SUM+C(M,N)*A(JT,NR)*(LL(LT)*A(KT,NR)+LL(KT)*A(LT,NR))
   14 CONTINUE
   13 CONTINUE
   12 CONTINUE
      SS(IT,NR)=SUM
   11 CONTINUE
   10 CONTINUE
C     PRINT FOR EACH VELOCITY THE PARTICLE MOTION AND
C     ENERGY FLOW DIRECTIONS AND FINALLY THE VELOCITY
      DO 15 NR=1,3
      WRITE(6,205)V(NR)
      CALL DIRECT(A,NR,210,1E-4)
      CALL DIRECT(SS,NR,211,1E-4)
   15 CONTINUE
    4 P=P+PS
    3 T=T+TS
      GOTO 23
C     CONTROL NOW RETURNS TO THE BEGINNING OF THE PROGRAM.
C     A NEW VALUE OF THE CONTROL INTEGER IS USED FOLLOWS.
C     TO END THE RUN OF THE PROGRAM, IC=0
C     TO ENABLE NEW RANGES OF THETA AND PHI TO BE SELECTED, IC=1
C     TO RUN THE PROGRAM FOR ANOTHER MATERIAL, IC=2
 1000 CONTINUE
      STOP
      END
      SUBROUTINE VELOCT(RHO,T,P,C,V,LL,A,NOT,IFAIL)
C     RHO IS THE DENSITY,T AND P ARE THE POLAR COORDINATES
C     THETA AND PHI IN RADIANS C IS A 6*6 ARRAY OF THE
C     ELASTIC CONSTANTS,V IS AN ARRAY OF THE 3 VELOCITIES
C     AND LL IS AN ARRAY OF THE 3 DIRECTION COSINES AND A IS
C     A 3*3 ARRAY OF THE EIGENVECTORS
      REAL LL
      DIMENSION NOT(3,3),C(6,6),G(3,3),A(3,3),LL(3),V(3),D(3),E(3)
```

VELOCITY PROGRAM

```
C     CALCULATE DIRECTION COSINES OF PROPAGATION DIRECTION
      LL(1)=SIN(T)*COS(P)
      LL(2)=SIN(T)*SIN(P)
      LL(3)=COS(T)
C     SET UP CHRISTOFFEL MATRIX
      DO 1 JS=1,3,1
      DO 2 KS=1,3,1
      SUMS=0.0
      DO 3 IS=1,3,1
      DO 4 LS=1,3,1
      MS=NOT(IS,JS)
      NS=NOT(KS,LS)
      SUMS=SUMS+LL(IS)*LL(LS)*C(MS,NS)
    4 CONTINUE
    3 CONTINUE
      G(JS,KS)=SUMS
      IF(JS.NE.KS) G(KS,JS)=G(JS,KS)
    2 CONTINUE
    1 CONTINUE
      CALL F01AJF(3,2.0**(-218),G,3,D,E,A,3)
      CALL F02AMF(3,2.0**(-37),D,E,A,3,IFAIL)
      IF (IFAIL) 5,5,6
    5 DO 7 JS=1,3
    7 V(JS)=SQRT(D(JS)/RHO)
    6 RETURN
      END
      SUBROUTINE DIRECT(A,NR,JW,EPS)
C     FOR A(I,R) WHERE I=1,2,3 IN WHICH THE A(I,R) ARE NOT
C     NECESSARILY NORMALIZED THE CORRESPONDING SPHERICAL
C     POLAR CO-ORDINATES THETA AND PHI ARE CALCULATED.THETA
C     PHI AND THE NORMALIZED DIRECTION COSINES ARE PRINTED
C     JW CONTROLS THE WRITE STATEMENT THAT IS USED
      REAL L
      DIMENSION L(3),A(3,3)
      PI=3.14159265+
      Z=PI/180.0
      S=0.0
      DO 1 ID=1,3
    1 IF(ABS(A(ID,NR)).GT.S) S=ABS(A(ID,NR))
C     REDUCE RANGE OF DIRECTION COSINES TO 0 TO 1
      DO 2 ID=1,3
    2 L(ID)=A(ID,NR)/S
      S=0.0
      DO 3 ID=1,3
    3 S=S+L(ID)**2
      S=SQRT(S)
C     NORMALIZE DIRECTION COSINES
      DO 4 ID=1,3
    4 L(ID)=L(ID)/S
C     CALCULATE THETA AND PHI FROM NORMALIZED DIRECTION COSINES
      IF(ABS(L(3)).GT.EPS) GO TO 5
      THETA=PI/2.0
      GO TO 7
    5 IF(ABS(L(3)).LT.(1.0-EPS)) GO TO 6
      THETA=0.0
      PHI=0.0
```

VELOCITY PROGRAM

```
C     PHI HAS NO MEANING FOR THETA=0.0 THEREFORE SET PHI=0.0
      GO TO 15
    6 THETA=ATAN(SQRT(1.0-L(3)**2)/L(3))
C     NOTE THETA DEFINED 0.0 TO PI WHEREAS ATAN RANGE -PI/2 TO +PI/2
      IF (THETA.GT.0.0) GO TO 7
      THETA=PI+THETA
    7 PHI=0.0
      IF (L(1)) 13,8,9
    8 IF (L(2)) 10,15,11
    9 IF (L(2)) 12,14,14
   10 PHI=PI
   11 PHI=PHI+PI/2.0
      GO TO 15
   12 PHI=PI
   13 PHI=PHI+PI
   14 PHI=PHI+ATAN(L(2)/L(1))
   15 CONTINUE
      THETA=THETA/2
      PHI=PHI/2
      IF (JW-210) 16,16,17
   16 WRITE(6,210)THETA,PHI,L(1),L(2),L(3)
      GOTO 18
   17 WRITE(6,211)THETA,PHI,L(1),L(2),L(3)
  210 FORMAT(1H ,"PARTICLE MOTION",15X,5F10.4)
  211 FORMAT(1H ,"ENERGY FLOW",19X,5F10.4)
   18 CONTINUE
      RETURN
      END
```

DATA FOR THE VELOCITY PROGRAM

```
2
CALCIUM MOLYBDATE DATA FROM B W JAMES J APPL PHYS VOL 45 3201 (1974)
4.255
143.92  68.61  48.43  0.0   0.0   12.72
 68.61 143.92  48.43  0.0   0.0  -12.72
 48.43  48.43 125.86  0.0   0.0    0.0
  0.0    0.0    0.0  39.91  0.0    0.0
  0.0    0.0    0.0   0.0  39.91   0.0
 12.72 -12.72   0.0   0.0   0.0   46.07
90.0  95.0  1
 0.0  90.0  6
0
```

RESULTS FROM THE VELOCITY PROGRAM

CALCIUM MOLYBDATE DATA FROM B W JAMES J APPL PHYS VOL 45 3201 (1974)

DENSITY =   4.2550 *1000 KGM-3

ELASTIC CONSTANTS GNM-2
```
CIJ MATRIX  143.9200    68.6100    48.4300     0.0000     0.0000    12.7200
CIJ MATRIX   68.6100   143.9200    48.4300     0.0000     0.0000   -12.7200
CIJ MATRIX   48.4300    48.4300   125.8600     0.0000     0.0000     0.0000
CIJ MATRIX    0.0000     0.0000     0.0000    39.9100     0.0000     0.0000
CIJ MATRIX    0.0000     0.0000     0.0000     0.0000    39.9100     0.0000
CIJ MATRIX   12.7200   -12.7200     0.0000     0.0000     0.0000    46.0700
```

THETA RANGE 90.0000 TO 95.0000 IN   1 STEPS
PHI RANGE   0.0000 TO 90.0000 IN   6 STEPS

	THETA	PHI	L1	L2	L3
PROPAGATION DIRECTION	90.0000	0.0000	1.0000	0.0000	-0.0000
MODE VELOCITY 3.063 KM/S					
PARTICLE MOTION	0.0000	0.0000	0.0000	-0.0000	1.0000
ENERGY FLOW	90.0000	360.0000	1.0000	-0.0000	-0.0000
MODE VELOCITY 3.232 KM/S					
PARTICLE MOTION	90.0000	97.2868	-0.1268	0.9919	0.0000
ENERGY FLOW	90.0000	328.9673	0.8569	-0.5155	-0.0000
MODE VELOCITY 5.849 KM/S					
PARTICLE MOTION	90.0000	7.2868	0.9919	0.1268	-0.0000
ENERGY FLOW	90.0000	10.4099	0.9835	0.1807	-0.0000
PROPAGATION DIRECTION	90.0000	15.0000	0.9659	0.2588	-0.0000
MODE VELOCITY 2.749 KM/S					
PARTICLE MOTION	90.0000	289.5796	0.3351	-0.9422	0.0000
ENERGY FLOW	90.0000	347.2489	0.9753	-0.2207	-0.0000
MODE VELOCITY 3.063 KM/S					
PARTICLE MOTION	0.0000	0.0000	0.0000	0.0000	1.0000
ENERGY FLOW	90.0000	15.0000	0.9659	0.2588	-0.0000
MODE VELOCITY 6.091 KM/S					
PARTICLE MOTION	90.0000	199.5796	-0.9422	-0.3351	0.0000
ENERGY FLOW	90.0000	21.1168	0.9328	0.3603	-0.0000
PROPAGATION DIRECTION	90.0000	30.0000	0.8660	0.5000	-0.0000
MODE VELOCITY 2.597 KM/S					
PARTICLE MOTION	90.0000	298.8262	0.4822	-0.8761	-0.0000
ENERGY FLOW	90.0000	38.6282	0.7812	0.6243	-0.0000

RESULTS FROM THE VELOCITY PROGRAM

```
MODE VELOCITY   3.063 KM/S
PARTICLE MOTION              0.0000     0.0000    0.0000    0.0000    1.0000
ENERGY FLOW                 90.0000    30.0000    0.8660    0.5000   -0.0000

MODE VELOCITY   6.157 KM/S
PARTICLE MOTION             90.0000   208.8262   -0.8761   -0.4822    0.0000
ENERGY FLOW                 90.0000    28.4542    0.8792    0.4765   -0.0000


PROPAGATION DIRECTION       90.0000    45.0000    0.7071    0.7071   -0.0000

MODE VELOCITY   2.919 KM/S
PARTICLE MOTION             90.0000   308.7462    0.6259   -0.7799   -0.0000
ENERGY FLOW                 90.0000    77.4021    0.2181    0.9759   -0.0000

MODE VELOCITY   3.063 KM/S
PARTICLE MOTION              0.0000     0.0000    0.0000   -0.0000    1.0000
ENERGY FLOW                 90.0000    45.0000    0.7071    0.7071   -0.0000

MODE VELOCITY   6.011 KM/S
PARTICLE MOTION             90.0000   218.7462   -0.7799   -0.6259    0.0000
ENERGY FLOW                 90.0000    36.4858    0.8040    0.5946   -0.0000


PROPAGATION DIRECTION       90.0000    60.0000    0.5000    0.8660   -0.0000

MODE VELOCITY   3.063 KM/S
PARTICLE MOTION              0.0000     0.0000   -0.0000    0.0000    1.0000
ENERGY FLOW                 90.0000    60.0000    0.5000    0.8660   -0.0000

MODE VELOCITY   3.416 KM/S
PARTICLE MOTION             90.0000   323.6264    0.8052   -0.5930    0.0000
ENERGY FLOW                 90.0000    85.2760    0.0824    0.9966   -0.0000

MODE VELOCITY   5.743 KM/S
PARTICLE MOTION             90.0000   233.6264   -0.5930   -0.8052    0.0000
ENERGY FLOW                 90.0000    50.5144    0.6359    0.7718   -0.0000


PROPAGATION DIRECTION       90.0000    75.0000    0.2588    0.9659   -0.0000

MODE VELOCITY   3.063 KM/S
PARTICLE MOTION              0.0000     0.0000    0.0000    0.0000    1.0000
ENERGY FLOW                 90.0000    75.0000    0.2588    0.9659   -0.0000

MODE VELOCITY   3.592 KM/S
PARTICLE MOTION             90.0000   346.9417    0.9741   -0.2259   -0.0000
ENERGY FLOW                 90.0000    67.5301    0.3822    0.9241   -0.0000

MODE VELOCITY   5.635 KM/S
PARTICLE MOTION             90.0000   256.9417   -0.2259   -0.9741    0.0000
ENERGY FLOW                 90.0000    78.0494    0.2071    0.9783   -0.0000
```

Physics Programs
Edited by A. D. Boardman
© 1980 John Wiley & Sons Ltd.

CHAPTER 8

A Computer-assisted Tutorial in Time-independent Non-degenerate Perturbation Theory

D. J. MARTIN

1. INTRODUCTION

Probably the most widely used approximate method in quantum mechanics is that known as perturbation theory. The idea of the following exercise is for the student to calculate the energies of a particle in a particular form of one-dimensional potential well, first with the computer program—which uses a numerical iterative technique to give very accurate results—and then by employing first- and second-order perturbation theory. The student has the incentive of trying to achieve agreement with the computer—rather as the theoretician seeks to achieve results in conformity with experiments.

The form of the potential—an infinite potential well with an extra potential (the perturbation) in part of the well (see Figure 1)—is chosen so that the Schrödinger equation can be solved numerically to a high degree of accuracy and that the perturbation theory calculation is relatively straightforward.

2. BASIC QUANTUM MECHANICS

2.1 Introduction

The central theoretical problem in treating any microscopic system—in which quantum-mechanical effects are significant—is that of finding the solution to the appropriate Schrödinger equation. For a single particle of mass M_p in a time-independent scalar field $V(\mathbf{r})$ the (time-independent) Schrödinger equation is

$$\left[-\frac{\hbar^2}{2M_p}\nabla^2 + V(\mathbf{r})\right]\psi(\mathbf{r}) = E\psi(\mathbf{r}), \tag{1}$$

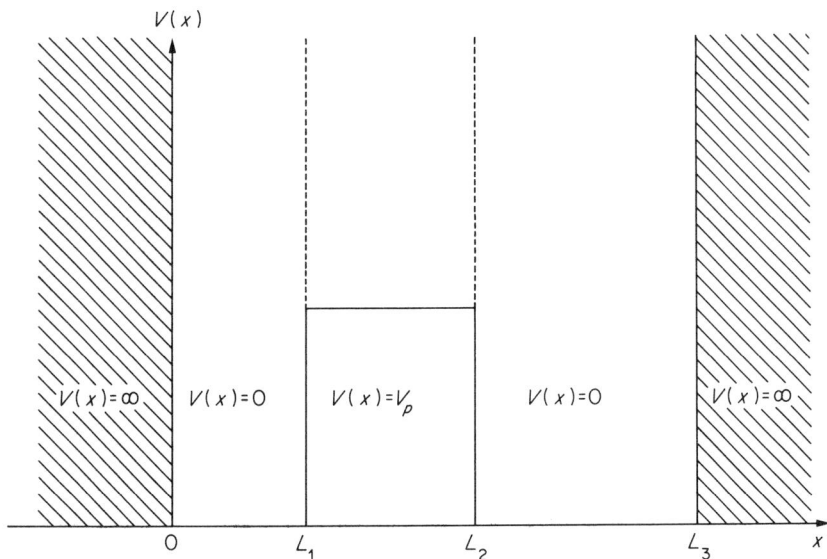

Figure 1. The form of the perturbed infinite potential well. V_p can be >0 or <0

where $\hbar = h/2\pi$, h is Planck's constant ($\hbar = 1.056 \times 10^{-34}$ J s) and ∇^2 is the Laplacian operator (sometimes written Δ) which in Cartesian coordinates is

$$\nabla^2 = \left\{\frac{\partial^2}{\partial x^2} + \frac{\partial^2}{\partial y^2} + \frac{\partial^2}{\partial z^2}\right\}, \qquad (2)$$

$\psi(\mathbf{r})$ is the wave function of the particle. If $\psi(\mathbf{r})$ is known then all possible information about the system can be obtained because the probability density at $\mathbf{r}$ is $\psi^*(\mathbf{r}) \cdot \psi(\mathbf{r})$, where $\psi^*(\mathbf{r})$ is the complex conjugate of $\psi(\mathbf{r})$. Here, E is the energy of the particle but not all values of E are necessarily possible. In fact it can be shown that if the particle is in a bound state (i.e. $\psi(\mathbf{r}) \to 0$ for $r \to \infty$) then E can only take discrete values, i.e. the energy is quantized. For a given $V(\mathbf{r})$ there are generally a number of different $\psi(\mathbf{r})$ and corresponding energies E which will be distinguished with the subscript n in the form $\psi_n(\mathbf{r})$ and E_n. The expression $[-(\hbar^2/2M_p)\nabla^2 - V(\mathbf{r})]$ is called the Hamiltonian ($\mathcal{H}$) and the time-independent Schrödinger equation can be written as $\mathcal{H}\psi_n(\mathbf{r}) = E_n\psi_n(\mathbf{r})$.

Typically M_p and $V(\mathbf{r})$ are known so that it remains to solve equation (1) for $\psi_n(\mathbf{r})$ and to compare E_n with the results of experiment. There are very few problems for which an exact analytical solution has been found for

$\psi_n(\mathbf{r})$. Indeed, in the majority of cases it is necessary to accept an approximate solution where, of course, approximate should not be confused with inaccurate (i.e. $\sin(x) \approx x$ with high accuracy provided $x \ll 1$.)

Probably the approximate method most widely used in quantum theory is that known as perturbation theory. This method is applicable if the Hamiltonian consists of two parts

$$\mathcal{H} = \mathcal{H}_0 + \mathcal{H}_p, \tag{3}$$

where $\mathcal{H}_p$ has only a small effects and is called the perturbation and where the Schrödinger equation, without $\mathcal{H}_p$, can be solved and all the possible $\psi_n^{(0)}(\mathbf{r})$ and corresponding energies $E_n^{(0)}$ can be found. (The (0) superscript on $\psi_n^{(0)}(\mathbf{r})$ and $E_n^{(0)}$ is used to indicate that they correspond to the unperturbed Hamiltonian $\mathcal{H}_0$.) In the absence of the perturbation $\mathcal{H}_p$ equation (1) is therefore:

$$\mathcal{H}_0 \psi_n^{(0)}(\mathbf{r}) = E_n^{(0)} \psi_n^{(0)}(\mathbf{r}). \tag{4}$$

The simplest form of perturbation theory applies when:

(1) $\mathcal{H}_0$ and $\mathcal{H}_p$ do not vary with time; and
(2) the solutions for the unperturbed system are non-degenerate, i.e. for a given energy $E_n^{(0)}$ there is a unique $\psi_n^{(0)}(\mathbf{r})$.

In this exercise we will confine our attention to quantized energy levels—which simplifies the notation. In fact, once the basic ideas have been grasped, the extension of perturbation theory to more general cases, where degeneracy occurs and/or $\mathcal{H}_p$ varies with time, is relatively simple.

2.2 Time-independent non-degenerate perturbation theory

In order to apply perturbation theory we must know $\psi_n^{(0)}(\mathbf{r})$ and $E_n^{(0)}$—the unperturbed wave functions and energies. Then, even if we cannot solve the Schrödinger equation in the presence of the perturbation $\mathcal{H}_p$, i.e.

$$(\mathcal{H}_0 + \mathcal{H}_p)\psi_n(\mathbf{r}) = E_n \psi_n(\mathbf{r}), \tag{5}$$

perturbation theory enables us to make an accurate estimate of $\psi_n(\mathbf{r})$ and E_n in terms of $\psi_n^{(0)}(\mathbf{r})$ and $E_n^{(0)}$, provided that the effects of $\mathcal{H}_p$ are small.

A full explanation and justification of perturbation theory is beyond the scope of this chapter (the subject is discussed in most intermediate-level textbooks on quantum theory), but the basis of the method is to express both E_n and $\psi_n(\mathbf{r})$—the perturbed energy and the perturbed wave function—as power series in what might be called the 'strength' of $\mathcal{H}_p$ (which we will write as λ_p, a dimensionless number), i.e.

$$E_n = \varepsilon_0 + \lambda_p / \varepsilon_1 + \lambda_p^2 / \varepsilon_2 + \ldots, \tag{6}$$
$$\psi_n(\mathbf{r}) = \chi_0(\mathbf{r}) + \lambda_p / \chi_1(\mathbf{r}) + \lambda_p^2 / \chi_2(\mathbf{r}) + \ldots.$$

Remembering that for $\mathcal{H}_p = 0$ we know the energy ($= E_n^0$) and the wave function ($= \psi_n^{(0)}(\mathbf{r})$) we can rewrite equation (6) as

$$E_n = E_n^{(0)} + \lambda_p/\varepsilon_1 + \lambda_p^2/\varepsilon_2 + \ldots, \quad (7)$$
$$\psi_n(\mathbf{r}) = \psi_n^{(0)}(\mathbf{r}) + \lambda_p/\chi_1(\mathbf{r}) + \lambda_p^2/\chi_2(\mathbf{r}) + \ldots.$$

The second term on the right-hand side of equation (7) is called the first-order correction to the energy or wave function, the third term is called the second-order correction, etc. In principle, perturbation theory enables us to evaluate all the terms in these infinite series and hence to find E_n and $\psi_n(\mathbf{r})$ exactly. However, the complexity of evaluating the terms increases with their order and, in practice, the calculation is limited to the first few terms. Nevertheless, provided λ_p is small, only the first few terms in the series will be significant, and the results for E_n and $\psi_n(\mathbf{r})$ will be accurate.

2.3 The first-order correction to the energy

If $\mathcal{H}_p$ is very small then $\psi_n(\mathbf{r})$ will be almost identical to $\psi_n^{(0)}(\mathbf{r})$. The first-order correction to the energy is found by assuming that any change in the wave function is negligible. The change in the energy will then be

$$\lambda_p \varepsilon_1 = E_n^{(1)} = E_n \approx E_n^{(0)} = \frac{\int_{\text{all space}} \psi_n^{(0)*}(\mathbf{r}) \mathcal{H}_p \psi_n^{(0)}(\mathbf{r}) \, d\mathbf{r}}{\int_{\text{all space}} \psi_n^{(0)*}(\mathbf{r}) \psi_n^{(0)}(\mathbf{r}) \, d\mathbf{r}} \quad (8)$$

which is the usual expression for the average value of an observable. (If you are unfamiliar with this type of expression then consider the following argument. Suppose $\mathcal{H}_p$ is an extra contribution to the potential, say $V_p(\mathbf{r})$. If the wave function is $\psi_n^{(0)}(\mathbf{r})$ then the probability density at $\mathbf{r}$ is $\psi_n^{(0)*}(\mathbf{r})\psi_n^{(0)}(\mathbf{r})$. The expression given above for the first-order correction to the energy is simply the probability density at $\mathbf{r}$, multiplied by the change in potential energy at $\mathbf{r}$, integrated, i.e. averaged, over all space, and divided by the total probability of the particle being anywhere in space.) It is convenient to choose $\psi_n^{(0)}(\mathbf{r})$ such that

$$\int_{\text{all space}} \psi_n^{(0)*}(\mathbf{r}) \psi_n^{(0)}(\mathbf{r}) \, d\mathbf{r} = 1 \quad (9)$$

and it will be assumed in the following that this is always the case. (The wave functions are then said to be normalized.) Then, to first order, equation (8) becomes

$$E_n \approx E_n^{(0)} + \int_{\text{all space}} \psi_n^{(0)*}(\mathbf{r}) \mathcal{H}_p \psi_n^{(0)}(\mathbf{r}) \, d\mathbf{r} \quad (10)$$

and, since $E_n^{(0)}$, $\psi_n^{(0)}(\mathbf{r})$, and $\mathcal{H}_p$ are known, the energy can be found to first order, provided that the integral can be evaluated.

2.4 The first-order correction to the wave function and the second-order correction to the energy

As explained above, the first-order correction to the energy is derived from the uncorrected ('zeroth-order') wave function. The second-order correction to the energy is derived from the first-order corrected wave function, and so on. The first-order correction to the wave function consists of an appropriate linear combination of all the *other* unperturbed wave functions, i.e. to first order:

$$\psi_n(\mathbf{r}) = \psi_n^{(0)}(\mathbf{r}) + \sum_{\substack{m \neq n \\ m=1}}^{\infty} a_m \psi_m^{(0)}(\mathbf{r}). \qquad (11)$$

It is possible to show that the coefficients a_m, which determine the amount of wave function $\psi_m^{(0)}(\mathbf{r})$ to be added, are given by

$$a_m = \frac{\int_{\text{all space}} \psi_m^{(0)*}(\mathbf{r}) \mathcal{H}_p \psi_n^{(0)}(\mathbf{r}) \, d\mathbf{r}}{(E_n^{(0)} - E_m^{(0)})}, \qquad (12)$$

and the resultant second-order correction to the energy is given by

$$\lambda_p^2 \varepsilon_2 = E_n^{(2)} = E_n - E_n^{(0)} - E_n^{(1)} = \sum_{\substack{m \neq n \\ m=1}}^{\infty} \frac{\left| \int_{\text{all space}} \psi_m^{(0)*}(\mathbf{r}) \mathcal{H}_p \psi_n^{(0)}(\mathbf{r}) \, d\mathbf{r} \right|^2}{(E_n^{(0)} - E_m^{(0)})}. \qquad (13)$$

The full expression for the perturbed energy, correct to second order is therefore

$$E_n = E_n^{(0)} + \int_{\text{all space}} \psi_n^{(0)*}(\mathbf{r}) \mathcal{H}_p \psi_n^{(0)}(\mathbf{r}) \, d\mathbf{r}$$

$$+ \sum_{\substack{m \neq n \\ m=1}}^{\infty} \frac{\left| \int_{\text{all space}} \psi_m^{(0)*}(\mathbf{r}) \mathcal{H}_p \psi_n^{(0)}(\mathbf{r}) \, d\mathbf{r} \right|^2}{(E_n^{(0)} - E_m^{(0)})}. \qquad (14)$$

Note that the first-order term for the energy only involves a single integral; its evaluation is therefore much simpler than that of the second-order term which is an infinite series of integrals involving all the unperturbed states.

3. A PARTICLE IN A ONE-DIMENSIONAL INFINITE POTENTIAL WELL

3.1 Unperturbed potential well

In one dimension the Schrödinger equation for a particle of mass M_p in a region $0 < x < L_3$, say, where $V(x) = 0$, is

$$\frac{-\hbar^2}{2M_p}\frac{d^2}{dx^2}\psi_n^{(0)}(x) = E_n^{(0)}\psi_n^{(0)}(x). \tag{15}$$

Because $\hbar$ is very small, quantum effects are usually only important in systems with atomic dimensions and masses. (A mass of 1 kg in an infinite potential well of width 1 m would have a ground state energy of 5.5×10^{-63} J, which corresponds to a r.m.s. velocity of 3.3×10^{-34} m s^{-1}; an electron of mass 9.11×10^{-31} kg in an infinite potential well of width 0.1 nm—typical of atomic dimensions—would have a ground state energy of 1.19×10^{-17} J = 75 eV, which corresponds to a r.m.s. velocity of 5.1×10^6 m s^{-1}.) Rather than enter numbers such as 9.11×10^{-31} (the mass of an electron in kg) into the computer for M_p we will use the atomic system of units in the following.

If lengths are expressed as multiples of the Bohr radius (1 Bohr radius = $4\pi\varepsilon_0\hbar^2/m_e e^2 = 5.29 \times 10^{-11}$ m) and energies and potentials are expressed as multiples of the Rydberg (1 Rydberg = − the ground state energy of the electron in the hydrogen atom = $m_e e^4/32\pi\varepsilon_0^2\hbar^2 = 2.19 \times 10^{-18}$ J = 13.6 eV) then equation (1), in dimensionless form, is

$$\left[-\frac{1}{M_p}\nabla^2 + V(\mathbf{r})\right]\psi(\mathbf{r}) = E\psi(\mathbf{r}), \tag{16}$$

where $M_p' = M_p/m_e$. For the one-dimensional infinite potential well (see Figure 2) equation (16) becomes (dropping the prime on M_p')

$$-\frac{1}{M_p}\frac{d^2}{dx^2}\psi_n^{(0)}(x) = E_n^{(0)}\psi_n^{(0)}(x). \tag{17}$$

The general solution to (17) is

$$\psi_n^{(0)}(x) = A\,e^{ikx} + B\,e^{-ikx}, \tag{18}$$

where $k^2 = E_n^{(0)}M_p$.

We must impose the boundary conditions that $\psi_n^{(0)}(x) \to 0$, both for $x \to 0$ and $x \to L_3$, because the particle would require an infinite energy to penetrate the region of infinite potential. Thus $B = -A$ and $\psi_n^{(0)}(x) =$

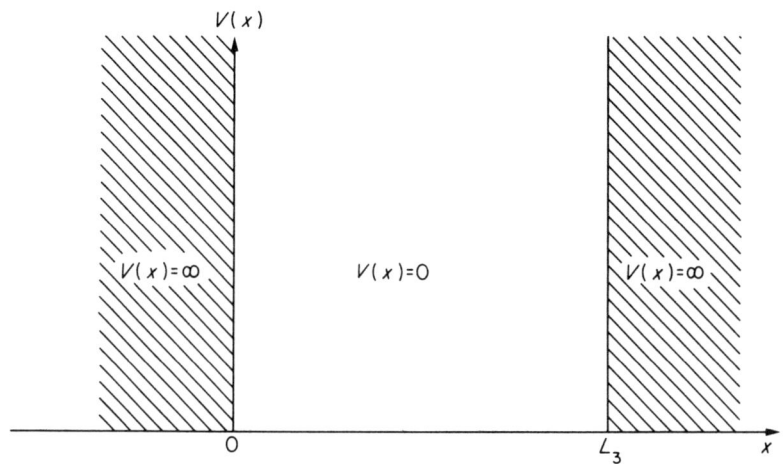

Figure 2. The unperturbed infinite potential well

$C \sin kx$ with $C = 2iA$. But $\psi_n^{(0)}(L_3) = C \sin(kL_3) = 0$, hence

$$k = \frac{n\pi}{L_3}, \quad n = 1, 2, 3, \ldots,$$

$$E_n^{(0)} = \frac{k^2}{M_p} = \frac{n^2\pi^2}{M_p L_3^2},$$

$$\psi_n^{(0)}(x) = C \sin\left(\frac{n\pi x}{L_3}\right). \tag{19}$$

These are the results for the unperturbed energies and wave functions (see Figure 3). The expressions given earlier for first- and second-order perturbation theory required normalized wave functions so C must be adjusted to ensure this. This is done quite simply by requiring that

$$\int_0^{L_3} C^* \sin\left(\frac{n\pi x}{L_3}\right) C \sin\left(\frac{n\pi x}{L_3}\right) dx = C^2 \left| \frac{x}{2} - \frac{L_3}{4n\pi} \sin\frac{2n\pi x}{L_3} \right|_0^{L_3} = C^2 \frac{L_3}{2} = 1, \tag{20}$$

i.e.

$$\psi_n^{(0)}(x) = \sqrt{\frac{2}{L_3}} \sin\left(\frac{n\pi x}{L_3}\right) \quad n = 1, 2, 3, \ldots \tag{21}$$

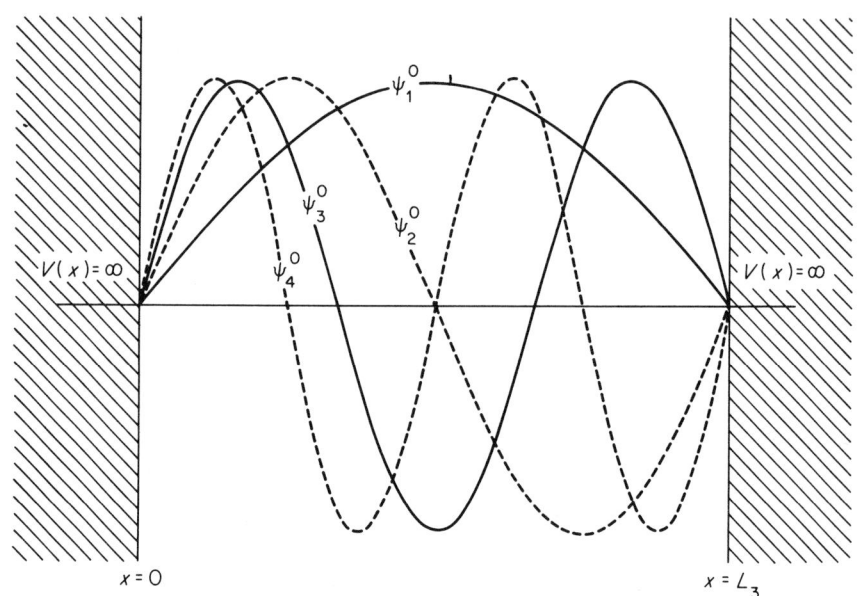

Figure 3. The unperturbed wave functions.

The case $V_p = 0$ should be tried, when running the computer program, detailed in section 4, to check that the computed energies agree with the standard expression

$$E_n^{(0)} = \frac{n^2 \pi^2}{M_p L_3^2}, \qquad n = 1, 2, 3, \ldots \tag{22}$$

for your chosen values of M_p and L_3.

3.2 The first-order correction to the energy for a particle in an infinite one-dimensional potential well with a perturbation V_p

The general expression for the first-order correction to the nth energy level is

$$E_n^{(1)} = \int_{\text{all space}} \psi_n^{(0)*}(\mathbf{r}) \mathcal{H}_p \psi_n^{(0)}(\mathbf{r}) \, d\mathbf{r}. \tag{23}$$

The $\psi_n^{(0)}(\mathbf{r})$ are the normalized unperturbed wave functions found above. The integration is carried out over all space where $\mathcal{H}_p$ is finite—in this case between L_1 and L_2. $\mathcal{H}_p$ itself is the extra term in the Hamiltonian—here simply a constant V_p. Equation (23) therefore becomes

$$E_n^{(1)} = \int_{L_1}^{L_2} \sqrt{\frac{2}{L_3}} \sin\left(\frac{n\pi x}{L_3}\right) V_p \sqrt{\frac{2}{L_3}} \sin\left(\frac{n\pi x}{L_3}\right) dx$$

$$= \frac{2V_p}{L_3} \int_{L_1}^{L_2} \sin^2\left(\frac{n\pi x}{L_3}\right) dx$$

$$= \frac{V_p}{L_3}(L_2 - L_1) - \frac{V_p}{2n\pi}\left\{\sin\left(\frac{2\pi n L_2}{L_3}\right) - \sin\left(\frac{2\pi n L_1}{L_3}\right)\right\}. \quad (24)$$

As V_p increases this energy will increase—as we would expect. However, the rate of increase will be different for different energy levels, because the probability of the particle being between L_1 and L_2 (and therefore 'seeing' V_p) will change with n. For example, if L_1 and L_2 are near the centre of the well the perturbation will affect the $n = 1$ level (in which the particle has a high probability of being near the centre) much more than the $n = 2$ level (where the particle has a low probability of being near the centre).

3.3 The second-order correction to the energy for a particle in an infinite one-dimensional potential well with a perturbation V_p

The general expression for the second-order correction to the energy of the nth level is

$$E_n^{(2)} = \sum_{\substack{m \neq n \\ m=1}}^{\infty} \frac{\left|\int_{\text{all space}} \psi_m^{(0)*}(\mathbf{r}) \mathcal{H}_p \psi_n^{(0)}(\mathbf{r}) d\mathbf{r}\right|^2}{(E_n^{(0)} - E_m^{(0)})} = \sum_{\substack{m \neq n \\ m=1}}^{\infty} E_{mn}^{(2)} \quad (25)$$

In the present case

$$I_{mn} = \int_{\text{all space}} \psi_m^{0*}(\mathbf{r}) \mathcal{H}_p \psi_n^{(0)}(\mathbf{r}) d\mathbf{r}$$

$$= \int_{L_1}^{L_2} \sqrt{\frac{2}{L_3}} \sin\left(\frac{m\pi x}{L_3}\right) V_p \sqrt{\frac{2}{L_3}} \sin\left(\frac{n\pi x}{L_3}\right) dx$$

$$= \frac{2V_p}{L_3} \int_{L_1}^{L_2} \sin\left(\frac{m\pi x}{L_3}\right) \sin\left(\frac{n\pi x}{L_3}\right) dx$$

$$= \frac{V_p}{L_3}\left\{\frac{\sin(\alpha L_2) - \sin(\alpha L_1)}{\alpha} - \frac{\sin(\beta L_2) - \sin(\beta L_1)}{\beta}\right\}, \quad (26)$$

where $\alpha = \pi(n-n)/L_3$, $\beta = \pi(n+n)/L_3$ and $|m| \neq |n|$.

4. THE COMPUTER PROGRAM

4.1 Basic features

The computer program calculates, to very high accuracy, the four lowest energy states of a particle of mass M_p, in a one-dimensional infinite potential well with an extra constant potential V_p in part of it. The program allows you to choose the particle mass M_p, the width of the finite potential well ($=L_3$), the position where the potential V_p starts ($=L_1$) and ends ($=L_2$), and the potential V_p itself. The basic procedure used to obtain the exact results is outlined below.

For $0 < x < L_1$ and $L_2 < x < L_3$, $V(x) = 0$ and

$$\psi(x) = A \exp(ik_1 x) + B \exp(-ik_1 x), \qquad k_1 = \sqrt{M_p E}. \tag{27}$$

Similarly, for $L_1 < x < L_2$, $V(x) = V_p$ and

$$\psi(x) = A' \exp(ik_2 x) + B' \exp(-ik_2 x), \qquad k_2 = \sqrt{M_p(E - V_p)}. \tag{28}$$

Here, $\psi(0) = 0$ is a necessary boundary condition, but this still leaves the assignment of $[d\psi(x)/dx]_{x=0}$. It is chosen to be unity, which is a compromise value resulting in $\psi(x)$ and $d\psi(x)/dx$ being of reasonable size over the whole system. (Note that in this computer method of solution, $\psi(x)$ is not, and need not, be normalized. Also this choice of $d\psi(x)/dx$ is reasonable only provided V_p has not a very extreme value.)

If E is then given $\psi(x)$ and $d\psi(x)/dx$ can be matched at L_1 and L_2 and hence $\psi(L_3)$ can be found. The subroutine PL3FEN, which finds $\psi(L_3)$ in this way, also returns the number of nodes of $\psi(x)$. The eigenvalues of the system will correspond to those values of E for which $\psi(L_3) = 0$. The problem is therefore equivalent to the determination of the roots of the complex determinantal transcendental equation which can be set up by conventional methods. However, the method adopted here is thought to possess some pedagogical advantages. An example of the variation of $\psi(L_3)$ with energy is shown in Figure 4.

The eigenvalues are first estimated using perturbation theory. Then, with these as a guide, by repeated doubling and halving, two energies, one above and one below the true nth energy level, are found where the number of nodes are $n-1$ and n respectively. The function EFPL30 is next employed to determine an accurate zero of $\psi(L_3)$ using a combination of the methods of linear extrapolation, linear interpolation, and bisection.

4.2 Running the program

Input the position where the perturbation starts (L_1) and ends (L_2), the width of the infinite potential well (L_3) and the mass of the particle (M_p).

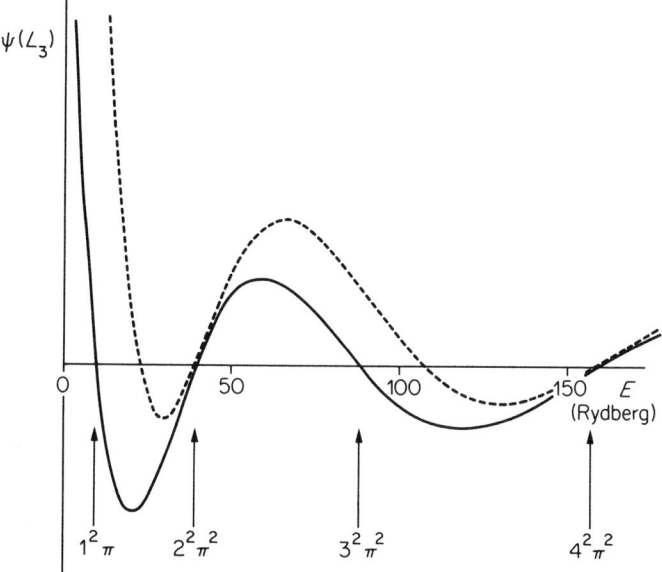

Figure 4. The variation of $\psi(L_3)$ with E, as returned by PL3FEN for $L_3 = 1.0$, $M_p = 1.0$. The solid curve shows the results for $V_p = 0$; the values of E for which $\psi(L_3) = 0$ coincide with the theoretical values for the energy: $E_n^0 = \pi^2 n^2 / M_p L_3^2 = \pi^2 n^2$. The dashed curve shows the result for $L_1 = 0.49$, $L_2 = 0.51$, $V_p = 500.0$. As we would expect from first-order perturbation theory, the ground state and second excited state energies are modified much more than the energies of the first and third excited states.

Rather than calculating the energy at a single value of V_p the program allows a range of values of V_p to be entered by reading the lowest required value of V_p (VP1)—which can be ≤ 0 or > 0, the number of different values of V_p required (NVP) and the amount by which V_p is to be increased (DVP). Choose these parameters so that the case $V_p = 0$ is included.

An approximate graph of the results can be plotted on a line printer; the last input should be YES or NO as required.

The computer will set out the results for the four lowest energies E_1, E_2, E_3, and E_4 as a function of V_p, in a table.

Remembering that the point of the exercise is to compare these computed results with perturbation theory, check that the chosen range of V_p is such that the energies change significantly (so that you can make the comparison) but not drastically (or otherwise V_p will be too large for perturbation theory to apply).

The program also gives a table of $\sin(n\pi L_1/L_3)$ and $\sin(n\pi L_2/L_3)$ for $n = 1, 2, \ldots 14$ which can be used to evaluate $E_n^{(1)}$, $E_n^{(2)}$ using equations (24) and (26).

4.3 Comparing the exact computer results with those from perturbation theory

Having run the program for chosen values of M_p, L_1, L_2, and L_3, the results can be compared with those of perturbation theory in the following manner:

(1) Plot the results of the program as a graph of E_n against V_p.
(2) Calculate the unperturbed energies and check that they coincide with the computer-generated results for $V_p = 0$.
(3) Use equation (24) to work out the first-order correction to the energy for the four lowest straight levels. A plot of these results will give four straight lines on your graph.
(4) The second-order correction to the energy is considerably more complicated to evaluate. It is therefore suggested that the calculation only be made for the ground state, i.e. evaluate equation (26) for $n = 1$, $m \neq 1$ but consider only the first few terms in the second-order correction to the energy. You will find (as is generally the case) that these decrease quite rapidly because $|E_n^0 - E_m^0|$ increases rapidly with $|n - m|$, and so the series converges quickly. Plot the second-order correction to the ground state energy on the graph.
(5) Can you make any general comments about the conditions for the validity of perturbation theory?

4.4 A Typical session

If the chosen parameters are:

L_1 (the position where the perturbation starts) = 0.4 ⎫
L_2 (the position where the perturbation ends) = 0.5 ⎬ (Bohr radii)
L_3 (the width of the infinite potential well) = 1.0 ⎭
M_p (the mass of the particle) = 1.0 (electron masses)
VP1 (the lowest value of the perturbation potential) = −150.0 ⎫
DVP (the amount by which the perturbing potential is to be increased) = 10.0 ⎬ (Rydbergs)
NVP (the required number of values of the perturbing potential) = 31 ⎭

and an approximate graph of the results on the line printer is requested, the results of the computer program are

A COMPUTER-ASSISTED TUTORIAL

```
TYPE THE WIDTH OF THE INFINITE POTENTIAL WELL, I.E. L3
1.0
TYPE THE POSITION WHERE THE PERTURBATION STARTS I.E. L1
0.4
TYPE THE POSITION WHERE THE PERTURBATION ENDS I.E. L2
0.5
TYPE THE MASS OF THE PARTICLE I.E. MP
1.0
TYPE THE LOWEST VALUE OF THE PERTURBING POTENTIAL VP
-150.0
TYPE THE REQUIRED NUMBER OF VALUES OF THE PERTURBING POTENTIAL
31
TYPE THE AMOUNT BY WHICH THE PERTURBING POTENTIAL IS TO BE INCREASED
10.0
```

RESULTS OF COMPUTER PROGRAM TO CALCULATE THE ENERGY OF A PARTICLE
IN A PERTURBED INFINITE POTENTIAL WELL. THE WELL WIDTH IS 0.1000E 01
THE PERTURBATION STARTS AT 0.4000E 00 AND ENDS AT 0.5000E 00
        THE PARTICLE MASS IS 0.1000E 01

VP	E1	E2	E3	E4
-0.1500E 03	-0.3776E 02	0.3658E 02	0.7090E 02	0.1484E 03
-0.1400E 03	-0.3344E 02	0.3673E 02	0.7175E 02	0.1489E 03
-0.1300E 03	-0.2927E 02	0.3689E 02	0.7264E 02	0.1495E 03
-0.1200E 03	-0.2526E 02	0.3706E 02	0.7359E 02	0.1500E 03
-0.1100E 03	-0.2140E 02	0.3722E 02	0.7458E 02	0.1506E 03
-0.1000E 03	-0.1771E 02	0.3740E 02	0.7563E 02	0.1512E 03
-0.9000E 02	-0.1419E 02	0.3757E 02	0.7674E 02	0.1518E 03
-0.8000E 02	-0.1083E 02	0.3776E 02	0.7789E 02	0.1524E 03
-0.7000E 02	-0.7650E 01	0.3795E 02	0.7910E 02	0.1530E 03
-0.6000E 02	-0.4640E 01	0.3815E 02	0.8036E 02	0.1537E 03
-0.5000E 02	-0.1803E 01	0.3835E 02	0.8167E 02	0.1543E 03
-0.4000E 02	0.8630E 00	0.3856E 02	0.8303E 02	0.1550E 03
-0.3000E 02	0.3359E 01	0.3878E 02	0.8443E 02	0.1557E 03
-0.2000E 02	0.5690E 01	0.3901E 02	0.8586E 02	0.1564E 03
-0.1000E 02	0.7858E 01	0.3924E 02	0.8733E 02	0.1572E 03
0.0000E 00	0.9870E 01	0.3948E 02	0.8883E 02	0.1579E 03
0.1000E 02	0.1173E 02	0.3973E 02	0.9034E 02	0.1587E 03
0.2000E 02	0.1345E 02	0.3998E 02	0.9187E 02	0.1595E 03
0.3000E 02	0.1503E 02	0.4024E 02	0.9340E 02	0.1603E 03
0.4000E 02	0.1648E 02	0.4050E 02	0.9493E 02	0.1611E 03
0.5000E 02	0.1782E 02	0.4077E 02	0.9645E 02	0.1620E 03
0.6000E 02	0.1904E 02	0.4105E 02	0.9795E 02	0.1628E 03
0.7000E 02	0.2015E 02	0.4132E 02	0.9944E 02	0.1637E 03
0.8000E 02	0.2117E 02	0.4160E 02	0.1009E 03	0.1646E 03
0.9000E 02	0.2210E 02	0.4188E 02	0.1023E 03	0.1655E 03
0.1000E 03	0.2296E 02	0.4216E 02	0.1037E 03	0.1664E 03
0.1100E 03	0.2373E 02	0.4245E 02	0.1051E 03	0.1673E 03
0.1200E 03	0.2444E 02	0.4273E 02	0.1064E 03	0.1683E 03
0.1300E 03	0.2509E 02	0.4300E 02	0.1077E 03	0.1692E 03
0.1400E 03	0.2568E 02	0.4328E 02	0.1089E 03	0.1701E 03
0.1500E 03	0.2623E 02	0.4355E 02	0.1101E 03	0.1711E 03

```
DO YOU WANT TO PLOT OUT THESE RESULTS AS AN APPROXIMATE GRAPH
ON THE TELETYPE.    TYPE YES OR NO.
YES
```

252 PHYSICS PROGRAMS

APPROXIMATE GRAPH OF THE ENERGIES OF THE GROUND STATE AND FIRST THREE EXCITED STATES AS A FUNCTION OF THE PERTURBATION VP. VP IS PLOTTED ON THE HORIZONTAL AXIS AND VARIES FROM -0.1500E 03 TO 0.1500E 03
THE ENERGY IS PLOTTED ON THE VERTICAL AXIS AND VARIES FROM 0.1711E 03 TO -0.3776E 02

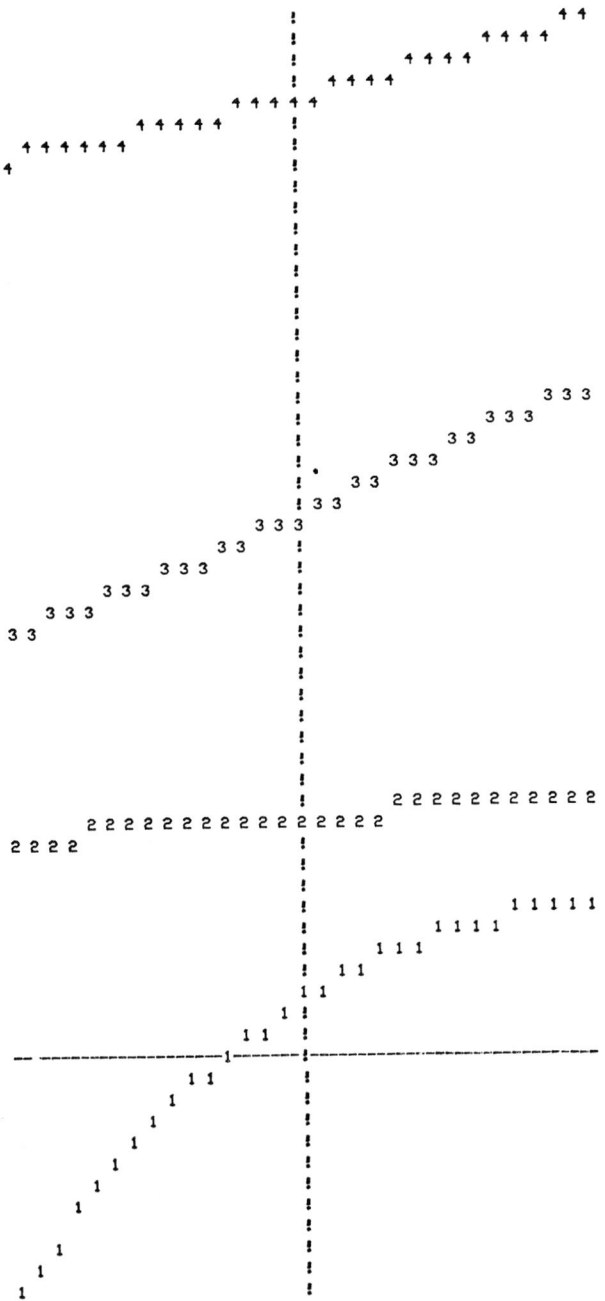

```
DO YOU WANT TO RE-RUN THE PROGRAM WITH THE SAME VALUES FOR L1, L2, L3
AND MP BUT WITH DIFFERENT VALUES FOR THE PERTURBING POTENTIAL VP.
TYPE YES OR NO
NO

    N         SIN(N*PI*L1/L3)      SIN(N*PI*L2/L3)

    1           0.951056            1.000000
    2           0.587786            0.000001
    3          -0.587784           -1.000000
    4          -0.951057           -0.000001
    5          -0.000002            1.000000
    6           0.951055            0.000004
    7           0.587788           -1.000000
    8          -0.587782           -0.000003
    9          -0.951058            1.000000
   10          -0.000004            0.000006
   11           0.951055           -1.000000
   12           0.587792           -0.000009
   13          -0.587777            1.000000
   14          -0.951059            0.000006
PROGRAM FINISHED
```

5. DISCUSSION

The unperturbed energies (in Rydbergs), for $V_p = 0$, are

$$E_n^{(0)} = \frac{n^2 \pi^2}{M_p L_3^2} \quad (29)$$

Equation (29), for the chosen parameters, gives $E_1^{(0)} = 9.87$, $E_2^{(0)} = 39.48$, $E_3^{(0)} = 88.83$, $E_4^{(0)} = 157.91$.

The first-order correction to the energy is given by equation (24). Thus, for the chosen parameters, the corrections to the first four levels are:

$n =$	1	2	3	4
$\dfrac{\text{(First-order correction)}}{V_p} =$	0.194	0.024	0.150	0.077

The second-order correction to the energy of the nth level is given by equation (25), which can be evaluated using equation (26). For the example under discussion, values of I_{mn}/V_p are given in Table 1 while $E_{mn}^{(2)}/V_p^2$ values are given in Table 2. The ranges of n and m are restricted to $1 \leq n \leq 4$ and $1 \leq m \leq 10$.

Table 1. Values of I_{mn}/V_p

	$m=1$	2	3	4	5	6	7	8	9	10
n										
1	—	0.059	−0.169	−0.107	0.126	0.136	−0.074	−0.141	0.023	0.125
2	0.059	—	−0.048	−0.043	0.028	0.052	−0.005	−0.050	−0.016	0.039
3	−0.169	−0.048	—	0.088	−0.117	−0.113	0.076	0.120	−0.036	−0.111
4	−0.107	−0.043	0.088	—	−0.053	−0.094	0.013	0.091	0.025	−0.072

Table 2. Values of $E^{(2)}_{mn}/V_p^2$

	$m=1$	2	3	4	5	6	7	8	9	10	Total
n											
1	—	-1.19×10^{-4}	-3.63×10^{-4}	-7.79×10^{-5}	-6.72×10^{-5}	-5.34×10^{-5}	-1.15×10^{-5}	-3.21×10^{-5}	-6.93×10^{-7}	-1.61×10^{-5}	-7.40×10^{-4}
2	1.19×10^{-4}	—	-4.69×10^{-5}	-1.57×10^{-5}	-3.91×10^{-6}	-8.66×10^{-6}	-6.43×10^{-8}	-4.30×10^{-6}	-3.26×10^{-7}	-1.60×10^{-6}	$+3.74\times10^{-5}$
3	3.63×10^{-4}	4.69×10^{-5}	—	-1.12×10^{-4}	-8.66×10^{-5}	-4.77×10^{-5}	-1.45×10^{-5}	-2.66×10^{-5}	-1.71×10^{-6}	-1.37×10^{-5}	$+1.07\times10^{-4}$
4	7.79×10^{-5}	1.57×10^{-5}	1.12×10^{-4}	—	-3.21×10^{-5}	-4.43×10^{-5}	-4.97×10^{-7}	-1.76×10^{-5}	-9.58×10^{-7}	-6.27×10^{-6}	$+1.03\times10^{-4}$

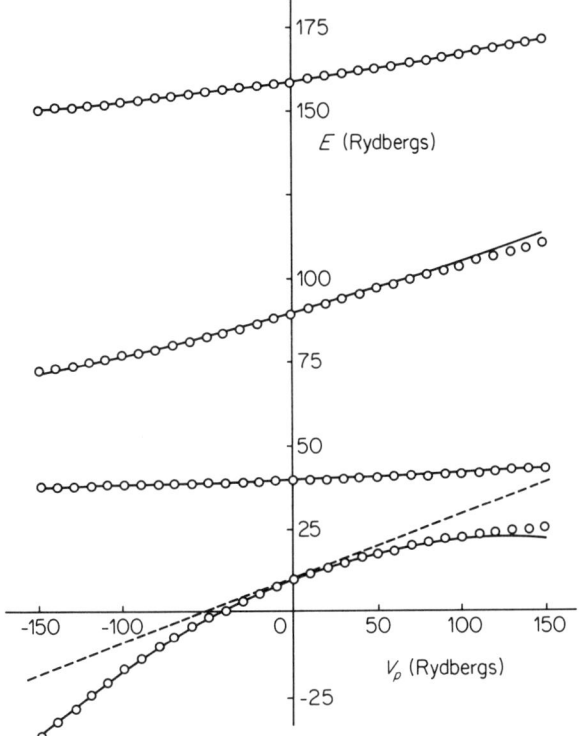

Figure 5. The results of the computer program (circles) for $L_1=0.4$, $L_2=0.5$, $L_3=1.0$, $M_p=1.0$ compared with the perturbation theory calculation. The solid curves are the results for first- plus second-order perturbation theory. The dashed line for the ground state shows the result for first-order perturbation theory alone

Figure 5 shows a comparison between the results of the computer program (the circles) and perturbation theory. The solid curves are the results for first- plus second-order perturbation theory corrections. The dashed line for the ground state shows the result for first-order perturbation theory alone. It can be seen that, even with the inclusion of the second-order correction, the results are starting to deviate from the (more accurate) computer results at the largest values of $|V_p|$. For these values of V_p the amount of admixture of other states into the unperturbed eigenstate is not small—which is the situation in which we would indeed expect second-order perturbation theory to break down.

Though they correspond to situations far outside the range of validity of perturbation theory the results for $V_p \gg E_n^0$ and $V_p \to -\infty$ are of some interest.

For $V_p \gg E_n^0$ the situation tends towards a pair of isolated infinite potential

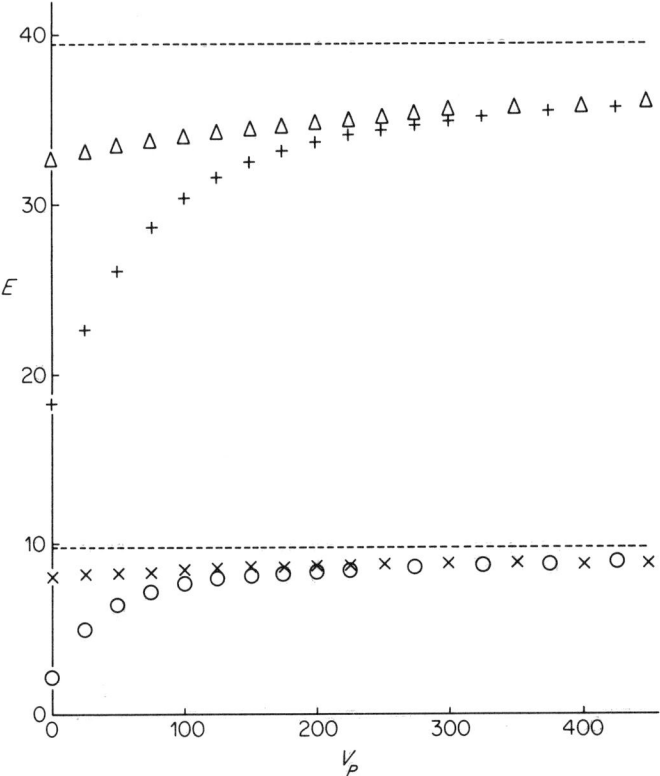

Figure 6. The energy levels for $V_p \gg E_n^0$. $L_1 = 1.0$, $L_2 = 1.2$, $L_3 = 2.2$, $M_p = 1.0$. The dashed lines indicate the ground state and first excited state energies for the particle in an infinite potential well of width $L_1 = L_3 - L_2$

wells, of width L_1 and $L_3 - L_2$. It will be found for this case that the energies tend towards the (constant) values for the infinite potential wells with these widths. Furthermore, if $L_1 = L_3 - L_2$ these levels are degenerate. This kind of situation is shown in Figure 6.

For $V_p \to -\infty$ the situation tends towards an infinite potential well of width $L_2 - L_1$ with the zero of energy shifted to V_p. Hence

$$E_n \to V_p + \frac{n^2 \pi^2}{M_p (L_2 - L_1)^2}.$$

An example of this behaviour is shown in Figure 7.

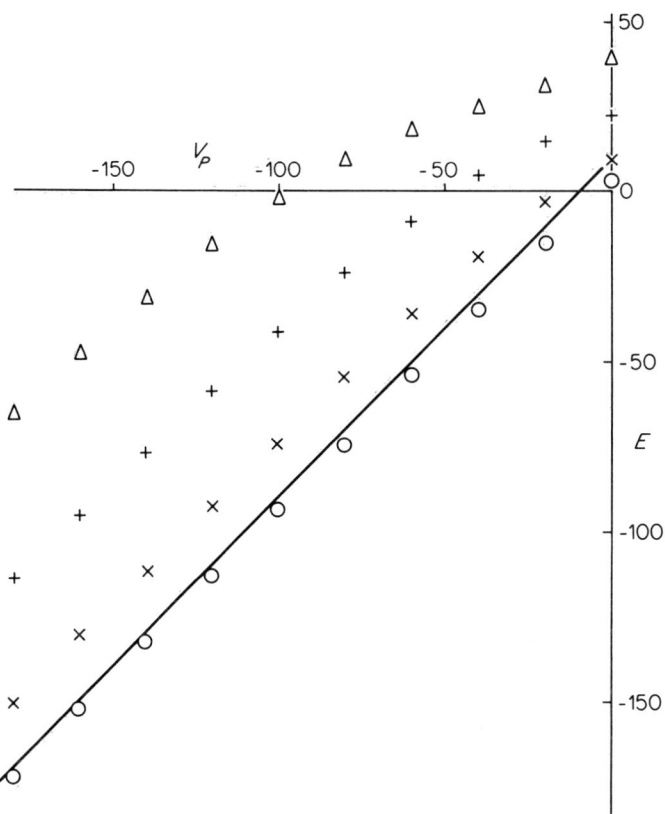

Figure 7. The energy levels for $V_p \to -\infty$ and $L_1 = 0.5$, $L_2 = 1.5$, $L_3 = 2.0$, $M_p = 1.0$. The straight line corresponds to $E = V_p + \pi^2/M_p(L_2 - L_1)^2$ and it can be seen that the ground state energy tends asymptotically to this value for $V_\mu \to -\infty$

MAIN PROGRAM

```
C  THIS PROGRAM USES AN ITERATIVE TECHNIQUE TO FIND THE GROUND STATE AND
C   FIRST THREE EXCITED STATES OF A PARTICLE IN AN INFINITE POTENTIAL
C   WELL WITH AN ADDITIONAL PERTURBING POTENTIAL IN PART OF
C   THE WELL.
      COMMON L1,L2,L3,VP,MP,N,PI
      DIMENSION SL1(15),SL2(15),HDV(5,10),EV(10),EC2(5),EP(5),E(51,4),
     1 PLOT(120),NAME(40)
      REAL L1,L2,L3,MP
      LOGICAL BOOL
      DATA Y,YN,PLOT1,PLOT2,PLOT3,PLOT4,PLOT5,PLOT6,PLOT7/4HYES ,4HNO  ,
     1 4H1    ,4H2    ,4H3    ,4H4    ,4H    ,4H-    ,4H!  /
C  THE WIDTH OF THE INFINITE POTENTIAL WELL (=L3) AND THE POSITION WHERE
C   THE PERTURBATION STARTS (=L1) AND ENDS (=L2) ARE READ IN.
      PI=ATAN(1.0)*4.0
C***********************************************************************
C
C    THE PROGRAM AS PRESENTED IS FOR USE ON AN INTERACTIVE SYSTEM.
C    FREE FORMAT IS USED FOR INPUT  -  F0.0 AND I0  -  WHICH MAY NOT
C    BE AVAILABLE ON SOME SYSTEMS.  ACCORDINGLY THE MATERIAL UP TO THE
C    NEXT SET OF *******'S MAY NEED SOME MODIFICATION.  IT IS NECESSARY
C    TO READ IN AT THIS POINT IN THE PROGRAM:
C       L1 (THE POSITION WHERE THE PERTURBATION STARTS)
C       L2 (THE POSITION WHERE THE PERTURBATION ENDS)
C       L3 (THE WIDTH OF THE INFINITE POTENTIAL WELL)
C       MP (THE MASS OF THE PARTICLE)
C
    2 CONTINUE
      WRITE(1,100)
  100 FORMAT(53HTYPE THE WIDTH OF THE INFINITE POTENTIAL WELL,I.E. L3)
      READ(1,1)L3
    1 FORMAT(F0.0)
      IF(L3)3,3,4
    3 CONTINUE
      WRITE(1,102)
  102 FORMAT(27HL3 MUST BE GREATER THAN 0.0)
      GOTO 2
    4 CONTINUE
      WRITE(1,106)
  106 FORMAT(55HTYPE THE POSITION WHERE THE PERTURBATION STARTS I.E. L1)
      READ(1,1)L1
      IF(L1.GE.0.0.AND.L1.LT.L3)GOTO 6
      WRITE(1,108)
  108 FORMAT(30HL1 MUST LIE BETWEEN 0.0 AND L3)
      GOTO 4
    6 CONTINUE
      WRITE(1,110)
  110 FORMAT(53HTYPE THE POSITION WHERE THE PERTURBATION ENDS I.E. L2)
      READ(1,1)L2
      IF(L2.GT.L1.AND.L2.LE.L3)GOTO 8
      WRITE(1,112)
  112 FORMAT(29HL2 MUST LIE BETWEEN L1 AND L3)
      GOTO 6
    8 CONTINUE
C NEXT THE MASS OF THE PARTICLE (=MP) IS INPUT.
      WRITE(1,114)
  114 FORMAT(37HTYPE THE MASS OF THE PARTICLE I.E. MP)
```

```
      MAIN PROGRAM
          READ(1,1)MP
          IF(MP)9,9,10
        9 CONTINUE
          WRITE(1,116)
      116 FORMAT(49HTHE MASS OF THE PARTICLE MUST BE GREATER THAN 0.0)
          GOTO 8
       10 CONTINUE
C***********************************************************************
C AT THIS POINT IN THE PROGRAM THE MAIN TERMS IN THE FIRST AND SECOND
C   ORDER PERTURBATION THEORY CORRECTIONS TO THE ENERGIES OF THE 5
C   LOWEST ENERGY STATES ARE COMPUTED. THEY ARE USED TO GUIDE THE
C   SUBSEQUENT ITERATIVE SEARCH FOR THE EIGENVALUES. THE RESULTS ARE
C   STORED AND SOME OF THEM ARE OUTPUT AT THE END OF THE PROGRAM AS
C   AN AID TO THE STUDENT.
          DO 12 I=1,15
          SL1(I)=SIN(FLOAT(I)*PI*L1/L3)
          SL2(I)=SIN(FLOAT(I)*PI*L2/L3)
       12 CONTINUE
          DO 14 N=1,5
          HDV(N,N)=(L2-L1)/L3-(SL2(2*N)-SL1(2*N))/(2.0*PI*FLOAT(N))
          NP=N+1
          DO 16 M=NP,10
          MMN=M-N
          MPN=M+N
          HDV(N,M)=(SL2(MMN)-SL1(MMN))/FLOAT(MMN)
          HDV(N,M)=(HDV(N,M)-(SL2(MPN)-SL1(MPN))/FLOAT(MPN))/PI
       16 CONTINUE
       14 CONTINUE
          DO 18 N=2,5
          NM=N-1
          DO 20 M=1,NM
          HDV(N,M)=HDV(M,N)
       20 CONTINUE
       18 CONTINUE
C COMPUTE THE UNPERTURBED ENERGIES (=EU(N))
          DO 22 N=1,10
          EU(N)=PI*PI*FLOAT(N)*FLOAT(N)/(MP*L3*L3)
       22 CONTINUE
C AS YET THE PERTURBING POTENTIAL HAS NOT BEEN READ IN. HOWEVER,
C   WE WISH TO BE ABLE TO EXPRESS THE ENERGY, UP TO SECOND ORDER,AS:
C                E = EU(N) + VP*HDV(N,N) + VP*VP*EC2(N) .    THUS.
          DO 24 N=1,5
          EC2(N)=0.0
          DO 26 M=1,10
          IF(M-N)25,26,25
       25 CONTINUE
          EC2(N)=EC2(N)+(HDV(N,M)*HDV(N,M))/(EU(N)-EU(M))
       26 CONTINUE
       24 CONTINUE
       28 CONTINUE
C THE VALUES OF THE PERTURBATION (=VP) ARE READ IN
C***********************************************************************
C
C    FURTHER INPUT IS REQUIRED HERE
C       VP1 (THE LOWEST VALUE OF THE PERTURBING POTENTIAL VP)
C       DVP (THE AMOUNT BY WHICH THE PERTURBING POTENTIAL IS TO BE INCREASED)
```

MAIN PROGRAM

```
C      NVP (THE REQUIRED NUMBER OF VALUES OF THE PERTURBING POTENTIAL)
C            NOTE THAT NVP IS AN INTEGER VARIABLE  -  THE OTERS ARE REAL
C
       WRITE(1,118)
   118 FORMAT(52HTYPE THE LOWEST VALUE OF THE PERTURBING POTENTIAL VP)
       READ(1,1)VP1
    30 CONTINUE
       WRITE(1,120)
   120 FORMAT(52HTYPE THE REQUIRED NUMBER OF VALUES OF THE PERTURBING ,
      1 9HPOTENTIAL)
       READ(1,29)NVP
    29 FORMAT(I0)
       IF(NVP.GT.0.AND.NVP.LT.52)GOTO 32
       WRITE(1,122)
   122 FORMAT(53HTHE REQUIRED NUMBER MUST BE POSITIVE AND LESS THAN 52)
       GOTO 30
    32 CONTINUE
       WRITE(1,124)
   124 FORMAT(55HTYPE THE AMOUNT BY WHICH THE PERTURBING POTENTIAL IS TO ,
      1 12HBE INCREASED)
       READ(1,1)DVP
       WRITE(1,208)
       WRITE(1,208)
C************************************************************************
       WRITE(1,200)
   200 FORMAT(1H ,45HRESULTS OF COMPUTER PROGRAM TO CALCULATE THE ,
      1 20HENERGY OF A PARTICLE)
       WRITE(1,202)L3
   202 FORMAT(1H ,49HIN A PERTURBED INFINITE POTENTIAL WELL. THE WELL ,
      1 9HWIDTH IS ,E11.4)
       WRITE(1,204)L1,L2
   204 FORMAT(1H ,27HTHE PERTURBATION STARTS AT ,E11.4,13H AND ENDS AT ,
      1 E11.4)
       WRITE(1,206)MP
   206 FORMAT(1H ,15X,21HTHE PARTICLE MASS IS ,E11.4)
C SET OUT THE HEADING FOR THE RESULTS
       WRITE(1,208)
   208 FORMAT(1H )
       WRITE(1,210)
   210 FORMAT(1H ,6X,2HVP,12X,2HE1,12X,2HE2,12X,2HE3,12X,2HE4)
       WRITE(1,208)
       DO 34 I=1,NVP
       VP=VP1+FLOAT(I-1)*DVP
C COMPUTE THE ENERGY OF THE 5 LOWEST STATES USING 2ND ORDER PERTURBATION
       DO 36 N=1,5
       EP(N)=EU(N)+VP*HDV(N,N)+VP*VP*EC2(N)
    36 CONTINUE
C  SORT THE EP(N) INCASE THE PERTURBATION HAS PRODUCED A CROSS OVER
C   OF THE LEVELS
    39 CONTINUE
       DO 37 N=1,4
       IF(EP(N).LT.EP(N+1))GOTO 37
       EPN=EP(N+1)
       EPNP=EP(N)
       EP(N)=EPN
       EP(N+1)=EPNP
```

MAIN PROGRAM

```
   37 CONTINUE
      IF(EP(1).GT.EP(2).OR.EP(2).GT.EP(3).OR.EP(3).GT.EP(4).
     1 OR.EP(4).GT.EP(5))GOTO 39
C  THE PROGRAM NOW STARTS TO SEARCH FOR VALUES OF ENERGY SUCH THAT
C  THE WAVE FUNCTION VANISHES AT L3.
C  IN THE FIRST PART OF THE SEARCH FOR THE N'TH EIGENSTATE BY
C  USING A BISECTION TECHNIQUE THE PPOGRAM FINDS TWO ENERGIES, ONE ABOVE
C  AND ONE BELOW THE TRUE N'TH ENERGY LEVEL, WHERE THE NUMBER OF NODES
C  ARE N-1 AND N RESPECTIVELY. THESE ENERGIES ARE THEN USED AS STARTING
C  POINTS FOR THE LINEAR INTERPOLATION ROOT FINDING FUNCTION.
      EF=AMIN1(0.0,UP)
      CALL PL3FEN(EF,PL3F,NODES)
      DO 38 N=1,4
      ET=(EP(N)+EP(N+1))*0.5
      GAP=ABS(ET-EF)
      IF(GAP.EQ.0.0)GAP=EU(1)
      ET=EF+GAP
      BOOL=.FALSE.
      DO 40 J=1,500
      CALL PL3FEN(ET,PL3T,NODES)
      PL3S=PL3T
      IF(NODES.GT.N)GOTO 76
      IF(NODES.LT.N)GOTO 78
      ES=ET
      GOTO 44
   76 CONTINUE
      BOOL=.TRUE.
      ET=ET-GAP
      GOTO 42
   78 CONTINUE
      ET=ET+GAP
   42 CONTINUE
      IF(BOOL)GAP=GAP*0.5
      IF(.NOT.BOOL)GAP=GAP*2.0
   40 CONTINUE
      WRITE(1,214)N
  214 FORMAT(1H0,49H WARNING - THE PROGRAM FAILED TO FIND A CORRECT  ,
     1 9HSTARTING ,/,26HPOINT FOR THE SEARCH WITH ,I4,6H NODES)
      ES=ET
   44 CONTINUE
C THE ACCURACY FOUND BY EFPL30 IS SET BY THE PARAMETERS EAC AND FAC.
      EAC=EU(1)*1.0E-5
      FAC=1.0E-5
      E(I,N)=EFPL30(EF,ES,PL3F,PL3S,EAC,FAC)
      IF(N-4)45,38,38
   45 CONTINUE
      EF=ET
      PL3F=PL3T
   38 CONTINUE
      WRITE(1,130)UP,E(I,1),E(I,2),E(I,3),E(I,4)
  130 FORMAT(1H ,5(E11.4,3X))
   34 CONTINUE
C  THE OPTION IS NOW PROVIDED OF PLOTTING A GRAPH OF E AGAINST UP ON THE
C  PRINTER.
C************************************************************************
C
```

MAIN PROGRAM

```
C      THE OPTION IS PROVIDED OF PLOTTING A GRAPH ON THE LINE PRINTER.
C      IF   YES   OR   NO   IS ENTERD AS REQUIRED IT CAN BE READ IN A
C      FORMAT AND COMPARED WITHE THE VARIABLES Y(=4HYES ,) AND
C      YN(=4HNO  ,) SET IN THE DATA STATEMENT. IF A GRAPH IS NOT REQUIRED
C      CONTROL CAN BE TRANSFERED TO STATEMENT 62
C
       WRITE(1,208)
       WRITE(1,170)
   170 FORMAT(//44HDO YOU WANT TO PLOT OUT THESE RESULTS AS AN,
      1 17HAPPROXIMATE GRAPH,/35H ON THE TELETYPE.   TYPE YES OR NO.)
    64 CONTINUE
       READ(1,171)YON
   171 FORMAT(A4)
       IF(YON.EQ.Y)GOTO 60
       IF(YON.EQ.YN)GOTO 62
       WRITE(1,174)
   174 FORMAT(1H ,40HRESPONSE NOT RECOGNISED.  TYPE YES OR NO)
       GOTO 64
    60 CONTINUE
C***************************************************************************
C      THE MAXIMUM WIDTH OF THE GRAPH IS SET BY THE PARAMETER IWG. THIS
C      IS CHOSEN SUCH THAT IWG-1 IS A MULTIPLE OF NVP-1
       IWG=70
       GAP=FLOAT(IWG-1)/FLOAT(NVP-1)
       IGAP=IFIX(AINT(GAP))
       IWG=(NVP-1)*IGAP+1
       DDVP=ABS(DVP/FLOAT(IGAP))
C      FIND THE MAXIMUM AND MINIMUM VALUES OF THE ENERGY
       EMIN=E(1,1)
       EMAX=E(1,1)
       DO 66 N=1,4
       DO 68 I=2,NVP
       IF(E(I,N).LT.EMIN)EMIN=E(I,N)
       IF(E(I,N).GT.EMAX)EMAX=E(I,N)
    68 CONTINUE
    66 CONTINUE
       IF(EMIN.GT.0.0)EMIN=0.0
C      THE HEIGHT OF THE GRAPH IS SET BY THE PARAMETER IHG
       IHG=IWG
       SCALE=(EMAX-EMIN)/FLOAT(IHG-1)
C      WRITE OUT HEADING FOR THE GRAPH
       VP2=VP1+FLOAT(NVP-1)*DVP
       WRITE(1,208)
       WRITE(1,218)
   218 FORMAT(1H ,48HAPPROXIMATE GRAPH OF THE ENERGIES OF THE GROUND ,
      1 21HSTATE AND FIRST THREE)
       WRITE(1,220)
   220 FORMAT(1H ,49HEXCITED STATES AS A FUNCTION OF THE PERTURBATION ,
      1 20HVP. VP IS PLOTTED ON)
       WRITE(1,222)VP1,VP2
   222 FORMAT(1H ,36HTHE HORIZONTAL AXIS AND VARIES FROM ,E11.4,4H TO ,
      1 E11.4)
       WRITE(1,224)EMAX
   224 FORMAT(1H ,47HTHE ENERGY IS PLOTTED ON THE VERTICAL AXIS AND ,
      1 12HVARIES FROM ,E11.4)
       WRITE(1,226)EMIN
```

MAIN PROGRAM

```
  226 FORMAT(1H ,10X,3HTO ,E11.4)
      WRITE(1,208)
C  PLOT GRAPH
      DO 70 J=1,IHG
      ETS=EMAX-SCALE*(FLOAT(J)-1.5)
      EBS=EMAX-SCALE*(FLOAT(J)-0.5)
      DO 72 I=1,NVP
      K1=(I-1)*IGAP+1
      K2=I*IGAP
      DO 74 K=K1,K2
      PLOT(K)=PLOT5
      IF(ETS.GE.0.0.AND.EBS.LT.0.0)PLOT(K)=PLOT6
      VP=VP1+DVP*FLOAT(K-1)/FLOAT(IGAP)
      IF(VP.LE.(DDVP/2.0).AND.VP.GT.(-DDVP/2.0))PLOT(K)=PLOT7
   74 CONTINUE
      IF(E(I,1).LE.ETS.AND.E(I,1).GT.EBS)PLOT(K1)=PLOT1
      IF(E(I,2).LE.ETS.AND.E(I,2).GT.EBS)PLOT(K1)=PLOT2
      IF(E(I,3).LE.ETS.AND.E(I,3).GT.EBS)PLOT(K1)=PLOT3
      IF(E(I,4).LE.ETS.AND.E(I,4).GT.EBS)PLOT(K1)=PLOT4
   72 CONTINUE
      WRITE(1,178)(PLOT(K),K=1,IWG)
  178 FORMAT(1H ,120A1)
   70 CONTINUE
   62 CONTINUE
C***********************************************************************
C
C     THE OPTION IS NOW PROVIDED  -  WITH AN INTERACTIVE SYSTEM - OF TRYING
C         DIFFERENT SETS OF VP'S.
C
      WRITE(1,132)
  132 FORMAT(/54HDO YOU WANT TO RE-RUN THE PROGRAM WITH THE SAME VALUES,
     1 14HFOR L1, L2, L3,/41HAND MP BUT WITH DIFFERENT VALUES FOR THE ,
     2 24HPERTURBING POTENTIAL VP.,/15H TYPE YES OR NO)
   46 CONTINUE
      READ(1,171)YON
      IF(YON.EQ.Y)GOTO 28
      IF(YON.EQ.YN)GOTO 48
      WRITE(1,136)
  136 FORMAT(39HRESPONSE NOT RECOGNISED. TYPE YES OR NO)
      GOTO 46
   48 CONTINUE
C***********************************************************************
C  OUTPUT SUPPLEMENTARY INFORMATION TO AID STUDENT'S PERTURBATION
C     CALCULATION.
      WRITE(1,208)
      WRITE(1,228)
  228 FORMAT(1H ,5X,1HN,6X,15HSIN(N*PI*L1/L3),6X,15HSIN(N*PI*L2/L3))
      WRITE(1,208)
      WRITE(1,140)(I,SL1(I),SL2(I),I=1,14)
  140 FORMAT(4X,I2,8X,F10.6,11X,F10.6)
      WRITE(1,312)
  312 FORMAT(1H ,16HPROGRAM FINISHED)
      STOP
      END
```

SUBROUTINE PL3FE

```
      SUBROUTINE PL3FEN(E,PL3FE,NODES)
C  SUBROUTINE TO RETURN THE VALUE OF THE WAVE FUNCTION AT L3
C  AND THE NUMBER OF NODES BETWEEN 0 AND L3
C  GIVEN THE ENERGY E, THE PERTURBATION VP AND THE DIMENSIONS.
C  THE DIMENSIONS (=L1,L2,L3), PERTURBING POTENTIAL (=VP) AND
C  PARTICLE MASS (=MP) ARE ALL TRANSFERED IN THE COMMON BLOCK.
C  THE MNEMONICS USED FOR THESE ARE AS IN THE MAIN PROGRAM.
      COMMON L1,L2,L3,VP,MP,N,PI
      REAL L1,L2,L3,MP,K1,K2
C  IF VP = 0 THE CALCULATION OF THE WAVE FUNCTION AT
C  L3 IS SIMPLE.
      IF(VP.NE.0.0)GOTO 1
      IF(E)5,7,3
    7 CONTINUE
      PL3FE=L3
      NODES=0
      GOTO 24
    3 CONTINUE
      K1=SQRT(MP*E)
      PL3FE=SIN(K1*L3)/K1
      NODES=IFIX(L3*K1/PI)
      GOTO 24
    5 CONTINUE
      NODES=0
      ALPHA=SQRT(-MP*E)
      F1=EXP(ALPHA*L3)
      F2=1.0/F1
      PL3FE=(F1-F2)/(2.0*ALPHA)
      GOTO 24
    1 CONTINUE
      NODES=0
C  CALCULATE THE WAVE FUNCTION AT L1 (=PL1) AND ITS SLOPE AT L1 (=DPL1)
      IF(E)2,4,6
    2 CONTINUE
      ALPHA=SQRT(-MP*E)
      F1=EXP(ALPHA*L1)
      F2=1.0/F1
      PL1=(F1-F2)/(2.0*ALPHA)
      DPL1=(F1+F2)/2.0
      GOTO 8
    4 CONTINUE
      PL1=L1
      DPL1=1.0
      GOTO 8
    6 CONTINUE
      K1=SQRT(MP*E)
      PL1=SIN(K1*L1)/K1
      DPL1=COS(K1*L1)
      NODES=IFIX(L1*K1/PI)
    8 CONTINUE
C  CALCULATE THE WAVE FUNCTION AT L2 (=PL2) AND ITS SLOPE AT L2 (=DPL2)
      IF(E-VP)10,12,14
   10 CONTINUE
      BETA=SQRT(-MP*(E-VP))
      F1=EXP(BETA*(L1-L2))
      A=PL1-DPL1/BETA
```

```
      SUBROUTINE PL3FE
          B=PL1+DPL1/BETA
          PL2=A*F1/2.0+B/(2.0*F1)
          DPL2=BETA*(-A*F1/2.0+B/(2.0*F1))
          IF(PL1.EQ.0.0)GOTO 16
          IF(PL2.NE.0.0)GOTO 11
          NODES=NODES+1
          GOTO 16
       11 CONTINUE
          IF((PL1/PL2).LT.0.0)NODES=NODES+1
          GOTO 16
       12 CONTINUE
          A=DPL1
          B=PL1-A*L1
          PL2=A*L2+B
          DPL2=A
          IF(PL1.EQ.0.0)GOTO 16
          IF(PL2.NE.0.0)GOTO 13
          NODES=NODES+1
          GOTO 16
       13 CONTINUE
          IF((PL1/PL2).LT.0.0)NODES=NODES+1
          GOTO 16
       14 CONTINUE
          K2=SQRT(MP*(E-UP))
          IF(L1.NE.0.0)GOTO 15
          PL2=SIN(K2*L2)/K2
          DPL2=COS(K2*L2)
          NODES=NODES+IFIX(L2*K2/PI)
          GOTO 16
       15 CONTINUE
          DELTA=ATAN(K2*PL1/DPL1)-K2*L1
          A=PL1/SIN(K2*L1+DELTA)
          PL2=A*SIN(K2*L2+DELTA)
          DPL2=K2*A*COS(K2*L2+DELTA)
          NODES=NODES+IFIX((L2*K2+DELTA)/PI+1.0)-IFIX((L1*K2+DELTA)/PI+1.0)
       16 CONTINUE
    C  CALCULATE THE WAVE FUNCTION AT L3 (=PL3FE)
          IF(E)18,20,22
       18 CONTINUE
          F1=EXP(ALPHA*L2)
          B=(PL2+DPL2/ALPHA)/(2.0*F1)
          A=PL2*F1-B*F1*F1
          F1=EXP(ALPHA*L3)
          PL3FE=A/F1+B*F1
          IF(PL2.EQ.0.0)GOTO 24
          IF(PL3FE.NE.0.0)GOTO 17
          NODES=NODES+1
          GOTO 24
       17 CONTINUE
          IF((PL2/PL3FE).LT.0.0)NODES=NODES+1
          GOTO 24
       20 CONTINUE
          A=DPL2
          B=PL2-A*L2
          PL3FE=A*L3+B
          IF(PL2.EQ.0.0)GOTO 24
```

```
      SUBROUTINE PL3FE
            IF(PL3FE.NE.0.0)GOTO 19
            NODES=NODES+1
            GOTO 24
        19 CONTINUE
            IF((PL2/PL3FE).LT.0.0)NODES=NODES+1
            GOTO 24
        22 CONTINUE
            IF(L2.NE.L3)GOTO 26
            PL3FE=PL2
            GOTO 24
        26 CONTINUE
            DELTA=ATAN(K1*PL2/DPL2)-K1*L2
            A=PL2/SIN(K1*L2+DELTA)
            PL3FE=A*SIN(K1*L3+DELTA)
            NODES=NODES+IFIX((L3*K1+DELTA)/PI+1.0)-IFIX((L2*K1+DELTA)/PI+1.0)
        24 CONTINUE
            RETURN
            END
```

```
FUNCTION EFPL30
      FUNCTION EFPL30(EF,ES,PL3F,PL3S,EAC,FAC)
C   THIS FUNCTION LOCATES THE VALUE OF THE ENERGY =EFPL30 WHEN THE WAVE
C   FUNCTION AT L3 IS ZERO IN THE INTERVAL EF TO ES BY A COMBINATION OF
C   THE METHODS OF LINEAR EXTRAPOLATION, INTERPOLATION AND BISECTION.
C   THE ACCURACY IS SPECIFIED BY THE PARAMETERS EAC AND FAC.
      COMMON L1,L2,L3,VP,MP,N,PI
      REAL L1,L2,L3,MP
      LOGICAL SWITCH
      SWITCH=.FALSE.
      EA=EF
      EB=ES
      PL3A=PL3F
      PL3B=PL3S
      EX=PL3S*(ES-EF)/(PL3S-PL3F)
      ET=ES-EX
      DO 2 I=1,1000
      CALL PL3FEN(ET,PL3T,NODES)
      IF(ABS(PL3T).GT.(L3*FAC))GOTO 16
      EFPL30=ET
      GOTO 14
   16 CONTINUE
C   TEST IF ET LIES OUTSIDE PREVIOSLY FOUND VALUES
      IF(ET.GE.EB.OR.ET.LE.EA)GOTO 4
C   TEST WHETHER EXTRAPOLATION WOULD INVOLVE DIVISION BY 0
      IF(PL3T.EQ.PL3S)GOTO 4
C   RESET EA AND EB
      IF(NODES.EQ.N)GOTO 10
      EA=ET
      PL3A=PL3T
      GOTO 12
   10 CONTINUE
      EB=ET
      PL3B=PL3T
   12 CONTINUE
C   USE LINEAR EXTRAPOLATION
      EF=ES
      PL3F=PL3S
      ES=ET
      PL3S=PL3T
      EX=PL3S*(ES-EF)/(PL3S-PL3F)
      IF((ABS(EX)).LT.EAC)GOTO 6
      IF(ET.EQ.0.0)GOTO 18
      IF(ABS(EX/ET).LT.FAC)GOTO 6
   18 CONTINUE
      ET=ES-EX
      GOTO 2
    4 CONTINUE
      IF(SWITCH)GOTO 8
C   USE LINEAR INTERPOLATION FROM THE TWO CLOSEST PREVIOUS VALUES
      EX=PL3A*(EA-EB)/(PL3A-PL3B)
      ET=EA-EX
      SWITCH=.TRUE.
      IF(ABS(EX).GT.EAC)GOTO 2
      IF(ET.EQ.0.0)GOTO 2
      IF(ABS(EX/ET).GT.FAC)GOTO 2
   20 CONTINUE
```

```
      FUNCTION EFPL30
          EFPL30=ET
          GOTO 14
        8 CONTINUE
C     ALTERNATIVELY, USE BISECTION FROM TWO CLOSEST PREVIOUS VALUES
          EX=(EA-EB)*0.5
          ET=(EA+EB)*0.5
          SWITCH=.FALSE.
          IF(ABS(EX).GT.EAC)GOTO 2
          IF(ET.EQ.0.0)GOTO 22
          IF(ABS(EX/ET).LT.FAC)GOTO 6
       22 CONTINUE
          EFPL30=ET
          GOTO 14
        2 CONTINUE
          WRITE(1,100)N
      100 FORMAT(1H ,44HWARNING : PROGRAM DOES NOT FIND THE CORRECT ,I1,
         1 3H TH)
          WRITE(1,102)VP
      102 FORMAT(1H ,20HEIGENSTATE FOR VP = ,E12.5)
          EFPL30=ES
          GOTO 14
        6 CONTINUE
          EFPL30=ES-EX
       14 CONTINUE
          RETURN
          END
```

Physics Programs
Edited by A. D. Boardman
© 1980 John Wiley & Sons Ltd.

CHAPTER 9

Simulation of Phonon Dispersion Curves and Density of States

G. J. KEELER

1. INTRODUCTION

In order to study the vibrational properties of crystalline solids, it is necessary to know in detail the frequency dependence of the normal modes of vibration of the crystal lattice.

Neutron-scattering measurements and other experimental observations provide overwhelming evidence that the normal modes are quantized, with energy $\hbar\omega$, and these quantized vibrations are referred to as phonons. An understanding of both the microscopic properties of the phonons, and macroscopic properties related to thermal vibrations (such as the specific heat and optical properties of insulating materials), requires a knowledge of the phonon dispersion curves and density of states.

In spite of the fundamental role played by the density states, it is very difficult to measure directly, and is almost invariably computed. Experimental data will normally give information on the phonon dispersion curves, and these will then be used to determine the interatomic force constants, by comparing the experimental curves with those calculated from the force constants. Even when these have been determined, it is by no means simple to calculate the density of states analytically, but modern computing methods have proved an ideal tool for solving the problem numerically.

2. LINEAR ATOMIC CHAIN

2.1 Dispersion curves

Before discussing dispersion curves in detail, it is worth pointing out that although calculation of specific heats, for example, requires a quantum-mechanical treatment, phonon dispersion curves can be calculated from a purely classical treatment.

Many of the salient features of lattice vibrations can be most clearly illustrated by consideration of a linear chain of atoms, as shown in Figure 1.

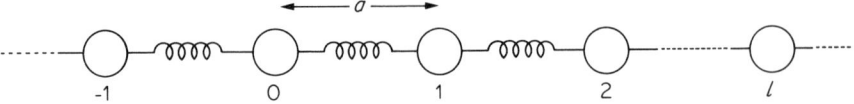

Figure 1. Linear chain of identical atoms, shown schematically with interatomic forces between nearest neighbours only.

We suppose that there are interactions between each pair of atoms in the chain (although Figure 1 has for simplicity been drawn to suggest forces between nearest neighbours only). We shall assume that the forces obey Hooke's law, i.e. they are linear in the relative displacements of the atoms (the 'harmonic approximation'), and we can then define interatomic force constants γ_l such that if u_l denotes the displacement of the lth atom, then the force on the atom at the origin due to the displacement u_l is

$$F_0 = +\gamma_l(u_l - u_0). \tag{1}$$

Summing over all atoms (l +ve and −ve), the equation of motion of the atom at the origin is

$$m\frac{\partial^2 u_0}{\partial t^2} = \sum_{l \neq 0} \gamma_l(u_l - u_0). \tag{2}$$

If we look for a solution for u_0 in the form of a wave of frequency ω and wave number $k(=2\pi/\lambda)$ travelling in the x-direction, this will have the general form

$$u(x) = A e^{i(\omega t - kx)}. \tag{3}$$

However, we need only consider displacements at actual atomic sites, so if the lattice spacing is a,

$$u_l = A e^{i(\omega t - kla)},$$

giving

$$-m\omega^2 A e^{i\omega t} = \sum_{l \neq 0} \gamma_l A \{e^{i(\omega t - kla)} - e^{i\omega t}\}. \tag{4}$$

Since by symmetry $\gamma_l = \gamma_{-l}$, we may rewrite this, after cancellation, as

$$m\omega^2 = \sum_{l > 0} 2\gamma_l(1 - \cos kla). \tag{5}$$

Consider the simple example of nearest-neighbour interactions only ($\gamma_1 = \gamma$, $\gamma_2 = \gamma_3 = \ldots = 0$). Then equation (5) gives

$$\omega^2(k) = \frac{2\gamma}{m}(1 - \cos ka), \quad \text{or} \quad \omega(k) = 2\sqrt{\frac{\gamma}{m}} |\sin \tfrac{1}{2}ka|. \tag{6}$$

SIMULATION OF PHONON DISPERSION CURVES AND DENSITY OF STATES 271

Thus the ω–k relationship, which we call the *dispersion curve*, is periodic in k as shown in Figure 2. However, let us consider the physical significance of this periodicity by looking at the relative motion of two successive atoms:

$$\frac{u_1}{u_0} = \frac{A e^{i(\omega t - ka)}}{A e^{i\omega t}} = e^{-ika}. \tag{7}$$

Thus a range of values of k of $2\pi/a$ covers all possible values of u_1/u_0. Since k must be allowed both positive and negative values to represent waves propagating in either direction, the range of independent values of k is

$$-\frac{\pi}{a} \leq k \leq \frac{\pi}{a}, \tag{8}$$

and this is called the *first Brillouin zone*.

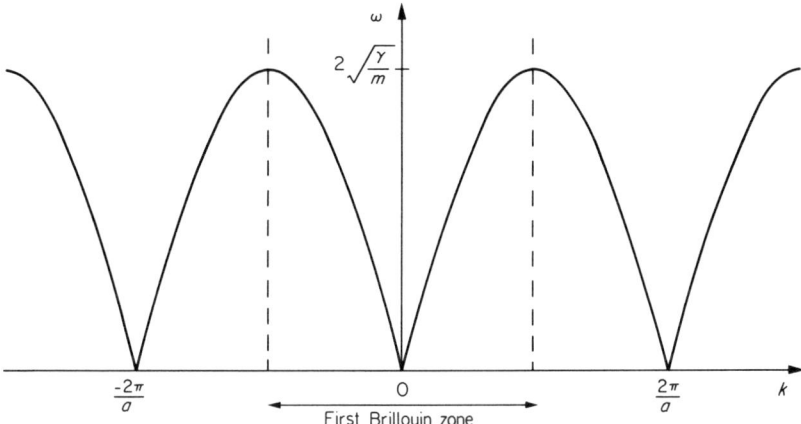

Figure 2. Plot of ω versus k (phonon dispersion curve) for a linear chain of atoms with interactions between nearest neighbours only.

2.2 Density of states

The above treatment assumes that all values of k are possible. However, physically the chain must be bounded, so that we must apply appropriate boundary conditions. The particular choice of boundary condition affects only the fine details of the density of states, and it is sufficient to consider the most commonly utilized boundary condition, which is for the solution to be periodic over a very large number of lattice spacings, N. Thus $u_l = u_{l+N}$, i.e.

$$e^{-ikla} = e^{-ik(l+N)a} \quad \text{or} \quad e^{-ikNa} = 1.$$

Hence, $kNa = 2n\pi$, where n is any integer, or

$$k = 0, \pm\frac{2\pi}{aN}, \pm\frac{2\pi}{a}\frac{2}{N}, \ldots, \pm\frac{2\pi\frac{1}{2}N}{a\ N}, \qquad (9)$$

i.e. there are approximately N possible values of k, evenly distributed throughout k-space with an interval $2\pi/Na$. Thus the *density of states in k-space*, $W(k)$—that is, the number of possible modes of vibration per unit interval in k—is a constant given by $2\pi W(k)/Na = 1$ or $W(k) = Na/2\pi$.

Of more interest, however, is the density of states as a function of frequency, $D(\omega)$. Now the number of states in a small frequency interval $d\omega$ will be

$$D(\omega)\,d\omega = W(k)\,dk = W(k)\frac{dk}{d\omega}\,d\omega; \qquad (10)$$

therefore

$$D(\omega) = \frac{W(k)}{d\omega/dk}. \qquad (11)$$

For the one-dimensional linear chain with nearest-neighbour interactions, it is easy to show that equation (11) becomes

$$D(\omega) = \frac{2N}{\pi(\omega_0^2 - \omega^2)^{\frac{1}{2}}}, \qquad \omega_0 = \sqrt{\frac{4\gamma}{m}}. \qquad (12)$$

The factor 2 arises because ω is always positive, so the negative and positive regions of k-space both contribute to the range $d\omega$.

2.3 Extension to all-neighbour forces

The simple form of dispersion curve shown in Figure 2 is a result of assuming nearest-neighbour forces only. Extension to further neighbours is quite simple in one dimension. For instance, if $\gamma_1 \neq 0$ and $\gamma_2 \neq 0$, equation (5) gives

$$\omega^2 = \frac{2}{m}(\gamma_1 + \gamma_2 - \gamma_1\cos ka - \gamma_2\cos 2ka). \qquad (13)$$

Figure 3 illustrates the case $\gamma_1 = \gamma_2$.

2.4 More than one atom per unit cell

Considerable complication occurs when the atoms in the chain are not all equivalent. Variations may occur in the masses, the force constants, and the atomic spacing. However, the essential features can be illustrated if we

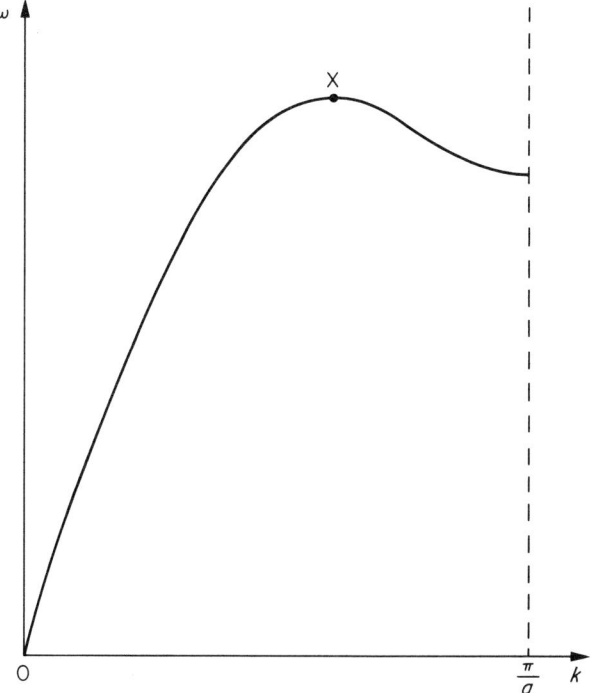

Figure 3. Phonon dispersion curve (first Brillouin zone, $k>0$) for a linear chain of atoms with equal force constants for nearest- and second-nearest-neighbour interactions. X labels the point where the phonon group velocity is zero

assume even atomic spacing, a single, nearest-neighbour force constant γ, but different atomic masses, as shown in Figure 4.

The lattice spacing b is now the distance between like atoms (i.e. $b=2a$), since the *unit cell* now contains two atoms. Thus we have two equations of motion, for the two types of atom (note that the suffix on u labels atom

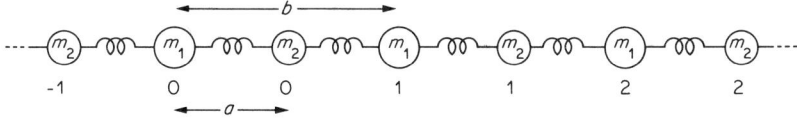

Figure 4. Linear chain of atoms of two species, having different masses but identical spacing and force constants. Although the distance apart is still a, the unit cell is of size $b=2a$.

position, whereas the suffix on m and the bracketed number label the species of atom):

$$m_1 \frac{\partial^2 u_0(1)}{\partial t^2} = \gamma\{u_0(2) - u_0(1)\} + \gamma\{u_{-1}(2) - u_0(1)\}$$
$$= \gamma\{u_0(2) + u_{-1}(2) - 2u_0(1)\}; \quad (14)$$

$$m_2 \frac{\partial^2 u_0(2)}{\partial t^2} = \gamma\{u_1(1) + u_0(1) - 2u_0(2)\}. \quad (15)$$

We now use, in equations (14) and (15), the plane wave solution

$$u_l(1) = A e^{i(\omega t - klb)},$$
$$u_l(2) = B e^{i[\omega t - k(l+\frac{1}{2})b]}. \quad (16)$$

This gives

$$-m_1\omega^2 A = -2\gamma\{A - B\cos\tfrac{1}{2}kb\};$$
$$-m_2\omega^2 B = -2\gamma\{B - A\cos\tfrac{1}{2}kb\}. \quad (17)$$

The condition for equation (17) to have non-trivial solutions for A and B is that the determinant of coefficients of A and B must vanish, i.e.

$$\begin{vmatrix} 2\gamma - m_1\omega^2 & -2\gamma\cos\tfrac{1}{2}kb \\ -2\gamma\cos\tfrac{1}{2}kb & 2\gamma - m_2\omega^2 \end{vmatrix} = 0, \quad (18)$$

giving

$$\omega^2 = \gamma\left(\frac{1}{m_1} + \frac{1}{m_2}\right) \pm \gamma\left[\left(\frac{1}{m_1} + \frac{1}{m_2}\right)^2 - \frac{4\sin^2\tfrac{1}{2}kb}{m_1 m_2}\right]^{\frac{1}{2}}. \quad (19)$$

We can write this as

$$\omega^2 = \tfrac{1}{2}\omega_0^2[1 \pm \sqrt{1 - C\sin^2\tfrac{1}{2}kb}] \quad (20)$$

where $\omega_0^2 = 2\gamma(1/m_1 + 1/m_2)$ and C is commonly called a coupling coefficient, defined as

$$C = \frac{4m_1 m_2}{(m_1 + m_2)^2} = \frac{4\rho}{(1+\rho)^2}, \quad \rho = \frac{m_1}{m_2}. \quad (21)$$

The dispersion curve is now as shown in Figure 5 (note that had we plotted ω^2 rather than ω, the curves would have been symmetrical about $\tfrac{1}{2}\omega_0^2$). The upper and lower branches of the curve are referred to as the optic and acoustic branches respectively, since this describes the way the corresponding phonons can be created in the long-wavelength limit.

We would expect that as $m_1 \to m_2$, the above case should go over smoothly to the single atom case, and a quick check of equation (19) above will confirm this, but the dispersion curve would appear to be quite different. However, the anomaly is resolved if we remember that $b = 2a$, and when

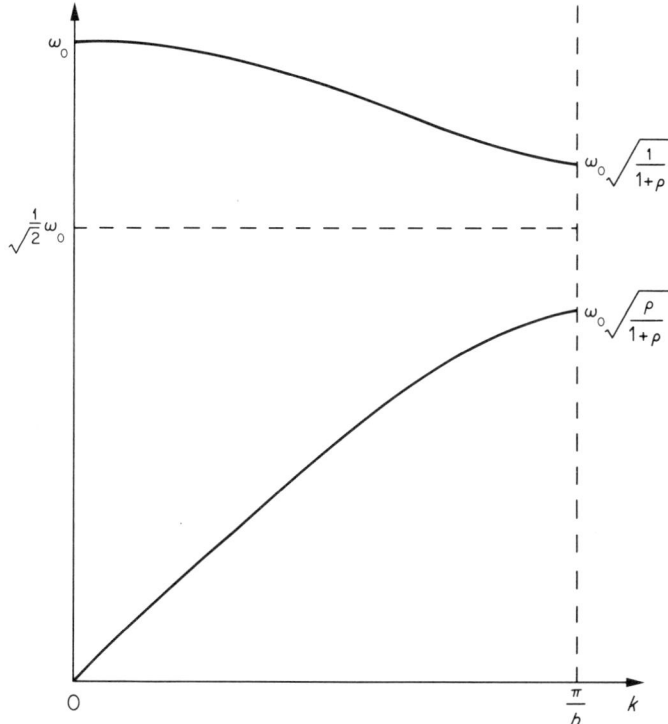

Figure 5. Phonon dispersion curve for a diatomic linear chain, with atoms of different masses m_1 and m_2 (where $\rho = m_1/m_2$, m_1 being the smaller mass) but identical spacing and force constants (between nearest neighbours only)

$m_1 \neq m_2$, the Brillouin zone is halved in size. Thus for the limiting case $m_1 = m_2$ the upper curve has been artificially 'folded back' from that part of the Brillouin zone where $\pi/2a < k \leq \pi/a$, as shown in Figure 6.

3. THREE-DIMENSIONAL CRYSTAL LATTICE

3.1 Normal mode frequencies in three dimensions

The biggest problem in generalizing the previous treatment to three dimensions is that the force constants become considerably more complicated, because the force on each atom due to displacement of its neighbours is a vector. Consider first how equation (1) for the force in one dimension may be rewritten:

$$F_0 = \sum_{l \neq 0} \gamma_l (u_l - u_0) = \sum_{l \neq 0} \gamma_l u_l - \sum_{l \neq 0} \gamma_l u_0. \tag{22}$$

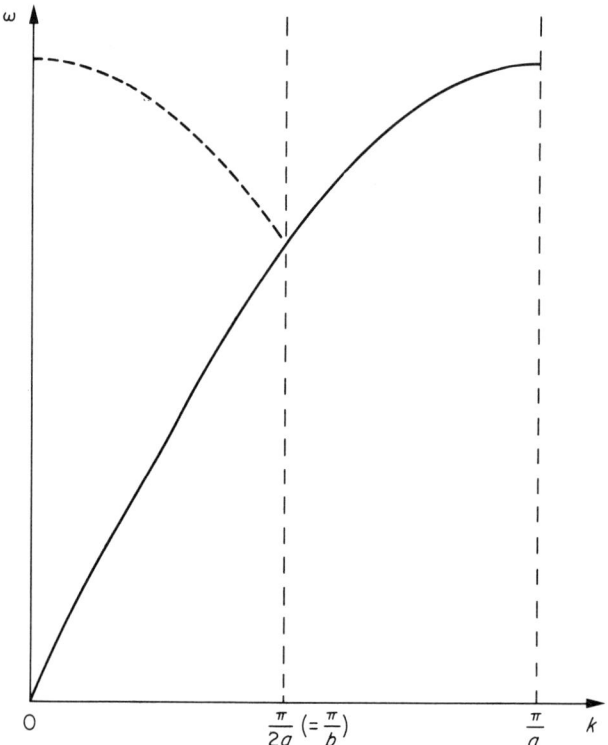

Figure 6. The true phonon dispersion curve for a monatomic linear chain (solid line) and when treated as the limiting case $m_1 = m_2$ for a diatomic linear chain (broken line), in which case the Brillouin zone is artifically reduced in size

If we define a special quantity γ_0 as $\gamma_0 = -\sum_{l \neq 0} \gamma_l$, F_0 reduces to

$$F_0 = \sum_{l=-\infty}^{+\infty} \gamma_l u_l. \tag{23}$$

The physical significance of γ_0 is simply that it represents the restoring force constant on the atom at the origin due to a displacement of itself.

In three dimensions, equation (23) can be generalized to

$$\mathbf{F}_0 = -\sum_l \mathbf{\Phi}_l \cdot \mathbf{u}_l, \tag{24}$$

(the minus sign is introduced simply for convenience, because $\mathbf{\Phi}$ then represents the second derivative of the equilibrium crystal potential energy) and the problem, which must be solved for the particular type of crystal structure under consideration, is to relate the tensor $\boldsymbol{\Phi}$ to the various force constants between neighbouring atoms.

If we use α and β as coefficients to represent the Cartesian components of the vectors, the force equation (24) can be written

$$F_{0,\alpha} = -\sum_{l}\sum_{\beta=1}^{3} \Phi_{l,\alpha\beta} u_{l,\beta}, \qquad \alpha = 1, 2, 3. \tag{25}$$

There are now three equations of motion for each atom, corresponding to the three degrees of freedom:

$$m\frac{\partial^2}{\partial t^2} u_{0,\alpha} = -\sum_{l}\sum_{\beta=1}^{3} \Phi_{l,\alpha\beta} u_{l,\beta} \tag{26}$$

and the three-dimensional form of a plane wave is

$$u_{l,\alpha}(\mathbf{k}) = A_{\alpha} \exp[i(\omega t - \mathbf{k} \cdot \mathbf{r}_l)], \tag{27}$$

for a wave travelling in the direction of $\mathbf{k}$, where $\mathbf{r}_l$ is the position vector of the lth atom. The precise form of $\mathbf{r}_l$ will depend on the crystal structure.

To facilitate the generalization to a unit cell containing more than one atom, it is convenient at this point to make two new definitions:

$$B_{\alpha} = m^{\frac{1}{2}} A_{\alpha}, \qquad \mathscr{D}_{\alpha\beta} = \frac{1}{m} \sum_{l} \Phi_{l,\alpha\beta} \exp(-i\mathbf{k} \cdot \mathbf{r}_l). \tag{28}$$

Using these quantities, substitution of equation (27) into the equations of motion (26) gives

$$\omega^2 B_{\alpha} = \sum_{\beta=1}^{3} \mathscr{D}_{\alpha\beta} B_{\beta}, \qquad \alpha = 1, 2, 3, \tag{29}$$

or in matrix form:

$$\omega^2 \mathbf{B} = \mathscr{D} \cdot \mathbf{B}. \tag{30}$$

This is a typical eigenvalue problem, and the solutions for ω^2 are the roots of the determinantal equation

$$\text{Det}(\mathscr{D} - \omega^2 \mathbf{I}) = 0. \tag{31}$$

The matrix $\mathscr{D}$ is commonly called the *dynamical matrix*, and the determinant is called the *secular determinant*.

It is interesting to note that, as might be expected, for the case of $\mathbf{k}$ in a symmetry direction the solutions (i.e. the normal modes of vibration) correspond to one longitudinal wave and two transverse waves (often degenerate). For non-symmetry directions, however, the normal modes do not necessarily have such a simple form.

3.2 Calculation of dynamical matrix for a BCC lattice

To proceed further it is necessary to specify the type of lattice and force constants to be included.

We shall consider the case of a body-centred cubic lattice which can be described by

$$\mathbf{r}_l = a\mathbf{l}, \qquad \mathbf{l} = (l_1, l_2, l_3), \tag{32}$$

where l_1, l_2, and l_3 are either all odd or all even integers.

We shall assume for simplicity that the interatomic force constants are purely radial, and that only nearest- and second-nearest-neighbour interactions are significant, with force constants γ and $R\gamma$, so that R represents the ratio of second-nearest to nearest force constants. We need to calculate the matrices $\mathbf{\Phi}_l$ for each nearest and second-nearest neighbour ($l = 1$–14), and for $l = 0$.

If the atom 0 lies at the centre of a cube of side $2a$, then the eight nearest neighbours lie on the corners of the cube, along the [111], type directions, as shown in Figure 7. Consider atom 1, lying on the [111]-direction. Referring back to equation (25), we see that $-\Phi_{l,\alpha\beta}$ is the force on atom 0 in the α-direction resulting from unit displacement of atom l in the β-direction.

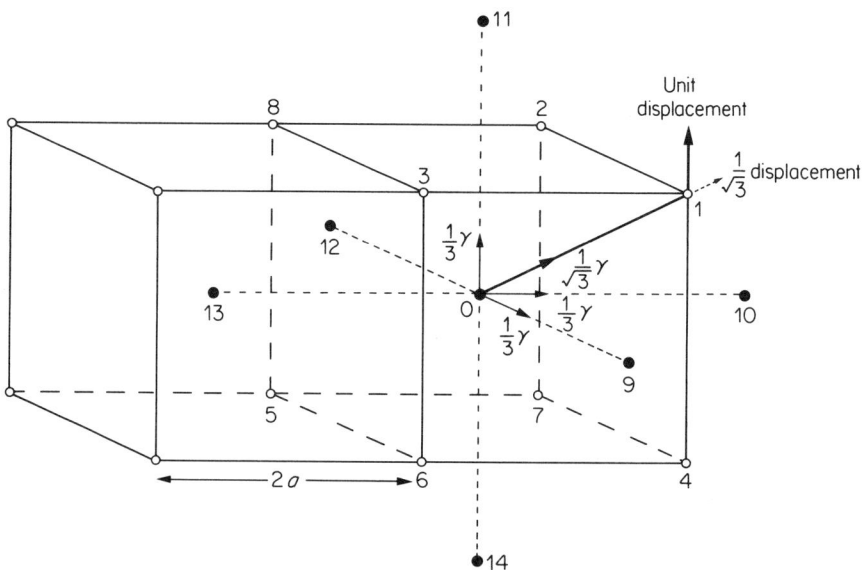

Figure 7. The BCC lattice, showing two of the unit cells, and all the nearest and second-nearest neighbours of the atom at the centre of the right-hand cell. This also represents the two-atom CsCl structure, where the two species are then represented by the solid and open circles respectively

Let us consider as an example the unit displacement of atom 1 in the x-, y- or z-direction. In each case this results in a radial component of displacement of $1/\sqrt{3}$, and since only radial forces are being considered, the resultant force on atom 0 is $(1/\sqrt{3})\gamma$. If we in turn resolve this force into its three Cartesian components, these can be written as $F_{0,x} = F_{0,y} = F_{0,z} = \frac{1}{3}\gamma$. Thus each Cartesian component of force due to unit displacement along any Cartesian axis is $\frac{1}{3}\gamma$, and this means that $-\Phi_{1,\alpha\beta} = \frac{1}{3}\gamma$ for all α and β. Thus

$$-\Phi_1 = \tfrac{1}{3}\gamma \begin{pmatrix} 1 & 1 & 1 \\ 1 & 1 & 1 \\ 1 & 1 & 1 \end{pmatrix}. \tag{33}$$

Exactly the same will be true for atom 5, in the $[\bar{1}\bar{1}\bar{1}]$-direction, i.e.

$$-\Phi_5 = -\Phi_1. \tag{34}$$

For the remaining atoms, along $[\bar{1}11]$-, $[1\bar{1}1]$-, and $[11\bar{1}]$-directions and their opposite neighbours, the matrices are respectively

$$-\Phi_2 = -\Phi_6 = \tfrac{1}{3}\gamma \begin{pmatrix} 1 & -1 & -1 \\ -1 & 1 & 1 \\ -1 & 1 & 1 \end{pmatrix}, \quad -\Phi_3 = -\Phi_7 = \tfrac{1}{3}\gamma \begin{pmatrix} 1 & -1 & 1 \\ -1 & 1 & -1 \\ 1 & -1 & 1 \end{pmatrix},$$

$$-\Phi_4 = -\Phi_8 = \tfrac{1}{3}\gamma \begin{pmatrix} 1 & 1 & -1 \\ 1 & 1 & -1 \\ -1 & -1 & 1 \end{pmatrix}. \tag{35}$$

The second-nearest neighbours lie on the six faces of a cube of side $4a$. Thus for atom 9, lying along the $[100]$-direction, a unit displacement in the x-direction gives a force $R\gamma$ in that direction on the atom at 0, and orthogonal displacements have no effect. Thus the only non-zero matrix elements are

$$-\Phi_{9,xx} = -\Phi_{12,xx} = -\Phi_{10,yy} = -\Phi_{13,yy} = -\Phi_{11,zz} = -\Phi_{14,zz} = R\gamma. \tag{36}$$

Finally we must calculate Φ_0. Now this is simply the force on atom 0 due to unit displacement of itself. Thus a unit displacement of 0 along $0x$ is equivalent to a displacement of all the neighbours along $-0x$. Adding the components of force due to all the other atoms, we find

$$-\Phi_{0,xx} = -\tfrac{8}{3}\gamma - 2R\gamma, \quad -\Phi_{0,xy} = 0, \quad -\Phi_{0,xz} = 0, \text{ etc.} \tag{37}$$

Thus

$$-\Phi_0 = -2\gamma(\tfrac{4}{3} + R) \begin{pmatrix} 1 & 0 & 0 \\ 0 & 1 & 0 \\ 0 & 0 & 1 \end{pmatrix}. \tag{38}$$

We are now in a position to calculate $\mathscr{D}$ from equation (28), i.e.

$$\mathscr{D}_{\alpha\beta} = \sum_l \frac{1}{m} \Phi_{l,\alpha\beta} \exp(-i\mathbf{k} \cdot \mathbf{r}_l). \tag{39}$$

Thus, for example, the contributions to $m\mathscr{D}_{xx}$ are, using Figure 7, from atoms 1 and 5:

$$-\tfrac{1}{3}\gamma[\exp\{-i(k_xa+k_ya+k_za)\}+\exp\{i(k_xa+k_ya+k_za)\}],$$

from atoms 2 and 6:

$$-\tfrac{1}{3}\gamma[\exp\{-i(-k_xa+k_ya+k_za)\}+\exp\{i(-k_xa+k_ya+k_za)\}],$$

from atoms 3 and 7:

$$-\tfrac{1}{3}\gamma[\exp\{-i(k_xa-k_ya+k_za)\}+\exp\{i(k_xa-k_ya+k_za)\}],$$

from atoms 4 and 8:

$$-\tfrac{1}{3}\gamma[\exp\{-i(k_xa+k_ya-k_za)\}+\exp\{i(k_xa+k_ya-k_za)\}],$$

from atoms 9 and 12:

$$-R\gamma[\exp(-ik_x \cdot 2a)+\exp(ik_x \cdot 2a)],$$

from atom 0:

$$+2\gamma(\tfrac{4}{3}+R).$$

Hence

$$\mathscr{D}_{xx} = \frac{2\gamma}{m}(-\tfrac{4}{3}\cos k_xa \cos k_ya \cos k_za - R\cos 2k_xa + \tfrac{4}{3}+R). \tag{40}$$

Other elements of $\mathscr{D}$ can be just as readily obtained, for example,

$$\mathscr{D}_{xy} = \mathscr{D}_{yx} = +\tfrac{8}{3}\gamma \sin k_xa \sin k_ya \cos k_za. \tag{41}$$

Using definitions such as

$$c_x = \cos k_xa, \quad s_x = \sin k_xa, \quad c_{2x} = \cos 2k_xa, \tag{42}$$
$$\delta = 1 + \tfrac{3}{4}R - c_xc_yc_z,$$

the elements of $\mathscr{D}$ can be considerably simplified. The final result is

$$\mathscr{D}(\mathbf{k}) = \frac{8\gamma}{3m} \begin{pmatrix} \delta - \tfrac{3}{4}Rc_{2x} & s_xs_yc_z & s_xc_ys_z \\ s_xs_yc_z & \delta - \tfrac{3}{4}Rc_{2y} & c_xs_ys_z \\ s_xc_ys_z & c_xs_ys_z & \delta - \tfrac{3}{4}Rc_{2z} \end{pmatrix}. \tag{43}$$

3.3 Two-atom unit cell

The principal difference in the equations, in the case of more than one atom per unit cell, is that we have three equations of motion for each of the atoms. If we allow l to label all atoms, of whichever sort (note this is the alternative procedure to the one adopted in section 2.4), then the force constant matrices will have the same form as before, but it will be necessary to label Φ and $\mathscr{D}$, and of course the masses m and the amplitudes $\mathbf{B}$, with the type of atom.

The reason for introducing $\mathbf{B}$ and $\mathscr{D}$ originally is that they must now take the form:

$$\mathbf{B}(\kappa) = m_\kappa^{\frac{1}{2}} \mathbf{A}(\kappa), \quad \mathscr{D}(\kappa, \kappa') = (m_\kappa m_{\kappa'})^{-\frac{1}{2}} \sum_l \Phi_l(\kappa, \kappa') \exp(-i\mathbf{k} \cdot \mathbf{r}_l), \quad (44)$$

where κ and κ' label the n different atoms in the unit cell. The equations of motion are

$$\omega^2 \mathbf{B}(\kappa) = \mathscr{D}(\kappa, \kappa) \cdot \mathbf{B}(\kappa) + \sum_{\kappa' \neq \kappa} \mathscr{D}(\kappa, \kappa') \cdot \mathbf{B}(\kappa'). \quad (45)$$

Equations (45) can be written more compactly as

$$\omega^2 \mathscr{B} = \mathscr{D}' \cdot \mathscr{B}, \quad (46)$$

where $\mathscr{D}'$ is a $3n \times 3n$ matrix and $\mathscr{B}$ is a column vector whose elements are $B_x(1)$, $B_y(1)$, $B_z(1)$, $B_x(2)$, etc.

Again we need to specify the type of lattice before proceeding any further. One of the simplest types of crystal structure having two atoms per unit cell is the CsCl structure, whose atoms form a BCC lattice having one type of atom at the cube centre and the other type on the corners of the cube, as shown in Figure 7. We shall assume as before that all the forces are radial and also that the force constants are the same for the two species of atom, so that we need consider only the difference in masses for the two types of atom. (In practice of course it would be physically unrealistic to consider only two force constants for an ionic crystal such as CsCl where the forces are quite long ranged.)

The matrices Φ will have exactly the same form as before, except that now they must also be labelled with the species of atom, and the first eight will couple different types whereas the next six couple identical atoms. Thus some matrices will now be zero, as follows: $\Phi_l(1, 1) = \Phi_l(2, 2) = 0$, $l = 1\text{–}8$; $\Phi_l(1, 2) = \Phi_l(2, 1) = 0$, $l = 9\text{–}14$; and of course $\Phi_0(1, 2) = \Phi_0(2, 1) = 0$ by definition, and $\Phi_0(1, 1) = \Phi_0(2, 2) = (\frac{8}{3}\gamma + 2\gamma R)\mathbf{I}$, as before.

If we remember that the ratio of masses is $\rho = m_1/m_2$, and use the notation of (42), we obtain for $\mathscr{D}'$

$$\mathscr{D}'(\mathbf{k}) = \tfrac{8}{3}\gamma(m_1 m_2)^{-\frac{1}{2}} \begin{pmatrix} \rho^{-\frac{1}{2}}(1+r_x) & 0 & 0 & -c_x c_y c_z & s_x s_y c_z & s_x c_y s_z \\ 0 & \rho^{-\frac{1}{2}}(1+r_y) & 0 & s_x s_y c_z & -c_x c_y c_z & c_x s_y s_z \\ 0 & 0 & \rho^{-\frac{1}{2}}(1+r_z) & s_x c_y s_z & c_x s_y s_z & -c_x c_y c_z \\ -c_x c_y c_z & s_x s_y c_z & s_x c_y s_z & \rho^{\frac{1}{2}}(1+r_x) & 0 & 0 \\ s_x s_y c_z & -c_x c_y c_z & c_x s_y s_z & 0 & \rho^{\frac{1}{2}}(1+r_y) & 0 \\ s_x c_y s_z & c_x s_y s_z & -c_x c_y c_z & 0 & 0 & \rho^{\frac{1}{2}}(1+r_z) \end{pmatrix},$$

$$(47)$$

where the r's arise from the second-nearest-neighbour forces, and are defined by $r_x = \frac{3}{4}R(1-c_{2x})$, etc.

3.4 Density of states

In order to make an analytical calculation of the density of states in three dimensions, it is necessary to integrate $(\nabla_\mathbf{k}\omega)^{-1}$ (the three-dimensional equivalent of equation (11)) over a constant frequency surface in **k**-space. Since this is not normally feasible, the usual approach is to compute the density of states numerically by the method first described by Walker.[1] This is usually referred to as the 'root sampling method', since it builds up the density of states by finding the roots of the secular equation at a large number of points in the Brillouin zone.

Just as in one dimension, the allowed values of **k** are evenly distributed in **k**-space, with a density such that there are N states in the first Brillouin zone. The simplest approach is to solve the secular determinant over a cubic mesh of points in **k**-space, and to plot a histogram of the computed frequencies. In practice, there are important refinements which involve interpolation between the calculated points,[2,3] but the principle remains the same.

In one dimension the edge of the first Brillouin zone was simply the point at which the wavelength was twice the lattice spacing and all higher values of k could be reflected back into the first Brillouin zone by simply subtracting a *reciprocal lattice vector*, $2\pi/a$.

In three dimensions, we need to be concerned with planes of atoms. (As a matter of fact the dispersion curve calculations of section 2.1 will also hold for vibrations of planes of atoms for waves in high symmetry directions, provided that the γ_l are reinterpreted as the interplanar force constants.)

For the two-atom unit cell of the CsCl structure, whose space lattice is simple cubic, the Brillouin zone is also a cube (but with its faces a perpendicular distance $\pi/2a$ from the origin since the unit cell has a side of $2a$). For a one-atom BCC lattice, on the other hand, the most widely spaced planes of atoms are the (110) planes, with spacing $\sqrt{2}a$, so that the limiting wavelength is $2\sqrt{2}a$, and (remembering that the wave vector is defined as $|\mathbf{k}| = 2\pi/\lambda$) the first Brillouin zone is that volume bounded by the 12 (110)-type planes, each a perpendicular distance $\pi/\sqrt{2}a$ from the origin.

It is not necessary to calculate frequencies over the whole Brillouin zone, of course. Just as in one dimension, we need only consider positive values of k_x, k_y, and k_z, which reduces calculations to the first octant. Moreover, symmetry reduces the calculations necessary still further. For a cubic lattice the x-, y-, and z-directions must be equivalent, so only one-third of the octant need be considered, and if we also remember that the (110) planes have mirror symmetry, this divides the unique portion of the Brillouin zone in half again, to only 1/48 of the whole zone (e.g. that part lying between the [100]-, [110]-, and [111]-directions).

It is worth noting at this point that for the BCC lattice the [111] dispersion curves do not repeat outside the Brillouin zone boundary, but only after twice this distance. The reason is that the (111) planes are closely spaced, and can support waves of very short wavelength (i.e. large wave vector). However, these do not lie within the first Brillouin zone because they can be reflected back into the zone (in fact, on to the zone boundary on the line joining the [$\bar{1}\bar{1}1$]- and [001]-directions) by subtracting a [110] reciprocal lattice vector.

3.5 Van Hove critical points

If we refer back to the analytical expression for the density of states in one dimension (equation (12)) we can see that there is a singularity at $\omega = \omega_0$. The reason for this is easy to understand if we re-examine Figure 2. The density of points in k-space is uniform, so that at any point on the curve where the slope approaches zero there will be a very large density of points in ω-space. For the simple model illustrated in Figure 2 this occurs at the Brillouin zone boundary, where $\omega = \omega_0$. However, any point on a dispersion curve where the slope (i.e. the group velocity of the phonons) is zero will create a singularity in the density of states—for instance the point X in Figure 3—and these are known as *Van Hove singularities*.

In three dimensions a similar feature occurs wherever $\nabla_\mathbf{k}(\omega) = 0$, except that whereas in equation (12) the effect was proportional to $(|\omega^2 - \omega_0^2|)^{-\frac{1}{2}}$, in three dimensions the corresponding contribution is proportional to the less drastic $(|\omega^2 - \omega_0^2|)^{\frac{1}{2}}$. The latter frequency dependence does not create a singularity at ω_0, merely a cusp or critical point in the curve as the extra contribution starts or ceases. Nevertheless, one of the most characteristic features of a three dimensional density of states curve is the series of Van Hove critical points, or discontinuities in the slope of the curve, and in many cases these can be associated with points of zero slope on the phonon dispersion curves in high symmetry directions.

4. COMPUTER PROGRAM

4.1 Computation of the density of states

In a normal experimental determination of the density of states, the actual measurements are made on the dispersion curves in symmetry directions in the crystal. The commonest method is probably inelastic neutron scattering, although the pioneering work by Walker[1] was based on X-ray diffuse scattering measurements.

The force constants (typically five to nine constants are determined, including some non-radial forces) are then found by solving the secular determinant along the symmetry directions and varying the constants to obtain the best fit to the experimental data.

For the purposes of the simulation program to be described here, this approach is reversed, and the user sets the ratio of the two force constants being considered. This value is then used to generate a theoretical set of dispersion curves so that the effect of the two constants on the dispersion curves can be examined.

The final determination of the density of states follows the normal experimental approach, using the values chosen for the force constants in the secular determinant and evaluating the determinant over a mesh of points in **k**-space to build up a histogram of the density of states.

4.2 Program details

Since numerical values are of less interest than the shape of the various curves, the program output consists entirely of curves, with arbitrary scales on the axes, plotted on the line printer (although it would be quite a simple task for the reader to alter the printing subroutines to have the output generated on a graph plotter). This means that it is unnecessary to specify values for the lattice spacing, force constant, or atomic mass, as it is only the ratios of the force constants and masses that affect the shape of the curves.

The program is based on the two subroutines D1 and D2, which set up the dynamical matrices for the one-atom and two-atom cases respectively. Each subroutine then calls a standard library subroutine to evaluate the secular determinant, whose eigenvalues are the squares of the frequencies of the normal modes of vibration. The program has been developed using the NAG subroutine F02AAF, which solves the simple eigenvalue problem $\mathbf{A} \cdot \mathbf{x} = \lambda \mathbf{x}$ for a real symmetric matrix. Any similar subroutine may be used instead, but it should be noted that the NAG subroutine returns the eigenvalues in ascending order. This fact is utilized in the main program, so if the subroutine chosen does not do this, it will be necessary to sort the values immediately after the subroutine call.

After reading in the necessary data (described in the next subsection) the program calculates the frequencies of the three normal modes at equal intervals of k, along the three symmetry directions [100], [110], and [111]. The subroutine PRDC then outputs the dispersion curves, in graphical form, on the line printer. The ω- and k-axes both have annotated scales with arbitrary units, but the respective scales are the same for all three sets of dispersion curves. Where the curves, or individual points on them, are degenerate, they are denoted by stars instead of crosses. For the one-atom case, the [111] dispersion curves do not repeat outside the Brillouin zone (as

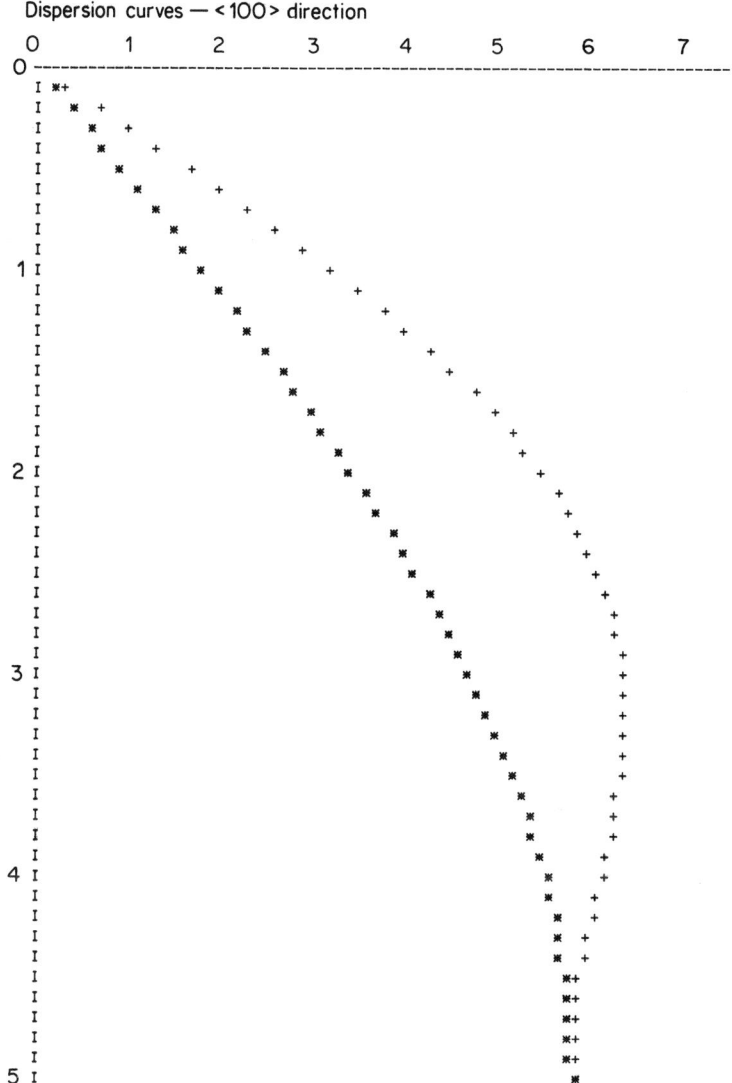

Figure 8. Specimen computer program output: the [100] phonon dispersion curve for a one-atom BCC crystal lattice. This particular curve is for the force constant ratio $R = 0.8$. DCXSC and DCYSC have been specified as 25 and 75 respectively; these values are smaller than would normally be specified for a line printer, but would be suitable for a terminal

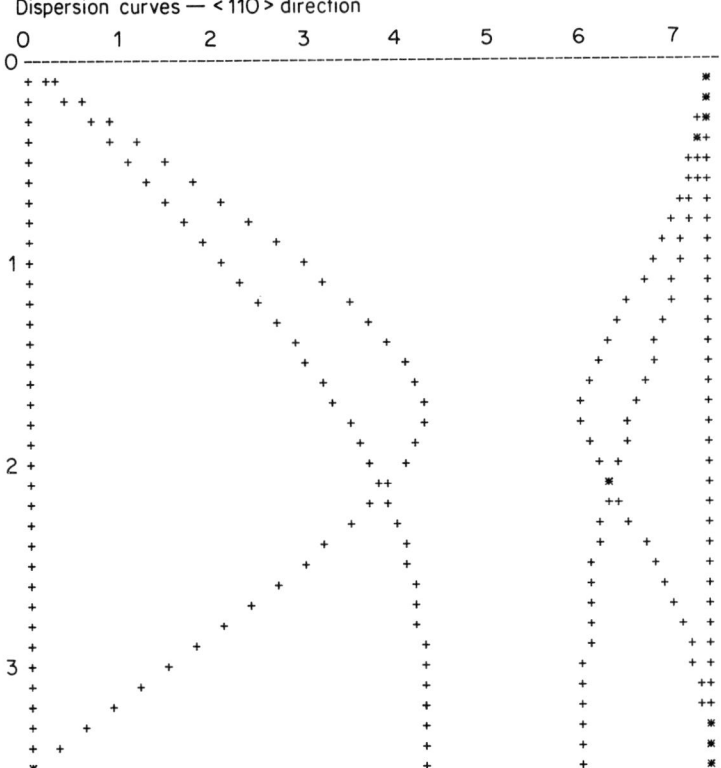

Figure 9. Specimen computer program output: the [110] phonon dispersion curve for a two-atom CsCl structure crystal. This particular curve is for a mass ratio $\rho = 0.5$ and force constant ratio $R = 0.0$. (Note that the latter choice has resulted in a soft mode in the [110]-directions)

was pointed out in section 3.4), so these curves are plotted beyond the zone boundary, which is marked by a broken line. Specimen curves for the one-atom and two-atom cases are illustrated in Figures 8 and 9 respectively.

The program then obtains values for the normal mode frequencies over a mesh of points in the irreducible portion of the Brillouin zone, giving only half weighting to points on the sides of the portion, since they are shared with the neighbouring portion. As the frequencies are calculated, they are immediately stored in histograms, which when completed form the densities of states for each of the modes of vibration. Instead of forming a simple histogram, each frequency value is made to contribute to the two frequency intervals centred above and below it, the contributions being weighted according to the closeness of the frequency value to the centre of the frequency interval. This weighting is a crude form of the refined calculations mentioned earlier, but it still results in a considerably smoother histogram for a given mesh size.

The subroutine PRDOS is used to plot the densities of states, again on the line printer. Three histograms, corresponding to the low-, medium-, and high-frequency normal modes, are output first. These do not necessarily correspond to any particular polarization, since the frequencies may cross over, and in any case the polarizations are not strictly longitudinal or transverse except in symmetry directions. For the two-atom case, acoustic and optic branches are combined in the same histogram since the frequencies do not overlap.

The total density of states is output last. The scales of the densities of states are again in arbitrary units, but the frequency scales do correspond to those of the dispersion curves, to facilitate comparison between them. Figure 10 illustrates typical output for the density of states, and a number of Van Hove critical points can clearly be seen.

4.3 Data for the program

Only six variables need to be specified as data. These are:
- DCXSC in I3 format
- DCYSC in I3 format

for each set of curves required:
- ATOMS in I3 format
- INT in I3 format
- R in F6.2 format

for the two-atom case only:
- RO in F6.2 format.

The function of these variables is as follows: DCXSC and DCYSC, which need to be specified only once, control the scale of the x- and y-axes of the output. The user may wish to experiment with these to obtain the most presentable size and relative scaling for the particular line printer (or terminal) in use. Once these are established, the values could be incorporated into the program.

ATOMS should be given the value 1 or 2 respectively to simulate the appropriate lattice—one-atom BCC or two-atom CsCl structure. The program will plot as many sets of curves as required and a value for ATOMS of zero must be specified to terminate the program.

INT sets the scale of the mesh for the density of states calculation. Obviously, the greater the number of mesh points the smoother the resulting histograms, but the number of points, and resulting computing time, will increase roughly as INT cubed. The user is left to set the size of the mesh

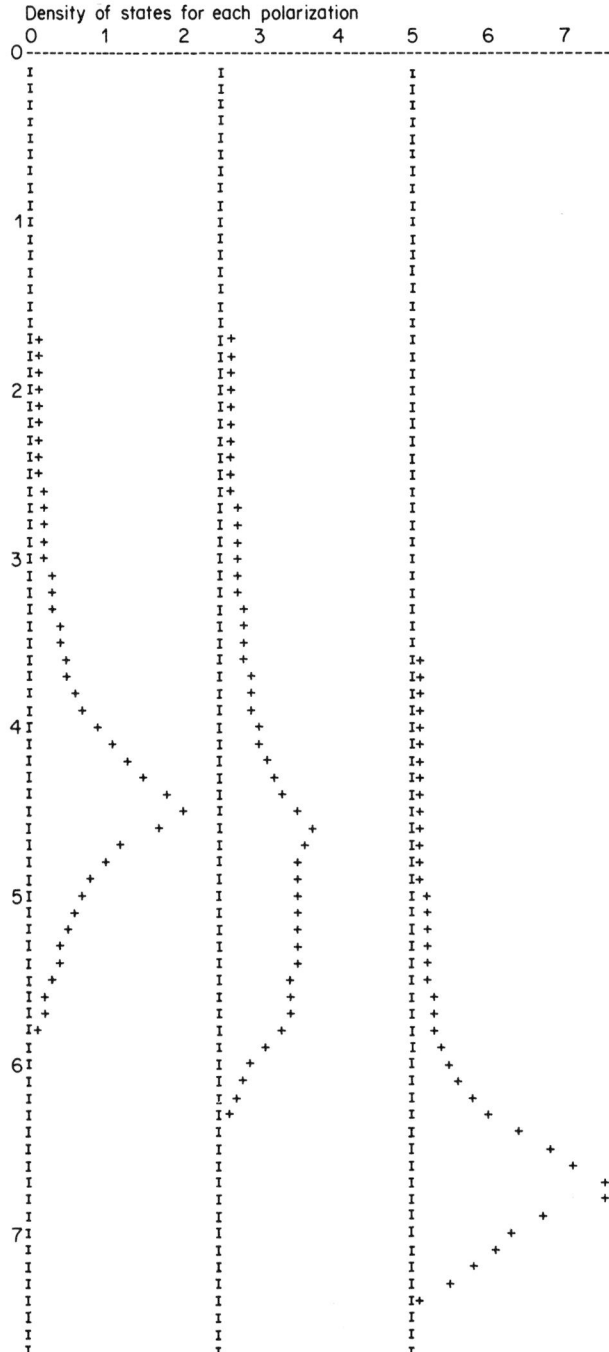

Figure 10. Specimen computer program output: the phonon densities of states. These particular curves are for a one-atom BCC lattice with $R = 0.8$ (a) Densities of states for the low-, medium-, and high-frequency polarizations; (b) total density of states

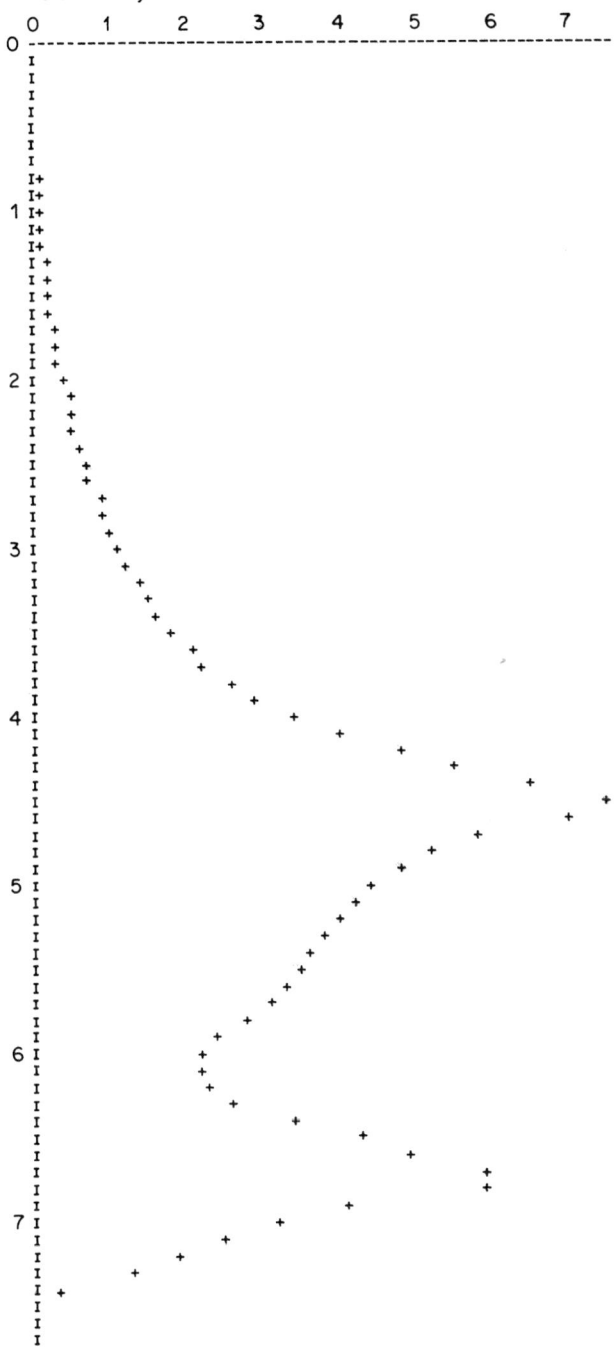

Figure 10b

according to the computing time available—a value of INT which gives reasonably smooth densities of states is 50.

R is the ratio of the second nearest to nearest-neighbour force constant, and any positive (or zero) value may be specified.

RO is only required for ATOMS = 2, and is the ratio of masses of the different species of atom. Only positive values are physically realistic, and to simulate all possibilities it is necessary only to investigate a range of values from 0 to 1.

5. EXERCISES FOR STUDENTS

5.1 Variation of parameters

The student can investigate the effect of varying the parameters R and RO for the two models. It is of particular interest to see the effects as the limiting values are approached.

In the limit as R becomes very large, the models approximate to two interpenetrating but non-interacting simple cubic lattices, and it will be seen that the second half of the [100] dispersion curve in the one-atom model becomes a reflection of the first half, because the larger simple cubic lattice has a Brillouin zone only half as long in the [100]-directions. Try and predict what the effect will be in the two-atom model.

Negative values of R (corresponding to a negative, or repulsive, second-neighbour force constant) are not disallowed in principle, but for the case of two radial force constants only it will be seen that the lattice becomes unstable at $R = 0.0$, as evidenced by the fact that one of the dispersion curves in the [110]-direction goes to zero. (A mode of vibration dropping to zero frequency is a characteristic of the onset of instability, and is often called a 'soft mode'.) This is because, without the stabilizing effect of (positive) second-neighbour forces, there is no resistance to shear of the (110) atom planes in the $[\bar{1}10]$-directions.

The results from the two models should be compared for the same values of R, and with $RO = 1.0$ for the two-atom model. This value of RO corresponds to a CsCl structure with identical atoms which therefore reduces to a one-atom BCC lattice. Why are the dispersion curves not identical?

5.2 Location of the Van Hove critical points

As many of the Van Hove critical points as possible should be located on the densities of states (a number can clearly be seen in Figure 10) for a variety of choices of parameters. It should be possible to associate most of the critical points with points of zero slope (including those at the zone boundary) on the dispersion curves. Why are there not a large number of critical points associated with zero slope positions elsewhere in the Brillouin zone?

5.3 Extension to face-centred cubic lattice

Readers with a reasonable programming ability should try to modify the program to perform the same calculations for a one-atom, face-centred cubic lattice (FCC). (The equivalent of a two-atom CsCl structure does not exist because there are four atoms to each cube, rather than two.) The only parts of the program which are affected are: the subroutine which calculates the dynamical matrix; the lengths of the dispersion curves; and also in consequence of the different Brillouin zone, the limits on the DO loops creating the mesh of points to calculate the densities of states.

The dynamical matrix for the FCC lattice is

$$\mathscr{D}(\mathbf{k}) = \frac{2\gamma}{m} \begin{pmatrix} \Delta + c_y c_z - Rc_{2x} & s_x s_y & s_x s_z \\ s_x s_y & \Delta + c_x c_z - Rc_{2y} & s_y s_z \\ s_x s_z & s_y s_z & \Delta + c_x c_y - Rc_{2z} \end{pmatrix}, \quad (48)$$

where $\Delta = 2 + R - (c_x c_y + c_y c_z + c_z c_x)$.

The first Brillouin zone is slightly more complicated, being bounded partly by six (100)-type planes a perpendicular distance π/a from the centre, and partly by eight (111)-type planes a perpendicular distance $\sqrt{3}\pi/2a$ from the centre. The conditions for a mesh point to lie within the zone will therefore be $k_x + k_y + k_z \leq 3\pi/2a$, $k_x \leq \pi/a$ and $k_z \leq k_y \leq k_x$ (the last condition is imposed to restrict the mesh to the irreducible 1/48 of the Brillouin zone). The limits of the dispersion curves are hence $k_x = k_y = k_z = \pi/2a$ for the [111]-direction, $k_x = \pi/a$ for the [100]-direction, and $k_x = k_y = 3\pi/4a$ for the [110]-direction. However, in an analogous fashion to the [111] curves of the BCC lattice, the [110] curves do not repeat outside the zone boundary, but only after the point $k_x = k_y = \pi/a$.

REFERENCES

1. C. B. Walker, *Phys. Rev.*, **103**, 547 (1956).
2. G. Gilat and G. Dolling, *Phys. Lett.*, **8**, 304 (1964).
3. G. Gilat and L. J. Raubenheimer, *Phys. Rev.*, **144**, 390 (1966).

LATTICE DYNAMICS PROGRAM

```
      INTEGER ATOMS,INT,KX,KY,KZ,K,LIM,INT1,INT2,INT3,DIR,J,I,L,M,WINT,
     1.        IFAIL,DCYSC,DCXSC,DOSXSC,CH(120)
      REAL PI,R,WMAX,NWMAX,WT,WFRAC,QX,QY,QZ,W(3,6,200),NW(6,200),
     1     N(200),RO,NMAX
      COMMON CH
C
C     ATOMS MUST BE SET TO 1 FOR THE CASE OF THE SINGLE ATOM, B.C.C.
C     LATTICE AND 2 FOR THE CASE OF THE TWO ATOM CSCL LATTICE.
C     J IS THUS THE NUMBER OF BRANCHES OF THE DISPERSION CURVES.
C     R IS THE RATIO OF SECOND-NEAREST TO NEAREST NEIGHBOUR FORCE
C     CONSTANTS. IN THE LATTER CASE, RO IS THE RATIO OF ATOMIC MASSES.
C     R MUST BE .GE. ZERO FOR THE MODELS CHOSEN, AND RO MUST BE .GT.
C     ZERO. VALUES OF RO .LE. ONE WILL COVER ALL EVENTUALITIES.
C     INT IS AN INTEGER, THE NUMBER OF POINTS IN K-SPACE IN EACH
C     DIRECTION TO BE USED IN THE CALCULATIONS OF THE DENSITY OF
C     STATES.
C
      IFAIL=0
      PI=4.0*ATAN(1.0)
      READ(1,1001)DCXSC
C
C     DCXSC CONTROLS THE SCALE OF THE K AXIS OF THE DISPERSION
C     CURVES. 40 IS SUGGESTED AS A SUITABLE VALUE, BUT IN ANY CASE IT
C     SHOULD NOT EXCEED 57 UNLESS THE SIZE OF ARRAY W IS CHANGED.
C
      READ(1,1001)DCYSC
C
C     DCYSC CONTROLS THE SCALE OF THE W AXIS OF ALL THE CURVES
C     AND THE Y AXIS OF THE DENSITIY OF STATES CURVES.
C     117 IS SUGGESTED AS A SUITABLE VALUE FOR A 120 CHARACTER
C     LINE PRINTER, BUT IN ANY CASE IT MUST BE AT LEAST 3 LESS
C     THAN THE AVAILABLE NUMBER OF CHARACTERS, IT MUST BE
C     DIVISIBLE BY 3, AND IT SHOULD NOT EXCEED 120 UNLESS THE
C     DIMENSION OF CH AND THE FORMAT STATEMENTS WHERE IT IS
C     PRINTED ARE ALTERED ACCORDINGLY.
C
  100 READ(1,1001)ATOMS
 1001 FORMAT(I3)
      IF (ATOMS.EQ.0) GOTO 199
      READ(1,1001)INT
      READ(1,1002)R
 1002 FORMAT(F6.2)
      WRITE(2,1003)
 1003 FORMAT(//////)
      WRITE(2,1005)
 1005 FORMAT(48H DISPERSION CURVES AND DENSITIES OF STATES FOR A)
      IF (ATOMS.EQ.2) GOTO 101
      J=3
      WRITE(2,1006)
 1006 FORMAT(33H MONATOMIC B.C.C. CRYSTAL LATTICE)
  101 CONTINUE
      IF (ATOMS.EQ.1) GOTO 102
      J=6
      READ(1,1002)RO
      WRITE(2,1007)
 1007 FORMAT(35H DIATOMIC CSCL TYPE CRYSTAL LATTICE)
```

SIMULATION OF PHONON DISPERSION CURVES AND DENSITY OF STATES

LATTICE DYNAMICS PROGRAM

```
  102 CONTINUE
      WRITE(2,1008)R
 1008 FORMAT(42H RATIO OF INTERATOMIC FORCE CONSTANTS = 1:,F8.4)
      IF (ATOMS.EQ.2) WRITE(2,1009)RO
 1009 FORMAT(28H RATIO OF ATOMIC MASSES = 1:,F8.4)
      WMAX=0.0
C
C     INTEGER COUNTERS K, KX, KY, KZ WILL BE USED TO CONTROL LOOPS OVER
C     MESHES IN K-SPACE. QX, QY, QZ WILL BE USED FOR THE ACTUAL VALUES
C     OF THE PRODUCT KA. THUS FOR THE ATOMS=1 CASE, THE MAXIMUM VALUE
C     OF EACH Q WILL BE PI ALONG THE <100> DIRECTIONS, AND PI/2 FOR THE
C     <110> AND <111> DIRECTIONS. HOWEVER, SINCE IN THE <111> DIRECTIONS
C     THE DISPERSION CURVES DO NOT REPEAT OUTSIDE THE BRILLOUIN ZONE
C     THESE ARE ALLOWED TO RUN TO THE REPEAT DISTANCE PI, AND THE EDGE
C     OF THE BRILLOUIN ZONE IS MARKED BY A BROKEN LINE ON THE PRINT-OUT.
C        FOR ATOMS=2 THE MAXIMUM VALUE OF EACH Q IS PI/2.
C
      INT1=DCXSC*2/ATOMS
      DO 103 K=1,INT1,1
      QX=FLOAT(K)*PI/FLOAT(INT1*ATOMS)
      QY=0.0
      QZ=0.0
      IF (ATOMS.EQ.1) CALL D1(1,K,QX,QY,QZ,R,W,IFAIL)
      IF (ATOMS.EQ.2) CALL D2(1,K,QX,QY,QZ,R,RO,W,IFAIL)
      IF (IFAIL.EQ.1) GOTO 199
      IF (WMAX.LT.W(1,J,K)) WMAX=W(1,J,K)
C
C     THE SUBROUTINE F02AAF IN THE SUBROUTINES D1 AND D2 RETURNS THE
C     EIGENVALUES OF W**2 IN ASCENDING ORDER, SO FINDING THE MAXIMUM
C     VALUE OF W(DIR,J,K) FOR ALL DIR AND K IS SUFFICIENT TO FIND THE
C     MAXIMUM FREQUENCY.
C
  103 CONTINUE
      INT2=IFIX(FLOAT(DCXSC)*SQRT(2.0))
      DO 104 K=1,INT2,1
      QX=FLOAT(K)*PI*0.5/FLOAT(INT2)
      QY=QX
      QZ=0.0
      IF (ATOMS.EQ.1) CALL D1(2,K,QX,QY,QZ,R,W,IFAIL)
      IF (ATOMS.EQ.2) CALL D2(2,K,QX,QY,QZ,R,RO,W,IFAIL)
      IF (IFAIL.EQ.1) GOTO 199
      IF (WMAX.LT.W(2,J,K)) WMAX=W(2,J,K)
  104 CONTINUE
      INT3=IFIX(FLOAT(DCXSC)*SQRT(3.0)*2.0/FLOAT(ATOMS))
      DO 105 K=1,INT3,1
      QX=FLOAT(K)*PI/FLOAT(INT3*ATOMS)
      QY=QX
      QZ=QX
      IF (ATOMS.EQ.1) CALL D1(3,K,QX,QY,QZ,R,W,IFAIL)
      IF (ATOMS.EQ.2) CALL D2(3,K,QX,QY,QZ,R,RO,W,IFAIL)
      IF (IFAIL.EQ.1) GOTO 199
      IF (WMAX.LT.W(3,J,K)) WMAX=W(3,J,K)
  105 CONTINUE
      WRITE(2,1010)
 1010 FORMAT(////37H DISPERSION CURVES - <100> DIRECTION)
      CALL PRDC(INT1,1,J,DCYSC,W,WMAX)
```

LATTICE DYNAMICS PROGRAM

```
      WRITE(2,1011)
 1011 FORMAT(//37H  DISPERSION CURVES - <110> DIRECTION)
      CALL PRDC(INT2,2,J,DCYSC,W,WMAX)
      WRITE(2,1012)
 1012 FORMAT(//37H  DISPERSION CURVES - <111> DIRECTION)
      CALL PRDC(INT3,3,J,DCYSC,W,WMAX)
C
C     THE IRREDUCIBLE PORTION OF THE BRILLOUIN ZONE IS SPECIFIED BY
C     THE CONDITION KZ<KY<KX. THE BRILLOUIN ZONE BOUNDARY FOR ATOMS=2
C     (SIMPLE CUBIC) IS SIMPLY A CUBE OF SIDE PI/2. THUS ALLOWING VALUES
C     OF QX UP TO PI/2 TERMINATES ON THE BRILLOUIN ZONE BOUNDARY.
C        THE BRILLOUIN ZONE FOR ATOMS=1 (F.C.C. RECIPROCAL LATTICE) IS
C     A (110) PLANE, I.E. IT IS SPECIFIED BY QX+QY=PI. THUS ALLOWING
C     QX VALUES UP TO PI, AND SPECIFYING QY<QX AND ALSO QY<(PI-QX)
C     TERMINATES ON THE ZONE BOUNDARY. IT IS NECESSARY TO GIVE POINTS ON
C     THE ZONE BOUNDARY, OR THE BOUNDARIES OF THE IRREDUCIBLE PORTION,
C     REDUCED WEIGHTING OF ONE HALF. HOWEVER, IF KZ IS GIVEN HALF
C     INTEGER VALUES, TWO OF THE THREE SIDES, AND ALL THE EDGES OF THE
C     PORTION ARE AVOIDED, WHICH SIMPLIFIES THE POINTS OF REDUCED
C     WEIGHTING TO THOSE FOR QY=QX OR (PI-QX), AND QX=PI/2 (THE LATTER
C     ONLY MATERIALLY AFFECTS THE ATOMS=2 CASE).
C
      DO 106 WINT=1,200,1
         N(WINT)=0
         DO 106 I=1,6,1
            NW(I,WINT)=0
  106 CONTINUE
      DOSXSC=0
      INT2=INT
      IF (ATOMS.EQ.2) INT2=INT/2
      DO 107 KX=1,INT2,1
         QX=FLOAT(KX)*PI/FLOAT(INT)
         LIM=KX
         IF (KX.GT.INT/2) LIM=INT-KX
C
C     SETS LIM AS THE APPROPRIATE BOUNDARY FOR KY.
C
         DO 107 KY=1,LIM,1
            QY=FLOAT(KY)*PI/FLOAT(INT)
            DO 107 KZ=1,KY,1
               QZ=(FLOAT(KZ)-0.5)*PI/FLOAT(INT)
               IF (ATOMS.EQ.1) CALL D1(1,1,QX,QY,QZ,R,W,IFAIL)
               IF (ATOMS.EQ.2) CALL D2(1,1,QX,QY,QZ,R,RO,W,IFAIL)
               IF (IFAIL.EQ.1) GOTO 199
C
C     IT IS UNNECESSARY THIS TIME TO FILL ALL ELEMENTS OF THE ARRAY W
C     SINCE THE VALUES WILL IMMEDIATELY BE USED TO INCREMENT THE
C     APPROPRIATE ARRAY ELEMENTS OF THE DENSITY OF STATES.
C
               WT=1.0
               IF (KY.EQ.LIM.OR.KX.EQ.INT2) WT=0.5
               DO 107 I=1,J,1
                  W(1,I,1)=FLOAT(DCYSC-1)*W(1,I,1)/WMAX
                  WINT=IFIX(W(1,I,1))
                  IF (DOSXSC.LT.WINT+2) DOSXSC=WINT+2
C
```

SIMULATION OF PHONON DISPERSION CURVES AND DENSITY OF STATES

LATTICE DYNAMICS PROGRAM

```
C       THE SCALE OF THE X AXIS (WHICH IS W IN THIS CASE) IS MADE THE
C       SAME AS THE Y AXIS (ALSO W) OF THE DISPERSION CURVES.
C         THE SCALE LENGTH OF THE X AXIS IS SET TO THE LARGEST VALUE OF
C       FREQUENCY ANYWHERE IN THE BRILLOUIN ZONE IN CASE THERE ARE ANY
C       FREQUENCIES GREATER THAN WMAX, WHICH IS THE GREATEST FREQUENCY
C       IN ANY SYMMETRY DIRECTION.
C
                WFRAC=W(1,I,1)-FLOAT(WINT)
                NW(I,WINT+1)=NW(I,WINT+1)+(1.0-WFRAC)*WT
                NW(I,WINT+2)=NW(I,WINT+2)+WFRAC*WT
C
C       TO GENERATE A SMOOTHER DENSITY OF STATES, RATHER THAN PLOT A
C       SIMPLE HISTOGRAM EACH CALCULATED FREQUENCY IS DIVIDED BETWEEN
C       THE TWO HISTOGRAM POINTS ABOVE AND BELOW, WEIGHTED ACCORDING TO
C       HOW CLOSE THE FREQUENCY IS TO EITHER POINT.
C
  107 CONTINUE
      NWMAX=0.0
      NMAX=0.0
      DO 108 WINT=1,200,1
         DO 109 I=1,3,1
            IF (ATOMS.EQ.2) NW(I,WINT)=NW(I,WINT)+NW(I+3,WINT)
            N(WINT)=N(WINT)+NW(I,WINT)
            IF (NWMAX.LT.NW(I,WINT)) NWMAX=NW(I,WINT)
  109    CONTINUE
         IF (NMAX.LT.N(WINT)) NMAX=N(WINT)
  108 CONTINUE
      CALL PRDOS(NW,NWMAX,N,NMAX,DCYSC,DOSXSC)
      GOTO 100
  199 STOP
      END




      SUBROUTINE D1(DIR,K,QX,QY,QZ,R,W,IFAIL)
      INTEGER DIR,K,I,J,IFAIL
      REAL QX,QY,QZ,R,DIAG,W(3,6,200),D(3,3),S(3),C(3),W2(3),
     1     WKSPC(3)
C
C       THIS SUBROUTINE SETS UP THE 3*3 DYNAMICAL MATRIX FOR THE ONE ATOM
C       UNIT CELL, AND USES A STANDARD LIBRARY SUBROUTINE F02AAF TO FIND
C       THE EIGENVALUES OF THE SECULAR DETERMINANT. ANY SIMILAR SUB-
C       ROUTINE WHICH FINDS THE EIGENVALUES OF A REAL SYMMETRIC MATRIX
C       MAY BE USED INSTEAD, AND THE EIGENVECTORS ARE NOT REQUIRED.
C
      C(1)=COS(QX)
      C(2)=COS(QY)
      C(3)=COS(QZ)
      S(1)=SIN(QX)
      S(2)=SIN(QY)
      S(3)=SIN(QZ)
      DIAG=1.0+0.75*R-C(1)*C(2)*C(3)
      D(1,1)=DIAG-0.75*R*COS(2.0*QX)
      D(2,2)=DIAG-0.75*R*COS(2.0*QY)
```

LATTICE DYNAMICS PROGRAM

```
      D(3,3)=DIAG-0.75*R*COS(2.0*QZ)
      D(1,2)=S(1)*S(2)*C(3)
      D(2,3)=C(1)*S(2)*S(3)
      D(3,1)=S(1)*C(2)*S(3)
      D(2,1)=D(1,2)
      D(3,2)=D(2,3)
      D(1,3)=D(3,1)
      CALL F02AAF(D,3,3,W2,WKSPC,0)
C
C     THE PARAMETERS OF F02AAF(D,N,N,W2,WKSPC,IFAIL) ARE AS FOLLOWS:
C     D IS THE DYNAMICAL MATRIX, OF DIMENSION NXN, W2 IS THE ARRAY
C     OF EIGENVALUES (THE SQUARES OF THE NORMAL MODE FREQUENCIES),
C     WKSPC IS AN ARRAY OF DIMENSION N PROVIDED FOR WORKING SPACE
C     AND IFAIL IS AN INTEGER DETERMINING THE MODE OF ACTION IN
C     THE EVENT OF FAILURE OF THE SUBROUTINE. THE VALUE 0 CAUSES
C     TERMINATION OF THE PROGRAM   WHENEVER AN ERROR IS DETECTED.
C
      DO 501 I=1,3,1
         IF(ABS(W2(I)).LT.0.00001) W2(I)=0.0
C
C     ROUNDING ERRORS MAY PRODUCE SMALL NEGATIVE VALUES IN
C     THE EIGENVALUES WHEN THE CORRECT VALUES ARE ZERO
C
         IF(W2(I).LT.0.0) GOTO 502
         W(DIR,I,K)=SQRT(W2(I))
  501 CONTINUE
      RETURN
  502 WRITE(2,5001)
 5001 FORMAT(42H SUBROUTINE D1 PRODUCES NEGATIVE FREQUENCY)
      IFAIL=1
      RETURN
      END


      SUBROUTINE D2(DIR,K,QX,QY,QZ,R,RO,W,IFAIL)
      INTEGER DIR,K,I,J,IFAIL
      REAL QX,QY,QZ,R,RO,ROOTRO,W(3,6,200),D(6,6),S(3),C(3),
     1     W2(6),WKSPC(6)
C
C     THIS SUBROUTINE IS SIMILAR TO D1, EXCEPT THAT IT SOLVES THE 6*6
C     MATRIX FOR THE TWO ATOM UNIT CELL.
C
      C(1)=COS(QX)
      C(2)=COS(QY)
      C(3)=COS(QZ)
      S(1)=SIN(QX)
      S(2)=SIN(QY)
      S(3)=SIN(QZ)
      ROOTRO=SQRT(RO)
      DO 801 I=1,6,1
         DO 801 J=1,6,1
            D(I,J)=0.0
  801 CONTINUE
```

SIMULATION OF PHONON DISPERSION CURVES AND DENSITY OF STATES

LATTICE DYNAMICS PROGRAM

```
      D(1,1)=ROOTRO*(1.0+0.75*R-0.75*R*COS(2.0*QX))
      D(2,2)=ROOTRO*(1.0+0.75*R-0.75*R*COS(2.0*QY))
      D(3,3)=ROOTRO*(1.0+0.75*R-0.75*R*COS(2.0*QZ))
      D(4,4)=(1.0+0.75*R-0.75*R*COS(2.0*QX))/ROOTRO
      D(5,5)=(1.0+0.75*R-0.75*R*COS(2.0*QY))/ROOTRO
      D(6,6)=(1.0+0.75*R-0.75*R*COS(2.0*QZ))/ROOTRO
      DO 802 I=1,3,1
         D(I,I+3)=-C(1)*C(2)*C(3)
         D(I+3,I)=D(I,I+3)
  802 CONTINUE
      D(1,5)=S(1)*S(2)*C(3)
      D(1,6)=S(1)*C(2)*S(3)
      D(2,6)=C(1)*S(2)*S(3)
      D(5,1)=D(1,5)
      D(2,4)=D(1,5)
      D(4,2)=D(1,5)
      D(6,1)=D(1,6)
      D(3,4)=D(1,6)
      D(4,3)=D(1,6)
      D(6,2)=D(2,6)
      D(3,5)=D(2,6)
      D(5,3)=D(2,6)
      CALL F02AAF(D,6,6,W2,WKSPC,0)
      DO 803 I=1,6,1
         IF (ABS(W2(I)).LT.0.00001) W2(I)=0.0
C
C     ROUNDING ERRORS MAY PRODUCE SMALL NEGATIVE VALUES IN
C     THE EIGENVALUES WHEN THE CORRECT VALUES ARE ZERO
C
         IF (W2(I).LT.0.0) GOTO 804
         W(DIR,I,K)=SQRT(W2(I))
  803 CONTINUE
      RETURN
  804 WRITE(2,8001)
 8001 FORMAT(42H SUBROUTINE D2 PRODUCES NEGATIVE FREQUENCY)
      IFAIL=1
      RETURN
      END


      SUBROUTINE PRDC(INT,DIR,J,DCYSC,W,WMAX)
      INTEGER INT,DIR,J,I,K,L,M,KMAX,XAXIS,YAXIS,SPACE,CROSS,SUB,
     1        DCYSC,STAR,PT(6),CH(120),NOUGHT,ONE,TWO,THREE,
     2        FOUR,FIVE,SIX,SEVEN,EIGHT,NINE,DCYSC2,SCALE
      REAL WMAX,W(3,6,200)
      COMMON CH
      DATA XAXIS,YAXIS,SPACE,CROSS,STAR/1HI,1H-,1H ,1H+,1H*/
      DATA NOUGHT,ONE,TWO,THREE,FOUR/1H0,1H1,1H2,1H3,1H4/
      DATA FIVE,SIX,SEVEN,EIGHT,NINE/1H5,1H6,1H7,1H8,1H9/
C
C     THIS SUBROUTINE PRODUCES A GRAPH OF THE DISPERSION CURVES ON
C     THE LINEPRINTER.
C
```

LATTICE DYNAMICS PROGRAM

```
      DCYSC2=DCYSC+2
      DO 601 L=1,DCYSC2,1
 601     CH(L)=SPACE
      CH(2)=NOUGHT
      CH(12)=ONE
      CH(22)=TWO
      CH(32)=THREE
      CH(42)=FOUR
      CH(52)=FIVE
      CH(62)=SIX
      CH(72)=SEVEN
      CH(82)=EIGHT
      CH(92)=NINE
      CH(102)=NOUGHT
      CH(112)=ONE
      WRITE(2,1601) (CH(K),K=1,DCYSC2)
      DO 602 L=2,DCYSC2,1
 602     CH(L)=YAXIS
      CH(1)=NOUGHT
      WRITE(2,1601) (CH(K),K=1,DCYSC2)
1601  FORMAT(1H ,120A1)
      DO 603 I=1,INT,1
         DO 604 L=1,DCYSC2,1
            CH(L)=SPACE
            IF(DIR.EQ.3.AND.J.EQ.3.AND.I.EQ.INT/2
     1         .AND.L.EQ.(L/2)*2) CH(L)=YAXIS
 604     CONTINUE
         CH(1)=XAXIS
         DO 605 L=1,J,1
            PT(L)=IFIX(W(DIR,L,I)*FLOAT(DCYSC-1)/WMAX+1.5)
            SUB=PT(L)
            IF (CH(SUB).EQ.CROSS.OR.CH(SUB).EQ.STAR) GOTO 606
            CH(SUB)=CROSS
            GOTO 605
 606        CH(SUB)=STAR
 605     CONTINUE
         SCALE=I/10
         IF(SCALE*10.EQ.I) GOTO 607
         WRITE(2,1602) (CH(K),K=1,DCYSC2)
1602     FORMAT(2H ,120A1)
         GOTO 603
 607     IF(SCALE.LT.10) GOTO 608
         SCALE=SCALE-10
         GOTO 607
 608     WRITE(2,1603)SCALE, (CH(K),K=1,DCYSC2)
1603     FORMAT(1H ,I1,120A1)
 603  CONTINUE
      RETURN
      END
```

LATTICE DYNAMICS PROGRAM

```
      SUBROUTINE PRDOS(NW,NWMAX,N,NMAX,DCYSC,DOSXSC)
      INTEGER I,L,CH(120),DCYSC,DOSXSC,XAXIS,YAXIS,SPACE,CROSS,
     1           INT2,SUB,NOUGHT,ONE,TWO,THREE,FOUR,FIVE,
     2           SIX,SEVEN,EIGHT,NINE,DCYSC2,SCALE,K
      REAL NW(6,200),NWMAX,N(200),NMAX
      COMMON CH
      DATA XAXIS,YAXIS,SPACE,CROSS/1HI,1H-,1H ,1H+/
      DATA NOUGHT,ONE,TWO,THREE,FOUR/1H0,1H1,1H2,1H3,1H4/
      DATA FIVE,SIX,SEVEN,EIGHT,NINE/1H5,1H6,1H7,1H8,1H9/
C
C     THIS SUBROUTINE PRODUCES ON THE LINEPRINTER GRAPHS OF THE DENSITY
C     OF STATES FOR EACH POLARIZATION, AND THE TOTAL DENSITY OF STATES.
C
      WRITE(2,1200)
 1200 FORMAT(//////41H  DENSITY OF STATES FOR EACH POLARIZATION)
      DCYSC2=DCYSC+2
      DO 201 L=1,DCYSC2,1
  201     CH(L)=SPACE
      CH(2)=NOUGHT
      CH(12)=ONE
      CH(22)=TWO
      CH(32)=THREE
      CH(42)=FOUR
      CH(52)=FIVE
      CH(62)=SIX
      CH(72)=SEVEN
      CH(82)=EIGHT
      CH(92)=NINE
      CH(102)=NOUGHT
      CH(112)=ONE
      WRITE(2,1201) (CH(K),K=1,DCYSC2)
      DO 202 L=2,DCYSC2,1
  202     CH(L)=YAXIS
      CH(1)=NOUGHT
      WRITE(2,1201) (CH(K),K=1,DCYSC2)
 1201 FORMAT(1H ,120A1)
      DOSXSC=DOSXSC+2
      DO 203 I=1,DOSXSC,1
          DO 204 L=1,DCYSC2,1
  204         CH(L)=SPACE
          INT2=DCYSC/3
          DO 205 L=1,3,1
              SUB=IFIX(NW(L,I+1)*FLOAT(INT2)/NWMAX+1.5)+INT2*(L-1)
              CH(SUB)=CROSS
  205     CONTINUE
          CH(1)=XAXIS
          CH(INT2+1)=XAXIS
          INT2=2*INT2
          CH(INT2+1)=XAXIS
          SCALE=I/10
          IF(SCALE*10.EQ.I) GOTO 206
          WRITE(2,1202) (CH(K),K=1,DCYSC2)
 1202     FORMAT(2H ,120A1)
          GOTO 203
  206     IF(SCALE.LT.10) GOTO 207
          SCALE=SCALE-10
```

LATTICE DYNAMICS PROGRAM

```
         GOTO 206
207      WRITE(2,1203)SCALE, (CH(K),K=1,DCYSC2)
1203     FORMAT(1H ,I1,120A1)
 203 CONTINUE
     WRITE(2,1300)
1300 FORMAT(//////25H   TOTAL DENSITY OF STATES)
     DO 301 L=1,DCYSC2,1
 301     CH(L)=SPACE
     CH(2)=NOUGHT
     CH(12)=ONE
     CH(22)=TWO
     CH(32)=THREE
     CH(42)=FOUR
     CH(52)=FIVE
     CH(62)=SIX
     CH(72)=SEVEN
     CH(82)=EIGHT
     CH(92)=NINE
     CH(102)=NOUGHT
     CH(112)=ONE
     WRITE(2,1301) (CH(K),K=1,DCYSC2)
     DO 302 L=2,DCYSC2,1
 302     CH(L)=YAXIS
     CH(1)=NOUGHT
     WRITE(2,1301) (CH(K),K=1,DCYSC2)
1301 FORMAT(1H ,120A1)
     DO 303 I=1,DOSXSC,1
         DO 304 L=1,DCYSC2,1
 304         CH(L)=SPACE
         SUB=IFIX(N(I+1)*FLOAT(DCYSC)/NMAX+1.5)
         CH(SUB)=CROSS
         CH(1)=XAXIS
         SCALE=I/10
         IF(SCALE*10.EQ.I)  GOTO 305
         WRITE(2,1302) (CH(K),K=1,DCYSC2)
1302     FORMAT(2H  ,120A1)
         GOTO 303
 305     IF(SCALE.LT.10) GOTO 306
         SCALE=SCALE-10
         GOTO 305
 306     WRITE(2,1303)SCALE, (CH(K),K=1,DCYSC2)
1303     FORMAT(1H ,I1,120A1)
 303 CONTINUE
     RETURN
     END
```

Physics Programs
Edited by A. D. Boardman
© 1980 John Wiley & Sons Ltd.

CHAPTER 10

Electron Energy Bands in a One-dimensional Periodic Potential

R. D. CLARKE and D. J. MARTIN

1. INTRODUCTION

A particularly demanding area encountered in any course on solid-state physics is that of the energies of electrons in crystals.[1] Familiarity with this material is basic to a proper understanding of electrical phenomena in metals and semiconductors; perhaps the most significant difficulty that arises is that any realistic treatment necessitates the use of numerical methods. Commonly, however, only qualitative arguments or rather unrealistic models, such as the Kronig–Penney model,[2] are presented. The computer program presented here enables the user to investigate electron energies in a system where the form, magnitude, and period of the potential can be specified by the user. The results can be compared with those derived from approximate analytic treatments for certain ranges of the parameters. Exercises of this kind can supplement more conventional presentations and give students some familiarity with the basic methods employed in band structure calculations.

Any system involving particles will exhibit quantum-mechanical features if the de Broglie wavelength associated with the momentum of the particles is of the same order of magnitude or greater than a typical length over which the potential acting on the particles changes significantly. It is easy to show that this is the case for conduction electrons in a solid by the following considerations.

In the case of metals a typical free electron density (N) is $\sim 4 \times 10^{28}\,\text{m}^{-3}$. An ideal Fermi–Dirac gas of this density would have kinetic energy (E) per particle $\sim \hbar^2(3\pi^2 N)^{\frac{2}{3}}/2m_e \sim 7 \times 10^{-19}\,\text{J}$, a velocity $(v) \sim \sqrt{2E/m_e} \sim 1.2 \times 10^6\,\text{m s}^{-1}$, and a corresponding wavelength $\lambda = h/m_e v \sim 6 \times 10^{-10}\,\text{m}$. For a semiconductor $E \sim k_B T \sim 10^{-21}\,\text{J}$, so $v \sim 10^5\,\text{m s}^{-1}$ and $\lambda \sim 7 \times 10^{-9}\,\text{m}$.

The potential acting on the electrons will vary significantly over the interatomic spacing $\sim 3 \times 10^{-10}$ m. It follows that quantum-mechanical methods are essential to tackle the problem.

The computer program presented here solves the Schrödinger equation to an accuracy of ~ 1 per cent for electrons in a one-dimensional potential, the form, magnitude, and period of which can be set by the user.

This study was limited to the one-dimensional case because it is designed as an aid to learning and deliberately avoids the numerical and conceptual complications of the three-dimensional case. It should, however, be noted that there are in fact some systems such as 'KCP' ($K_2Pt(CN)_4Br_{0.3} \cdot 3H_2O$) where the electron motion is in reality effectively confined to one dimension.[3]

One of the most significant factors influencing the behaviour of electrons in a typical metal or semiconductor is the fact that the potential to which they are subject is *periodic*. The reason for the periodicity is that the atoms are arranged in a regular crystal lattice. (Amorphous—randomized—metals and semiconductors exist but a rather different theoretical approach is then necessary.) The potential is periodic over many thousands of atoms, even if the material as a whole is polycrystalline.

2. ELECTRONS IN A PERIODIC POTENTIAL

2.1 Brillouin zones

It is not difficult to appreciate that the periodic potential due to the regular crystal lattice will affect the electron energies, and indeed to see in general terms what the effect will be. Electrons in free space have kinetic energy $E = \frac{1}{2}m_e v^2$. Knowing that for electrons in a solid quantum effects are important we can use the de Broglie relationship to express this as

$$E = \frac{h^2}{2m_e \lambda^2} \quad \text{or} \quad E = \frac{\hbar^2 k^2}{2m_e}$$

where $k (= 2\pi/\lambda)$ is called the wave number. The corresponding wave function in free space is $\psi(x) = \exp(ikx)$. In the case of a crystalline solid, states are specified by n (the band index) and k (the 'crystal momentum').

For a weak potential of period a, $E = \hbar^2 k^2 / 2m_e$ except when k is close to $n\pi/a$ ($n = \pm 1, \pm 2, \ldots$)—the Brillouin zone boundaries. The reason why deviations always occur in these regions is that a wave function of the form $\psi(x) = \exp(ikx)$ represents a *travelling* wave in the $+x$-direction, of wavelength $2\pi/k$. From the condition for Bragg reflection: $2a \sin \theta = n\lambda$ (where a is the interplane spacing) we might expect the travelling electron wave to be strongly affected in our one-dimensional case (for which $\sin \theta = 1$) when $2a = n\lambda$, i.e. $2a = n \cdot 2\pi/k$, i.e. $k = n\pi/a$. The electron waves cannot propagate at the zone boundaries and a discontinuity occurs in the E–k

ELECTRON ENERGY BANDS IN A ONE-DIMENSIONAL PERIODIC POTENTIAL 303

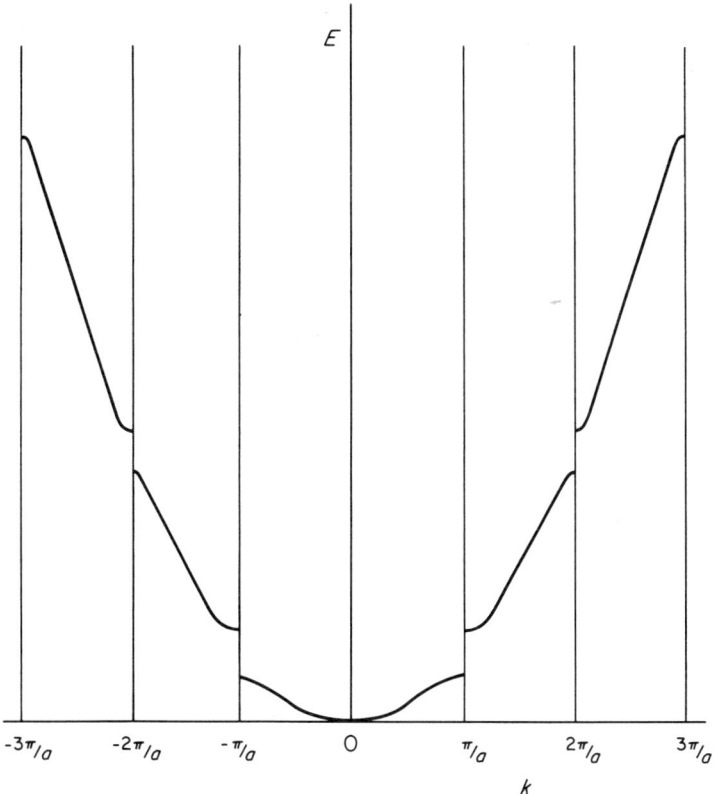

Figure 1. A typical E–k relationship for a weak potential. Discontinuities in E occur at $k = n\pi/a$ ($n = \pm 1, \pm 2, \ldots$)

relationship (see Figure 1). The region for $0 < |k| < \pi/a$ is called the first Brillouin zone, $\pi/a < |k| < 2\pi/a$ the second Brillouin zone, etc.

While these general arguments give a qualitative picture, a determination of the details of the variation of E with k depends on solving the Schrödinger equation for the particular periodic potential involved.

2.2 Bloch functions

The time-independent Schrödinger equation for an electron in one dimension is

$$\left[-\frac{\hbar^2}{2m_e} \frac{d^2}{dx^2} + V(x) \right] \psi(x) = E\psi(x). \tag{1}$$

If we use atomic units (see Chapter 9) and measure energies in Rydbergs and distances in Bohr radii then this is equivalent to setting $\hbar = 1$, $m_e = \frac{1}{2}$. Hence for an electron equation (1) takes the form:

$$\left[-\frac{d^2}{dx^2} + V(x)\right]\psi(x) = E\psi(x), \tag{2}$$

where $V(x)$, the potential, is periodic with period a, i.e. $V(x) = V(x + ma)$ and m is an integer.

It might appear that $\psi(x)$ should also be periodic; in fact the reality is more complicated. The probability density $\psi^*(x)\psi(x)$ is indeed periodic with period a, but this can still hold if $\psi(x)$ itself is equal to the product of a function which is periodic ($u(x)$ say) and a complex quantity whose product with its own complex conjugate is unity. The general form of such a complex quantity must be $\exp(if[x, k])$, i.e. $\psi(x) = \exp(if[x, k])u(x)$ and

$$\begin{aligned}\psi^*(x)\psi(x) &= \exp(-if[x, k])u^*(x)\exp(if[x, k])u(x) \\ &= u^*(x)u(x),\end{aligned} \tag{3}$$

which shows that the probability density is of period a. We know that for $V(x) \to 0$, $\psi(x) \to \exp(ikx)$ and so the possibility suggests itself that $\exp(if[x, k]) = \exp(ikx)$ and $\psi(x) = \exp(ikx)u_k(x)$, where the k subscript on $u_k(x)$ implies a dependence of the periodic function on k. Wave functions of this form are known as Bloch functions. (Bloch first established this result in the present connection,[4] on the basis of the periodic potential and 'periodic boundary condition'). One consequence of the form of the Bloch functions is that there are states at $k + 2n\pi/a$ ($n = \pm 1, \pm 2, \ldots$) with the same energy as a state at k. Consider the state $\psi_k(x) = \exp(ikx)u_k(x)$ and the state at $k' = k + 2n\pi/a$;

$$\begin{aligned}\psi_{k'}(x) &= \exp(ik'x)u_{k'}(x) \\ &= \exp(ikx)\exp(i2n\pi x/a)u_{k'}(x),\end{aligned} \tag{4}$$

where $\exp(i2n\pi x/a)$ is of period a so that $\exp(i2n\pi x/a)u_{k'}(x)$ is of period a. A possible form for this function, which will certainly correspond to a solution of equation (1) is

$$\exp(i2n\pi x/a)u_{k'}(x) = u_k(x), \tag{5}$$

in which case

$$\psi_k(x) = \psi_{k'}(x), \tag{6}$$

i.e. the two states are the same.

There are, in consequence, three entirely equivalent ways in which the E-k relationship can be presented (see Figure 2):

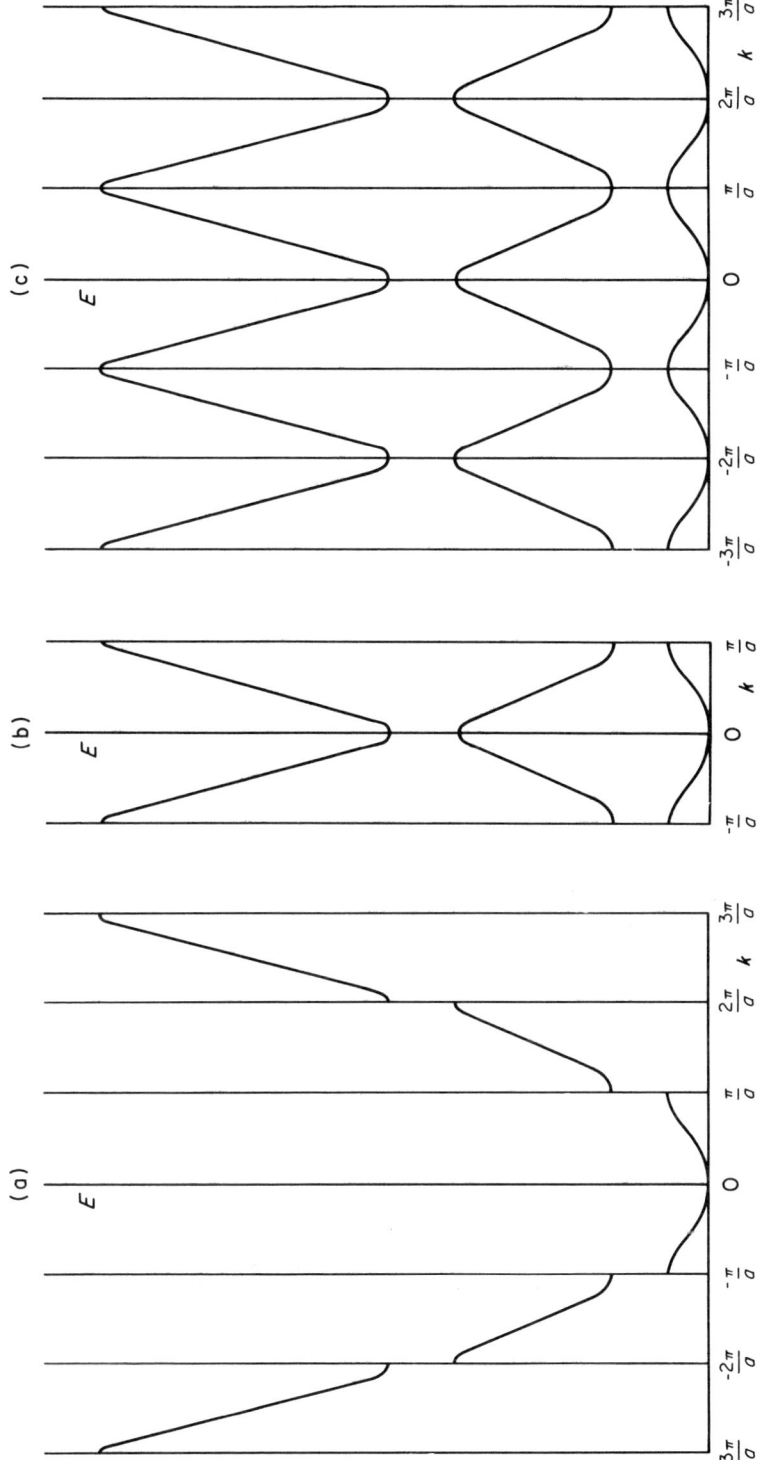

Figure 2. The three equivalent zone schemes: (a) the extended zone scheme; (b) the reduced zone scheme; and (c) the repeated or periodic zone scheme

305

(a) The extended zone scheme, in which E is a single-valued function of k and all the relevant Brillouin zones are shown.
(b) The reduced zone scheme, in which E is a multivalued function of k and only the first Brillouin zone is shown.
(c) The repeated or periodic zone scheme, in which E is a multivalued function of k, and all the relevant Brillouin zones are shown.

Furthermore, $E(k) = E(-k)$—the electron energy depends on λ but does not depend on whether the electron is travelling in the $+x$- or the $-x$-direction—so that it is really only necessary to present the results for $k > 0$. In this discussion, and in the graphs generated by the computer program, the reduced zone scheme is used with $0 < k < \pi/a$.

Substituting $\psi(x) = \exp(ikx) u_k(x)$ into equation (1) gives

$$-\frac{d^2 u_k(x)}{dx^2} - 2ik \frac{du_k(x)}{dk} + [k^2 + V(x)] u_k(x) = E u_k(x). \tag{7}$$

It is required to find a solution to this form of the Schrödinger equation for the appropriate $V(x)$. Exact analytic solutions are unobtainable but numerical techniques can be used to achieve accurate results.

2.3 The variational method

The approach used in most realistic band structure calculations employs the variational method. This method depends on the result that if the wave function of a system is expressed as a function of a set of N adjustable parameters (the variational parameters) $c_1, \ldots, c_l, \ldots, c_N$, i.e. $\psi = \psi(x, c_1, \ldots, c_l, \ldots, c_N)$ then the values of these parameters for which $\partial \langle E \rangle / \partial c_l = 0$ (where $\langle E \rangle$ is the expectation value for the energy) gives a 'best estimate' for the true wave function. In the form of the variational method most suitable for numerical applications (the Ritz variational method) the unknown wave function is expressed as a linear combination of N functions χ_l (the 'basis functions') which form a complete orthonormal set and the chosen c_l's are the coefficients of this series, i.e.

$$\psi(x) = \sum_{l=1}^{N} c_l \chi_l(x), \tag{8}$$

where

$$\int \chi_l^*(x) \chi_m(x) \, dx = \begin{cases} \text{constant, usually 1, for } l = m. \\ 0 \text{ for } l \neq m \end{cases} \tag{9}$$

Equation (9) is the definition of orthonormality. The constant is referred to as the 'norm' of the basis functions.

For bound states the limits of this integral (and subsequent integrations in this sub-section) are usually $-\infty$ to $+\infty$. For the non-bound states of interest in band structure work it is usual to carry out the integration over the unit cell ($0 \leq x \leq a$ here).

For $N \to \infty$ the series approaches an exact solution to the Schrödinger equation. In practice, of course, the series used is finite but is chosen to be sufficiently long that any errors due to truncation are negligibly small. Now $\langle E \rangle$ is given by

$$\langle E \rangle = \frac{\int \psi^*(x) \mathcal{H} \psi(x) \, dx}{\int \psi^*(x) \psi(x) \, dx}. \tag{10}$$

If equation (8) is used then

$$\langle E \rangle \int \sum_{l=1}^{N} c_l^* \chi_l^*(x) \sum_{m=1}^{N} c_m \chi_m(x) \, dx = \int \sum_{l=1}^{N} c_l^* \chi_l^*(x) \mathcal{H} \sum_{m=1}^{N} c_m \chi_m(x) \, dx. \tag{11}$$

The left-hand side can be simplified because of the orthonormality of the functions $\chi_l(x)$ and if we adopt the conventional abbreviation

$$H_{lm} = \int \chi_l^*(x) \mathcal{H} \chi_m(x) \, dx, \tag{12}$$

then

$$\langle E \rangle \sum_{l=1}^{N} c_l^* c_l = \sum_{l=1}^{N} \sum_{m=1}^{N} c_l^* c_m H_{lm}. \tag{13}$$

Taking the derivative with respect to c_l^* gives

$$\frac{\partial \langle E \rangle}{\partial c_l^*} \sum_{l=1}^{N} c_l^* c_l + \langle E \rangle c_l = \sum_{m=1}^{N} c_m H_{lm}. \tag{14}$$

But, according to the variational principle, the value of the c_l's, corresponding to a solution of the Schrödinger equation, occur for

$$\frac{\partial \langle E \rangle}{\partial c_l^*} = 0, \tag{15}$$

which from equation (14) gives

$$\sum_{m=1}^{N} c_m H_{lm} = \langle E \rangle c_l \quad (l = 1 \ldots N). \tag{16}$$

This constitutes a set of N simultaneous linear equations which have non-trivial solutions if

$$\begin{vmatrix} H_{11}-\langle E \rangle & H_{12} & \cdots\cdots\cdots\cdots & H_{1N} \\ H_{21} & H_{22}-\langle E \rangle & & \vdots \\ \vdots & & & \vdots \\ \vdots & & & \vdots \\ H_{N1} & \cdots\cdots\cdots\cdots & & H_{NN}-\langle E \rangle \end{vmatrix} = 0. \qquad (17)$$

This equation has N roots for $\langle E \rangle$ and can be solved by standard numerical techniques. The N roots constitute a 'best estimate' of the N lowest energy levels of the system, for the chosen basis functions. The corresponding values for the c_i's, and hence $\psi(x)$, can also be found.

A key issue is therefore the choice of a suitable set of basis functions $\chi(x)$ and the central problem of the subsequent analysis is the evaluation of

$$H_{lm} = \int \chi_l^*(x) \mathcal{H} \chi_m(x) \, dx. \qquad (18)$$

One noteworthy feature of the variational method is that the results for $\langle E \rangle$ are generally more accurate than the results for $\psi(x)$ because a first-order error in $\psi(x)$ only leads to a second-order error in $\langle E \rangle$.

2.4 The basis functions

To apply the Ritz variational method we must express $u_k(x)$ as a linear combination of complete orthonormal functions. The fact that $u_k(x)$ is periodic, of period a, suggests the possibility of expressing $u_k(x)$ as a Fourier series:

$$u_k(x) = \frac{a_0}{2} + \sum_{l=1}^{N} a_l \cos\left(\frac{2\pi l x}{a}\right) + \sum_{l=1}^{N} b_l \sin\left(\frac{2\pi l x}{a}\right). \qquad (19)$$

(The terms of a Fourier series form a complete orthonormal set.) This approach was adopted in the present work and is useful for a general-purpose treatment. In fact, as we shall see, there are better ways of expanding $u_k(x)$ for the case of realistic potentials.

ELECTRON ENERGY BANDS IN A ONE-DIMENSIONAL PERIODIC POTENTIAL 309

Rather than employing explicit cosine and sine terms it is more convenient to use $\chi_l(x) = \exp[i(2\pi l x/a)]$ and to express the Fourier series as

$$u_k(x) = \sum_{l=-N}^{+N} c_l \exp\left(\frac{i2\pi l x}{a}\right), \tag{20}$$

where the c_l's are in general complex. These functions are normalized to 'a' rather than 1 over the unit cell, but this factor subsequently cancels.

The potential $V(x)$ is also periodic and can likewise be expanded as a Fourier series:

$$V(x) = \frac{d_0}{2} + \sum_{j=1}^{N} d_j \cos\left(\frac{2\pi j x}{a}\right) + \sum_{j=1}^{N} e_j \sin\left(\frac{2\pi j x}{a}\right). \tag{21}$$

Many crystals possess what is called an 'inversion centre' and if such a point is chosen as the origin of coordinates then $V(\mathbf{r}) = V(-\mathbf{r})$. Symmetry of this kind leads to a considerable reduction in the computing involved and for all the potentials investigated here we can choose an origin such that $V(x) = V(-x)$. Now $\cos(ax) = \cos(-ax)$ but $\sin(ax) = -\sin(-ax)$ so, if we make this choice of origin, all the e_j's are zero. It is again more convenient to write

$$V(x) = \sum_{j=-N}^{+N} f_j \exp\left(\frac{i2\pi j x}{a}\right). \tag{22}$$

Because the sine terms are absent

$$f_j = f_{-j} = \frac{d_j}{2}, \tag{23}$$

and, since $V(x)$ is real, all the f_j's are real.

The Hamiltonian operator takes the form

$$\begin{aligned}\mathcal{H} &= -\frac{d^2}{dx^2} - 2ik\frac{d}{dx} + k^2 + V(x) \\ &= -\frac{d^2}{dx^2} - 2ik\frac{d}{dx} + k^2 + \sum_{j=-N}^{+N} f_j \exp\left(\frac{i2\pi j x}{a}\right).\end{aligned} \tag{24}$$

Hence

$$H_{lm} = \frac{1}{a}\int_0^a \exp\left(\frac{-i2\pi l x}{a}\right)\left[\frac{4\pi^2 m^2}{a^2} + \frac{4\pi m k}{a}\right.$$

$$\left. + k^2 + \sum_{j=-N}^{N} f_j \exp\left(\frac{i2\pi j x}{a}\right)\right] \exp\left(\frac{i2\pi m x}{a}\right) dx. \tag{25}$$

The integration is over the unit cell and the factor of $1/a$ cancels with the norm of the basis functions. Because of the orthogononality of the basis functions over the unit cell, equation (25) becomes

$$H_{lm} = \begin{cases} f_{l-n} & (l \neq m) \\ \left(k + \dfrac{2\pi m}{a}\right)^2 + f_0 & (l = m). \end{cases} \quad (26)$$

Hence, using equation (23) the matrix Hamiltonian is

$$\mathcal{H} = \begin{bmatrix} \left(k - \dfrac{2N\pi}{a}\right)^2 + \dfrac{d_0}{2} & \cdots & & & & \dfrac{d_{2N}}{2} \\ & \left(k - \dfrac{2\pi}{a}\right)^2 + \dfrac{d_0}{2} & \dfrac{d_1}{2} & \dfrac{d_2}{2} & & \\ & \dfrac{d_1}{2} & k^2 + \dfrac{d_0}{2} & \dfrac{d_1}{2} & & \\ & \dfrac{d_2}{2} & \dfrac{d_1}{2} & \left(k + \dfrac{2\pi}{a}\right)^2 + \dfrac{d_0}{2} & & \\ \dfrac{d_{2N}}{2} & & & \cdots & \left(k + \dfrac{2N\pi}{a}\right)^2 + \dfrac{d_0}{2} \end{bmatrix}. \quad (27)$$

The task therefore reduces to Fourier analysing $V(x)$, setting up the Hamiltonian matrix $\mathcal{H}$ (27) and then employing numerical techniques to determine the energies (diagonalization) and, if required, the corresponding wave functions.

2.5 The periodic potential

In a real crystal the periodic potential arises from the Coulomb interaction of the electron with all of the atomic nuclei and all of the other electrons. In the computer program presented here the user can choose between a number of potentials which have been selected principally with a view to their heuristic value rather than to their similarity to real crystal potentials. The potentials are of period a which is set by the user. The origin is an inversion centre, i.e. $V(x) = V(-x)$ so that it is only necessary to specify them for $0 \leq x \leq a/2$. The potentials are (see Figure 3):

(i) A rectangular potential

$$V(x) = \begin{cases} 0 & 0 \leq x < \left(\dfrac{a}{2} - \dfrac{b}{2}\right) \\ V_0 & \left(\dfrac{a}{2} - \dfrac{b}{2}\right) < x \leq \dfrac{a}{2}. \end{cases} \quad (28)$$

ELECTRON ENERGY BANDS IN A ONE-DIMENSIONAL PERIODIC POTENTIAL

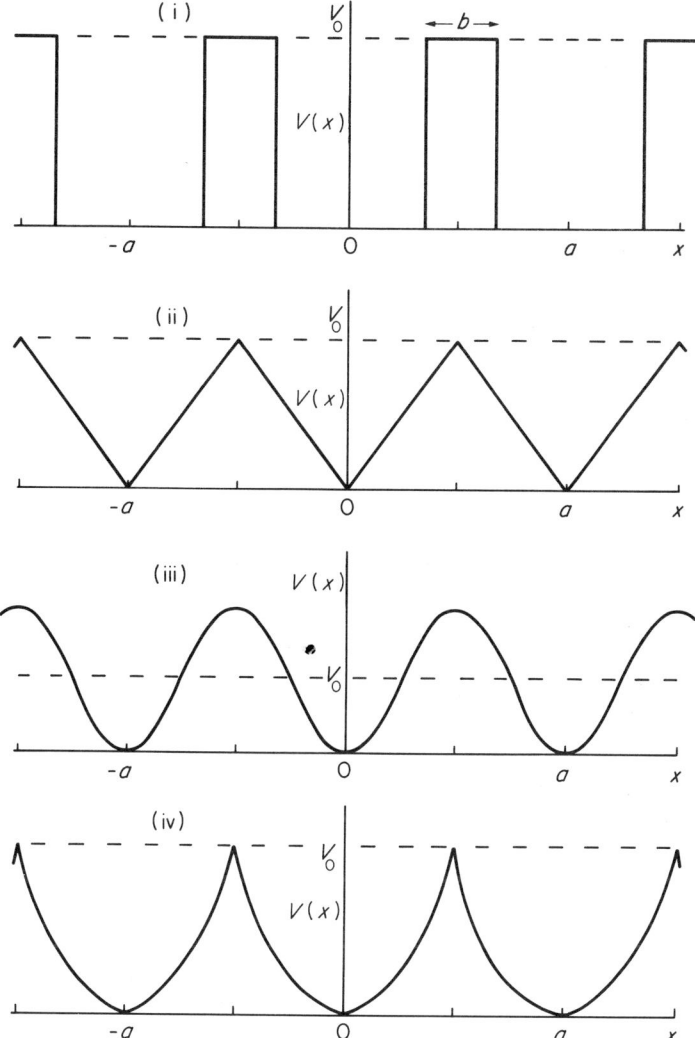

Figure 3. The potential options: (i) the rectangular potential; (ii) the sawtooth potential; (iii) the cosine potential; and (iv) the harmonic potential. (An arbitrary potential can also be set up by linear interpolation)

(ii) A 'sawtooth' potential

$$V(x) = 2V_0 \frac{x}{a}, \qquad 0 < x < \frac{a}{2}. \tag{29}$$

This form is used so that $V(x = a/2) = V_0$.

(iii) A cosine potential

$$V(x) = V_0 \left\{ 1 - \cos\left(\frac{2\pi x}{a}\right) \right\}. \tag{30}$$

(iv) A 'harmonic' potential

$$V(x) = 4V_0 \frac{x^2}{a^2} \quad 0 < x < \frac{a}{2}, \tag{31}$$

so that again $V(x = a/2) = V_0$.

(v) Finally the user can set up an arbitrary potential by linear interpolation specifying the number of points, their x-values, and the potential at each point.

2.6 The core state approximation

For $V_0 a \to \infty$ the probability of elections moving from atom to atom falls and their probability density is only significant where $V(x)$ takes its lowest values. For potentials (i), (iii), and (iv) under these conditions the effective potential tends towards that for several well-known situations and the computed results approach the corresponding values.

This is equivalent to the result, which is exploited in full-scale band structure calculations, that the lowest-lying electron levels—the 'core states'—have practically the same energies and wave functions in the solid as in the isolated atom. The core state electrons are bound to one nucleus and do not contribute to electrical conduction in the material.

A necessary condition for the validity of the core state approximation can be derived from first-order perturbation theory. If $\phi_{\text{atom}}(x)$ E_{atom} and $U_{\text{atom}}(x)$ are the isolated 'atomic' wave function, energy, and potential respectively, then it is required that the energy change in the solid be small, i.e.

$$\left| \int_{-\infty}^{+\infty} \phi^*_{\text{atom}}(x) [V(x) - U_{\text{atom}}(x)] \phi_{\text{atom}}(x) \, dx \right| \ll |E_{\text{atom}}|. \tag{32}$$

For $V_0 a \to \infty$ the potentials (28) to (31) lead to the following limiting energies and normalized wave functions in atomic units (for $-a/2 < x < a/2$):

(i) For $V_0 > 0$ and $V_0 a \to \infty$ the rectangular potential (28) tends towards an infinite potential well of width $(a-b)$.

$$E_n \to \frac{\pi^2 n^2}{(a-b)^2}, \qquad n = 1, 2, 3, \ldots, \tag{33}$$

$$\psi_n \to \frac{1}{\sqrt{a-b}} \cos\left(\frac{n\pi x}{a-b}\right), \qquad n = 1, 3, \ldots, \tag{34}$$

$$\psi_n \to \frac{1}{\sqrt{a-b}} \sin\left(\frac{n\pi x}{a-b}\right), \qquad n = 2, 4, \ldots. \tag{35}$$

The situation could be more realistically approximated by a finite potential well of depth V_0, though this problem does not have an analytic solution.[5,6]

(ii) There is no analytic solution for the core state of the sawtooth potential.

(iii) In the case of the cosine potential (30), $V(x)$ is lowest for $x \to 0$ and in this region

$$V_0\left[1-\cos\left(\frac{2\pi x}{a}\right)\right] \to V_0\left[1-\left(1-\frac{4\pi^2 x^2}{2!a^2}+\ldots\right)\right] \quad (36)$$

$$\approx V_0 2\pi^2 \frac{x^2}{a^2}.$$

The situation is therefore essentially equivalent to that for the harmonic potential.

(iv) The harmonic potential (31) will give results for $V_0 a \to \infty$ corresponding to the quantum-mechanical harmonic oscillator, i.e.

$$E_n \to (n+\tfrac{1}{2})\frac{4}{a}\sqrt{V_0}, \quad (37)$$

$$\psi_1(x) \to \left(\frac{4V_0^{\frac{3}{8}}}{a^2\pi^2}\right)^{\frac{1}{8}} \exp\left(-\sqrt{V_0}\cdot\frac{x^2}{a}\right), \quad (38)$$

$$\psi_2(x) \to \left(\frac{2^{\frac{5}{4}}V_0^{\frac{3}{8}}}{a^{\frac{3}{4}}\pi^{\frac{1}{4}}}\right) x \exp\left(-\sqrt{V_0}\frac{x^2}{a}\right), \quad (39)$$

$$\psi_3(x) \to \frac{1}{2^{\frac{1}{4}}}\left(\frac{V_0}{a^2\pi^2}\right)^{\frac{1}{8}}\left(\frac{4}{a}\sqrt{V_0}x^2-1\right)\exp\left(-\sqrt{V_0}\frac{x^2}{a}\right). \quad (40)$$

In all cases, as $V_0 a$ is increased, the lowest energy state will approach the core state limit soonest because, having less energy, it is more closely confined to the regions where $V(x)$ is low.

As an example of the application of the core state approximation the data points in Figure 4 shows the results of a series of runs of the computer program for the energies at $k = 0$ employing the cosine potential for a wide range of values of V_0. The solid lines show the energies of the three lowest energy states of an electron as predicted by the core state approximation (i.e. $E_n = (n+\tfrac{1}{2})(2\pi/a)\sqrt{2V_0}$), for the cosine potential. Note that agreement deteriorates for low V_0, particularly for the third band.

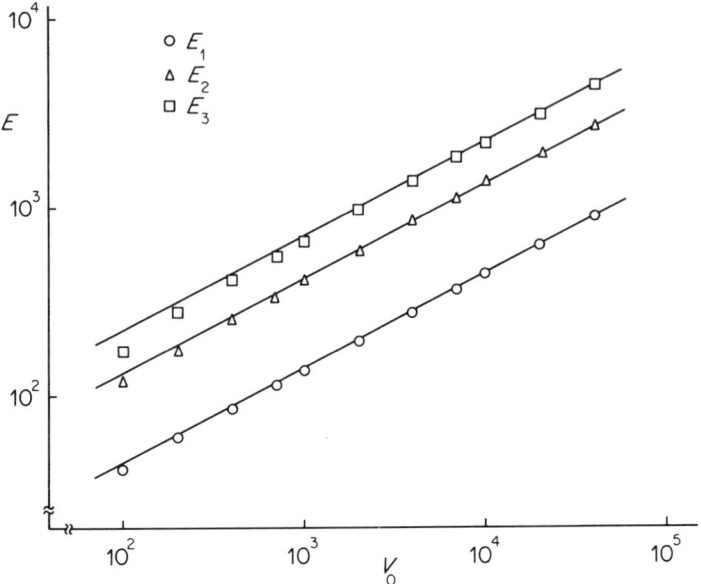

Figure 4. The results of the computer program employing the cosine potential with a period of 1.0 showing the energies at $k=0$ as a function of V_0: $E_1(\bigcirc)$, $E_2(\triangle)$, and $E_3(\square)$. The straight lines show the predictions of the core state approximation

The computer program also finds the probability density $\psi^*(x)\psi(x)$. In Figure 5 the data points show the computed results for $k=0$ with the cosine potential; the solid curves are the predictions of the core state approximation (which are simple analytic expressions). Agreement is poorest at larger $|x|$ where the difference between

$$V_0\left\{1-\cos\left(\frac{2\pi x}{a}\right)\right\} \quad \text{and} \quad V_0\left(\frac{2\pi^2 x^2}{a^2}\right)$$

is greatest.

2.7 The nearly-free electron approximation

The opposite extreme of very low values of $V_0 a$ can be treated using the nearly-free approximation. If $V(x)=0$ then $\psi(x)=\exp(ikx)$ and $E=k^2$. If $V(x)$ is small we can treat it as a perturbation and from the usual expression

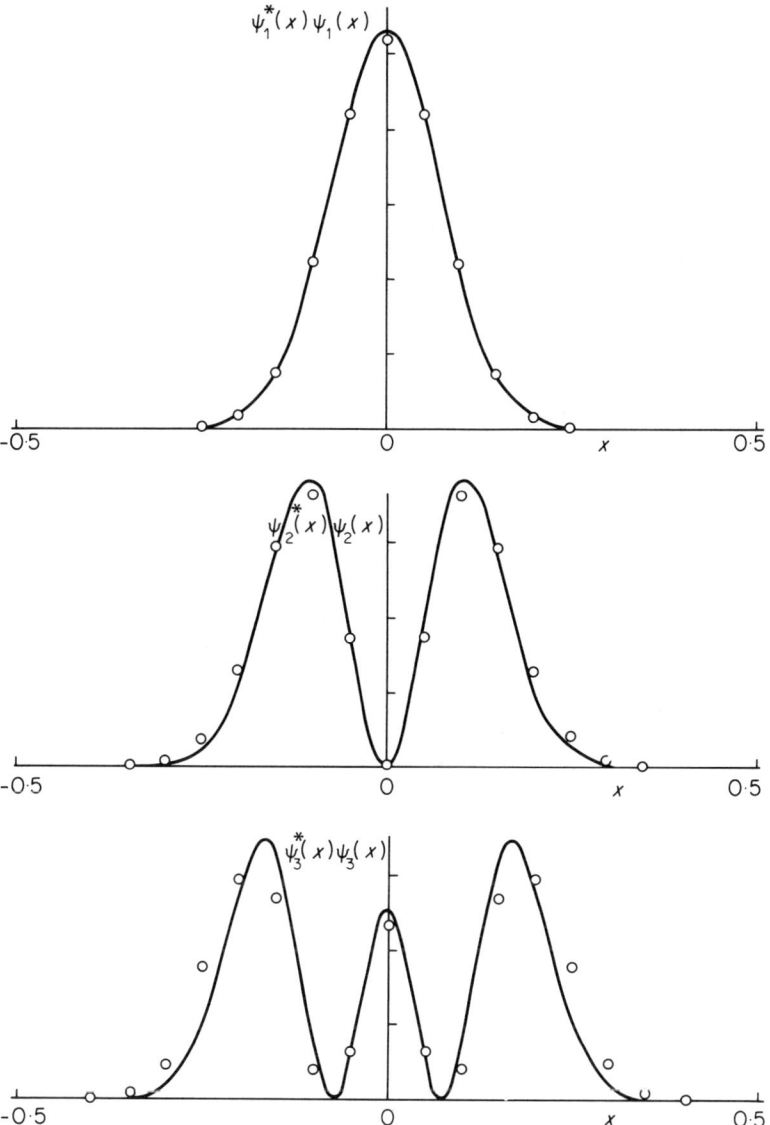

Figure 5. The circles are the results of the computer program for the probability density when the cosine potential is employed with a period of 1.0, $V_0 = 400$ and $k = 0$. The curves show the predictions of the core state approximation

for non-degenerate perturbation theory, up to second order,

$$E = k^2 + \frac{1}{a}\int_0^a \psi_k^*(x)V(x)\psi(x)\,dx + \frac{1}{a^2}\int_{-\infty \atop k'\neq k}^{+\infty} \frac{\left|\int_0^a \psi_{k'}^*(x)V(x)\psi_k(x)\,dx\right|^2}{(k^2-k'^2)}\,dk'$$

$$= k^2 + \frac{1}{a}\int_0^a \exp(-ikx)\sum_{j=-N}^{+N} f_j \exp\left(\frac{i2\pi jx}{a}\right)\exp(ikx)\,dx$$

$$+ \frac{1}{a^2}\int_{-\infty \atop k'\neq k}^{+\infty} \frac{\left|\int_0^a \exp(-ik'x)\sum_{j=-N}^{+N} f_j \exp\left(\frac{i2\pi jx}{a}\right)\exp(ikx)\,dx\right|^2}{(k^2-k'^2)}\,dk'. \quad (41)$$

The factors $1/a$ and $1/a^2$ arise because the norm of the wave functions over the unit cell is a. Because the exponential functions are orthogonal over the unit cell,

$$E = k^2 + f_0 + \sum_{j=-N \atop j\neq 0}^{+N} \frac{f_j^2}{\left\{k^2 - \left(k - \frac{2\pi j}{a}\right)^2\right\}}. \quad (42)$$

Equation (42) is accurate except when there is degeneracy, or near degeneracy, between the states at k and at $k - 2\pi j/a$, i.e.

$$|k| \approx \left|k - \frac{2\pi j}{a}\right|, \quad (j\neq 0), \quad (43)$$

leading to

$$k \approx -k + \frac{2\pi j}{a}, \quad (44)$$

which implies that

$$k \approx \frac{\pi j}{a} \quad (j = \pm 1, \pm 2, \ldots), \quad (45)$$

as will be the case in the region of the zone boundaries. We then need to consider the explicit form of the wave functions.

As an example of the approach consider the lowest energy band. The unperturbed wave function $\psi(x) = \exp(ikx)$. The perturbation $V(x)$ will 'mix in' states with wave functions

$$\exp\left\{i\left(k - \frac{2\pi j}{a}\right)x\right\} \quad j = \pm 1, \pm 2 \ldots$$

but, close to $k = \pi/a$ (the zone boundary) the main contribution will be due to the state $\exp\{i(k - 2\pi/a)x\}$ that is degenerate with the unperturbed state at the zone boundary, i.e.

$$\psi(x) \approx \alpha \exp(ikx) + \beta \exp\left\{i\left(k - \frac{2\pi}{a}\right)x\right\}, \tag{46}$$

where α and β are to be determined. The Schrödinger equation takes the form:

$$\left[-\frac{d^2}{dx^2} + \sum_{j=-N}^{+N} f_j \exp\left(\frac{i2\pi j}{a}\right)\right]\left(\alpha \exp(ikx) + \beta \exp\left\{i\left(k - \frac{2\pi}{a}\right)x\right\}\right)$$
$$= E\left(\alpha \exp(ikx) + \beta \exp\left\{i\left(k - \frac{2\pi}{a}\right)x\right\}\right), \tag{47}$$

that becomes

$$(k^2 - E)\alpha \exp(ikx) + \left[\left(k - \frac{2\pi}{a}\right)^2 - E\right]\beta \exp\left\{i\left(k - \frac{2\pi}{a}\right)x\right\}$$
$$+ \sum_{j=-N}^{+N} f_j \exp\left(\frac{i2\pi j}{a}\right)\left[\alpha \exp(ikx) + \beta \exp\left\{i\left(k - \frac{2\pi}{a}\right)x\right\}\right] = 0. \tag{48}$$

Since this equation is true for all x the coefficients of all the exponential terms must vanish. The coefficient of $\exp(ikx)$ is

$$(k^2 - E)\alpha + f_0\alpha + f_1\beta = 0, \tag{49}$$

and of $\exp\left\{i\left(k - \frac{2\pi}{a}\right)x\right\}$ is

$$\left[\left(k - \frac{2\pi}{a}\right)^2 - E\right]\beta + f_0\beta + f_{-1}\alpha = 0. \tag{50}$$

The other exponential terms should strictly be incorporated into similar equations involving the other (negligible) admixed wave functions. Eliminating α/β from the two equations above and substituting for f_0 and $f_1(=f_{-1})$ gives

$$\left[k^2 + \frac{d_0}{2} - E\right]\left[\left(k - \frac{2\pi}{a}\right)^2 + \frac{d_0}{2} - E\right] - \frac{d_1^2}{4} = 0, \tag{51}$$

which has the roots

$$E = k^2 + \frac{d_0}{2} + \frac{2\pi}{a}\left[\frac{\pi}{a} - k \pm \sqrt{\left(\frac{\pi}{a} - k\right)^2 + \left(\frac{ad_1}{4\pi}\right)^2}\right]. \tag{52}$$

The lower root applies to the lowest energy band. The higher root applies to the second energy band close to the first zone boundary.

The energy gap at the first zone boundary ($k = \pi/a$) is therefore $|d_1|$, and quite generally the energy gap at the jth zone boundary is $|d_j|$.

This result can be readily obtained by diagonalizing the degenerate two-dimensional submatrix of the Hamiltonian matrix (27) appropriate at the jth zone boundary, i.e.

$$\begin{bmatrix} \dfrac{\pi^2 j^2}{a^2} + \dfrac{d_0}{2} & \dfrac{d_j}{2} \\ \dfrac{d_j}{2} & \dfrac{\pi^2 j^2}{a^2} + \dfrac{d_0}{2} \end{bmatrix}$$

Hence

$$E(k = \pi j/a) = \frac{\pi^2 j^2}{a^2} + \frac{d_0}{2} \pm \frac{d_j}{2}. \tag{53}$$

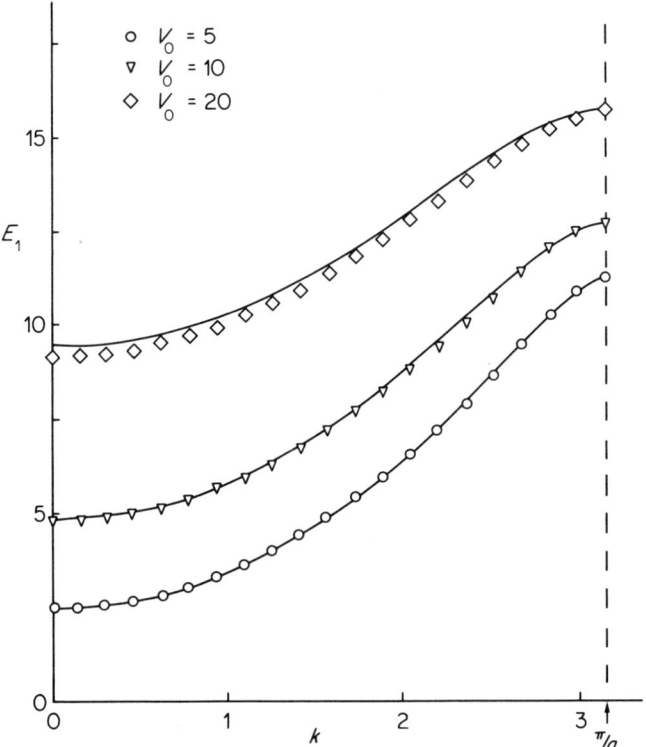

Figure 6. The results for the E–k relationship of the lowest band found by the computer program with a sawtooth potential of period 1.0 for $V_0 = 5$(○), $V_0 = 10$(▽) and $V_0 = 20$(◇). The solid curves show the predictions of the nearly-free electron approximation

ELECTRON ENERGY BANDS IN A ONE-DIMENSIONAL PERIODIC POTENTIAL 319

Figure 6 shows a comparison between the results of the computer program (the data points) and the nearly-free electron approximation (the solid curves) for the E–k relationship of the lowest band for three values of V_0 with the sawtooth potential. In this case agreement deteriorates for *high* V_0. Figure 7 shows the results of a series of program runs for the energy gap at the first zone boundary (the circles) as a function of V_0, for the sawtooth potential. The solid line shows the prediction of the nearly-free electron approximation, i.e. $4V_0/\pi^2$. Agreement decreases at large V_0.

It is of some interest to determine the coefficients α and β, and hence the wave function, especially at the zone boundary. Now using equations (23) and (49)

$$\alpha\left(k^2 + \frac{d_0}{2} - E\right) = -\beta\frac{d_1}{2}, \qquad (54)$$

and, at the zone boundary.

$$E = k^2 + \frac{d_0}{2} \pm \frac{d_1}{2}. \qquad (55)$$

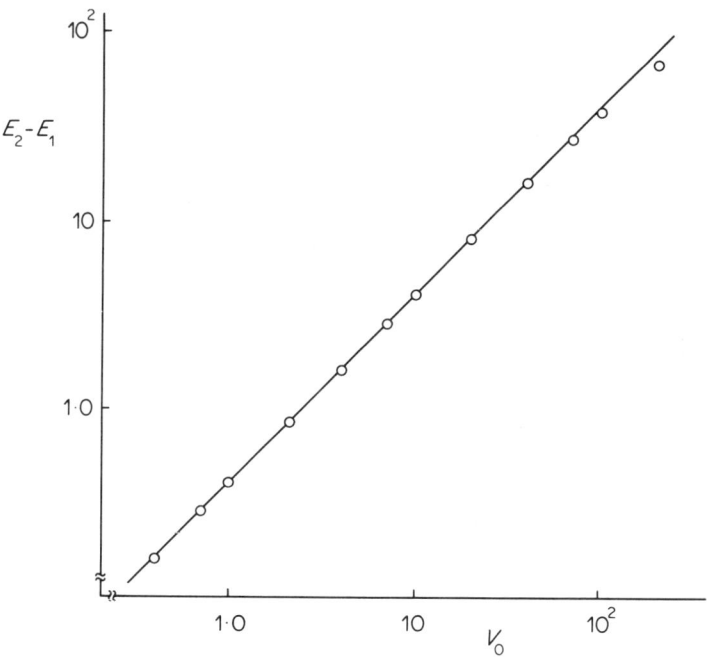

Figure 7. The energy gap at the first zone boundary: $E_2(k = \pi/a) - E_1(k = \pi/a)$ as a function of V_0. The circles show the results of the computer program for a sawtooth potential of period 1.0. The solid line shows the prediction of the nearly-free electron approximation

If $d_1 < 0$, which is the case here for $V_0 > 0$, then for the lowest energy band the upper sign in equation (55) is taken and

$$\alpha\left(-\frac{d_1}{2}\right) = -\beta\left(\frac{d_1}{2}\right). \tag{56}$$

The normalization requires $\alpha^2 + \beta^2 = 1$ so that for the lowest band, at the zone boundary,

$$\psi(x) = \frac{1}{\sqrt{2}}\left[\exp\left(\frac{i\pi x}{a}\right) + \exp\left(-\frac{i\pi x}{a}\right)\right]$$

$$= \sqrt{2}\cos\left(\frac{\pi x}{a}\right). \tag{57}$$

For the second band at $k = \pi/a$, a similar analysis gives

$$\psi(x) = i\sqrt{2}\sin\left(\frac{\pi x}{a}\right). \tag{58}$$

Thus, for both bands, the wave functions at the zone boundary correspond to standing waves (due to Bragg reflection) rather than travelling waves. Furthermore, the probability density is high in the region of low potential for the lower energy band and high in the region of high potential for the higher band.

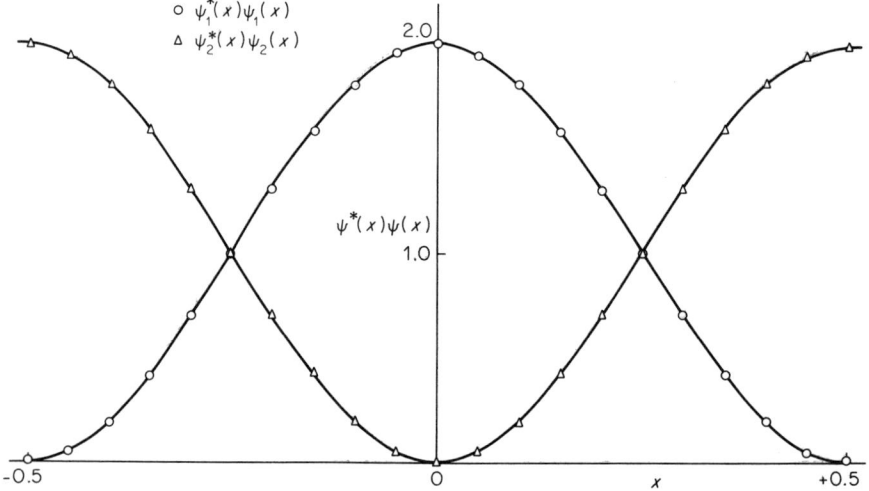

Figure 8. The results of the computer program for the probability density at $k = \pi/a$ of the first (○) and the second band (△) for a sawtooth potential of period 1.0 with $V_0 = 1.0$. The solid curves show the predictions of the nearly-free electron approximation. Note that for the lower energy band the probability density is high where $V(x)$ is low; for the higher energy band the probability density is high where $V(x)$ is high

In Figure 8 the computed results for the probability density (the data points) are compared with the above results for the nearly-free electron approximation. The potential used was the sawtooth potential.

In view of the rather large magnitude of the Fourier components of the Coulomb potential it might be thought that the nearly-free electron approximation would be of little value in practice, and indeed, for many years, it was not clear why it gives reasonable results in many cases. The reason is that, because the 'free' electron wave functions are orthogonal to the core states, the Schrödinger equation can be transformed so that the potential can be replaced by a 'pseudopotential'[7] which is weaker than the true Coulomb potential and which has Fourier components of smaller magnitude which are rapidly convergent.

2.8 The tight-binding approximation

If $V_0 a$ is large, but not sufficiently large for the core state approximation to apply, we might expect that a good approximation to the true wave function would be a linear combination of atomic wave functions $\phi_{atom}(x)$ on each site. Bloch[4] suggested a linear combination of the form:

$$\psi_k(x) = A \sum_{n=-\infty}^{+\infty} \exp(ikna)\phi_{atom}(x - na), \qquad (59)$$

where $\phi_{atom}(x - na)$ is an atomic wave function centred on the site at na, and A is a normalizing factor.

These functions satisfy the Bloch condition $\psi_k(x) = \exp(ikx)u_k(x)$, where $u_k(x)$ is of period a. The suggested linear combination implies

$$u_k(x) = \exp(-ikx) A \sum_{n=-\infty}^{+\infty} \exp(ikna)\phi_{atom}(x - na). \qquad (60)$$

Hence

$$u_k(x + ma) = \exp\{-ik(x + ma)\} A \sum_{n=-\infty}^{+\infty} \exp(ikna)\phi_{atom}(x + ma - na)$$

$$= \exp(-ikx) A \sum_{n=-\infty}^{+\infty} \exp(ik[n-m]a)\phi_{atom}(x - [n-m]a), \qquad (61)$$

and, since the sum over n runs from $-\infty$ to $+\infty$,

$$u_k(x + ma) = \exp(-ikx) A \sum_{n=-\infty}^{+\infty} \exp(ikna)\phi_{atom}(x - na)$$

$$= u_k(x). \qquad (62)$$

Now equation (10) is

$$\langle E \rangle = \frac{\int_0^a \psi^*(x)\mathcal{H}\psi(x)\,dx}{\int_0^a \psi^*(x)\psi(x)\,dx} = \frac{N}{D}. \tag{63}$$

The denominator, using equation (60), is

$$D = \int_0^a \psi^*(x)\psi(x)\,dx$$

$$= A^*A \int_0^a \sum_{n=-\infty}^{+\infty} \exp(-ika)\phi^*_{atom}(x-na) \sum_{m=-\infty}^{+\infty} \exp(ikma)\phi_{atom}(x-ma)\,dx. \tag{64}$$

If we write $l = n - m$ equation (64) becomes

$$D = A^*A \sum_{n=-\infty}^{+\infty} \sum_{l=-\infty}^{+\infty} \exp(-kla) \int_0^a \phi^*_{atom}(x-na)\phi_{atom}(x+la-na)\,dx. \tag{65}$$

Summing over n and then integrating over the interval $0 \leq x \leq a$, i.e. over the unit cell, is equivalent to integrating over all space giving

$$D = A^*A \sum_{l=-\infty}^{+\infty} \exp(-ikla) \int_{-\infty}^{+\infty} \phi^*_{atom}(x)\phi_{atom}(x+la)\,dx. \tag{66}$$

If we write

$$S_l = \int_{-\infty}^{+\infty} \phi^*_{atom}(x)\phi_{atom}(x+la)\,dx, \tag{67}$$

equation (66) simplifies to

$$D = A^*A \sum_{l=-\infty}^{+\infty} \exp(-ikla) S_l. \tag{68}$$

(Note that S_0 is simply the norm of $\phi_{atom}(x)$.)

Major simplification occurs because, when the tight-binding approximation is appropriate, the atomic wave functions $\phi_{atom}(x)$ are highly localized, and S_l is significant only for $l = 0$ (the dominant term) and for $l = \pm 1$.

The numerator in the expression for $\langle E \rangle$ is

$$N = \int_0^a \psi^*(x) \mathcal{H} \psi(x) \, dx$$

$$= AA^* \int_0^a \sum_{n=-\infty}^{+\infty} \exp(-ikna) \phi^*_{\text{atom}}(x-na) \mathcal{H} \sum_{m=-\infty}^{+\infty} \\ \times \exp(ikma) \phi_{\text{atom}}(x-ma) \, dx, \quad (69)$$

and, by a similar argument, if we define

$$H_l = \int_{-\infty}^{+\infty} \phi^*_{\text{atom}}(x) \mathcal{H} \phi_{\text{atom}}(x+la) \, dx, \quad (70)$$

equation (69) becomes

$$N = A^*A \sum_{l=-\infty}^{+\infty} \exp(-ikla) H_l. \quad (71)$$

Now

$$\mathcal{H} \phi_{\text{atom}}(x+la) = \left[-\frac{d^2}{dx^2} + V(x) \right] \phi_{\text{atom}}(x+la)$$

$$= \left[-\frac{d^2}{dx^2} + U_{\text{atom}}(x+la) \right] \phi_{\text{atom}}(x+la)$$

$$+ [V(x) - U_{\text{atom}}(x+la)] \phi_{\text{atom}}(x+la), \quad (72)$$

(here $U_{\text{atom}}(x+la)$ is the atomic potential centred on the atom at $-la$). Therefore,

$$\mathcal{H} \phi_{\text{atom}}(x+la) = E_{\text{atom}} \phi_{\text{atom}}(x+la) + [V(x) - U_{\text{atom}}(x+la)] \phi_{\text{atom}}(x+la), \quad (73)$$

so that

$$H_l = \int_{-\infty}^{+\infty} \phi^*_{\text{atom}}(x) \mathcal{H} \phi_{\text{atom}}(x+la) \, dx$$

$$= E_{\text{atom}} S_l + g_l, \quad (74)$$

where

$$g_l = \int_{-\infty}^{+\infty} \phi^*_{\text{atom}}(x) [V(x) - U_{\text{atom}}(x+la)] \phi_{\text{atom}}(x+la) \, dx. \quad (75)$$

The g_l's are referred to as overlap integrals. Hence

$$\langle E \rangle = \frac{\sum\limits_{l=-\infty}^{+\infty} \exp(-ikla) \cdot (E_{\text{atom}} S_l + g_l)}{\sum\limits_{l=-\infty}^{+\infty} \exp(-ikla) S_l}$$

$$= E_{\text{atom}} + \frac{\sum\limits_{l=-\infty}^{+\infty} \exp(-ikla) g_l}{\sum\limits_{l=-\infty}^{+\infty} \exp(-ikla) S_l}, \quad (76)$$

and the only significant contributions will come from terms for small $|l|$.

Figure 9 shows some of the functions involved in the integrals which give S_l and g_l for the case of the lowest energy band with the harmonic potential. For $V_0 a \to \infty$ the only significant terms are for $l = 0$ and

$$\langle E \rangle \approx E_{\text{atom}} + \frac{g_0}{S_0}. \quad (77)$$

The first term is the core state approximation. The second term is the correction to it predicted by first-order perturbation theory.

For somewhat lower values of $V_0 a$ the $|l| = 1$ terms will also be significant.

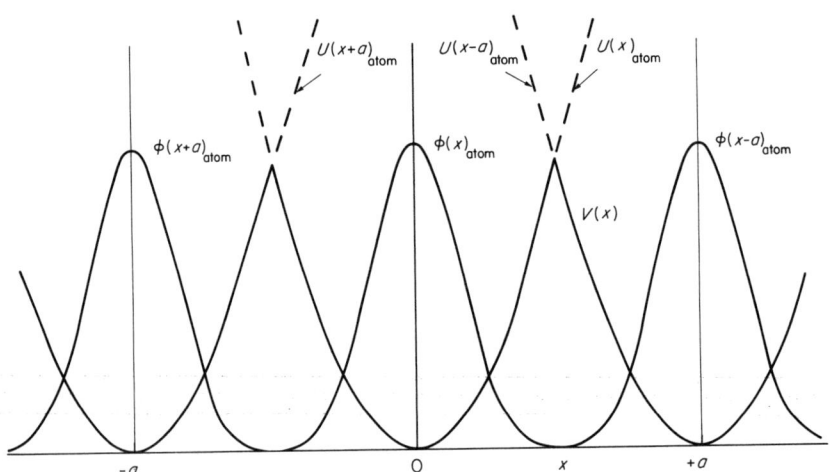

Figure 9. Some of the functions involved in calculating S_l and g_l for the lowest energy band in the tight-binding approximation with the harmonic potential

ELECTRON ENERGY BANDS IN A ONE-DIMENSIONAL PERIODIC POTENTIAL

Usually the term in $g_{\pm 1}$ is significant before that in $S_{\pm 1}$ and, since $g_1 = g_{-1}$,

$$\langle E \rangle \approx E_{\text{atom}} + \frac{g_0}{S_0} + \frac{g_1}{S_0} [\exp(ika) + \exp(-ika)]$$

$$= E_{\text{atom}} + \frac{g_0}{S_0} + \frac{2g_1}{S_0} \cos(ka), \qquad (78)$$

and, if $\phi_{\text{atom}}(x)$ is normalized to unity,

$$\langle E \rangle = E_{\text{atom}} + g_0 + 2g_1 \cos(ka). \qquad (79)$$

Thus, in the tight-binding approximation,

$$E(k=0) - E(k=\pi/a) = 4g_1 \quad \text{and} \quad E(k=\pi/2a) = E_{\text{atom}} + g_0.$$

The tight-binding approximation can be applied to the following potentials employed in the computer program:

(i) The cosine potential where, for the lowest energy band,

$$g_0 = V_0 \left(\frac{2V_0}{a^2}\right)^{\frac{1}{4}} \int_{-\infty}^{+\infty} \exp\left(-\frac{x^2 \pi}{a}\sqrt{2V_0}\right) \left[1 - \cos\left(\frac{2\pi x}{a}\right) - \frac{2\pi^2 x^2}{a^2}\right] dx$$

$$= V_0 \left[1 - \frac{\pi}{a\sqrt{2V_0}} - \exp\left(-\frac{\pi}{a\sqrt{2V_0}}\right)\right], \qquad (80)$$

and

$$g_1 = V_0 \left(\frac{2V_0}{a^2}\right)^{\frac{1}{4}} \int_{-\infty}^{+\infty} \exp\left(-\frac{x^2 \pi}{a}\sqrt{\frac{V_0}{2}}\right) \left[1 - \cos\left(\frac{2\pi x}{a}\right) - \frac{2\pi^2}{a^2}(x+a)^2\right]$$

$$\times \exp\left(-\frac{[x+a]^2 \pi}{a}\sqrt{\frac{V_0}{2}}\right) dx$$

$$= V_0 \exp\left(-\frac{a\pi}{2}\sqrt{\frac{V_0}{2}}\right)\left[1 - \frac{\pi^2}{2} - \frac{\pi}{a\sqrt{2V_0}} + \exp\left(-\frac{\pi}{a\sqrt{2V_0}}\right)\right]. \qquad (81)$$

(ii) The harmonic potential. For the lowest band

$$g_0 \approx \frac{8}{a^2}\left(\frac{4V_0}{a^2\pi^2}\right)^{\frac{1}{4}} \int_{a/2}^{+\infty} \exp\left(-2\sqrt{V_0}\frac{x^2}{a}\right)[V_0(x-a)^2 - V_0 x^2] dx$$

$$= 4V_0\left[1 - \Phi(V_0^{\frac{1}{4}}\sqrt{a})\right] - \frac{4\sqrt{2}V_0^{\frac{3}{4}}}{\sqrt{\pi a}} \exp\left(-\sqrt{V_0}\frac{a}{2}\right), \qquad (82)$$

$$g_1 \approx \frac{4}{a^2}\left(\frac{4V_0}{a^2\pi^2}\right)^{\frac{1}{4}} \int_{-a/2}^{+a/2} \exp\left(-\frac{\sqrt{V_0}x^2}{a}\right)$$

$$\times [V_0 x^2 - V_0(x+a)^2] \exp\left(-\sqrt{V_0}\frac{(x+a)^2}{a}\right) dx$$

$$= 2\sqrt{2} \cdot \frac{V_0^{\frac{3}{4}}}{\sqrt{\pi a}} \exp\left(-\sqrt{V_0}\frac{a}{2}\right)[\exp(-2\sqrt{V_0}a) - 1]. \tag{83}$$

For the harmonic potential the probability integral

$$\Phi(z) = \sqrt{\frac{2}{\pi}} \int_0^z \exp(-t^2/2)\, dt \tag{84}$$

is required. This integral cannot be expressed in closed form. It can be approximated by a series expansion, but for the large values of z typically involved, convergence is slow and it is more convenient to use tabulated values for $\Phi(z)$.

The tight-binding approximation cannot be applied to the rectangular potential if the 'atomic' wave functions are taken to be those corresponding to an infinite potential well because in that case $g_0 = g_1 = \ldots = 0$. It can be used if the atomic states are taken as those appropriate for a finite potential well, and an approach along these lines has been presented by Wetsel[6] though the overlap integrals for the finite potential well cannot be expressed as analytic functions but involve the solution of a transcendental equation.

In Figure 10 the computed results for the variation of E with k of the lowest band (the data points) are compared with the predictions of the tightbinding approximation (the solid curves) for the harmonic potential. Agreement deteriorates as V_0 increases. In Figure 11 the circles show the computed results for $E(k = \pi/a) - E(k = 0)$ as a function of V_0 with the harmonic potential. The solid curve shows $4g_1$, i.e. the energy difference predicted by the tight-binding approximation.

2.9 The effective mass

There are situations, particularly those involving the dynamic behaviour of the electrons, when the $E-k$ relationship is not the most useful form in which to present the results of a band structure calculation. The wave function $\psi(x) = \exp(ikx)u_k(x)$ extends over all space; if we want to represent the motion of a 'localized' electron the uncertainty principle indicates that we must build up a wave-packet with a spread of k-values. The appropriate velocity is then the group velocity (V_G) which is equal to the derivative of

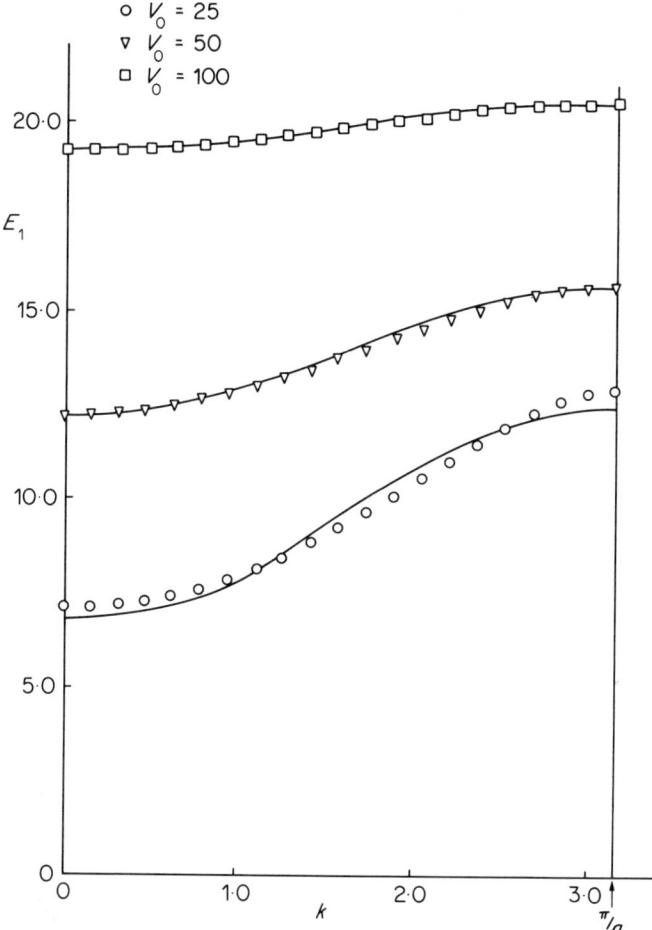

Figure 10. The results of the computer program for the lowest energy band with the harmonic potential of period 1.0 for $V_0 = 25(\bigcirc)$, $V_0 = 50(\triangledown)$, and $V_0 = 100(\square)$. The solid curves show the predictions of the tight-binding approximation

the angular frequency (ω) with respect to k

$$V_G = \frac{\partial \omega}{\partial k} = \frac{\partial E}{\partial k} \tag{85}$$

($E = \omega$, since $\hbar = 1$ in the atomic system of units).

If a force F acts on this electron wave-packet in the $+x$-direction then

$$FV_G = \frac{\partial E}{\partial t}, \tag{86}$$

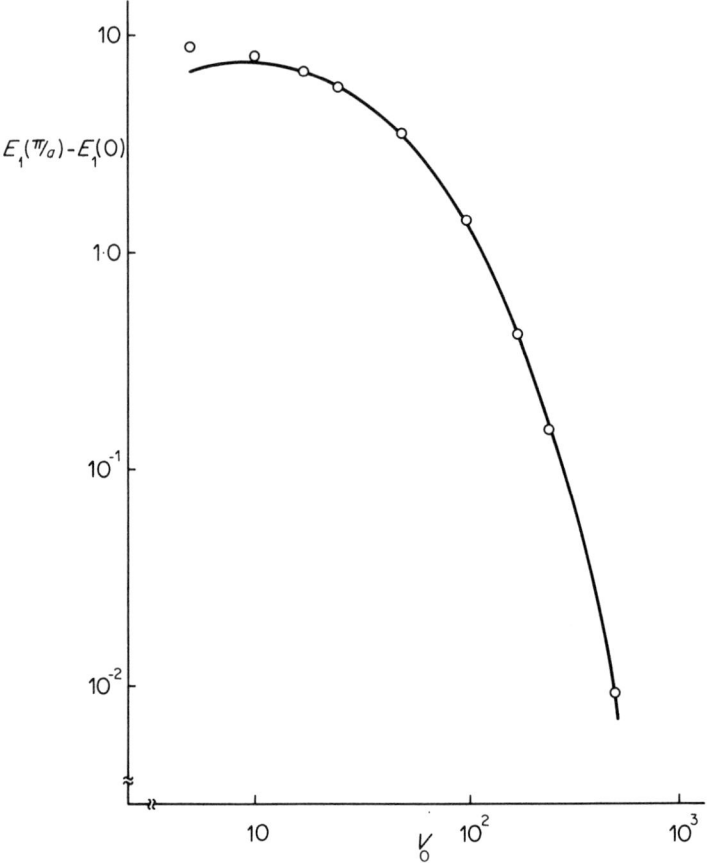

Figure 11. The circles show the results computed for $E(k = \pi/a) - E(k = 0)$ of the lowest energy band with a harmonic potential of period 1.0, as a function of V_0. The solid curve shows $4g_1$ which is the energy difference predicted by the tight-binding approximation

and

$$\begin{aligned}\frac{\partial V_G}{\partial t} &= \frac{\partial}{\partial t}\left(\frac{\partial E}{\partial k}\right) = \frac{\partial}{\partial k}\left(\frac{\partial E}{\partial t}\right) \\ &= \frac{\partial}{\partial k}(FV_G) = \frac{\partial}{\partial k}\left(F\frac{\partial E}{\partial k}\right) \\ &= F\frac{\partial^2 E}{\partial k^2}.\end{aligned} \qquad (87)$$

Comparing this result with Newton's second law of motion for a particle of mass M, i.e. $\partial V/\partial t = (1/M)F$ we see that the motion of the electron wave-packet can be described by the semi-classical concept of an effective mass $(m^*) = 1/(\partial^2 E/\partial k^2)$.

We should perhaps observe that there are situations[8,9] where a different definition of effective mass, $1/m^* = (1/k)\partial E/\partial k$, appears to be appropriate.

For the lowest band $\partial^2 E/\partial k^2 > 0$ near $k = 0$ but $\partial^2 E/\partial k^2 < 0$ near $k = \pi/a$. Rather than refer to a negative effective mass it is conventional to take $m^* = 1/|\partial^2 E/\partial k^2|$ but to refer to electrons (with electronic charge $-e$) when $\partial^2 E/\partial k^2 > 0$ and holes (with charge $+e$)—regarded as the *absence* of an electron from a band—for $\partial^2 E/\partial k^2 < 0$.

It should be noted that, since in the atomic system of units the mass of an electron $m_e = \frac{1}{2}$,

$$\frac{m^*}{m_e} = 2 \bigg/ \left|\frac{\partial^2 E}{\partial k^2}\right| \quad \text{(in atomic units).} \tag{88}$$

It is easy to see that equation (88) is plausible in the two extremes of completely free electrons and the core state approximation. For completely free electrons $E = k^2$, $\partial^2 E/\partial k^2 = 2$ and $m^* = m_e$. In the case of the core state approximation (where the electrons are not free to travel through the crystal) E is independent of k, $\partial^2 E/\partial k^2 \to 0$, and $m^* \to \infty$.

If the nearly-free electron approximation or the tight-binding approximation apply, then the appropriate analytic expressions for $E(k)$ can be transformed to analytic expressions for m^*/m_e as a function of k.

The seemingly perverse behaviour of holes can be understood if we recall that at the zone boundaries Bragg reflection leads to wave functions which are standing waves. Consequently, a wave-packet at a boundary has $V_G = 0$ and cannot propagate. For the lowest energy band then, at $k = 0$ the wave-packet has zero velocity; as k and E increase its velocity increases initially, however, when $k = \pi/a$ the velocity has fallen again to zero. Consequently, there are regions of the energy bands approaching the zone boundaries where an *increase* in energy leads to a *decrease* in velocity—the acceleration is in the opposite direction to the force. The concept of holes is used to describe this unexpected behaviour.

Having found E at $k = 0$, $\pi/20a$, $\pi/10a$, ..., π/a the program estimates $\partial^2 E/\partial k^2$ at these k-values from the results for $E_{k-\Delta k}$, E_k, and $E_{k+\Delta k}$. By Taylor's series:

$$E_{k+\Delta k} = E_k + \frac{\Delta k}{1!}\frac{\partial E}{\partial k} - \frac{\Delta k^2}{2!}\frac{\partial^2 E}{\partial k^2} + \cdots. \tag{89}$$

Hence

$$\left(\frac{\partial^2 E}{\partial k^2}\right)_k \approx \frac{1}{\Delta k^2}[E_{k+\Delta k} + E_{k-\Delta k} - 2E_k]. \tag{90}$$

The program displays the results for m^*/m_e as a function of k, printing a letter E (electron) when $\partial^2 E/\partial k^2 > 0$ and a letter H (hole) when $\partial^2 E/\partial k^2 < 0$.

2.10 The probability density

The variation of E with k (or related variables) is usually the most important result of a band structure calculation. There are, however, situations when the actual wave functions or related quantities are of interest. In particular the total electron probability density, which can be investigated by X-ray and neutron diffraction techniques, according to equation (3) involves the product $u_k^*(x)u_k(x)$. Now

$$u_k(x) = \sum_{l=-N}^{+N} c_l \exp\left(\frac{i2\pi l x}{a}\right), \tag{91}$$

and for systems with an inversion centre, the Hamiltonian matrix is real. Consequently the c_l's are real, and

$$u_k^*(x) = \sum_{m=-N}^{+N} c_m \exp\left(-\frac{i2\pi m x}{a}\right). \tag{92}$$

Hence

$$\begin{aligned}
u_k^*(x)u_k(x) &= \sum_{m=-N}^{+N} c_m \exp\left(-\frac{i2\pi m x}{a}\right) \sum_{l=-N}^{+N} c_l \exp\left(\frac{i2\pi l x}{a}\right) \\
&= \sum_{m=-N}^{+N} \sum_{l=-N}^{+N} c_m c_l \exp\left(i2\pi[l-m]\frac{x}{a}\right) \\
&= \sum_{m=-N}^{+N} \sum_{l=-N}^{+N} c_m c_l \cos\left(2\pi[l-m]\frac{x}{a}\right) \\
&= 1 + 2 \sum_{m=-N}^{+N} \sum_{l=m+1}^{+N} c_m c_l \cos\left(2\pi[l-m]\frac{x}{a}\right),
\end{aligned} \tag{93}$$

since $\sum_{m=-N}^{+N} c_m^2 = 1$ and $u_k^*(x)u_k(x)$ is real.

The computer program finds the probability densities in the range $-a/2 < x < a/2$ for a value of k specified by the user. The results can be compared with the predictions of the core state approximation or the nearly-free electron approximation when they are appropriate.

As is generally the case when the variational method is employed, the results for the probability density can be expected to be less accurate than the results for the energy.

3. THE COMPUTER PROGRAM

The program allows the user to choose the potential from the set of even periodic functions described in section 2.5. The first section prints questions about the potential. The program as written is for use on an interactive system. This section would need to be modified for use in batch mode. The user's responses define the potential function by setting the integer variable NPOT and the value of constants, such as the period and the maximum value of the potential.

The subroutine FRANCS(NTERMS, NDATA, DC, A) is then called to find the coefficients of the cosine terms in the Fourier series. The first parameter NTERMS ($=33$) is the number of coefficients required. NDATA ($=$NPTS) is used only when the interpolated potential is employed. For that particular case FRANCS( ) employs trapezoidal integration and NDATA specifies the number of subdivisions required. The fourth parameter DC is the constant term in the Fourier series for $V(x)$, i.e. $d_0/2$. The fifth parameter is the array A( ) which is set by FRANCS( ) to the Fourier coefficients; $A(1) = d_1$, $A(2) = d_2, \ldots, (A33) = d_{33}$. In the case of all except the interpolated potential the Fourier coefficients are determined analytically.

Next $K(=k)$ is set. It is increased in 20 steps from 0 to π/a. Because the true matrix for the Hamiltonian is of infinite dimension in practice it has to be truncated, which introduces some error into the results. In an attempt to maintain this error less than ~ 1 per cent the dimension of the Hamiltonian (NAR) is set initially to either 10 or 11. The energies are then determined by the subroutine ENERGY( ) and NAR increased by 2. ENERGY( ) is called again and if the fractional changes in the results for the energy are less than 3×10^{-3} then they are printed out. If one or more of the results has changed by more than 3×10^{-3}, NAR is increased by a further 8 and ENERGY( ) called again.

This process is continued till NAR is 32 or 33. At this point the computing time taken becomes significant, so rather than increasing NAR further, the results are printed out with an accuracy warning. If higher accuracy were to be required the program could be modified by increasing the upper limit on NAR.

The subroutine ENERGY(K, E, EV, ND, N, DC, A, H, W, BOOL) is used to find the energy eigenvalues and also (if BOOL is set.FALSE.) the eigenvectors.

The parameters set on entry are: $K(=k)$, ND – the first dimension of H( ) (the Hamiltonian) as specified in ENERGY( ), N is the order of H( ), DC and A( ) are the Fourier coefficients described above and BOOL is a logical variable. If BOOL is set .TRUE. the NAG[10] library routine F02AAF()—which calculates eigenvalues—is called. If BOOL is set .FALSE. the NAG routine F02ABF( )—which calculates eigenvalues and eigenvectors—is called.

ENERGY( ) sets up the Hamiltonian matrix H( ). The input variables common to F02AAF( ) and F02ABF( ) are H( ), ND, N (as described above), and IFAIL which specifies what will happen if a failure occurs during a call to these routines. Both F02AAF and F02ABF require H( ) to be real and symmetric. The array W( ) is used by the NAG routines as working space. On exit from F02AAF or F02ABF E( ) contains the eigenvalues *in ascending order*. On exit from F02ABF the array EV( ) contains the *normalized* eigenvectors in columns corresponding to the eigenvalues. By calling ENERGY( ) for varying values of k the program calculates and prints out results for the E–k diagram, showing the three lowest energy bands.

The user has the option of plotting these results on the line printer. The plot is done by the subroutine GRAPH(Y, NP, NG, X). Y( ) contains the energy values and X( ) the corresponding k-values. If a more sophisticated graph-plotting routine were to be available to the reader GRAPH( ) could be replaced.

The user may also calculate the approximate effective mass as a function of k and the probability density function as a function of x (x being increased in 20 steps from $-a/2$ to $+a/2$), for a chosen value of k. For the probability density function calculation ENERGY( ) is called with BOOL = .FALSE. so that F02ABF( ) is now called. The potential is also found as a function of x by the function V(Y) and printed out with the corresponding probability density.

The calculations of the probability density and the effective mass will be less accurate than the eigenvalue calculations, especially the effective mass results which are only intended to give a qualitative picture of the variation of m^*/m_e with k. In both cases there is the option of plotting the results graphically.

Finally the user has the option of either stopping or of changing the potential and initiating a further set of calculations.

Because of the necessary truncation of the Hamiltonian matrix there is always some error. It is found that significant errors (accompanied by a warning message) generally occur in the extreme core-state limit. This is only to be expected because the very large potential values involved generate large—and non-negligible—Fourier terms, even for the higher terms in the series. For example, the jth coefficient for the harmonic potential is

$$d_j = \frac{2V_0 a \cos(j\pi)}{(j\pi)^2}$$

The magnitude of these Fourier terms decreases as j increases but increases with increasing $V_0 a$.

3.1 A typical session

If the rectangular potential is used with:

$$V_0 = 5.0 \qquad \text{(in Rydbergs)}$$

ELECTRON ENERGY BANDS IN A ONE-DIMENSIONAL PERIODIC POTENTIAL 333

Period $(a) = 1.5$ (in Bohr radii)

Width of rectangle $(b) = 0.5$ (in Bohr radii)

and if computation of the effective mass and the probability density are requested then the results of the computer program are:

```
PROGRAM TO CALCULATE FIRST THREE ENERGY LEVELS OF
AN ELECTRON SUBJECT TO A GIVEN PERIODIC POTENTIAL.
ALL INPUT AND OUTPUT IS IN ATOMIC UNITS
I.E. THE UNIT OF DISTANCE IS ONE BOHR RADIUS
THE UNIT OF ENERGY IS ONE RYDBERG

WHAT SORT OF POTENTIAL DO YOU WANT
TYPE 1 FOR RECTANGULAR POTENTIAL
TYPE 2 FOR SAWTOOTH POTENTIAL
TYPE 3 FOR COSINE POTENTIAL
TYPE 4 FOR HARMONIC POTENTIAL
TYPE 5 FOR INTERPOLATED POTENTIAL
1
RECTANGULAR POTENTIAL
INPUT HEIGHT OF RECTANGLE AS REAL NUMBER
5.0
INPUT WIDTH OF RECTANGLE AS REAL NUMBER
0.5
INPUT PERIOD AS REAL NUMBER
1.5

        PERIOD         1.5

         K         E1         E2         E3
      0.0000     1.4499    18.4804    20.0641
      0.1047     1.4603    18.1049    20.4621
      0.2094     1.4917    17.4000    21.2342
      0.3142     1.5440    16.6373    22.1089
      0.4189     1.6171    15.8717    23.0315
      0.5236     1.7109    15.1179    23.9872
      0.6283     1.8253    14.3814    24.9709
      0.7330     1.9600    13.6649    25.9800
      0.8378     2.1147    12.9698    27.0133
      0.9425     2.2890    12.2974    28.0700
      1.0472     2.4824    11.6486    29.1497
      1.1519     2.6940    11.0245    30.2523
      1.2566     2.9226    10.4265    31.3775
      1.3614     3.1666     9.8564    32.5252
      1.4661     3.4233     9.3169    33.6951
      1.5708     3.6884     8.8125    34.8874
      1.6755     3.9548     8.3501    36.1018
      1.7802     4.2107     7.9419    37.3383
      1.8850     4.4355     7.6082    38.5968
      1.9897     4.5970     7.3814    39.8768
      2.0944     4.6571     7.2995    41.1096

DO YOU WANT TO PLOT THESE RESULTS
TYPE YES OR NO
YES
```

E-K DIAGRAM SHOWING FIRST THREE ENERGY LEVELS

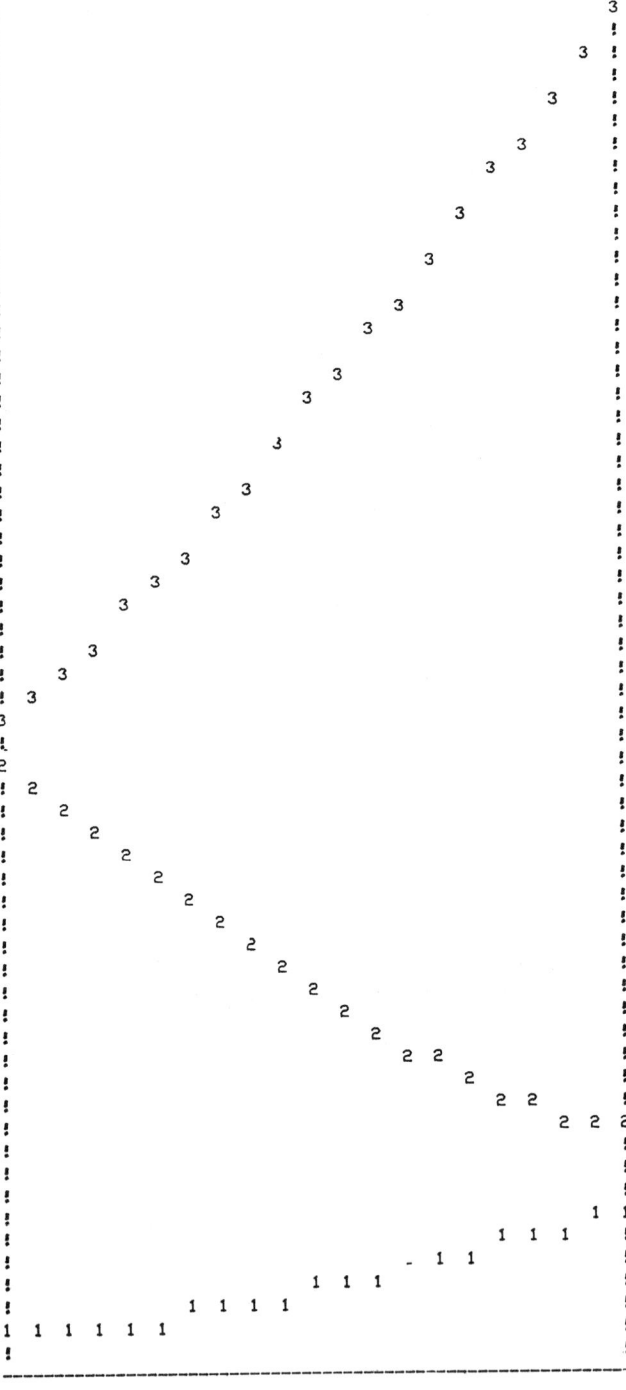

ELECTRON ENERGY BANDS IN A ONE-DIMENSIONAL PERIODIC POTENTIAL

```
DO YOU WANT TO CALCULATE THE EFFECTIVE MASS
 TYPE YES OR NO
YES


THE MASS OF A FREE ELECTRON IS 1.0
```

K	M1	M2	M3
0.0000	1.05 E	0.03 H	0.03 E
0.1047	1.05 E	0.07 H	0.06 E
0.2094	1.05 E	0.38 H	0.21 E
0.3142	1.05 E	7.40 H	0.46 E
0.4189	1.06 E	1.86 E	0.66 E
0.5236	1.07 E	1.27 E	0.79 E
0.6283	1.08 E	1.10 E	0.86 E
0.7330	1.10 E	1.02 E	0.91 E
0.8378	1.12 E	0.97 E	0.93 E
0.9425	1.15 E	0.93 E	0.96 E
1.0472	1.20 E	0.89 E	0.96 E
1.1519	1.28 E	0.84 E	0.97 E
1.2566	1.43 E	0.79 E	0.98 E
1.3614	1.73 E	0.71 E	0.98 E
1.4661	2.60 E	0.63 E	0.99 E
1.5708	15.80 E	0.52 E	0.99 E
1.6755	2.07 H	0.41 E	0.99 E
1.7802	0.71 H	0.29 E	1.00 E
1.8850	0.35 H	0.21 E	1.02 E
1.9897	0.22 H	0.15 E	0.47 H
2.0944	0.18 H	0.13 E	0.01 H

```
DO YOU WANT TO PLOT THESE RESULTS
 TYPE YES OR NO
NO
```

```
DO YOU WANT TO CALCULATE THE PROBABILITY DENSITY
FUNCTIONS FOR THE FIRST THREE ENERGY STATES
TYPE YES OR NO
YES
 INPUT A VALUE OF K AS A REAL NUMBER
2.0944
```

X	PD1	PD2	PD3	V
-0.7500	-0.0000	1.7479	-0.0000	5.0000
-0.6750	0.0410	1.7247	0.3807	5.0000
-0.6000	0.1640	1.6579	1.2313	5.0000
-0.5250	0.3712	1.5536	1.9127	5.0000
-0.4500	0.6541	1.3909	1.8729	0.0000
-0.3750	0.9788	1.1409	1.0989	0.0000
-0.3000	1.3072	0.8310	0.2388	0.0000
-0.2250	1.6075	0.5166	0.0327	0.0000
-0.1500	1.8471	0.2462	0.6556	0.0000
-0.0750	2.0014	0.0641	1.5750	0.0000
0.0000	2.0552	-0.0000	2.0028	0.0000
0.0750	2.0014	0.0641	1.5750	0.0000
0.1500	1.8471	0.2462	0.6556	0.0000
0.2250	1.6075	0.5166	0.0327	0.0000
0.3000	1.3072	0.8310	0.2388	0.0000
0.3750	0.9788	1.1409	1.0989	0.0000
0.4500	0.6541	1.3909	1.8729	0.0000
0.5250	0.3712	1.5536	1.9127	5.0000
0.6000	0.1640	1.6579	1.2313	5.0000
0.6750	0.0410	1.7247	0.3807	5.0000
0.7500	-0.0000	1.7479	-0.0000	5.0000

```
DO YOU WANT TO PLOT THESE RESULTS
 TYPE YES OR NO
 YES
```

ELECTRON ENERGY BANDS IN A ONE-DIMENSIONAL PERIODIC POTENTIAL

PROBABILITY DENSITY VS X FOR FIRST THREE STATES.
THE POTENTIAL IS SHOWN AS 4

```
                    DO YOU WANT TO CALCULATE ANOTHER SET OF
                    PROBABILITY DENSITY FUNCTIONS
                    TYPE YES OR NO
                 NO



                    DO YOU WANT TO TRY ANOTHER POTENTIAL
                    TYPE YES OR NO
                 NO
                 END OF RUN
```

3.2 Discussion

The program can be used in a variety of ways and the accompanying documentation tailored to suit a variety of levels of user, but a particularly instructive procedure is to concentrate on one form of potential and first to perform a series of runs, varying V_0 over a wide range whilst keeping the period constant. It is then relatively simple for the cases of the rectangular potential, the cosine potential, and the harmonic potential to compare graphically the energy difference between the first and second energy bands at $k = \pi/a$ (and/or between the second and third bands at $k = 0$) as a function of V_0 with the predictions of both the core state approximation, and the nearly-free electron approximation. In this way an estimate can be made of the value of V_0 below which the nearly free electron approximation is accurate, and the value of V_0 above which the core state approximation holds. A second criterion that aids in identifying which approximation, if any, is appropriate is the form of the $E-k$ relationship itself. If the energy gaps at the zone boundaries are small then the nearly-free electron approximation is likely to hold; if E is virtually independent of k the core state approximation probably applies. Having established the respective regions of validity and the extent of the intermediate region, more detailed comparisons involving the variation of E with k and the probability density can then be made.

Comparison with the tight-binding approximation is rather more complicated. It is recommended that the procedure described above first be used to identify the region of applicability of the core state approximation; the region of validity of the tight-binding approximation extends to somewhat lower values of V_0. In Figure 12 the data points show the results for a typical series of runs as suggested, employing the rectangular potential. The energy gaps predicted by the nearly-free electron approximation and the core state approximation are indicated by the solid lines. From the graphs it is clear that the nearly-free electron approximation was valid for the choice

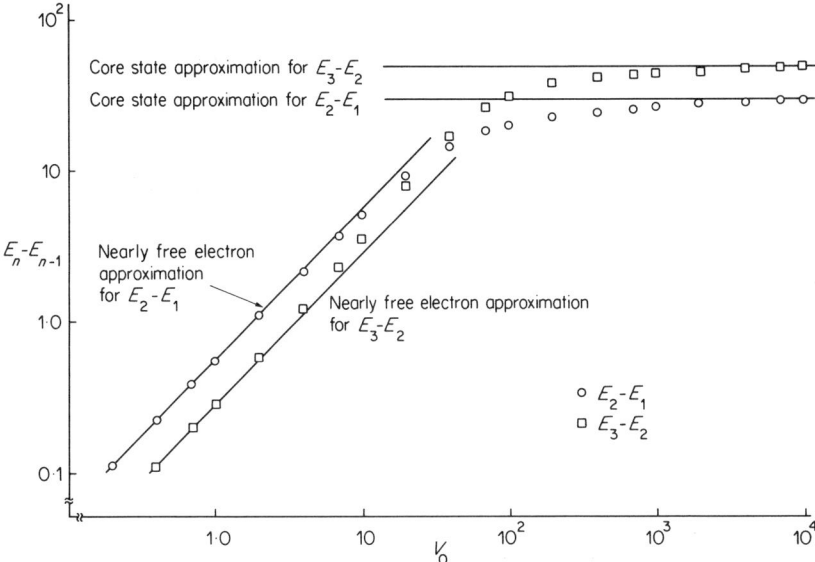

Figure 12. The computed results for $E_2(k = \pi/a) - E_1(k = \pi/a)$ (○) and $E_3(k = 0) - E_2(k = 0)$ (□) as a function of V_0 for a rectangular potential of period 1.5 and with a rectangle width (b) of 0.5. The solid lines show the predictions of the nearly-free electron approximation (for low V_0) and the core state approximation (for high V_0)

of parameters employed in section 3.2. This conclusion is consistent with the variation of E with k and the form of the probability density found in that case.

4. FURTHER CALCULATIONS

4.1 Estimating the potential

In the present program the potential can be varied widely in form, magnitude, and period. In a full-scale calculation for a particular material it is chosen to conform as closely as possible to the actual potential. It should, however, be noted that treating the problem as that of finding the eigenstates of a single electron in a potential due to the nuclei and the other electrons is itself an approximation. In principle the true wave function should embrace all the interacting particles. Fortunately, it appears that the results for the important physical properties are generally satisfactory using the single-electron approximation.

The principal contributions to the potential are taken to be:

(i) A long-range Coulomb potential due to each of the atomic nuclei screened by their associated core state electrons less the band electrons.

(ii) A more intense short-range Coulomb potential due to the fact that the nuclear charge is less fully screened by the core state electrons close to the nucleus. The degree of screening is estimated from the probability density found for the core state electrons in quantum-mechanical calculations for the isolated atom.

(iii) An 'exchange' interaction with the core state electrons.

(iv) A Coulomb interaction with all of the *other* band electrons. This contribution to the potential is estimated by assuming that the interaction with the band electrons is equivalent to that produced by a charge density proportional to the sum of the band electron probability densities. The procedure adopted is iterative; on the basis of an assumed potential the Schrödinger equation for the band electrons is solved and hence their combined probability density estimated. This leads in general to a somewhat different result for the potential than was assumed originally. On the basis of the revised potential the Schrödinger equation is again solved, the cycle being continued until no significant change in any of the parameters of interest occurs. This type of procedure is referred to as a self-consistent calculation.

4.2 The three-dimensional case

In a normal crystalline conductor the band electrons are free to move in three dimensions. The major difference introduced by this elaboration into the formalism of band calculations is that several quantities which can be treated as scalars in a one-dimensional model have to be recognized as vectors in the three-dimensional case. This is obviously the case for the variable used to represent position in the crystal; x has been used here for the one-dimensional case, $\mathbf{r}$ is generally used in the three-dimensional case. Similarly, the scalar k becomes the vector $\mathbf{k}$. The one-dimensional free electron wave function $\psi(x) = \exp(ikx)$ becomes $\psi(\mathbf{r}) = \exp(i\mathbf{k} \cdot \mathbf{r})$ and specifies the value at $\mathbf{r}$ of a wave function of wavelength $\lambda = 2\pi/|\mathbf{k}|$ propagating in the direction of the vector $\mathbf{k}$ (the wave vector). One speaks of '$\mathbf{k}$-space' or 'reciprocal space' ($\mathbf{k}$ has units L^{-1}).

The quantity corresponding to $2n\pi/a$, which arises as a consequence of Bragg reflections and determines the zone boundaries, must also be expressed as a vector in the three-dimensional case. The condition for Bragg reflection is $2a \sin \theta = n\lambda$ (where a is the spacing between the atomic *planes*), i.e.

$$2a \sin \theta = n\frac{2\pi}{|\mathbf{k}|},$$

so that

$$|\mathbf{k}| \sin \theta = \frac{n\pi}{a}.$$

Bragg reflection therefore occurs when the component of **k** in a direction normal to a lattice plane is $n\pi/a$. The change in the component of **k** normal to the plane is $2n\pi/a$. Vectors $\mathbf{K}_n$, of magnitude $2n\pi/a$ directed normally to the various lattice planes, of interplane spacing a, fill, in the three-dimensional case, the role occupied by $2n\pi/a$ in the one-dimensional case. These vectors $\mathbf{K}_n$ are referred to as reciprocal lattice vectors.

The most important new feature which occurs as a consequence of these considerations in the three-dimensional case is that the electron energy now depends not only on the magnitude but also on the direction of **k**. In consequence the effective mass is likewise dependent on direction and is expressed as a second-rank tensor.

4.3 Other basis functions

Since any realistic potential involves Coulomb interactions with the atomic nuclei a large number of terms are required to represent it accurately by a Fourier series of the kind used in the program presented here. Since the algorithm used to find the eigenvalues takes a computing time approximately proportional to the cube of the number of terms employed, this feature can render the accurate solution of the problem completely impracticable. For this reason alternative basis functions are generally used in full-scale band structure calculation. Three of the most widely used approaches are as follows.

(i) *The Orthogonalized Plane Wave (OPW) method*[11]

The core state electrons are of no real interest in a band structure calculation—their energies and wave functions are already known from calculations on the free atom and are unchanged in the solid. In the method of orthogonalized plane waves the core states are deliberately ignored by using the Schmidt orthogonalization procedure to construct basis functions which consist of a linear combination of plane waves and core states. The basis functions so formed are orthogonal to the core states.

(ii) *The Augmented Plane Wave (APW) method*[12]

The potential close to an atomic nucleus in a solid is nearly spherically symmetrical about the nucleus; well away from the nuclei it is nearly constant. In the augmented plane wave approach the radial part of the Schrödinger equation in the regions close to the nuclei is solved by numerical integration. The full solutions—the product of the radial solutions and the spherical harmonics—are matched at the surface of non-overlapping spheres, centred on each of the nuclei, to plane waves (plane waves being the solutions to the Schrödinger equation in a region of constant potential).

A set of plane waves, each differing in wave vector by a reciprocal lattice vector and each matched to the solutions close to the nucleus, are used as basis functions for the variational method.

(iii) *The scattering or Korringa, Kohn, and Rostoker (KKR) method*[13]

This method is allied to the phase-shift method used in scattering theory. The basis functions are made up of ingoing and outgoing spherical waves centred on the nuclei. In the most common form of this method the Green's function approach is employed to cast the Schrödinger equation in integral form.

For materials with $Z \geqslant 55$ relativistic corrections become important, and all of these methods can, if necessary, be modified to incorporate these effects.[14–16]

In principle, provided the basis functions form a complete set, the results of a band structure calculation should be independent of the choice of basis functions—provided enough are used. In practice there are always some deviations, though they are usually small.

4.4 Metals and semiconductors

The methods employed in band structure calculations for metals and semiconductors are quite similar; their different physical properties arise because a metal has one or more partially filled bands, a semiconductor at the absolute zero of temperature has energy bands that are either completely full (the valence band and the core states) or completely empty (the conduction band).

In the case of a metal, band structure calculations lead to an estimate of the density of states, the Fermi energy, and hence of the form of the Fermi surface—the energy surface in **k**-space which delineates the boundary between the filled and the empty electron states at the absolute zero. In a metal it is the electrons close to the Fermi surface that are responsible for the electronic transport properties.

In the case of a semiconductor the most important features determined by a band structure calculation are the energy gap between the valence band and the conduction band, the effective masses and the densities of states of the hole states close to the top of the valence band, and the electron states at the bottom of the conduction band. It is these states which lead to electronic conduction in semiconductors.

REFERENCES

1. A clear discussion of band structure calculations is given by L. Pincherle, *Electronic Energy Bands in Solids* (Macdonald, London, 1971).

2. R. de L. Kronig and W. G. Penney, *Proc. Roy. Soc. (London)*, **A130,** 499 (1931).
3. H. G. Schuster (Ed.) *One Dimensional Conductors*, (Springer-Verlag, New York, 1975).
4. F. Bloch, *Zeit. Phys.*, **52,** 555 (1928).
5. R. M. Eisberg, *Fundamentals of Modern Physics* (Wiley, New York, 1961).
6. G. C. Wetsel, Jr., *Am. J. Phys.*, **46,** 714 (1978).
7. M. L. Cohen, V. Heine, and D. Weaire, *Solid State Physics*, **24,** (1970).
8. R. Barrie, *Proc. Phys. Soc.*, **69,** 553 (1956).
9. T. C. Harman and J. M. Honig, *J. Phys. Chem. Solids*, **23,** 913 (1962).
10. NAG—Numerical Algorithms Group; Central Office: Oxford University Computing Laboratory, 13, Banbury Road, Oxford OX2 6NN, U.K.
11. T. O. Woodruff, *Solid State Physics*, **4,** 367 (1957).
12. T. L. Loucks, *Augmented Plane Wave Method* (Benjamin, New York, 1967).
13. G. C. Fletcher, *The Electron Band Theory of Solids*. Ch. 12 (North-Holland, London, 1971).
14. P. Soven, *Phys. Rev.*, **A137,** 1706 (1965).
15. T. L. Loukes, *Phys. Rev.*, **A139,** 1333 (1965).
16. Y. Onodera, M. Okazaki, and T. Inui, *J. Phys. Soc. Japan*, **21,** 2173 (1966).

ENERGY BANDS

```
C       MASTER SEGMENT OF PROGRAM.
C       THIS SEGMENT CONTAINS SECTIONS WHERE DATA
C       IS INPUT FROM A TERMINAL.
C       CHANGES WOULD BE NECESSARY TO RUN THE PROGRAM
C       IN BATCH MODE.
        DOUBLE PRECISION H(33,33),A(33),E(33),EF(33,33),W(33)
        COMMON AMP,WIDTH,PERIOD,PI,NPOT,FVAL(20),XVAL(20),NVAL
        REAL K,EL(21,3),KX(21),EM(21,3),ELEC(21,3),WF(21,4),XX(21)
        LOGICAL LOG,LOG1,LOG2,LOG3
        DATA YY,YN,EE,EH/4HYES ,4HNO  ,1HE,1HH/
        PI = 4.0*ATAN(1.0)
        WRITE(1,190)
        WRITE(1,191)
        WRITE(1,192)
        WRITE(1,193)
        WRITE(1,194)
      1 WRITE(1,200)
        WRITE(1,201)
        WRITE(1,202)
        WRITE(1,203)
        WRITE(1,204)
        WRITE(1,205)
        READ(1,100)NPOT
        IF(NPOT.EQ.1) GOTO 10
        IF(NPOT.EQ.2) GOTO 20
        IF(NPOT.EQ.3) GOTO 30
        IF(NPOT.EQ.4) GOTO 40
        IF(NPOT.EQ.5) GOTO 50
        WRITE(1,207)
        GOTO 1
C       DATA FOR RECTANGULAR POTENTIAL
     10 WRITE(1,210)
        WRITE(1,211)
        READ(1,101) AMP
        WRITE(1,212)
        READ(1,101) WIDTH
        WRITE(1,208)
        READ(1,101) PERIOD
        IF(PERIOD.GT.WIDTH) GOTO 70
        WRITE(1,213)
        GOTO 10
C       DATA FOR SAWTOOTH POTENTIAL
     20 WRITE(1,220)
        WRITE(1,221)
        READ(1,101)AMP
        WRITE(1,208)
        READ(1,101)PERIOD
        GOTO 70
C       DATA FOR COSINE POTENTIAL
     30 WRITE(1,230)
        WRITE(1,231)
        WRITE(1,241)
        READ(1,101)AMP
        WRITE(1,208)
        READ(1,101)PERIOD
        GOTO 70
```

ELECTRON ENERGY BANDS IN A ONE-DIMENSIONAL PERIODIC POTENTIAL

```
      ENERGY BANDS
C        DATA FOR HARMONIC POTENTIAL
   40 WRITE(1,240)
      WRITE(1,241)
      READ(1,101)AMP
      WRITE(1,208)
      READ(1,101)PERIOD
      AMP=AMP*4.0/(PERIOD*PERIOD)
      GOTO 70
C        DATA FOR INTERPOLATED POTENTIAL
   50 WRITE(1,250)
      WRITE(1,251)
      WRITE(1,252)
      WRITE(1,253)
   51 WRITE(1,271)
      WRITE(1,256)
      READ(1,102)NVAL
      IF(NVAL.LT.2) GOTO 51
      IF(NVAL.GT.20) GOTO 51
   52 WRITE(1,257)
      READ(1,101)FVAL(1)
      XVAL(1)=0.0
      DO 54 I=2,NVAL
      WRITE(1,258)
      READ(1,101)XVAL(I)
      IF(XVAL(I).GT.XVAL(I-1)) GOTO 53
      WRITE(1,270)
      GOTO 52
   53 WRITE(1,259)
      READ(1,101)FVAL(I)
   54 CONTINUE
      PERIOD=2.0*XVAL(NVAL)
      GOTO 70
   70 NPTS=100.0*PERIOD
C        CALCULATION OF RESULTS FOR E-K DIAGRAM
      CALL FRANCS(33,NPTS,DC,A)
      WRITE(1,300)PERIOD
      WRITE(1,301)
      DO 2 L=1,21
      K=0.05*PI*FLOAT(L-1)/PERIOD
      LOG=.FALSE.
      IF(L.LE.10)NAR=10
      IF(L.GT.10)NAR=11
   12 CONTINUE
      CALL ENERGY(K,E,EF,NAR,NAR,DC,A,H,W,.TRUE.)
      IF(LOG)GOTO 13
      F1=E(1)
      E2=E(2)
      E3=E(3)
      NAR=NAR+2
      LOG=.TRUE.
      GOTO 12
   13 CONTINUE
      F1=E(1)
      F2=E(2)
      F3=E(3)
      LOG=.FALSE.
```

ENERGY BANDS

```
      DE=F3-F1
      IF(DE.EQ.0.0)DE=F3
      NAR=NAR+8
      LOG1=.TRUE.
      LOG2=.TRUE.
      LOG3=.TRUE.
      ERES=ABS((F1-E1)/DE)
      IF(ERES.GT.3.0E-3)LOG1=.FALSE.
      ERES=ABS((F2-E2)/DE)
      IF(ERES.GT.3.0E-3)LOG2=.FALSE.
      ERES=ABS((F3-E3)/DE)
      IF(ERES.GT.3.0E-3)LOG3=.FALSE.
      IF(LOG1.AND.LOG2.AND.LOG3)GOTO 16
      IF(NAR.LE.31)GOTO 12
C     OUTPUT WARNING IF ONE OF THE ACCURACIES IS BELOW LIMITS
      IF(.NOT.LOG1.AND.LOG2.AND.LOG3)WRITE(1,400)
  400 FORMAT(41H *** WARNING ACCURACY OF E1 ON NEXT LINE ,
     1 13HUNCERTAIN ***)
      IF(LOG1.AND.(.NOT.LOG2).AND.LOG3)WRITE(1,401)
  401 FORMAT(41H *** WARNING ACCURACY OF E2 ON NEXT LINE ,
     1 13HUNCERTAIN ***)
      IF(LOG1.AND.LOG2.AND.(.NOT.LOG3))WRITE(1,402)
  402 FORMAT(41H *** WARNING ACCURACY OF E3 ON NEXT LINE ,
     1 13HUNCERTAIN ***)
      IF((.NOT.LOG1).AND.(.NOT.LOG2).AND.LOG3)WRITE(1,403)
  403 FORMAT(48H *** WARNING ACCURACY OF E1 AND E2 ON NEXT LINE ,
     1 13HUNCERTAIN ***)
      IF((.NOT.LOG1).AND.LOG2.AND.(.NOT.LOG3))WRITE(1,404)
  404 FORMAT(48H *** WARNING ACCURACY OF E1 AND E3 ON NEXT LINE ,
     1 13HUNCERTAIN ***)
      IF(LOG1.AND.(.NOT.LOG2).AND.(.NOT.LOG3))WRITE(1,405)
  405 FORMAT(48H *** WARNING ACCURACY OF E2 AND E3 ON NEXT LINE ,
     1 13HUNCERTAIN ***)
      IF((.NOT.LOG1).AND.(.NOT.LOG2).AND.(.NOT.LOG3))WRITE(1,406)
  406 FORMAT(52H *** WARNING ACCURACY OF E1, E2 AND E3 ON NEXT LINE ,
     1 13HUNCERTAIN ***)
   16 CONTINUE
      KX(L)=K
      DO 3 N=1,3
      EL(L,N)=E(N)
    3 CONTINUE
      WRITE(1,302) KX(L),(EL(L,N),N=1,3)
    2 CONTINUE
    4 WRITE(1,280)
      WRITE(1,291)
      READ(1,103) YA
      IF(YA.EQ.YN) GOTO 5
      IF(YA.NE.YY) GOTO 4
      WRITE(1,303)
      CALL GRAPH(EL,21,3,KX)
    5 WRITE(1,281)
      WRITE(1,291)
      READ(1,103) YA
      IF(YA.EQ.YN) GOTO 71
      IF(YA.NE.YY) GOTO 5
      WRITE(1,304)
```

ENERGY BANDS

```
      WRITE(1,305)
C     APPROXIMATE CALCULATION OF EFFECTIVE MASS
C     THE VALUE IS TRUNCATED AT 10**6 TIMES THE
C     MASS OF A FREE ELECTRON
      DK2=(KX(2)-KX(1))**2
      DO 7 N=1,3
      EM(1,N)=2.0*(EL(2,N)-EL(1,N))/DK2
      DO 6 L=2,20
      EM(L,N)=(EL(L+1,N)+EL(L-1,N)-2.0*EL(L,N))/DK2
    6 CONTINUE
      EM(21,N)=2.0*(EL(20,N)-EL(21,N))/DK2
    7 CONTINUE
      DO 8 L=1,21
      DO 9 N=1,3
      IF(EM(L,N).GE.0.0) ELEC(L,N)=EE
      IF(EM(L,N).LT.0.0) ELEC(L,N)=EH
      IF(ABS(EM(L,N)).GT.2.0E-6) GOTO 80
      EM(L,N)=2.0E-6
   80 EM(L,N)=2.0/ABS(EM(L,N))
    9 CONTINUE
      WRITE(1,310) KX(L),(EM(L,N),ELEC(L,N),N=1,3)
    8 CONTINUE
   81 WRITE(1,280)
      WRITE(1,291)
      READ(1,103) YA
      IF(YA.EQ.YN) GOTO 71
      IF(YA.NE.YY) GOTO 81
      WRITE(1,306)
      DO 82 L=1,21
      DO 82 N=1,3
      IF(EM(L,N).GT.50.0) EM(L,N)=50.0
   82 CONTINUE
      CALL GRAPH(EM,21,3,KX)
   71 WRITE(1,282)
      WRITE(1,283)
      WRITE(1,291)
      READ(1,103) YA
      IF(YA.EQ.YN) GOTO 78
      IF(YA.NE.YY) GOTO 71
   72 WRITE(1,284)
      READ(1,101) K
C     CALCULATION OF PROBABILITY DENSITY FUNCTIONS
      CALL ENERGY(K,E,EF,15,15,DC,A,H,W,.FALSE.)
      DO 73 I=1,21
      XX(I)=0.05*FLOAT(I-11)*PERIOD
      X=XX(I)
      WF(I,4)=V(X)
      DO 74 J=1,3
      WF(I,J)=PROB(J,EF,15,15,X)
   74 CONTINUE
   73 CONTINUE
      WRITE(1,307)
      DO 75 I=1,21
      WRITE(1,302) XX(I),(WF(I,J),J=1,4)
   75 CONTINUE
   76 WRITE(1,280)
```

ENERGY BANDS

```
      WRITE(1,291)
      READ(1,103) YA
      IF(YA.EQ.YN) GOTO 77
      IF(YA.NE.YY) GOTO 76
      WRITE(1,308)
      WRITE(1,309)
      PDMAX=WF(1,1)
      DO 771 I=1,21
      DO 771 J=1,3
      IF(WF(I,J).GT.PDMAX) PDMAX=WF(I,J)
  771 CONTINUE
      VMAX=ABS(WF(1,4))
      DO 772 I=2,21
      IF(ABS(WF(I,4)).GT.VMAX) VMAX=ABS(WF(I,4))
  772 CONTINUE
      IF(VMAX.LT.0.1*PDMAX) VMAX=0.5*PDMAX
      SCALE=1.5*PDMAX/VMAX
      DO 773 I=1,21
      WF(I,4)=WF(I,4)*SCALE
  773 CONTINUE
      CALL GRAPH(WF,21,4,XX)
   77 WRITE(1,285)
      WRITE(1,286)
      WRITE(1,291)
      READ(1,103) YA
      IF(YA.EQ.YN) GOTO 78
      IF(YA.NE.YY) GOTO 77
      GOTO 72
   78 WRITE(1,290)
      WRITE(1,291)
      READ(1,103) YA
      IF(YA.EQ.YN) GOTO 90
      IF(YA.NE.YY) GOTO 78
      GOTO 1
   90 WRITE(1,299)
      STOP
C     FORMATS FOR INTERACTIVE SECTIONS OF PROGRAM.
  100 FORMAT(I1)
  101 FORMAT(F10.4)
  102 FORMAT(I2)
  103 FORMAT(A4)
  190 FORMAT(1X,49HPROGRAM TO CALCULATE FIRST THREE ENERGY LEVELS OF)
  191 FORMAT(1X,50HAN ELECTRON SUBJECT TO A GIVEN PERIODIC POTENTIAL.)
  192 FORMAT(1X,39HALL INPUT AND OUTPUT IS IN ATOMIC UNITS)
  193 FORMAT(1X,44HI.E. THE UNIT OF DISTANCE IS ONE BOHR RADIUS)
  194 FORMAT(1X,33HTHE UNIT OF ENERGY IS ONE RYDBERG//)
  200 FORMAT(1X,34HWHAT SORT OF POTENTIAL DO YOU WANT)
  201 FORMAT(1X,32HTYPE 1 FOR RECTANGULAR POTENTIAL)
  202 FORMAT(1X,29HTYPE 2 FOR SAWTOOTH POTENTIAL)
  203 FORMAT(1X,27HTYPE 3 FOR COSINE POTENTIAL)
  204 FORMAT(1X,29HTYPE 4 FOR HARMONIC POTENTIAL)
  205 FORMAT(1X,33HTYPE 5 FOR INTERPOLATED POTENTIAL)
  207 FORMAT(1X,36HNUMBER INPUT MUST BE BETWEEN 1 AND 5)
  208 FORMAT(1X,27HINPUT PERIOD AS PEAL NUMBER)
  210 FORMAT(1X,21HRECTANGULAR POTENTIAL)
  211 FORMAT(1X,40HINPUT HEIGHT OF RECTANGLE AS REAL NUMBER)
```

ENERGY BANDS

```
  212 FORMAT(1X,39HINPUT WIDTH OF RECTANGLE AS REAL NUMBER)
  213 FORMAT(1X,46HPERIOD MUST BE GREATER THAN WIDTH OF RECTANGLE)
  220 FORMAT(1X,18HSAWTOOTH POTENTIAL)
  221 FORMAT(1X,39HINPUT HEIGHT OF SAWTOOTH AS REAL NUMBER)
  230 FORMAT(1X,37HCOSINE POTENTIAL I.E. PROPORTIONAL TO)
  231 FORMAT(1X,24H1.0-COS(2.0*PI*X/PERIOD))
  240 FORMAT(1X,49HHARMONIC POTENTIAL I.E. PROPORTIONAL TO X SQUARED)
  241 FORMAT(1X,48HINPUT CONSTANT OF PROPORTIONALITY AS REAL NUMBER)
  250 FORMAT(1X,22HINTERPOLATED POTENTIAL)
  251 FORMAT(1X,46HTHE POTENTIAL IS GIVEN BY LINEAR INTERPOLATION)
  252 FORMAT(1X,44HBETWEEN POINTS SPECIFIED OVER HALF A PERIOD.)
  253 FORMAT(1X,38HTHE PERIOD IS TWICE THE FINAL X VALUE.)
  256 FORMAT(1X,22HINPUT NUMBER OF POINTS)
  257 FORMAT(1X,43HINPUT AS REAL NUMBER THE POTENTIAL AT X=0.0)
  258 FORMAT(1X,33HINPUT NEXT X VALUE AS REAL NUMBER)
  259 FORMAT(1X,30HINPUT POTENTIAL AS REAL NUMBER)
  270 FORMAT(1X,42HEACH X VALUE MUST BE GREATER THAN LAST ONE)
  271 FORMAT(1X,46HTHE NUMBER OF POINTS MUST BE BETWEEN 02 AND 20)
  280 FORMAT(//33HDO YOU WANT TO PLOT THESE RESULTS)
  281 FORMAT(//43HDO YOU WANT TO CALCULATE THE EFFECTIVE MASS)
  282 FORMAT(//48HDO YOU WANT TO CALCULATE THE PROBABILITY DENSITY)
  283 FORMAT(1X,43HFUNCTIONS FOR THE FIRST THREE ENERGY STATES)
  284 FORMAT(1X,35HINPUT A VALUE OF K AS A REAL NUMBER)
  285 FORMAT(1X,39HDO YOU WANT TO CALCULATE ANOTHER SET OF)
  286 FORMAT(1X,29HPROBABILITY DENSITY FUNCTIONS)
  290 FORMAT(////36HDO YOU WANT TO TRY ANOTHER POTENTIAL)
  291 FORMAT(1X,14HTYPE YES OR NO)
  299 FORMAT(1X,10HEND OF RUN)
C     ************************************************
  300 FORMAT(//6X,6HPERIOD,F12.1)
  301 FORMAT(/7X,1HK,9X,2HE1,8X,2HE2,8X,2HE3)
  302 FORMAT(1X,5F10.4)
  303 FORMAT(//5X,45HE-K DIAGRAM SHOWING FIRST THREE ENERGY LEVELS//)
  304 FORMAT(//34HTHE MASS OF A FREE ELECTRON IS 1.0)
  305 FORMAT(/7X,1HK,10X,2HM1,9X,2HM2,9X,2HM3)
  306 FORMAT(//5X,45HABS(EFFECTIVE MASS) VS K IN FIRST THREE BANDS)
  307 FORMAT(//7X,1HX,9X,3HPD1,7X,3HPD2,7X,3HPD3,7X,1HV)
  308 FORMAT(//5X,48HPROBABILITY DENSITY VS X FOR FIRST THREE STATES.)
  309 FORMAT(1X,27HTHE POTENTIAL IS SHOWN AS 4//)
  310 FORMAT(1X,F10.4,3(F9.2,1X,A1))
      END
C
      SUBROUTINE ENERGY(K,E,EV,ND,N,DC,A,H,W,BOOL)
C
C        SUBROUTINE TO CALCULATE THE FIRST N ENERGY LEVELS
C        AND CORRESPONDING STATE VECTORS FOR GIVEN K AND
C        GIVEN PERIODIC POTENTIAL.
C        F02ABF IS A NAG LIBRARY ROUTINE THAT CALCULATES
C        THE EIGENVALUES AND EIGENVECTORS OF A REAL
C        SYMMETRIC MATRIX.
C        F02AAF IS A NAG LIBRARY ROUTINE THAT CALCULATES
C        JUST THE EIGENVALUES OF A REAL SYMMETRIC MATRIX.
C
      DOUBLE PRECISION E(N),H(ND,N),A(N),EV(ND,N),W(N)
      LOGICAL BOOL
      COMMON AMP,WIDTH,PERIOD,PI,NPOT,FVAL(20),XVAL(20),NVAL
```

ENERGY BANDS

```
      REAL K
      N1=N/2
      IF(2*N1.EQ.N) GOTO 1
      N2=(N+1)/2
      DO 2 I=1,N
      H(I,I)=(K-FLOAT(N2-I)*2.0*PI/PERIOD)**2+DC
    2 CONTINUE
      GOTO 3
    1 DO 4 I=1,N
      H(I,I)=(K-FLOAT(N1-I+1)*2.0*PI/PERIOD)**2+DC
    4 CONTINUE
    3 NM=N-1
      DO 5 I=1,NM
      IP=I+1
      DO 5 J=IP,N
      JMI=J-I
      H(I,J)=A(JMI)/2.0
      H(J,I)=H(I,J)
    5 CONTINUE
      IFAIL=0
      IF(BOOL)CALL F02AAF(H,ND,N,E,W,IFAIL)
      IF(.NOT.BOOL)CALL F02ABF(H,ND,N,E,EV,ND,W,IFAIL)
      RETURN
      END
C
      SUBROUTINE FRANCS(NTERMS,NDATA,DC,A)
C
C     SUBROUTINE TO FOURIER ANALYSE THE POTENTIAL.
C     FOR THE INTERPOLATED POTENTIAL THE COEFFICIENTS
C     ARE EVALUATED BY NUMERICAL INTEGRATION.
C     FOR THE OTHER POTENTIALS ANALYTIC EXPRESSIONS
C     FOR THE FOURIER COEFFICIENTS ARE USED.
C
      DOUBLE PRECISION A(NTERMS)
      COMMON AMP,WIDTH,PERIOD,PI,NPOT,FVAL(20),XVAL(20),NVAL
      REAL K
      IF(NPOT.EQ.1) GOTO 10
      IF(NPOT.EQ.2) GOTO 20
      IF(NPOT.EQ.3) GOTO 30
      IF(NPOT.EQ.4) GOTO 40
      GOTO 50
C     FOURIER COEFFICIENTS FOR RECTANGULAR POTENTIAL
   10 DC=WIDTH*AMP/PERIOD
      DO 11 M=1,NTERMS
      FM=M
      A(M)=2.0*COS(FM*PI)*AMP*SIN(FM*PI*WIDTH/PERIOD)/(FM*PI)
   11 CONTINUE
      RETURN
C     FOURIER COEFFICIENTS FOR SAWTOOTH POTENTIAL
   20 DC=AMP/2.0
      DO 21 M=1,NTERMS
      FM=M
      A(M)=2.0*AMP*(COS(FM*PI)-1.0)/(FM*PI)**2
   21 CONTINUE
      RETURN
C     FOURIER COEFFICIENTS FOR COSINE POTENTIAL
```

ELECTRON ENERGY BANDS IN A ONE-DIMENSIONAL PERIODIC POTENTIAL 351

ENERGY BANDS

```
   30 DC=AMP
      A(1)=-AMP
      DO 31 M=2,NTERMS
      A(M)=0.0
   31 CONTINUE
      RETURN
C       FOURIER COEFFICIENTS FOR HARMONIC POTENTIAL
   40 DC=AMP*PERIOD**2/12.0
      DO 41 M=1,NTERMS
      FM=M
      A(M)=AMP*PERIOD**2*COS(FM*PI)/(FM*PI)**2
   41 CONTINUE
      RETURN
C       FOURIER COEFFICIENTS BY NUMERICAL INTEGRATION
   50 K=2.0*PI/PERIOD
      FD=NDATA
      STEP=PERIOD/FD
      DC=V(0.0)
      DO 51 M=1,NTERMS
      A(M)=V(0.0)
   51 CONTINUE
      ND1=NDATA-1
      DO 52 I=1,ND1
      FI=I
      X=FI*STEP
      VX=V(X)
      DC=DC+VX
      DO 53 M=1,NTERMS
      FM=M
      A(M)=A(M)+COS(FM*K*X)*VX
   53 CONTINUE
   52 CONTINUE
      DC=DC/FD
      DO 54 M=1,NTERMS
      A(M)=A(M)*2.0/FD
   54 CONTINUE
      RETURN
      END
C
      FUNCTION V(Y)
C
C        FUNCTION TO EVALUATE THE POTENTIAL AT ANY POINT.
C        THE POTENTIAL IS ALWAYS AN EVEN FUNCTION.
C
      COMMON AMP,WIDTH,PERIOD,PI,NPOT,FVAL(20),XVAL(20),NVAL
      X=Y
      IF(X.LT.0.0) X=-X
      P2=PERIOD/2.0
      R=AMOD(X,PERIOD)
      IF(R.GT.P2) R=PERIOD-R
      IF(NPOT.EQ.1) GOTO 10
      IF(NPOT.EQ.2) GOTO 20
      IF(NPOT.EQ.3) GOTO 30
      IF(NPOT.EQ.4) GOTO 40
      GOTO 50
C       RECTANGULAR POTENTIAL
```

ENERGY BANDS

```
   10 B=(PERIOD-WIDTH)/2.0
      IF(R.LT.B) V=0.0
      IF(R.EQ.B) V=AMP/2.0
      IF(R.GT.B) V=AMP
      RETURN
C        SAWTOOTH POTENTIAL
   20 V=R*AMP/P2
      RETURN
C        COSINE POTENTIAL
   30 Q=2.0*PI/PERIOD
      V=AMP*(1.0-COS(Q*R))
      RETURN
C        HARMONIC POTENTIAL
   40 V=AMP*R**2
      RETURN
C        INTERPOLATED POTENTIAL
   50 NV1=NVAL-1
      DO 51 I=1,NV1
      IF(R.GE.XVAL(I).AND.R.LT.XVAL(I+1)) J=I
   51 CONTINUE
      GRAD=(FVAL(J+1)-FVAL(J))/(XVAL(J+1)-XVAL(J))
      V=FVAL(J)+GRAD*(R-XVAL(J))
      RETURN
      END
C
      FUNCTION PROB(L,EV,ND,N,X)
C
C     FUNCTION TO EVALUATE THE PROBABILITY DENSITY AT X
C     FOR GIVEN K AND GIVEN STATE VECTOR.
C
      DOUBLE PRECISION EV(ND,N)
      COMMON AMP,WIDTH,PERIOD,PI,NPOT,FVAL(20),XVAL(20),NVAL
      S=0.0
      ND1=ND-1
      DO 1 I=1,ND1
      I1=I+1
      DO 2 J=I1,ND
      F=2.0*PI*FLOAT(J-I)/PERIOD
      S=S+EV(I,L)*EV(J,L)*COS(F*X)
    2 CONTINUE
    1 CONTINUE
      PROB=1.0+2.0*S
      RETURN
      END
C
      SUBROUTINE GRAPH(Y,NP,NG,X)
C
C     SUBROUTINE TO PLOT UP TO FOUR FUCTIONS
C     ON THE SAME GRAPH.
C
      DIMENSION Y(NP,NG),X(NP),PLOT(120),PLT(7)
      DATA PLT(1),PLT(2),PLT(3),PLT(4),PLT(5),PLT(6),PLT(7)/
     1 4H1    ,4H2    ,4H3    ,4H4    ,4H     ,4H-    ,4H!   /
      IWG=70
      GAP=FLOAT(IWG-1)/FLOAT(NP-1)
      IGAP=IFIX(AINT(GAP))
```

ENERGY BANDS

```
      IWG=(NP-1)*IGAP+1
      YMIN=Y(1,1)
      YMAX=Y(1,1)
      DO 66 N=1,NG
      DO 68 I=1,NP
      IF(Y(I,N).LT.YMIN) YMIN=Y(I,N)
      IF(Y(I,N).GT.YMAX) YMAX=Y(I,N)
   68 CONTINUE
   66 CONTINUE
      IF(YMIN.GT.0.0) YMIN=0.0
      IHG=IWG
      SCALE=(YMAX-YMIN)/FLOAT(IHG-1)
      DO 70 J=1,IHG
      YTS=YMAX-SCALE*(FLOAT(J)-1.5)
      YBS=YMAX-SCALE*(FLOAT(J)-0.5)
      PLOT(1)=PLT(7)
      PLOT(IWG)=PLT(7)
      DO 72 I=1,NP
      K1=(I-1)*IGAP+1
      K2=I*IGAP
      DO 74 K=K1,K2
      IF(K.NE.1.AND.K.NE.IWG) PLOT(K)=PLT(5)
      IF(YTS.GE.0.0.AND.YBS.LT.0.0) PLOT(K)=PLT(6)
   74 CONTINUE
      DO 73 N=1,NG
      IF(Y(I,N).LE.YTS.AND.Y(I,N).GT.YBS) PLOT(K1)=PLT(N)
   73 CONTINUE
   72 CONTINUE
      WRITE(1,178)(PLOT(K),K=1,IWG)
  178 FORMAT(1H ,120A1)
   70 CONTINUE
   62 CONTINUE
      RETURN
      END
```

PART 4

Applied Physics

Physics Programs
Edited by A. D. Boardman
© 1980 John Wiley & Sons, Ltd.

CHAPTER 11

Computer Simulation of Hot Electron Behaviour in Semiconductors Using Monte Carlo Methods

A. D. BOARDMAN

1. INTRODUCTION

Transistors become less effective as the operation frequency increases. This is particularly true in the microwave (GHz) range of frequencies. This fact has stimulated, for many years, a great deal of research effort directed towards the development of devices that behave as microwave sources or act as microwave amplifiers.

In 1963, Gunn[1] discovered that the current through bulk gallium arsenide (GaAs) becomes unstable and fluctuates periodically at a microwave frequency, provided the applied bias field (cf. Figure 1) exceeds a certain critical value, thus confirming earlier theoretical work by Ridley and Watkins[2] and Hilsum.[3] It is now known as the Gunn effect. The word 'bulk' is significant because, in the bulk device shown in Figure 1, there are no gates or junctions. Instead of the electrons becoming warm, as they do in a normal transistor, they become hot and then use the peculiarities of the band structure to transfer from one part of momentum space to another.

For materials in which such transfers can take place, such as GaAs, the low field behaviour fits the familiar Ohm's law, but at strong electric fields a deviation from Ohm's law occurs. Therefore, in the variation of mean velocity v, against electric field F, as sketched in Figure 2, the differential mobility dv/dF is negative beyond a field value called the threshold bias. A simple argument shows that it is this possibility of voltage-controlled negative resistance in the bulk that leads to instability and hence to microwave current oscillations.

Consider the one-dimensional forms of the equation of continuity and

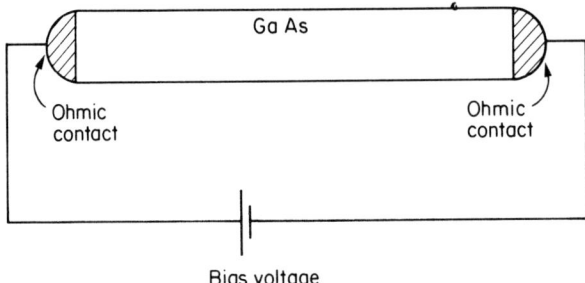

Figure 1. Slab of GaAs with a bias voltage applied through ohmic contacts. If the bias voltage exceeds a critical value, negative differential mobility occurs in the device

Poisson's equation, i.e. using x for position and t for time,

$$\frac{\partial n}{\partial t} - \frac{\partial}{\partial x}(nv(F)) = 0, \tag{1}$$

$$\frac{\partial F}{\partial x} = -\frac{e}{\varepsilon}(n - n_0), \tag{2}$$

where diffusion is neglected. Here, n is the number density of the electrons

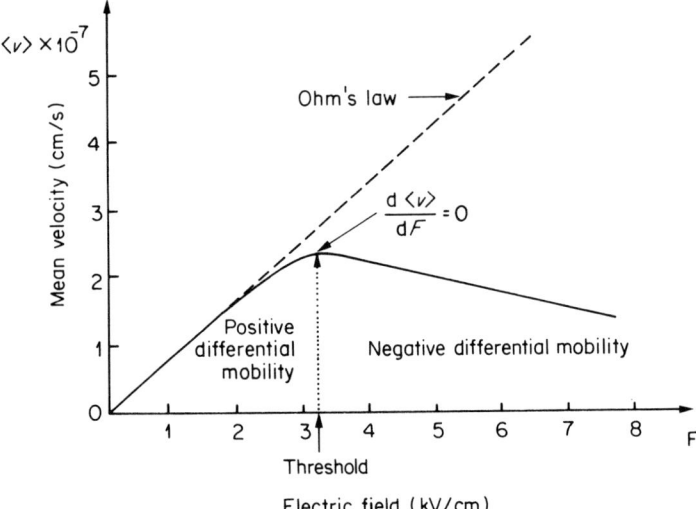

Figure 2. Sketch of the velocity-field characteristic of bulk GaAs. A threshold electric field ~ 3 kV/cm separates positive and negative differential mobility regimes. The deviation from Ohm's law is clearly evident

inside a semiconductor, $v(F)$ is the electron velocity, n_0 is the equilibrium density, F is the applied electric field, e is the electronic charge, and $\varepsilon = \kappa_0 \varepsilon_\infty$, where κ_0 is the permittivity of free space and ε_∞ is the high-frequency dielectric constant. Now suppose that a fluctuation in the electron density occurs that drives n_0 to $n = n_0 + n_1(x, t)$. In such an event the equilibrium values of velocity and electric field (v_0 and F_0) are driven to the new values $v = v_0 + v_1(x, t)$, $F = F_0 + F_1(x, t)$ so that

$$\frac{\partial n_1}{\partial t} - n_0 \frac{\partial v_1}{\partial x} - v_0 \frac{\partial n_1}{\partial x} = 0, \qquad (3)$$

$$\frac{\partial F_1}{\partial x} = -\frac{e}{\varepsilon} n_1, \qquad (4)$$

$\partial v_1/\partial x$ is $(\partial v/\partial F)_{F_0} \partial F_1/\partial x$, however, which means that equations (3) and (4) reduce to

$$\left[\frac{\partial}{\partial t} - v_0 \frac{\partial}{\partial x} + \frac{n_0 e}{\varepsilon} \left(\frac{\partial v}{\partial F}\right)_{F_0}\right] n_1 = 0 \qquad (5)$$

where the quantity $(\partial v/\partial F)_{F_0}$, the gradient of the velocity-field characteristic, is called the differential mobility μ_0. Note that it is not the same as v/F that is the more usually discussed mobility. If all the fluctuating quantities vary as $e^{i(kx - \omega t)}$, where k is a wave number and ω is an angular frequency, then the dispersion equation of the system is

$$\omega + k v_0 = -i \left(\frac{n_0 e}{\varepsilon}\right) \mu_0 \qquad (6)$$

From equation (6) it is seen that all fluctuations must vary as $\exp[(-n_0 e/\varepsilon)\mu_0 t]$. If, for any electric field, $\mu_0 < 0$ then the differential mobility is negative and the space charge n will grow. This is what is meant by instability. For real k the oscillation frequency is $k v_0$ and the rise or decay constant is $\text{Im}(\omega)$. If the lower limit of k is $\sim 2\pi/L$, where L is the length of the sample, then the oscillation frequency is $f \sim v_0/L$. Typically L is $\sim 10\,\mu\text{m}$ and v_0 is between 10^7 and 2×10^7 cm/s. These figures imply that $f \sim 10\text{--}20$ GHz.

The critical field at which the Gunn effect begins in GaAs is a little over 3 kV/cm. This means, for devices as thin as $\sim 10\,\mu\text{m}$, that the applied voltage need only be ~ 3 V. This is obviously a great advantage implying, as it does, the possibility of making hand-held microwave devices.

The transferred electron mechanism in GaAs can be used to make amplifiers and oscillators, but probably the most dramatic development is the use of the GaAs microwave generator to make miniature radar systems.[4] As can be seen from the rough calculations given earlier, these require, as a power source, only a simple torch battery. They are used to make burglar alarms, small portable radar torches for ground surveillance (giving ranges of ~25 to 625 m), navigation beacons, and hand-held speed torches for use by traffic police.

At the heart of all these devices lies the transferred electron effect and it is the computer simulation of this mechanism that this chapter deals with. The effect involves the movement of electrons that receive so much energy from the applied electric field that they become hot, in the sense that they have an energy far in excess of the thermal energy. It is necessary, therefore, to study the motion of electrons in an applied electric field and to study the mechanisms that can scatter them. This is a difficult task, especially if the conventional transport equations are solved. However, since a computer has to be used anyway it turns out that a simulation, using Monte Carlo techniques, has a number of advantages.

2. THE MONTE CARLO METHOD

The application of simulation techniques in theoretical physics and engineering serves two main purposes. The first is circumventing the difficulty of providing useful solutions to important non-linear equations such as the Boltzmann transport equation. Analytical solutions to these equations often involve drastic approximations and provide unrealistic results. The second purpose is to provide deeper insight into the physical mechanisms that such equations describe.

One simulation technique that has proved immensely powerful in electron transport theory is the Monte Carlo method.[5-8] This chapter is devoted to a description of that method and is illustrated by an application to the study of electron transport in non-degenerate semiconductors such as GaAs with electron densities of $\sim 10^{21}\,\mathrm{m}^{-3}$.

Many parameters of a physical system are governed by probability distributions. Therefore, if a mathematically random distribution is used to interrogate such distributions we can, in principle, generate the physical values of these parameters. This is, broadly speaking, the Monte Carlo method. In practice, of course, the physical distributions may be quite complex and difficult to manipulate even with a computer. The manipulations can be simplified by 'mapping' the complex distributions on to a simple pseudo-random distribution; the most convenient pseudo-random distribution is the uniform distribution, which is readily available on most computer systems.

In general, if $p(\phi)$ and $p(r)$ are the respective probability densities, associated with ϕ in the physical distribution and r in the pseudo-random distribution, then

$$\int_0^\phi p(\phi')\,\mathrm{d}\phi' = \int_0^r p(r')\,\mathrm{d}r' \tag{7}$$

In a uniform distribution $p(r) = 1$ so that (7) becomes

$$r = \int_0^\phi p(\phi') \, d\phi' \tag{8}$$

Hence, provided that this integral can be evaluated in a simple closed analytical form, inversion will yield a random value for the physical variable ϕ in terms of the uniformly distributed random number r. A simple example of this technique is the generation of the random flight times of a classical particle in a gas for the case when Γ, the total scattering rate, is constant. The probability of this particle travelling unimpeded for a time t and then being scattered at the end of this flight is

$$p(t) = \Gamma e^{-\Gamma t} \tag{9}$$

From equation (8)

$$r = \int_0^t \Gamma e^{-\Gamma t'} \, dt' = 1 - e^{-\Gamma t}, \tag{10}$$

so that the random flight times are given by

$$t = -\frac{1}{\Gamma} \ln(1-r) = -\frac{1}{\Gamma} \ln(r) \tag{11}$$

Note that, since r is uniformly distributed, $\ln(1-r)$ is equivalent to $\ln(r)$.

In effect, what happens in the calculation is that the cumulative probability $P(t)$ is calculated, i.e. the area under the probability curve up to a value t, and its value is then mapped on to a uniform random number distribution from which t is directly selected. This process is illustrated in Figures 3 and 4. Similar manipulations provide values for any other physical parameter

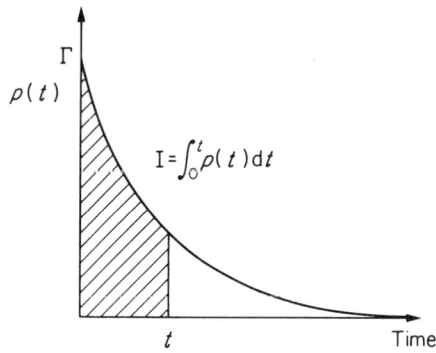

Figure 3. Simple probability density as a function of time

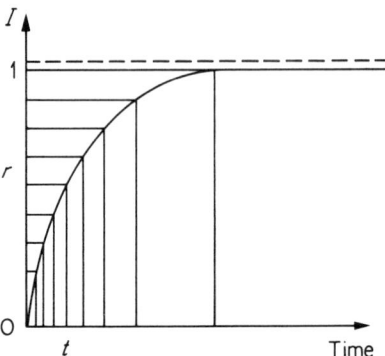

Figure 4. Cumulative probability obtained from Figure 3 mapped on to a uniformly distributed random distribution lying between 0 and 1

whose instantaneous value can be described by a probability density distribution, an example being the angular deviation of a particle in a scattering interaction.

3. ELECTRON MOTION

Before this simulation technique is applied to the problem of charge carrier transport in semiconducting solids, it is helpful to examine, briefly, the manner in which charge carriers move through a semiconductor and determine those features of the motion that are governed by probability distributions of the kind discussed in the previous section.

A conductor contains a large number of charge carriers and it is impossible to directly simulate such a many-body distribution of mutually interacting particles that are also being scattered by lattice vibrations. However, the actual system can be considered as a quasi-system of independent particles. This means that it is legitimate to monitor the history of a single particle, undergoing many lattice-scattering events, in order to generate the behaviour of a large set. In other words, the time spent by an electron in each part of momentum space is proportional to the electron distribution function.

In a simple semiconductor electrons or holes are regarded as free particles with an effective mass m^* and an energy $E(k)$ given by the parabolic dispersion law

$$E(\mathbf{k}) = \frac{\hbar^2 k^2}{2m^*}, \tag{12}$$

where $\hbar = h/2\pi$, $h = 6.625 \times 10^{-34}$ J s is Planck's constant and **k** labels the state of the particle. Thus $\hbar \mathbf{k}$ is interpreted as particle momentum and, in the case of electrons, m^* is usually considerably smaller than the free-electron mass $m_0 = 9.11 \times 10^{-31}$ kg. Many semiconductors are not as simple as this but, if they are not, it is sometimes possible, as in the case of GaAs, to provide a description in terms of two effective masses.

In the semiconductor to be simulated here, individual electrons are pictured as drifting under the action of an applied electric field through a lattice of atoms or ions that act as scattering agents. Under these circumstances an electron moves, under the influence of an external electric field, for a time dependent upon the total probability of scattering due to any of the interaction processes with the surrounding material.

The motion is most usefully described in 'momentum space'; here the particle moves at a constant rate between collisions and changes its momentum state from one value to another instantaneously during a collision, as illustrated in Figure 5.

The momentum space is really the space defined by $\hbar \mathbf{k}$ but, since $\hbar$ is a constant scaling factor, only the motion in **k**-space is considered. If the carriers are electrons with a charge $-e$ then a constant, uniform electric field $\mathbf{F} = (0, 0, -F)$ causes them to drift in the positive k_z direction and the equation of motion of the electron has a solution

$$\mathbf{k}_f(t) = \mathbf{k}_i - \frac{e\mathbf{F}}{\hbar} t, \tag{13}$$

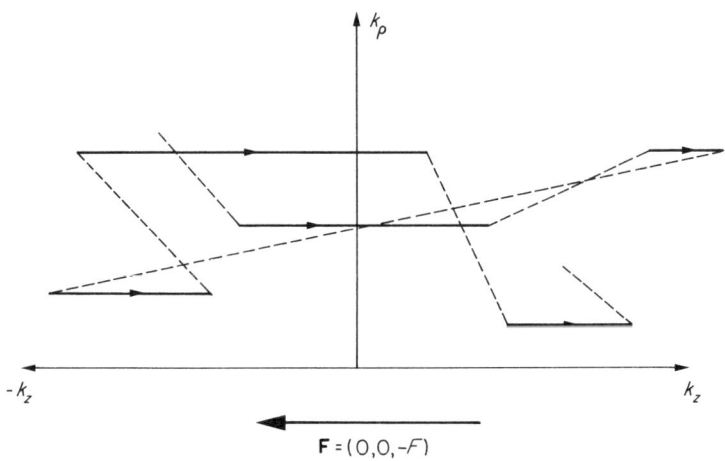

Figure 5. Electron motion in momentum space. The axes are $k_\rho = (k_x^2 + k_y^2)^{\frac{1}{2}}$, k_z, and the scaling factor $\hbar$ is omitted

where $\mathbf{k}_i$ is the value at the start of the electron flight, $\mathbf{k}_f(t)$ is the value at the end of the flight (drifting time), and t is the flight time.

Since the **k**-space of the electrons is symmetric about **F** we can use an equivalent two-dimensional (k_ρ, k_z)-space, where $k_\rho = (k_x^2 + k_y^2)^{\frac{1}{2}}$, instead of the three-dimensional (k_x, k_y, k_z)-space. Hence k_ρ remains fixed during an electron flight from an initial position (k_ρ, k_{zi}) to a final position (k_ρ, k_{zf}) and only the z-component of equation (13) needs to be considered, i.e.

$$k_{zf} = k_{zi} + \frac{eF}{\hbar} t. \tag{14}$$

The simulation is started by choosing an arbitrary starting position and is then allowed to proceed until many flight times have occurred between scattering events. If enough flight times are considered, i.e. if the electron path is long enough, then the starting position has no influence on the final results. A flight is terminated by a scattering process that, since there are several competing mechanisms, is selected by another Monte Carlo procedure. The scattering mechanism automatically specifies the electron energy after scattering; however, the determination of the new position of the electron in momentum space requires a knowledge of the scattering angle. Since this angle has a probability distribution that depends on the scattering probability, the scattering angle is also generated by a Monte Carlo method. This is described later.

4. THE ELECTRON-PHONON SCATTERING PROCESSES

Each process that can scatter an electron at the end of a free flight is characterized by a transition rate, $S_n(\mathbf{k}, \mathbf{k}')$, from the momentum state $\hbar \mathbf{k}$ to the momentum state $\hbar \mathbf{k}'$. Here the subscript n denotes an individual scattering process and can take values $n = 1, 2, ,,,, N$ if there are N possible processes. The total scattering rate from the state **k**, because of the nth process, is

$$\lambda_n(\mathbf{k}) = \int S_n(\mathbf{k}, \mathbf{k}') \, d\mathbf{k}' \tag{15}$$

Hence, the total scattering rate, due to all processes, is

$$\lambda(\mathbf{k}) = \sum_{n=1}^{N} \lambda_n(\mathbf{k}), \tag{16}$$

where **k** is a function of time (cf. equation (7). In practice the total scattering rates are only functions of $k = |\mathbf{k}|$ so, by virtue of equation (12), $\lambda(\mathbf{k})$ is easily transformed to $\lambda(E)$.

As an illustration of hot electron behaviour in semiconductors we consider the important material GaAs, whose energy-momentum band structure ($E(k)$ versus $\pm k$) is sketched in Figure 6, in the vicinity of the band extremities, along the ⟨100⟩ directions. There are energy minima, located at the zone centre, and at the edges, that we will call valleys. There is one central valley, with a small radius of curvature, where electrons have a low effective mass equal to 0.067 m_0, and there is a shallow satellite valley at the equivalent (100) Brillouin zone edges. By symmetry, however, two other, identical, satellite valleys are located at the (010) and (001) edges as well. Electrons associated with these satellite valleys have an effective mass equal to 0.35 m_0. Thus the conduction band has valley structure and, at zero applied electric field ($\mathbf{F}=0$) and room temperature (or less), the free electrons are located in the bottom regions of the central valley. The valence band is well separated from the conduction band so the transport of holes need not be considered.

In a non-degenerate sample of GaAs, the electron density is quite small ($\sim 10^{21}$ m^{-3}) so that the Pauli exclusion principle does not have to be taken into account. This is what we assume here although, within the framework of a Monte Carlo calculation, it is reasonably straightforward to include

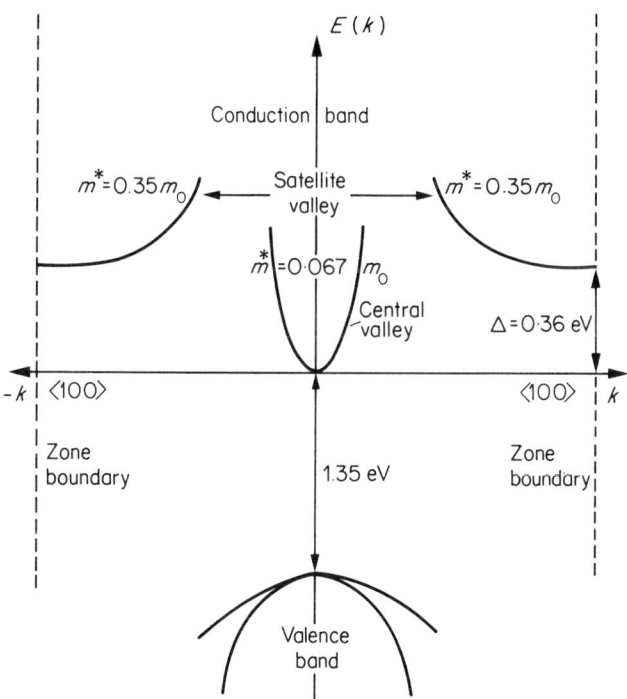

Figure 6. Sketch of the band structure of GaAs

degeneracy. A reasonable approximation, for a student exercise such as this, is to assume that both types of valleys are parabolic, i.e. that the electron energy minima are described by equation (12). Nevertheless the central valley of GaAs is really non-parabolic and, strictly speaking, equation (12) should be modified by the addition of a term of the form αk^4, where α is a constant.[6] This non-parabolicity, on its own, turns out to be quite important but it is, in fact, partially offset by a wave-number dependence that should be included in $S_n(\mathbf{k}, \mathbf{k}')$ if Bloch wave functions, as opposed to plane waves, are used. For these reasons it is a consistent, quite good approximation, to use parabolic valleys and a plane wave representation of the free-electron states.

If the applied electric field is zero the electrons are in equilibrium with the vibrating solid atoms which have a mean energy $\sim (k_B T/1.6) \times 10^{19}$ eV, where T is the absolute temperature of the semiconductor and k_B is Boltzmann's constant 1.38×10^{-23} J K^{-1}. For a material at 300 K the mean electron energy is therefore ~ 0.03 eV. The application of an electric field causes the electrons to cease to be in equilibrium with the solid lattice and to increase in energy until they acquire a temperature in excess of that of the solid. The electrons become hot, therefore, and occupy the upper regions of the central valley. At sufficiently high electric fields the electrons may become hot enough to transfer from the central valley to the satellite valleys and switch from being light electrons to being heavy electrons. This transfer is achieved through the agency of phonons that are energy quanta of the lattice vibrations and is a scattering event in the history of the electron. Similarly, transfers occur from the satellite valley back to the central valley and between the three equivalent satellite valleys.

All these intervalley scattering events and, in addition, intravalley scattering events that leave the electron in the same valley, involve phonons that have an energy $\hbar\omega$, where ω is the frequency of a lattice vibration. In general, an electron that experiences a scattering event due to an interaction with a phonon field has an ante-scattering energy $E(\mathbf{k})$ and a post-scattering energy $E(\mathbf{k}')$, where $\mathbf{k}$ and $\mathbf{k}'$ are the initial and final $\mathbf{k}$ states such that $E(\mathbf{k}') - E(\mathbf{k}) \pm \hbar\omega = 0$. If the intravalley scattering occurs, via polar optical phonons, then to a good approximation the electron either emits or absorbs a phonon during the scattering event according to the following rules:

Absorption: $\quad E(\mathbf{k}') = E(\mathbf{k}) + \hbar\omega_0$ \hfill (17a)

Emissions: $\quad E(\mathbf{k}') = E(\mathbf{k}) - \hbar\omega_0,$ \hfill (17b)

where ω_0 is a constant frequency. If only acoustic phonons are involved then, to a good approximation, the intravalley scattering can be assumed to be elastic with $E(\mathbf{k}') = E(\mathbf{k})$.

For GaAs the important intravalley scattering mechanism between states inside the central valley and between states inside a satellite valley is polar optical phonon scattering. Obviously, acoustic phonon scattering also takes place within the central and satellite valleys but it has a minor influence on the electron distribution.[6]

If scattering between the central and satellite valley occurs then the electrons must acquire at least an energy Δ before the transition becomes

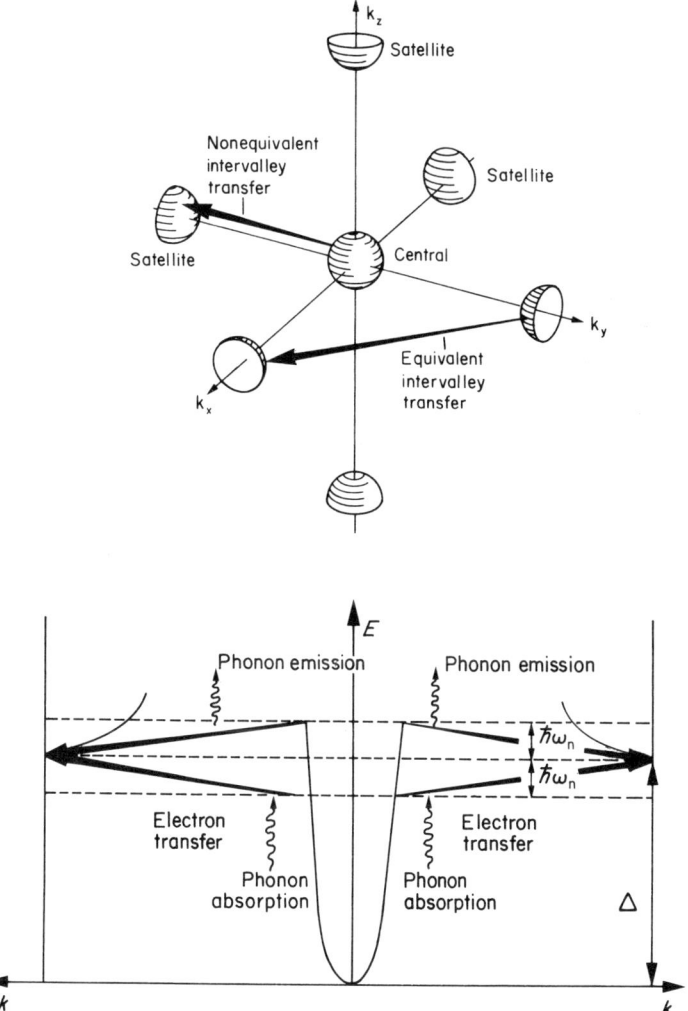

Figure 7. Intervalley transitions on GaAs. For simplicity only one-way transitions are shown. Obviously, the reverse transitions also occur with phonon emission or absorption changing to absorption or emission

Table 1. Scattering mechanisms in GaAs

Scattering mechanism	Transition rate $(\hbar/2\pi)S(\mathbf{k}',\mathbf{k})$	Total scattering rate $\lambda(\mathbf{k}) \equiv \lambda(E)$	Definitions
Acoustic phonon for either absorption or emission (intravalley)	$\dfrac{\hbar D_a^2 \|\mathbf{k}-\mathbf{k}'\| N_a \delta(A)}{(2\rho s)(2\pi)^3}$ $N_a \approx \dfrac{k_B T}{\hbar s \|\mathbf{k}-\mathbf{k}'\|}$	$\dfrac{(2m_{c,s}^*)^{\frac{3}{2}} k_B T D_a^2 E^{\frac{1}{2}}}{4\pi \rho s^2 \hbar^4}$ absorption or emission $A = E' - E = 0$	Central or satellite valley effective mass, $m_{c,s}^*$; density, ρ; velocity of sound, s; acoustic deformation potential, D_a; phonon occupation number, N_a.
Polar optical phonon (intravalley)	$\dfrac{2\pi e^2 \hbar \omega_0}{\|\mathbf{k}-\mathbf{k}'\|^2} \left(\dfrac{1}{\varepsilon_\infty} - \dfrac{1}{\varepsilon_0}\right) \dfrac{X}{4\pi\kappa_0 (2\pi)^3}$ $X = N_0 \delta(A_a)$ absorption $(N_0+1)\delta(A_e)$ emission $N_0 = [\exp(\hbar\omega_0/k_B T) - 1]^{-1}$	$\dfrac{Y e^2 m^{*\frac{1}{2}} \omega_0}{\sqrt{2}\hbar(4\pi\kappa_0)^2 E^{\frac{1}{2}}} \left(\dfrac{1}{\varepsilon_\infty} - \dfrac{1}{\varepsilon_0}\right) \ln \left\|\dfrac{E^{\frac{1}{2}} + E'^{\frac{1}{2}}}{E^{\frac{1}{2}} - E'^{\frac{1}{2}}}\right\|$ $Y = N_0$ absorption (N_0+1) emission $A_{a,e} = E' - E \mp \hbar\omega_0 = 0$	High frequency and static dielectric constants, ε_∞, ε_0; phonon occupation number, N_0; permittivity of free space, κ_0.

Scattering mechanism	Scattering rate		Notes
Equivalent intervalley (satellite ⇄ satellite)	$\dfrac{(Z-1)\hbar D_e^2 X}{(2\pi)^3 (2\rho\omega_e)}$ absorption $X = N_e \delta(A_a)$ $(N_e+1)\delta(A_e)$ emission $N_e = [\exp(\hbar\omega_e/k_B T)-1]^{-1}$	$\dfrac{(Z-1)m_s^{*\frac{3}{2}} D_e^2 E^{\frac{1}{2}} Y}{\sqrt{2}\pi\rho\omega_e \hbar^3}$ $Y = N_e$ absorption (N_e+1) emission $A_{a,e} = E' - E \mp \hbar\omega_e$	Intervalley deformation potential, D_e; number of equivalent valleys, Z (for GaAs, $Z=3$); phonon occupation number, N_e.
Non-equivalent intervalley ((a) central → satellite	(a) $\dfrac{3\hbar D_n^2 X}{2\rho\omega_n (2\pi)^3}$ $X = N_n \delta(A_a)$ $(N_n+1)\delta(A_e)$ emission	$\dfrac{3m_s^{*\frac{3}{2}} D_n^2 E^{\frac{1}{2}} Y}{\sqrt{2}\pi\rho\omega_n \hbar^3}$ $Y = N_n$ absorption (N_n+1) emission $A_{a,e} = E' - E + \Delta \mp \hbar\omega_n = 0$	Intervalley deformation potential D_n; phonon occupation number, N_n.
(b) satellite → central)	(b) $\dfrac{\hbar D_n^2 X}{2\rho\omega_n (2\pi)^3}$ $X = N_n \delta(A_a)$ absorption $(N_n+1)\delta(A_e)$ emission $N_n = [\exp(\hbar\omega_n/k_B T)-1]^{-1}$	$\dfrac{m_c^{*\frac{3}{2}} D_n^2 E^{\frac{1}{2}} Y}{\sqrt{2}\pi\rho\omega_n \hbar^3}$ $Y = N_n$ absorption (N_n+1) emission $A_{a,e} = E' - E - \Delta \mp \hbar\omega_n = 0$	

possible. However, in the satellite valley the electron energy is measured from its minimum so it is necessary to add Δ to the electron energy if scattering from the satellite to the central valley occurs. Such non-equivalent intervalley transitions are shown in Figure 7 and are summarized as:

Central $\rightarrow$ Satellite:

Absorption: $\qquad E_S(\mathbf{k}') = E_C(\mathbf{k}) - \Delta + \hbar\omega_n;$ (18a)

Emission: $\qquad E_S(\mathbf{k}') = E_C(\mathbf{k}) - \Delta - \hbar\omega_n.$ (18b)

Satellite $\rightarrow$ Central:

Absorption: $\qquad E_C(\mathbf{k}') = E_S(\mathbf{k}) + \Delta + \hbar\omega_n;$ (19a)

Emission: $\qquad E_C(\mathbf{k}') = E_S(\mathbf{k}) + \Delta - \hbar\omega_n$ (19b)

where ω_n is a constant frequency. The other transitions that must be considered take place between the satellite valleys and involve the emission or absorption of a phonon of energy $\hbar\omega_e$ where ω_e is also a constant frequency. The energy relationship for satellite $\rightleftarrows$ satellite transitions has the same form as equations (17a) and (17b).

The complete list[6] of electron-phonon scattering processes for plane wave states in parabolic valleys is given in Table 1. It should be noted that if V is the volume of the crystal then a factor $V/8\pi^3$ should appear in front of the integral of equation (15). This factor has been absorbed into the definition of $S_n(\mathbf{k}, \mathbf{k}')$. The formulae are presented without proof, but the references at the end of this chapter and a good textbook on solid-state physics for

Table 2. Data for GaAs[6]

Parameter	Value
Density	5.37 g/cm^3
Velocity of sound, s	5.22 × 10^5 cm/s
High frequency dielectric constant, ε_∞	10.82
Static dielectric constant ε_0	12.53
Polar optical phonon frequency, ω_0	5.37 × 10^{13} rad.S^{-1}
Equivalent intervalley phonon frequency, ω_e	4.54 × 10^{13} rad.S^{-1}
Non-equivalent intervalley phonon frequency, ω_n	4.54 × 10^{13} rad.S^{-1}
Acoustic deformation potential in central and satellite valleys	7 eV
Equivalent intervalley deformation potential, D_e	10^9 eV/cm
Non-equivalent intervalley deformation potential, D_n	10^9 eV/cm
Central valley effective mass, m_c^*	0.067 m_0
Satellite valley effective mass, m_s^*	0.35 m_0
Valley separation	0.36 eV

Note. If SI units are used to calculate $\lambda_n(E)$ then care must be exercised with quantities measured in eV/cm.

advanced students, will enable the reader to check them without too much difficulty. Each process is governed by an energy-conserving delta-function condition $\delta(A)$ in which we write $E' \equiv E(\mathbf{k}')$ and $E \equiv E(\mathbf{k})$. Data for GaAs that are required to evaluate these formula are given in Table 2. The computer program given at the end of this chapter, however, can be run with any set of data required, even if it does not correspond to a real substance.

5. COMPUTATION OF ELECTRON FLIGHT TIMES AND SCATTERING CHANNEL SELECTION

If $p(t)$ is the probability per unit time that an electron has a flight, of duration time t, terminated by some scattering process (i.e. has a drift time t in momentum space and is then scattered) then the solution of equation (8) in the form

$$r = \int_0^t p(t') \, dt'$$

will enable a random distribution of such flight times to be generated.

Now suppose that the electron drifts for a time t before being scattered and that this time consists of n tiny increments $\delta t_1, \delta t_2, \ldots, \delta t_n$. The probability of the electron being scattered, within the time interval δt_i, is $\lambda(k)\delta t_i$, where $\lambda(\mathbf{k})$ is the total scattering rate defined by equation (16). The probability that there will will not be any scattering during δt_i is thus $(1 - \lambda(\mathbf{k})\delta t_i)$. Therefore, the probability of an electron drifting in momentum space, for a time t, is

$$S(t) = \prod_{i=1}^{n} (1 - \lambda(\mathbf{k})\delta t_i) \tag{20}$$

so that

$$\log\{S(t)\} = \sum_{i=1}^{n} \log(1 - \lambda(\mathbf{k})\delta t_i). \tag{21}$$

However, since $\lambda(\mathbf{k})\delta t_i \ll 1$ equation (21) reduces to

$$\log\{S(t)\} = -\sum_{i=1}^{n} \delta t_i \lambda(\mathbf{k}) \tag{22}$$

which gives, immediately,

$$S(t) = \exp\left\{-\int_0^t \lambda(\mathbf{k}) \, dt'\right\} \tag{23}$$

where $\lambda(\mathbf{k})$ is a function of time through equation (13). The probability density $p(t)$ is therefore

$$p(t) = \lambda(\mathbf{k})\exp\left\{-\int_0^t \lambda(\mathbf{k})\,dt'\right\} \tag{24}$$

Using equation (24) and a uniformly distributed random number distribution gives

$$r = 1 - \exp\left\{-\int_0^t \lambda(\mathbf{k})\,dt'\right\}. \tag{25}$$

Equation (25) is very complicated and cannot, for the scattering mechanisms discussed in section 3, be solved analytically for t. There is no conceptual difficulty here, however, because, if necessary, the integrations could be performed by numerical quadrature to produce r and t in tubular form for for each value of electron energy. Such a process involves several subroutines and, spread over thousands of scattering events, involves a significant amount of computer time and store. It is also rather inconvenient to program.

A new technique for circumventing this difficulty has been found[5,6] and involves supplementing the real scattering processes of section 3 by the addition of a virtual scattering process that does not affect the state of the electron. Suppose a virtual scattering transition probability $S_V(\mathbf{k}', \mathbf{k})$ is defined as

$$S_V(\mathbf{k}', \mathbf{k}) = \lambda_V(\mathbf{k})\delta(\mathbf{k}-\mathbf{k}'), \tag{26}$$

where, since the electron state is unaltered, the electron distribution remains unaffected so that $\lambda_V(\mathbf{k})$ is quite arbitrary. We are then at liberty to choose

$$\lambda_V(\mathbf{k}) = \Gamma - \lambda - \lambda(\mathbf{k}), \tag{27}$$

where $\lambda(\mathbf{k})$ is the total real scattering rate and Γ is some constant sufficiently large to avoid negative probabilities. Thus $\lambda_V(\mathbf{k}) > 0$ for all values of $\mathbf{k}$ that could be involved in the situation and the new total scattering rate for the electron that now includes the virtual process is simply

$$\lambda_T(\mathbf{k}) = \lambda(\mathbf{k}) + \lambda_V(\mathbf{k}) = \Gamma. \tag{28}$$

Equation (25) now reduces to the elementary form

$$r = 1 - e^{-\Gamma t}, \tag{29}$$

giving the amazing result that

$$t = -\frac{1}{\Gamma}\ln(1-r) = -\frac{1}{\Gamma}\ln(r). \tag{30}$$

The time given by equation (30) is not usually the flight time between real scattering processes but is more likely to be the time between one virtual scattering process and another. During the simulation a flight time t_1, say, is generated by a certain random number in accordance with equation (30). A further random number is then used to determine which scattering process (i.e. scattering channel) is to be selected for possible termination of the flight. This selection normally uses random numbers lying between 0 and 1 so that all the scattering rates are usually normalized with Γ. Hence, if the electron, at the moment of scattering, has an energy E, $\lambda_n(E)/\Gamma$ are calculated and a uniformly distributed random number is used, in the manner of Figure 8, to select the appropriate scattering channel.

If a virtual process is encountered the flight is not terminated and the flight time t_1 is stored. A new flight time t_2 is then selected, in the same way

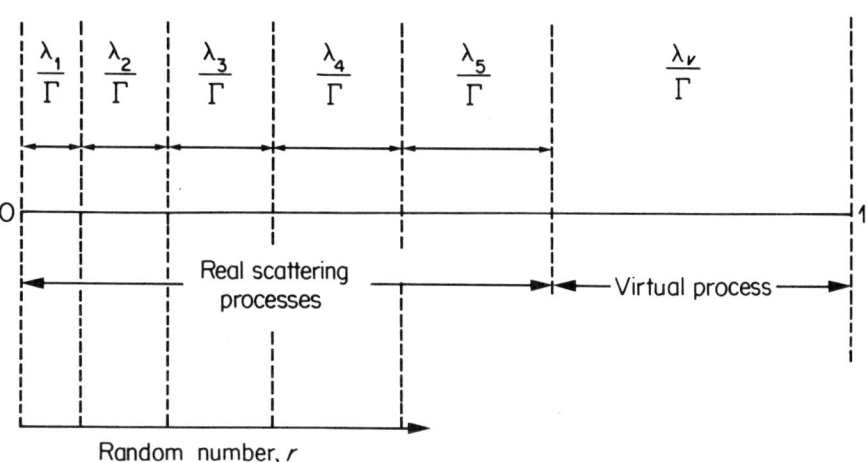

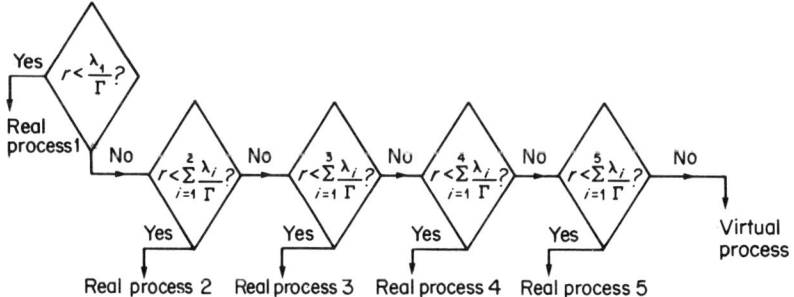

Figure 8. Scattering channel selection assuming that $N = 5$

as t_1, and another scattering channel is selected. If a virtual process is encountered, once again, then t_2 is added to t_1 and a further flight time t_3 is calculated. This process goes on until the flight is finally terminated by the selection of a real scattering mechanism. It is clearly advisable to have as few virtual processes as possible, but the penalty that has to be paid here is that many virtual processes occur. This means that we have to accumulate, say, M subflights t_i until the real flight time $t = \sum_{i=1}^{M} t_i$ is obtained. This penalty, in computing terms, is relatively minor compared to using equation (25) together with only real scattering mechanisms. Nevertheless, the number of virtual processes can be minimized by making Γ as small as possible, without making $\lambda_V < 0$. In the program given here we set $\Gamma = \{\sum_{n=1}^{N} \lambda_n(E)\}_{\max}$ that, for two-valley GaAs processes, happens to be $\sum_{n=1}^{N} \lambda_n(E_{\max})$, where $E_{\max} \sim 1$ eV is the maximum energy allowed during the simulation.

5. ANGLE SELECTIONS, MEAN VELOCITY, AND MEAN ENERGY

5.1 Angle selections

After a real scattering process an electron makes an instantaneous transition, in **k**-space, from a position specified by **k** to a new position specified by **k′**. All possible new positions **k′** are, of course, randomly distributed and are selected by means of uniformly distributed random numbers.

The simplest scattering events that occur involve the acoustic (within the quasi-elastic approximation used here) and intervalley processes. For all of these the transition probability $S_n(\mathbf{k}, \mathbf{k}')$ does not depend upon the angle between **k** and **k′** so that all wave vectors that satisfy the appropriate law of conservation of energy are equally probable. Such scattering processes are called randomizing, or isotropic, which means that the scattered electron has the same probability of being in any direction after a collision. Hence, all that needs to be known is the energy of the electron after it has been scattered.

Suppose the vector **k′** makes an angle θ with the k_z-axis direction then the probability density $p(\theta)$ is proportional to the number of available states on the circumference of a circle of radius

$$|\mathbf{k}'| \sin \theta = \left(\frac{2m^* E'}{\hbar^2}\right)^{\frac{1}{2}} \sin \theta$$

(cf. Figure 9a). The basic statement is that

$$p(\theta) = A \sin \theta, \tag{31}$$

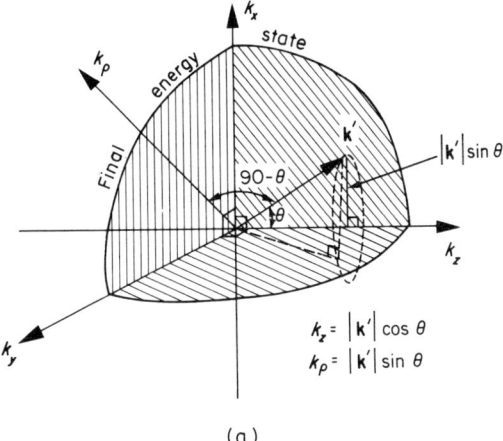

Figure 9a. Angle selection geometry for acoustic and intervalley processes

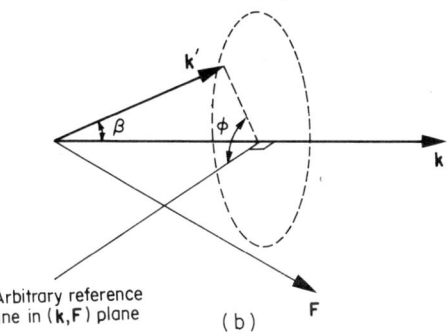

Figure 9b. Scattering of electron in state **k** to a state **k'** induced by optical polar phonons

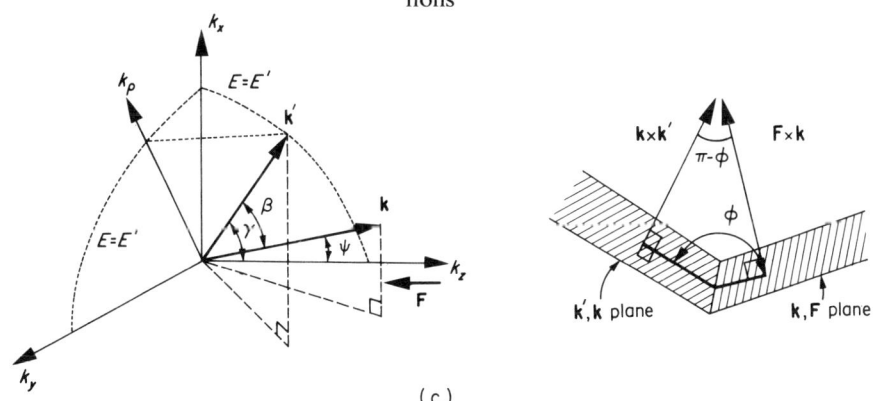

Figure 9c. Full geometry of electron scattering by optical polar phonons

where normalization requires that

$$A \int_0^\pi \sin\theta \, d\theta = 1. \tag{32}$$

The solution of equation (32) is $A = \frac{1}{2}$ and the θ distribution is obtained from a uniformly distributed random number set lying on the interval $(0, 1)$ by inverting

$$r = \int_0^\theta p(\theta') \, d\theta' = \tfrac{1}{2}(1 - \cos\theta). \tag{33}$$

The new state $\mathbf{k}'$ is therefore

$$\mathbf{k}' = (|\mathbf{k}'| \sin\theta, |\mathbf{k}'| \cos\theta) \tag{34}$$

The situation for polar optical phonon scattering is much more difficult because, as Table 1 shows, $S_n(\mathbf{k}, \mathbf{k}')$ depends upon the angle between $\mathbf{k}$ and $\mathbf{k}'$. Actually this transition rate describes a dipole scattering interaction between electrons and a lattice that has two types of atoms. This is a form of Coulombic scattering and is not isotropic since it emphasizes small angle deviations.

It is necessary to consider first the angle β between $\mathbf{k}$ and $\mathbf{k}'$ without reference to the coordinate system of the actual problem but using $\mathbf{k}$ as the reference direction, as shown in Figure 9b. The probability density $p(\beta)$, as in the previous mechanisms, must be proportional to $\sin\beta$, but it must be weighted, also, by the factor $|\mathbf{k} - \mathbf{k}'|^{-2}$ (cf. Table 1). This means that

$$p(\beta) = \frac{A \sin\beta}{E + E' - 2(EE')^{\frac{1}{2}} \cos\beta}, \tag{35}$$

where A is a constant, $|\mathbf{k} - \mathbf{k}'|^2$ has been replaced by $(2m^*/\hbar^2) \times (E + E' - 2(EE')^{\frac{1}{2}} \cos\beta)$ and the factor $2m^*/\hbar^2$ is absorbed into A. If $f = 2(EE')^{\frac{1}{2}}/(E^{\frac{1}{2}} - E'^{\frac{1}{2}})^2$ then it is easy to see that β is determined by

$$r = \frac{\int_{\cos\beta}^{1} \frac{dx}{(1 + f - fx)}}{\int_{-1}^{1} \frac{dx}{(1 + f - fx)}}, \tag{36}$$

where r is the usual random number. The integrations here are elementary and result in

$$r = \frac{\ln(1 + f - f \cos\beta)}{\ln(1 + 2f)}, \tag{37}$$

giving
$$\cos \beta = (1+f-(1+2f)^r)/f. \tag{38}$$

As is shown in Figure 9b all the possible positions of $\mathbf{k}'$ are the generators of a cone, about $\mathbf{k}$, of semi-angle β. If an arbitrary line is drawn on the base of this cone through $\mathbf{k}$ then the end point of $\mathbf{k}'$ that lies on the circular edge is in a position designated by an azimuthal angle ϕ for which $p(\phi) = 1$. The relationship of a uniformly distributed random number r to ϕ is $r = \phi/2\pi$.

The post-scattering coordinates (k'_ρ, k'_z) are

$$k'_\rho = |\mathbf{k}'| \sin \gamma, \tag{39a}$$
$$k'_z = |\mathbf{k}'| \cos \gamma, \tag{39b}$$

where γ is the angle between $\mathbf{k}'$ and $\mathbf{F}$ and γ is related to β and ϕ. The latter arises from the vector relationship, shown in Figure 9c,

$$(\mathbf{k}' \times \mathbf{k}) \cdot (\mathbf{k} \times \mathbf{F}) = (\mathbf{k}' \cdot \mathbf{k})(\mathbf{k} \cdot \mathbf{F}) - (\mathbf{k}' \cdot \mathbf{F})(\mathbf{k} \cdot \mathbf{k}) \tag{40}$$

This gives immediately

$$-\sin \beta \sin \psi \cos \phi = \cos \beta \cos \psi - \cos \gamma, \tag{41}$$

where ψ is the angle between $\mathbf{k}$, and $\mathbf{F}$. ϕ is the uniformly distributed random angle discussed above. Hence positive and negative values of $\cos \phi$ are equally possible. At the end of a flight the electron position is k_z and its energy is E so that $|\mathbf{k}| = (2m^*E)^{\frac{1}{2}}/\hbar$ and $\cos \psi = k_z/|\mathbf{k}|$.

5.2 Mean velocity and mean energy

The time spent by an electron in each part of $\mathbf{k}$-space is proportional to the electron distribution function and, in principle, this is computed and used to calculate useful mean values, such as the mean electron velocity and the mean electron energy. Indeed, since the Monte Carlo calculation described here is equivalent to an exact solution of the Boltzmann equation for a non-degenerate electron gas, it would seem that the accumulation of the visiting time of an electron in each element of $\mathbf{k}$-space is the primary computing objective.

This procedure, however, would require the establishment of a $\mathbf{k}$-space histogram with quite a fine mesh in order to obtain accurate answers. Fortunately, it is not necessary to do this, although such a histogram, with a reasonable mesh, will be calculated for qualitative purposes. The mean values of velocity and energy can be accumulated directly by monitoring each electron flight and then taking an average over all flights.

For the electron velocity this procedure is understood as follows. The velocity of an electron is

$$\mathbf{v} = \frac{1}{\hbar}\nabla_\mathbf{k} E(\mathbf{k}) = \frac{\hbar \mathbf{k}}{m^*} \qquad (42)$$

and, since the motion in $\mathbf{k}$-space along the k_z-axis is uniform, $p(k_z) = 1$. This means that, for all flights from starting positions k_{zi} to final positions k_{zf}, the mean drift velocity is defined as

$$v = \frac{\frac{\hbar}{m^*}\sum \int_{k_{zi}}^{k_{zf}} k_z \, dk_z}{\sum \int_{k_{zi}}^{k_{zf}} dk_z} = \frac{\sum \int_{k_{zi}}^{k_{zf}} \frac{1}{\hbar}\frac{\partial E}{\partial k_z} dk_z}{\sum \int_{k_{zi}}^{k_{zf}} dk_z} = \frac{\sum (E_f - E_i)}{\hbar D}, \qquad (43)$$

where the summation is over all electron flights that occur during the simulation and $1/D$ is the distribution function per unit length of the $\mathbf{k}$-space flight. The same reasoning leads to a mean energy E being defined as

$$E = \frac{\hbar^2}{2m^*} \frac{\sum \int_{k_{zi}}^{k_{zf}} (k_\rho^2 + k_z^2) \, dk_z}{\sum \int_{k_{zi}}^{k_{zf}} dk_z}. \qquad (44)$$

This technique avoids the setting up of a histogram mesh that is fine enough to ensure a convergent result. All that is required is the accumulation of contributions from each flight, thus making the simulation both accurate and very much easier to program.

Naturally v and E are accumulated for each valley and the mean values for the whole valley structure are then calculated by weighting each contribution according to the electron population ratio. The quantity D, when multiplied by $\hbar/eF$, gives the total time spent by the electron in a valley. If T_c is the time spent in the central valley and T_s the time spent in the satellite valleys then $N1 = T_c/(T_c + T_s)$ and $N2 = T_s/(T_c + T_s)$ give the relative electron population ratios of the valleys. The evaluation of the integrals in equations (43) and (44) leads to the following expressions for v and E:

$$v = \frac{\hbar}{2m^*D} \sum (k_{zf}^2 - k_{zi}^2) \qquad (45)$$

and

$$E = \frac{\hbar^2}{2m^*D} \sum \left\{ k_\rho^2 (k_{zf} - k_{zi}) + \frac{(k_{zf}^3 - k_{zi}^3)}{3} \right\}. \qquad (46)$$

6. COMPUTER PROGRAMS

6.1 Description

The optimum use of the programs, listed at the end of this chapter, is in an interactive mode on the type of computer terminal that has a visual display unit (VDU) with a graphics capability. In the interactive mode the program prompts the user to input data or to execute other actions as the calculation proceeds. If only an ordinary terminal is available the program can still be used by omitting the graphics package or suitably editing it to work off-line. This may require the deletion of specific reference to the VDU and feeding in the required data in a modified manner.

The main program is called CARLO and the graph plotting program is called GMONTE. The compiled, loaded, ready to run, binary versions are called *CARLO and *GMONTE. The files that are used are:

FILE	CONTENTS
MONTEF:	complete tables of final results
MDATA 1:	electric field values
MDATA 2:	mean velocities
MDATA 3:	mean energies
MDATA 4:	**k**-space histograms
MDATA 5:	input data
GRAPP:	test labels for choice of action

All these files are established as reading, writing, or reading/writing files by the instruction CALL SEARCH (a, 'FILE', b) where $a = 1$ denotes a file from which information is read, $a = 2$ denotes a file into which information is written, and $a = 3$ denotes the reading/writing mode. Here, b is an arbitrary file number. Since files are of variable length two actions are normally used to close them. The first

$$\text{CALL SEARCH } (8, 0, b)$$

truncates the file, while the second

$$\text{CALL SEARCH } (4, 0, b)$$

actually closes it. For reading and writing the VDU has device number 1 but reading out of and writing into the files uses $b+4$.

Subroutines CALL ZREALD (A), CALL ZINTRD (A) are used that halt the program until a piece of data or an instruction is entered or the computer is simply requested to proceed. Actually these are free-format instructions that enable data to be entered in free format into A or, if RETURN is activated treat A as dummy and pass on to the next instruction. CALL PAGE is a little subroutine that clears the VDU screen. RESUME

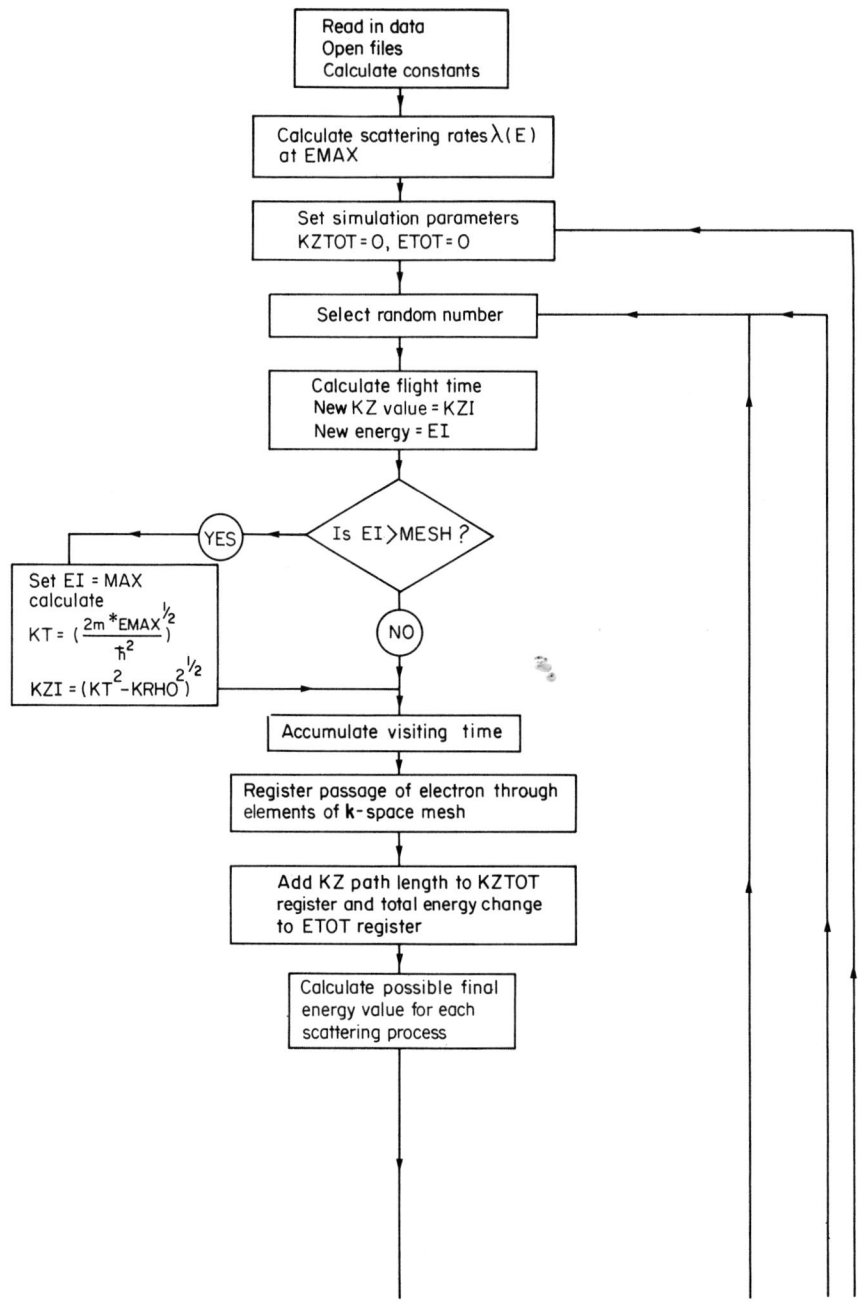

COMPUTER SIMULATION USING MONTE CARLO METHODS

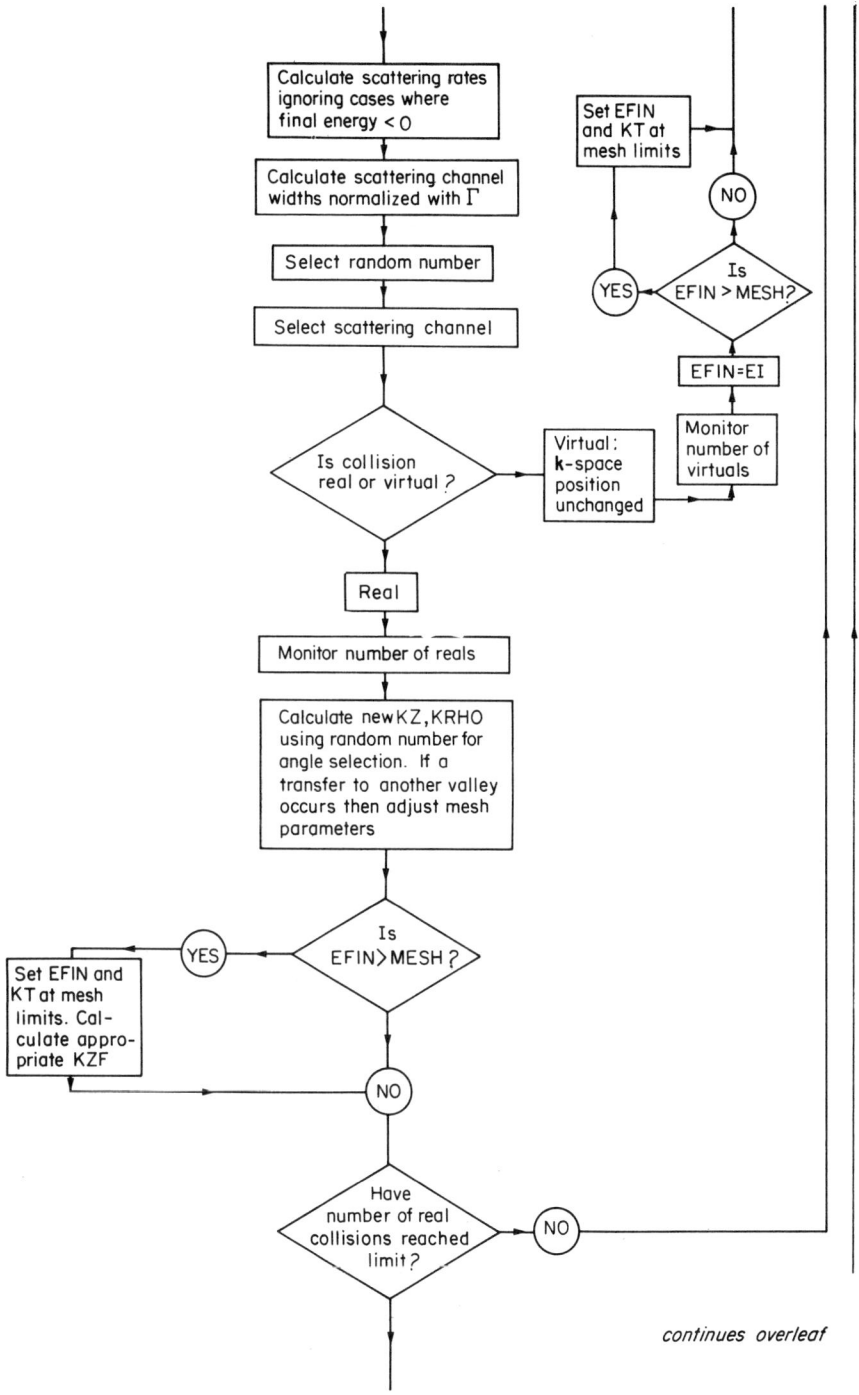

continues overleaf

continued from previous page

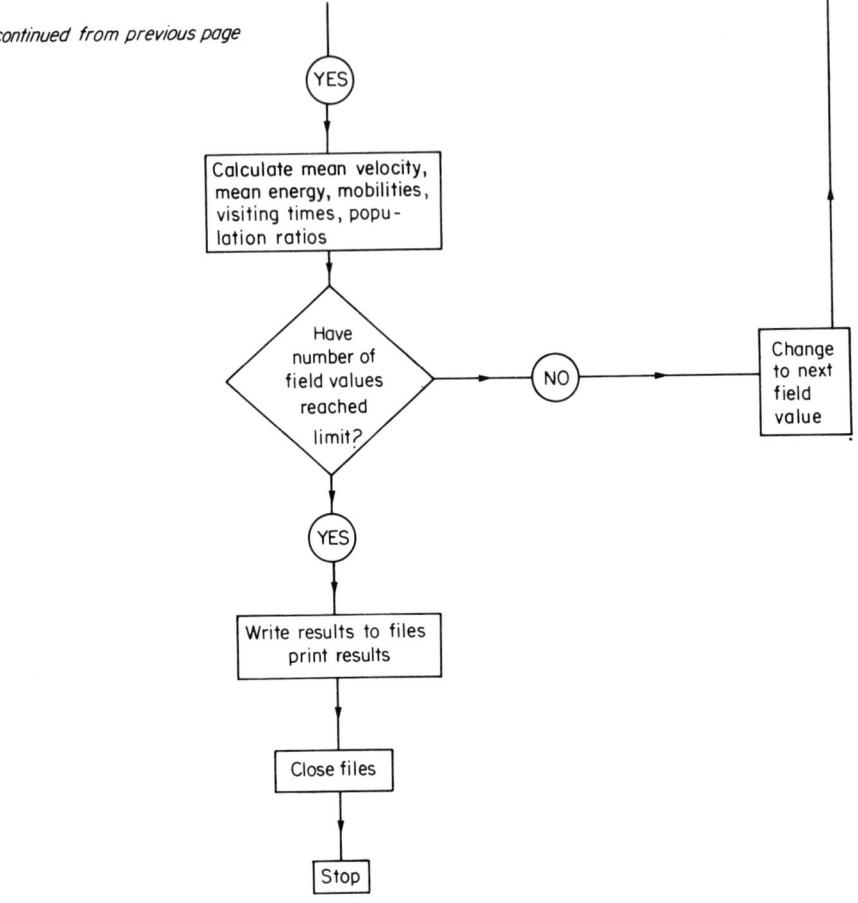

Figure 10. MONTE CARLO flow diagram

('*GMONTE') causes entry into the graph plotting program, RESUME ('*CARLO') causes entry into the main Monte Carlo program.

CARLO the main program actually does the Monte Carlo calculations, while GMONTE takes the results and displays them graphically. The organization of the main program can be seen in the CARLO flow diagram (Figure 10). In CARLO a virtual scattering process is referred to as self- or pseudo-scattering and the electrons are assumed to move in a **k**-space bounded by $E = E_{max}$, where E_{max} is called, for convenience, the 'mesh' size. Checks are made, after every flight and every scattering event, to see whether the electron has escaped from the 'mesh'. If it has done so it is placed on the edge of the mesh. This is done repeatedly, until it is scattered inwards again. CARLO produces, for each electric field value, a mean velocity and a mean energy, averaged over the two valleys, the visiting time, mean velocity, ordinary mobility, and **k**-space histogram for each valley.

The **k**-space histogram is the distribution function $f(\mathbf{k})$ in (k_ρ, k_z) space. It is obtained by dividing this space into reasonably fine meshes or boxes and computing the total electron visiting time for each box for a given number of real collisions; $f(\mathbf{k})$ is plotted, for a fixed k_ρ, along the k_z-axis and CARLO allows the possibility of plotting this at 21 different k_ρ levels. The k_z-value at the centre of each mesh element is calculated. The function $f(\mathbf{k})$ is then accumulated in the following way. As the electron progresses along the k_z-axis a box is filled by assigning 1 to it if an electron completely crosses it, but assigning only a fraction to it if it does not. This procedure is justified by the uniform motion in **k**-space; $f(\mathbf{k})$ is then displayed as the unnormalized accumulation in these boxes, for each electric field value.

CARLO also monitors the number of virtual or self-scattering processes, together with the number of times the electron energy exceeds E_{max}. The electron energy at the end of a free-flight is called EI, indicating that this is its initial value prior to being scattered. Its energy after scattering is called EFIN.

The graphics program GMONTE (Figure 11) is designed to produce a plot of the calculated mean velocity (in cm/sec) and of mean energy (in eV) against electric field (in kV/cm), and to produce the electron distribution functions for each electric field value. These plots are produced either at the VDU or off-line on a graph plotter as hard copies. The velocity-field plot, called the static characteristic, is compared by the program with experimental values, if any are available. These experimental values are the only interactive mode data requests GMONTE makes. After the theoretical curve is plotted the program is halted. Restarting causes the experimental curve to be superimposed.

All graph plotting instructions involve CALL FIXAXS(.....) and CALL FGPLT(.....). These, or their equivalents, will be instantly recognizable by most computer program advisory units. CALL T1OU(7) is only an embellishment being, in fact, the activation of a warning buzzer that is sounded to signify the completion of a line of results or a graph. The display of graphs is held static by using CALL CHAMOD, after each graph is completed, to put the console back into a numeric mode, and then using CALL ZREALD(A). This simply halts the program until return is entered at the VDU. In this way the graphs can be examined at leisure.

Finally, GMONTE allows three other possibilities (a) production of hard copies, (b) a return to CARLO to rerun the Monte Carlo program, (c) ending the program operation. All these features are illustrated in the GMONTE flow diagram.

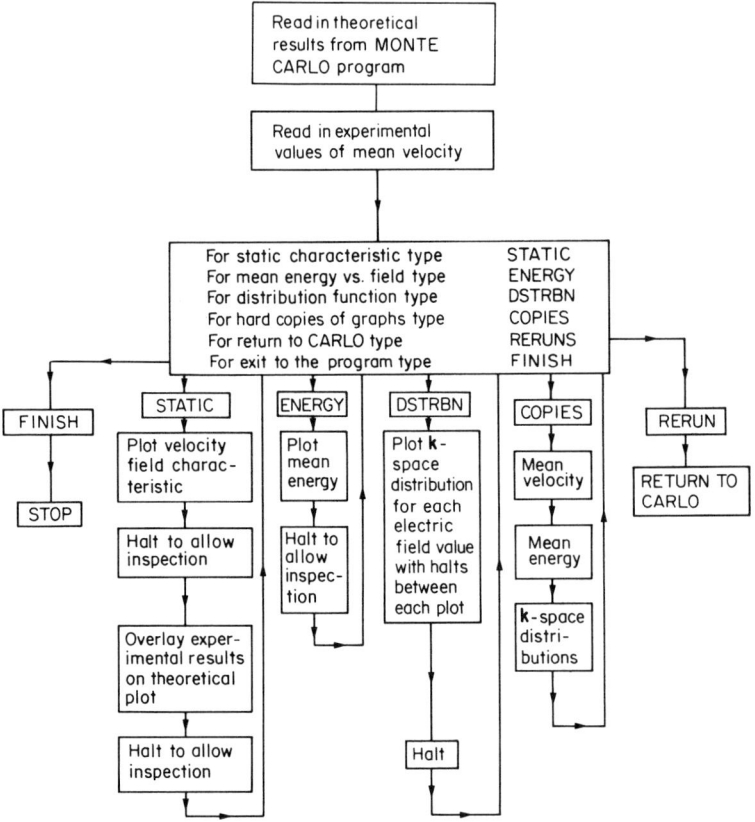

Figure 11. GMONTE flow diagram

6.2 Running the programs

The programs are quite straightforward to run since so many prompts are given to the user. The main program, CARLO, does not contain any graphics and can be run on any kind of terminal. The material data, specifying the semiconductor, are placed, for recall purposes, in the file MDATA5 and comprises, for GaAs, the values given in Table 2. On rerunning the program these data can be used again by typing OLD in response to the question FOR SAME MATERIAL DATA TYPE OLD, FOR NEW TYPE NEW.

An examination of the program listed soon reveals that most of the material parameters are fed in with scaled values since obvious powers of 10, and the free electron mass, are already built into it. Thus, for example, the deformation potential D_n is entered into the computer as 1.0. Furthermore, all energies are expressed in eV, wave numbers in cm^{-1}, and the scattering rates are calculated using cgs units.

The temperature of the solid, the maximum allowed energy and the maximum allowed number of real collisions are entered next. These quantities can be varied at will. However, for the example output, we have taken 300 K, 1 eV, and 2000. Such a temperature corresponds to room-temperature operation, but the temperature variation of the results should be investigated. An E_{max} of 1 eV is just about right, for this example, because we do not intend the electric field to exceed 10 kV/cm. Much higher fields may require a larger E_{max}. Also, too small a value would seriously, and artificially, distort the electron distribution causing the electron to be constantly leaving the 'mesh'. Too large a value would encourage occasional long flights that contribute very little towards the distribution function. Experimentation with E_{max} is, of course, a very interesting exercise. The program has a built-in limit to the number of real collisions (i.e. real scattering events) allowed. This can be changed quite easily. Indeed, it is only there to limit the time of a computer run. Obviously, accuracy is lost this way and the student would be well advised to run the program for a varying number of real collisions to check the rate of convergence. This is entertaining since a lot of good physics is learned through interpreting the shorter runs.

The next section of the data requests asks for the number of electric fields to be used, the values of these fields in units of $kV\,cm^{-1}$, and for which valley plots of the **k**-space distribution function are required. The number of electric fields used is a matter of choice; however, from the point of view of the time available, for experimentation a maximum number equal to seven is suggested. An initial trial run of the field values 1.0, 2.0, 3.0, 4.0, 5.0, 6.0, and 7.0 is recommended. Such a run will reveal, for GaAs, the negative differential mobility, with v against F having a peak near to 3.3 kV/cm. Then the peak area and any other region that looks interesting can be investigated more carefully.

After deciding to look at either the **k**-space distribution function of the central or satellite valley, a k_ρ level is selected. The population density, as measured by the content of the boxes, changes as each k_ρ is chosen and will fall off at very large k_ρ. Also, as a function of k_z, some form of rough Maxwellian distribution should be observed. If it is desired to simulate a simpler one-valley semiconductor then $\Delta = 0$ and the intervalley deformation potentials are set equal to zero.

For GaAs typical experimental values of mean velocity v cm/sec and electric field F kV/cm required by GMONTE are: (1.0, 0.85×10^7), (2.0, 1.63×10^7), (3.0, 2.15×10^7), (4.0, 2.00×10^7), (5.0, 1.80×10^7), (6.0, 1.60×10^7), (7.0, 1.45×10^7).

6.3 Output from the programs

After running CARLO the file MONTEF will contain some tables of final results and MDATA4 will contain the **k**-space histograms for the k_ρ level, and for each value of electric field, requested. In addition MDATA1, MDATA2, and MDATA3 will contain, respectively, a table of electric field values, mean velocities, and mean energies. The contents of MONTEF, for the GaAs data discussed earlier, using the parameters suggested above, are listed at the end of the chapter under the heading Monte Carlo results. In this listing the following definitions are used.

FIELD : electric field value.
VELOCITY : mean electron velocity, in cm/sec, averaged over both valleys.
GAMMA1 : $\Gamma = \Gamma_1$ for central valley.
GAMMA2 : $\Gamma = \Gamma_2$ for satellite valleys.
TIME IN 1 : time in seconds spent central valley.
TIME IN 2 : time in seconds spent satellite valleys.
MN. ENERGY: mean electron energy in eV averaged over both valleys.
SELF : number of virtual collisions.
MESH : number of times electron energy exceeded E_{max}.
VEL1 : average velocity of electrons in central valley.
VEL2 : average velocity of electrons in satellite valleys.
MOB1 : mobility = VEL1/FIELD of electrons in central valley.
MOB2 : mobility = VEL2/FIELD of electrons in satellite valleys.
N1 : fraction of electron population in central valley.
N2 : fraction of electron population in satellite valleys.

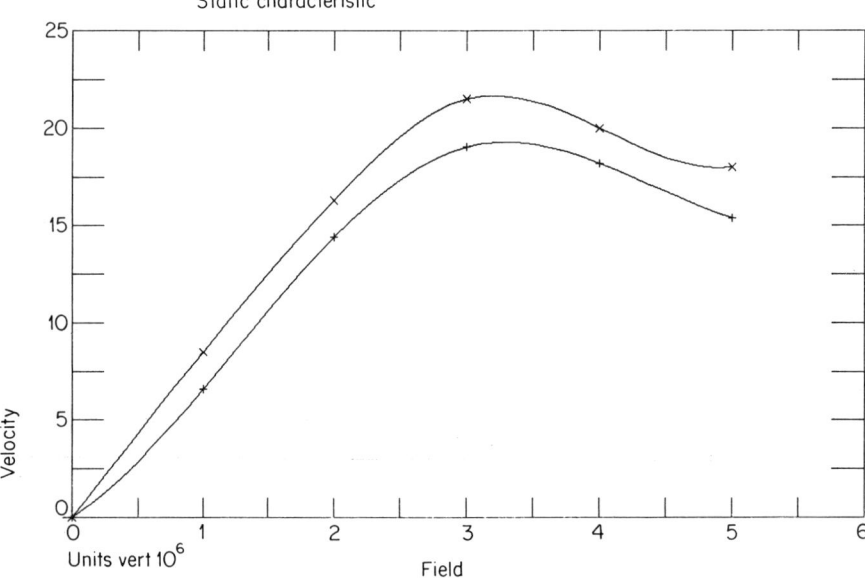

Figure 12. Computer-generated plot of mean velocity of electrons (cm/sec) as a function of applied electric field (kV/cm). + Monte Carlo simulation: 2000 real collisions; × experimental results

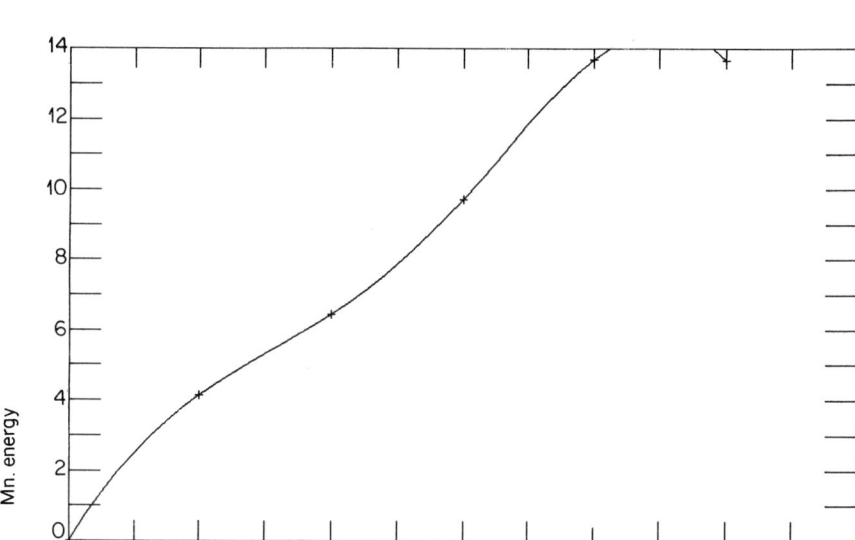

Figure 13. Computer-generated plot of electron mean energy (eV) as a function of applied electric field)kV/cm)

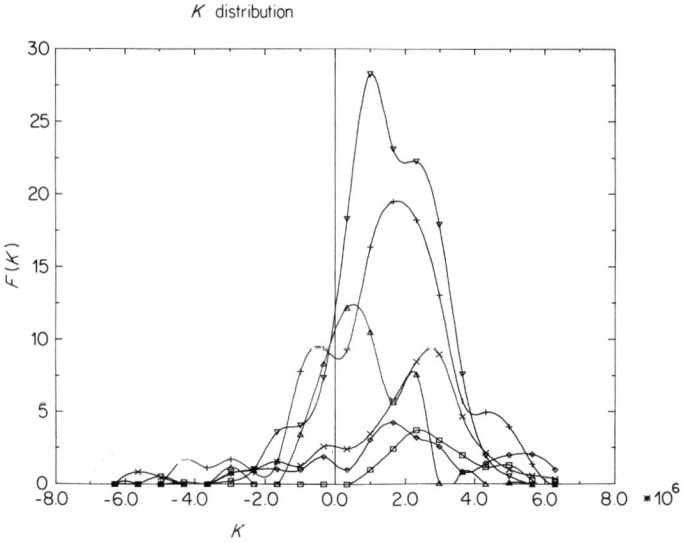

Figure 14. k-space distribution for different electric fields (F kV/cm). $\triangle$: $F=1$, ∇: $F=2$, $+$: $F=3$, $\times$: $F=4$, $\square$: $F=5$, $\diamond$: $F=6$

Figures 12, 13, and 14 show the computer-generated graphs (hard copies) obtained for 2000 real collisions. In Figure 12 it is seen that, even though not too much should be expected, for so few real collisions, the static characteristic agrees surprisingly well with experiment and is very similar to Figure 2. Note that the computer has drawn the graph through the origin even though this point was not calculated. Figure 13, the mean energy plot shows that the mean energy rises with the applied electric field. In this case also the computer has drawn the graph through the origin, but since the calculations began at 1 kV/cm this point should be ignored. In any case, since the electron distribution is Maxwellian, extremely high accuracy is needed before any displacement, caused by low applied electric fields, can be computed with any confidence.

The displaced nature of the Maxwellian distributions, at higher electric fields, is easily seen from an examination of the plots in Figure 14. Note that as the field gets higher the distribution flattens out as more electrons occupy the higher k_z regions.

7. EXERCISES AND PROJECT SUGGESTIONS

In the previous section the use of the computer program is illustrated with data representing GaAs. The number of real collisions is set at 2000, the maximum energy is set at 1 eV, and a temperature of 300 K is used. This set of data, however, leaves a lot of room for manoeuvre. Obviously, if GaAs is going to be studied then the material constants should not be altered too much, but the result of varying intervalley deformation potentials, say, can be studied as a means of understanding that, quite often, these are disposable parameters varied to fit any available experimental results. Varying the number of real collisions is an interesting exercise in order to see how the results converge and to try and see at what point the results begin to make physical sense. The physical sense of the results also allow an interpretation of the variation of E_{max} that was discussed earlier. Both large and small values should be tried.

The variation of temperature between, say, 77 and 500 K is quite an important exercise[9] and is extremely straightforward to do. The results of this exercise should reveal that, for the temperature range 77–500 K, the threshold field, for the onset of negative differential mobility, is rather insensitive to temperature changes, changing only from 3.1 kV/cm to 3.7 kV/cm. On the other hand the negative differential mobility reduces from $\sim 4200 \text{ cm}^2 \text{ V}^{-1} \text{ s}^{-1}$, at 77 K, to $\sim 1000 \text{ cm}^2 \text{ V}^{-1} \text{ s}^{-1}$ at 500 K. The physics of the behaviour of the threshold field can be understood by considering the condition that determines the threshold field F_T. This is

$$\left[\frac{d}{dF}(n_1 v_1)\right]_{F_{Th}} = 0, \qquad (47)$$

where n_1 and v_1 are the electron concentration and velocity in the central valley and the satellite valleys are assumed not to contribute. Equation (47) gives

$$\left[\left|\frac{dn_1}{dF}\right|\right]_{F_{Th}} = \left[\frac{n_1}{V_1}\frac{dV_1}{dF}\right]_{F_{Th}} \quad (48)$$

which contains all that is needed to explain the results provided the variations of n_1 and V_1 with F are obtained.

A more substantial exercise, or project, is to consider the introduction of an ionized impurity scattering mechanism.[9] This is omitted in CARLO so that the results it produces are valid for materials with a doping strength of less than $\sim 10^{15}$ impurities per cm^3 or for more heavily doped materials in electric fields in excess of a few thousand volts per centimetre. The effect of doping is to introduce ionized impurity centres that elastically scatter electrons. The electrons interact with these ions through a screened Coulomb potential and the inverse screening length, in the classical gas approximation, is

$$\Lambda^2 = ne^2/(4\pi\kappa_o k_B T\varepsilon), \quad (49)$$

where T is temperature, κ_0 is the permittivity of free space, ε is an average static dielectric constant, k_B is Boltzmann's constant, e is the electronic charge, and n is the electron density. Typically, even up to electron concentrations of 10^{17} cm^{-3}, Λ does not exceed $\sim 4 \times 10^5$ cm^{-1}.

The transition probability, for ionized impurity scattering, is

$$S(\mathbf{k},\mathbf{k}') = \frac{2\pi}{\hbar}|\langle\mathbf{k}'|V(\mathbf{r})|\mathbf{k}\rangle|^2\delta(E(\mathbf{k}') - E(\mathbf{k})), \quad (50)$$

where elastic scattering is assumed and the matrix element is simply the Fourier transform of a Brooks–Herring potential energy (assuming singly ionized impurities), i.e.

$$V(r) = \frac{-e^2}{4\pi\kappa_0\varepsilon}\frac{\exp(-\Lambda r)}{r} \quad (51)$$

and

$$\langle\mathbf{k}'|V(\mathbf{r})|\mathbf{k}\rangle = \frac{-e^2}{\kappa_0\varepsilon\sqrt{\Omega}}\frac{1}{(K^2 + \Lambda^2)}, \quad (52)$$

where Ω is the volume and $K = |\mathbf{k} - \mathbf{k}'|$. Therefore, including the factor $\Omega/8\pi^3$ and setting the ionized impurity density equal to n,

$$S(\mathbf{k},\mathbf{k}') = \frac{4e^4n}{\hbar(4\pi\kappa_0)^2\varepsilon^2(K^2 + \Lambda^2)^2} \quad (53)$$

and

$$\lambda(\mathbf{k}) = \frac{4\pi e^4 n 2\sqrt{2}m^{*\frac{1}{2}}E^{\frac{1}{2}}}{\hbar^2\varepsilon^2(r\pi\kappa_0)^2\Lambda^2[(\hbar^2\Lambda^2/2m^*) + 4E]}. \quad (54)$$

It is only when $E > 0.005$ eV that the electron distribution function is not basically spherically symmetric.[9] This fact, together with the fact that $\Lambda \leq 4 \times 10^5$ cm^{-1} for densities up to 10^{17} cm^{-3}, implies that $\hbar^2\Lambda^2/8mE \ll 1$. Hence,

$$\lambda(\mathbf{k}) \simeq \frac{\pi e^4 n 2\sqrt{2} m^{*\frac{1}{2}}}{\hbar^2 \varepsilon^2 (4\pi\kappa_0)^2 \Lambda^2 E^{\frac{1}{2}}} = \frac{2\sqrt{2}\pi e^2 m^{*\frac{1}{2}} k_B T}{\hbar^2 \varepsilon (4\pi\kappa_0) E^{\frac{1}{2}}} \qquad (55)$$

It may appear to be a little strange that λ is not a function of n, but it should be remembered that this result is only approximately true. Furthermore, the important point is that the distribution of scattering angles is affected by the size of the electron density. The probability density of scattering an electron through angle θ is proportional to the scattering matrix element and to $\sin\theta$, i.e.

$$p(\theta) = \frac{A \sin\theta}{(2k^2(1-\cos\theta) + \Lambda^2)^2}, \qquad (56)$$

where A is a constant. The application of the Monte Carlo principle[9] leads to

$$\cos\theta = 1 - \frac{2(1-r)}{1 - r(4k^2/\Lambda^2)}, \qquad (57)$$

where r is a uniformly distributed random number. The rest of the procedure is the same as it is for optical polar phonon scattering.

A word of warning is needed here because the ionized impurity scattering rate is an order of magnitude greater than the phonon rates. Scattering by phonons, however, has a much greater influence on the transport properties. Some inefficiency can be encouraged, therefore, if ionized impurity scattering is introduced without any de-weighting procedures. The straightforward method will work, of course, but the user should try to devise methods of self-consistently reducing λ and increasing θ.

Another interesting project would be to consider the combined effect of a constant magnetic field $\mathbf{B}$ and a constant electric field $\mathbf{F}$. The equation of motion in momentum space of an electron of charge e and crystal momentum $\hbar \mathbf{k}$ is

$$\hbar \frac{d\mathbf{k}}{dt} = e\mathbf{F} + \frac{e\hbar}{m^*} \mathbf{k} \times \mathbf{B}. \qquad (58)$$

For fields $\mathbf{F} = (F, 0, 0)$ and $\mathbf{B} = (0, 0, B)$. The solutions of this equation are[7]

$$k_x = \left(\frac{eE}{\omega\hbar} + k_y^{(0)}\right)\sin(\omega t) + k_x^{(0)}\cos(\omega t), \qquad (59)$$

$$k_y = \left(\frac{eE}{\omega\hbar} + k_y^{(0)}\right)\cos(\omega t) - k_x^{(0)}\sin(\omega t) - \frac{eE}{\omega\hbar}, \qquad (60)$$

where $(k_x^{(0)}, k_y^{(0)}$, and $k_z^{(0)})$ is the position of the electron at time $t=0$. These equations show that the electron rotates in a circular orbit in the k_x–k_y plane centred on the point $(0, -m^*E/\hbar B)$, with a frequency $\omega = eB/m^*$ (the cyclotron frequency) and radius $K = [k_x^{(0)2} + (k_y^{(0)} + m^*E/(\hbar B))^2]^{\frac{1}{2}}$.

Using the same accumulation ideas discussed earlier in this chapter, and measuring θ anticlockwise (in a coordinate system $k_x' = K\cos\theta$, $k_y' = K\sin\theta$) from some suitable reference line, say the k_x'-axis, summations over all flights gives[7]

$$v_y = \frac{\hbar}{m^*} \frac{\sum K(\cos\theta_1 - \cos\theta_2)}{\sum(\theta_2 - \theta_1)} - \frac{E}{B} \qquad (61)$$

$$v_x = \frac{\hbar}{m^*} \frac{\sum K(\sin\theta_2 - \sin\theta_1)}{\sum(\theta_2 - \theta_1)} \qquad (62)$$

where θ_1 is the initial angle and θ_2 is the final angle on any flight. These formulae arise because integrations over angular displacements are used now instead of over k_z as in equation (43), i.e. $\int_{\theta_1}^{\theta_2} d\theta K \cos\theta$, etc. The mean electron energy is therefore

$$E = \frac{\hbar^2}{2m^*} \frac{\sum \int_{\theta_1}^{\theta_2} \left\{ k_z^2 + k_x'^2 + \left(k_y' - \frac{eE}{\omega\hbar}\right)^2 \right\} d\theta}{\sum \int_{\theta_1}^{\theta_2} d\theta}. \qquad (63)$$

Therefore

$$E = \frac{\hbar^2}{2m^*} \frac{\sum \int_{\theta_1}^{\theta_2} \left\{ k_z^2 + K^2 + \left(\frac{eE}{\omega\hbar}\right)^2 - 2K\sin\theta\left(\frac{eE}{\omega\hbar}\right) \right\} d\theta}{\sum(\theta_2 - \theta_1)}$$

$$= \frac{\hbar^2}{2m^*\sum(\theta_2-\theta_1)} \left\{ \sum(k_z^2 + K^2)(\theta_2 - \theta_1) \right.$$
$$\left. - \frac{2eE}{\omega\hbar} \sum K(\cos\theta_1 - \cos\theta_2) \right\} + \tfrac{1}{2}m^*\left(\frac{E}{B}\right)^2. \qquad (64)$$

The mean value of a quantity A, averaged over two valleys and setting $S = \sum(\theta_2 - \theta_1)$, is

$$\langle A \rangle = \frac{m_1 S_1 \langle A \rangle_1 + m_2 S_2 \langle A \rangle_2}{m_1 S_1 + m_2 S_2}. \qquad (65)$$

The determination of the distribution of scattering angles is more difficult than in the zero magnetic field case. A scattering event taking the electron

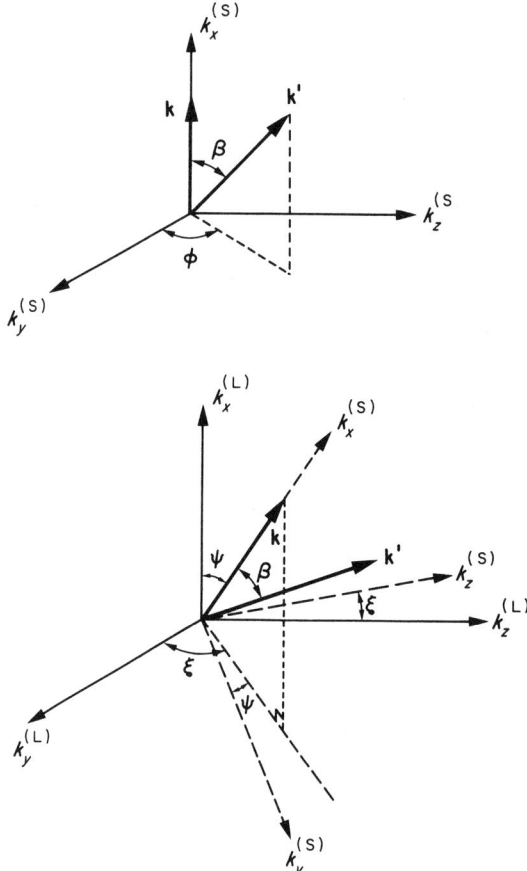

Figure 15. Scattering geometry of an electron in crossed electric and magnetic fields

from **k** to **k'** is shown in Figure 15. The components of **k'** are, in what we will call the scattering frame,

$$
\begin{aligned}
k_x^{(S)} &= k' \cos \beta, \\
k_y^{(S)} &= k' \sin \beta \cos \phi, \\
k_z^{(S)} &= k' \sin \beta \cos \phi.
\end{aligned}
\quad (66)
$$

If a laboratory frame is now defined then, as ia really seen from Figure 15, the vector $\mathbf{k}^{(L)}$ is related to $k^{(S)}$ through

$$
\begin{pmatrix} k_x^{(L)} \\ k_y^{(L)} \\ k_z^{(L)} \end{pmatrix} = \begin{pmatrix} \cos \psi & -\sin \psi & 0 \\ \sin \psi \cos \xi & \cos \psi \cos \xi & -\sin \xi \\ \sin \psi \sin \xi & \cos \psi \sin \xi & \cos \xi \end{pmatrix} \begin{pmatrix} k_x^{(S)} \\ k_y^{(S)} \\ k_z^{(S)} \end{pmatrix}, \quad (67)
$$

where $k_x^{(L)}$, $k_y^{(L)}$, and $k_z^{(L)}$ are the **k**-space coordinates of the electron after a scattering event and $k' = \sqrt{2mE'}/\hbar$; $E' = E_f \pm \hbar\omega$, where E_f is the electron energy at the end of a flight.

The angle β is found in the usual way and the angles ξ, ψ are found as follows. At the end of a flight, and before a scattering event occurs,

$$k_x^{(f)} = k \cos \psi,$$
$$k_y^{(f)} = k \sin \psi \cos \xi, \tag{68}$$
$$k_z^{(f)} = k \sin \psi \sin \xi$$

so that

$$\cos \psi = \frac{k_x^{(f)}}{k}, \quad \sin \xi = \frac{k_z^{(f)}}{\sqrt{k_x^{(f)2} + k_z^{(f)2}}}, \tag{69}$$

ϕ is a random angle and is part of the polar scattering. It is therefore determined in the same way, i.e. $\phi = 2\pi r$.

Acoustic and intervalley scattering must also be treated in three dimensions, but the results are quite straightforward. If the polar coordinates of an electron in **k**-space, after scattering, are θ, ϕ then $\cos \theta = 1 - 2r_1$, $\phi = 2\pi r_2$, where r_1 and r_2 are two random uniformly distributed numbers.

The student with all this information should now be in a position to modify the computer program given here and make a study of magnetic field effects, bearing in mind that the **k**-space histogram is no longer relevant since the system now has cylindrical symmetry. Many interesting quantities can be obtained, but perhaps the most interesting is that a complete

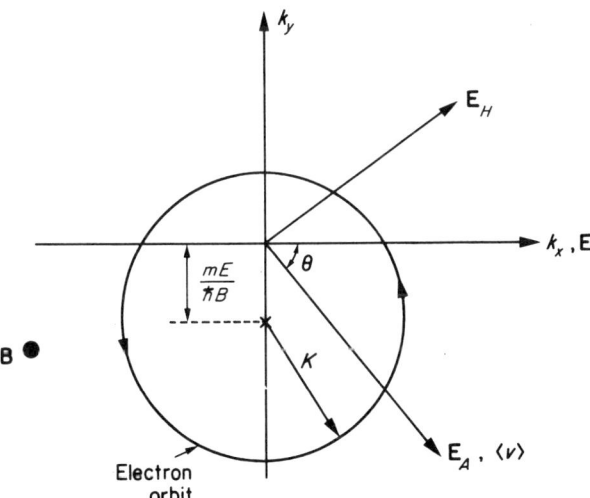

Figure 16. Schematic representation of the electron trajectory in k-space under the influence of crossed electric (E) and magnetic fields (B)

simulation of the Hall effect is possible. In Figure 16 it is seen that $E \cos \theta$ is interpreted as the applied field E_A and $E \sin \theta$ as the Hall field E_H. The Hall coefficient R_H is also given directly, the general definition of which is

$$R_H = \frac{E_H}{BJ}, \tag{70}$$

where J is the current density given by nev, n being the electron concentration. R_H is related to a scattering factor r through

$$R_H = \frac{r}{ne}, \tag{71}$$

so that

$$r = \frac{E_H}{Bv} = \frac{\mu_H(B)}{\mu_D(B)}, \tag{72}$$

where $\mu_D(B) = v/E_A$ is the drift mobility and $\mu_H(B) = E_H/BE_A$ is the Hall mobility.

These are some of the short-term and long-term exercises, and projects, that can be attempted. Obviously there are many others, such as the inclusion of non-parabolity[6] and accounting for the Pauli exclusion principle as the electron gas becomes degenerate.[10] These are not discussed here but the answers can be found by studying the references at the end of the chapter.

REFERENCES

1. J. B. Gunn, Sol. State Comm., **1,** 88 (1963).
2. B. K. Ridley and T. B. Watkins, Proc. Phys. Soc., **78,** 293 (1961).
3. C. Hilsum, Proc. I.R.E., **50,** 185 (1962).
4. P. J. Bulman, G. S. Hobson, and B. C. Taylor, *Transferred Electron Devices* (Academic Press, London and New York, 1972).
5. A. D. Boardman, W. Fawcett, and H. D. Rees, Sol. State Comm., **6,** 305 (1968).
6. W. Fawcett, A. D. Boardman, and S. Swain, J. Phys. Chem. Sol., **31,** 1963 (1970).
7. A. D. Boardman, W. Fawcett, and J. G. Ruch, Phys. Stat. Sol. (a), **4,** 133 (1971).
8. G. C. Aers, A. D. Boardman, and E. D. Isaac, Phys. Stat. Sol. (a), **36,** 357 (1976).
9. J. G. Ruch and W. Fawcett, J. App. Phys., **41,** 3843 (1970).
10. S. Bosi and C. Jacaboni, J. Phys. C., **9,** 315 (1976).

COMPUTER SIMULATION USING MONTE CARLO METHODS

MONTE CARLO RESULTS

GAMMA1= 0.7212E 14   GAMMA2= 0.7921E 14
TEMP=0.3000E 03 REALCOLSNS= 2000 MAXENERGY(EV)=1.00

FIELD	VELOCITY	TIME IN 1	TIME IN 2	MN.ENERGY	SELF	MESH
0.100E 01	0.659E 07	0.458E-09	0.000E 00	0.413E-01	30663	0
0.200E 01	0.144E 08	0.371E-09	0.864E-11	0.644E-01	25469	0
0.300E 01	0.190E 08	0.272E-09	0.242E-10	0.972E-01	19376	0
0.400E 01	0.182E 08	0.149E-09	0.502E-10	0.137E 00	12706	0
0.500E 01	0.154E 08	0.862E-10	0.615E-10	0.137E 00	8983	0
0.600E 01	0.122E 08	0.651E-10	0.679E-10	0.125E 00	8065	0

FIELD	VEL1	VEL2	MOB1	MOB2	N1	N2
0.100E 01	0.659E 07	0.000E 00	0.659E 04	0.000E 00	0.100E 01	0.615E-01
0.200E 01	0.147E 08	0.269E 07	0.734E 04	0.135E 04	0.977E 00	0.227E-01
0.300E 01	0.207E 08	0.116E 0	0.691E 04	0.385E 0	0.918E 00	0.817E-01
0.400E 01	0.240E 08	0.107E 07	0.600E 04	0.268E 03	0.748E 00	0.252E 00
0.500E 01	0.250E 08	0.194E 07	0.500E 04	0.388E 03	0.584E 00	0.416E 00
0.600E 01	0.236E 08	0.128E 07	0.394E 04	0.214E 03	0.490E 00	0.510E 00

```
      GRAPH PLOTTING:MONTE CARLO PROBLEM
C
C    GRAPH PLOTTING PROGRAM FOR MONTE CARLO CALCULATION
C    PLOTS GRAPHS OF DRIFT VELOCITY,MEAN ENERGY AND
C    K SPACE DISTRIBUTION FUNCTIONS ON VDU AND GRAPH PLOTTER
C
      REAL MNE1
      DIMENSION IT(21),F(10),VEL(10),MNE1(10),IT2(17)
     1   ,YINT(10,20),VEL1(10),IT3(13),XVAL(20),YVAL(20)
      COMMON/SCDTA/RV(160)
      DATA IT/21,'STATIC CHARACTERISTIC',5,'FIELD',8,'VELOCITY'/
      DATA IT2/11,'MEAN ENERGY',5,'FIELD',9,'MN ENERGY'/
      DATA IT3/14,'K DISTRIBUTION',1,'K',4,'F(K)'/
      CALL SEARCH(1,'GRAPPP',15)
C
C    READS TEST LABELS FOR CHOICE OF ACTION
C    FILE CONTAINS.
C                       STATIC
C                       ENERGY
C                       DSTRBN
C                       RERUNS
C                       COPIES
C
      READ(19,343)Y1
  343 FORMAT(A3)
      READ(19,343)Y2
      READ(19,343)Y3
      READ(19,343)Y4
      READ(19,343)Y5
      CALL SEARCH(4,0,15)
C
C    READ DATA FILES FROM PROGRAM CARLO FOR GRAPHICAL OUTPUT
C
      CALL SEARCH(1,'MDATA1',5)
      CALL SEARCH(1,'MDATA2',6)
      CALL SEARCH(1,'MDATA3',7)
      READ(9,801)NFG
  801 FORMAT(I2)
C
C    CALCULATES AXIS SIZES FOR GRAPHS
C
      NF=NFG
      FIX1=0.0
      FIX2=0.0
      CALL PAGE
      DO 100 I=1,NF
      READ(9,800)F(I)
      READ(10,800)VEL(I)
      IF(VEL(I).GT.FIX1) FIX1=VEL(I)
      READ(11,800)MNE1(I)
      IF(MNE1(I).GT.FIX2) FIX2=MNE1(I)
  800 FORMAT(E12.4)
C
C    INPUT OF EXPERIMENTAL VALUES FOR DRIFT VELOCITY TO
C    COMPARE WITH CALCULATED CURVE
C
      WRITE(1,910)F(I)
```

```
GRAPH PLOTTING:MONTE CARLO PROBLEM
  910 FORMAT(///,34HTYPE EXPERIMENTAL RESULT AT FIELD=,F4.2,2HKV)
      CALL ZREALD(VEL1(I))
      IF(VEL1(I).GT.FIX1) FIX1=VEL1(I)
  100 CONTINUE
      FIXX=F(NF)
C
C   INCLUDE ORIGIN AS POINT IN GRAPH
C
      NF1=NF+2
      DO 590 J=1,NF
      I=NF1-J
      K=NF-J+1
      VEL(I)=VEL(K)
      VEL1(I)=VEL1(K)
      MNE1(I)=MNE1(K)
      F(I)=F(K)
  590 CONTINUE
      F(1)=0.0
      VEL(1)=0.0
      VEL1(1)=0.0
      MNE1(1)=0.0
      NF=NF+1
C
C   CLOSE DATA FILES
C
      CALL SEARCH(4,0,5)
      CALL SEARCH(4,0,6)
      CALL SEARCH(4,0,7)
  888 CALL PAGE
C
C   MAKE SELECTION OF GRAPHS REQUIRED
C
      WRITE(1,288)
  288 FORMAT(///,10X,38HFOR STATIC CHARACTERISTIC TYPE STATIC///
     1,10X,38HFOR MEAN ENERGY VS. FIELD TYPE ENERGY///
     1,10X,38HFOR DISTRIBUTION FUNCTION TYPE DSTRBN///
     1,10X,38HFOR HARD COPIES OF GRAPHS TYPE COPIES///
     1,10X,38HTO RUN MONTE CARLO PROGRAM TYPE RERUNS///
     1,10X,38HFOR AN EXIT TO THE PROGRAM TYPE FINISH///)
      READ(1,343)X
      IF(X.EQ.Y1)GOTO 887
      IF(X.EQ.Y2)GOTO 886
      IF(X.EQ.Y3)GOTO 885
      IF(X.EQ.Y4)GOTO 883
      IF(X.EQ.Y5)GOTO 700
      GOTO 884
  700 CALL PAGE
C
C   CALLS GRAPH PLOTTER FOR HARD COPIES OF ALL GRAPHS.
C
C   "CALL A4" COMMAND SPECIFIES AN A4 FORMAT FOR GRAPHS.
C
      CALL A4
C
C   PLOTS DRIFT VELOCITY/FIELD GRAPHS
C
```

GRAPH PLOTTING:MONTE CARLO PROBLEM

```
      CALL FIXAXS(0.0,FIXX,0.0,FIX1)
      CALL FGPLT(F,VEL,NF,3,3,1,0,IT,6)
      CALL FGPLT(F,VEL1,NF,3,4,1,1,IT,6)
C
C   SIGNALS COMPLETION TO VDU AND SOUNDS BUZZER
C
      WRITE(1,701)
  701 FORMAT(////,25X,23HVELOCITY PLOT COMPLETED,/)
      CALL T10U(7)
C
C   PLOTS MEAN ENERGY/FIELD GRAPH
C
      CALL FIXAXS(0.0,FIXX,0.0,FIX2)
      CALL FGPLT(F,MNE1,NF,3,3,1,0,IT2,6)
C
C   SIGNALS COMPLETION TO VDU AND SOUNDS BUZZER.
C
      WRITE(1,702)
  702 FORMAT(/,25X,26HMEAN ENERGY PLOT COMPLETED,/)
      CALL T10U(7)
      KK=1
C
C   GOES TO LABEL 889 TO CALCULATE GRAPH AXES FOR
C   DISTRIBUTION PLOT KK=1 ENABLES RETURN TO LABEL 703.
C
      GOTO 889
  703 CALL SEARCH(4,0,8)
C
C   PLOTS K SPACE DISTRIBUTION FUNCTION AT EACH FIELD
C
      DO 705 I=1,NF
      DO 704 K=1,20
      YVAL(K)=YINT(I,K)
  704 CONTINUE
      JG=0
      IF(I.NE.1)JG=1
      CALL FGPLT(XVAL,YVAL,20,3,I,1,JG,IT3,6)
C
C   SIGNALS COMPLETION TO VDU AND SOUNDS BUZZER
C
      WRITE(1,706)I
  706 FORMAT(/,25X,19HDISTRIBUTION CURVE ,I2,10H COMPLETED,/)
      CALL T10U(7)
  705 CONTINUE
      GOTO 888
C
C   PLOTS VELOCITY/FIELD GRAPHS ON VDU
C
C   CALCULATED CURVE
C
  887 CALL T4010
      CALL FIXAXS(0.0,FIXX,0.0,FIX1)
      CALL FGPLT(F,VEL,NF,3,3,1,0,IT,5)
C
C   SOUNDS BUZZER AND WAITS FOR CARRIAGE RETURN
C
```

```
GRAPH PLOTTING:MONTE CARLO PROBLEM
      CALL CHAMOD
      CALL T10U(7)
      CALL ZREALD(A)
C
C   PLOTS EXPERIMENTAL VELOCITY/FIELD CURVE ON SAME GRAPH
C
      CALL FGPLT(F,VEL1,NF,3,4,1,1,IT,S)
C
C   SOUNDS BUZZER AND WAITS FOR CARRIAGE RETURN
C
      CALL CHAMOD
      CALL T10U(7)
      CALL ZREALD(A)
      GOTO 888
C
C   PLOTS MEAN ENERGY/FIELD CURVE ON VDU
C
  886 CALL FIXAXS(0.0,FIXX,0.0,FIX2)
      CALL FGPLT(F,MNE1,NF,3,3,1,0,IT2,S)
C
C   SOUNDS BUZZER AND WAITS FOR CARRIAGE RETURN
C
      CALL CHAMOD
      CALL T10U(7)
      CALL ZREALD(A)
      GOTO 888
C
C   CALCULATES AXIS SIZE FOR DISTRIBUTION FUNCTION PLOT
C
  885 KK=0
  889 CALL SEARCH(1,'MDATA4',8)
      NF=NFG
      DO 38 I=1,20
      READ(12,31)XVAL(I)
   31 FORMAT(E12.4)
   38 CONTINUE
      FIX3=0.0
      DO 56 I=1,NF
      DO 56 K=1,20
      READ(12,32)YINT(I,K)
      IF(YINT(I,K).GT.FIX3) FIX3=YINT(I,K)
   32 FORMAT(E12.4)
   56 CONTINUE
      CALL FIXAXS(XVAL(1),XVAL(20),0.0,FIX3)
      IF(KK.EQ.1)GOTO 703
      DO 37 I=1,NF
      DO 36 K=1,20
      YVAL(K)=YINT(I,K)
   36 CONTINUE
      JG=0
      IF(I.NE.1)JG=1
C
C   PLOTS DISTRIBUTION FUNCTIONS ON ONE GRAPH ON VDU.
C   AFTER EACH CURVE,SOUNDS BUZZER AND WAITS FOR
C   CARRIAGE RETURN
C
```

```
GRAPH PLOTTING:MONTE CARLO PROBLEM
      CALL FGPLT(XVAL,YVAL,20,3,I,1,JG,IT3,5)
      CALL CHAMOD
      CALL T10U(7)
      CALL ZREALD(A)
  37  CONTINUE
C
C   CLOSE DISTRIBUTION DATA FILE
C
      CALL SEARCH(4,0,8)
      GOTO 888
C
C   RESTARTS MONTE CARLO CALCULATION.
C
 883  CALL RESUME('*CARLO')
C
C         END PROGRAM HARD COPY GRAPHICAL OUTPUT STARTS AT THIS POINT
C
 884  CALL DEVFIN
      STOP
      END
```

COMPUTER SIMULATION USING MONTE CARLO METHODS

```
MONTE CARLO PROGRAM
C
C     A MONTE CARLO CALCULATION OF THE DRIFT VELOCITY,MEAN ENERGY
C     AND DISTRIBUTION FUNCTION OF HOT ELECTRONS IN A ONE OR TWO
C     VALLEY SEMICONDUCTOR.SCATTERING PROCESSES INCLUDE ACOUSTIC
C     ,POLAR OPTICAL AND INTERVALLEY PHONON SCATTERING.
C
      REAL KB,KRHO,KZF,KZTOT,KZI,KT,L,KZ,NPR1,NPR2,MNE,M
     1  ,NO,NE,NI,KMESH,KKI,KKZF,KKZI,KM,KR
      INTEGER NF,VV,VR,TT,V,SR,SS,K,GMAX,NRC
      DIMENSION KR(20),F(10),GAMMA(2),L(10),VTIM(2),EF(10)
     1  ,SCATT(10),VEL(10)
     1  ,ET(2),KZ(2),V1(10),V2(10),U1(10),U2(10),NPR1(10)
     2  ,NPR2(10),DK(2),KMESH(2,21,21)
      DATA Y/4HNEW /
C
C     OPEN DATA STORAGE FILES
C
      CALL SEARCH(2,'MONTEF',5)
      CALL SEARCH(2,'MDATA1',6)
      CALL SEARCH(2,'MDATA2',7)
      CALL SEARCH(2,'MDATA3',8)
      CALL SEARCH(2,'MDATA4',9)
      CALL SEARCH(3,'MDATA5',10)
C
C     FUNDAMENTAL CONSTANTS
C
      H=1.05459
      E=1.60219
      C=4.803213
      KB=1.38062
      M=9.109136
C
C     DATA RECALL OPTION OR NEW DATA INPUT
C
      CALL PAGE
      WRITE(1,47)
47    FORMAT(//,10X,43HMONTE CARLO CALCULATION OF DRIFT VELOCITIES,/
     1   ,10X,43H-------------------------------------------,//
     1   ,48HFOR SAME MATERIAL DATA TYPE OLD,FOR NEW TYPE NEW)
      READ(1,303)X
303   FORMAT(A3)
      IF(X.EQ.Y) GOTO 49
      READ(14,789)RHE
      READ(14,789)S
      READ(14,789)R1
      READ(14,789)R2
      READ(14,789)WO
      READ(14,789)WE
      READ(14,789)WI
      READ(14,789)THA
      READ(14,789)THE
      READ(14,789)THI
      READ(14,789)D
      READ(14,789)EM1
      READ(14,789)EM2
789   FORMAT(E10.4)
```

```
MONTE CARLO PROGRAM
      GOTO 46
49    WRITE(1,41)
41    FORMAT(1H ,30HTYPE MATERIAL DENSITY(GM.CM-3))
      CALL ZREALD(RHE)
      IF(RHE.EQ.0.0)RHE=1.0
      WRITE(1,42)
42    FORMAT(1H ,34HTYPE VELOCITY OF SOUND IN MATERIAL,
     1 23H(UNITS OF 10**5 CM.S-1))
      CALL ZREALD(S)
      IF(S.EQ.0.0)S=1.0
      WRITE(1,43)
43    FORMAT(1H ,39HTYPE HIGH FREQUENCY DIELECTRIC CONSTANT)
      CALL ZREALD(R1)
      WRITE(1,44)
44    FORMAT(1H ,38HTYPE LOW FREQUENCY DIELECTRIC CONSTANT)
      CALL ZREALD(R2)
      WRITE(1,1)
1     FORMAT(1H ,29HTYPE OPTICAL PHONON FREQUENCY,
     1 25H(UNITS 10**13 RAD.SEC.-1))
      CALL ZREALD(WO)
      WRITE(1,2)
2     FORMAT(1H ,44HTYPE EQUIVALENT INTERVALLEY PHONON FREQUENCY,
     1 28H(UNITS OF 10**13 RAD.SEC.-1))
      CALL ZREALD(WE)
      IF(WE.EQ.0.0)WE=1.0
      WRITE(1,3)
3     FORMAT(1H ,33HTYPE INTERVALLEY PHONON FREQUENCY,
     1 28H(UNITS OF 10**13 RAD.SEC.-1))
      CALL ZREALD(WI)
      IF(WI.EQ.0.0)WI=1.0
      WRITE(1,4)
4     FORMAT(1H ,39HTYPE ACOUSTIC DEFORMATION POTENTIAL(EV))
      CALL ZREALD(THA)
      WRITE(1,5)
5     FORMAT(1H ,45HTYPE EQUIV. INTERVALLEY DEFORMATION POTENTIAL,
     1 24H(UNITS OF 10**9 EV.CM-1))
      CALL ZREALD(THE)
      WRITE(1,6)
6     FORMAT(1H ,38HTYPE INTERVALLEY DEFORMATION POTENTIAL,
     1 24H(UNITS OF 10**9 EV.CM-1))
      CALL ZREALD(THI)
      WRITE(1,7)
7     FORMAT(1H ,26HTYPE VALLEY SEPARATION(EV))
      CALL ZREALD(D)
      WRITE(1,8)
8     FORMAT(1H ,34HTYPE CENTRAL VALLEY EFFECTIVE MASS)
      CALL ZREALD(EM1)
      WRITE(1,9)
9     FORMAT(1H ,36HTYPE SATELLITE VALLEY EFFECTIVE MASS)
      CALL ZREALD(EM2)
      WRITE(14,689)RHE
      WRITE(14,689)S
      WRITE(14,689)R1
      WRITE(14,689)R2
      WRITE(14,689)WO
      WRITE(14,689)WE
```

```
       MONTE CARLO PROGRAM
              WRITE(14,689)WI
              WRITE(14,689)THA
              WRITE(14,689)THE
              WRITE(14,689)THI
              WRITE(14,689)D
              WRITE(14,689)EM1
              WRITE(14,689)EM2
       689    FORMAT(E10.4)
       C
       C      FINAL DATA INPUT
       C
       46     CALL PAGE
              WRITE(1,10)
       10     FORMAT(1H ,32HTYPE TEMPERATURE(DEGREES KELVIN))
              CALL ZREALD(T)
              WRITE(1,21)
       21     FORMAT(1H ,23HTYPE MAXIMUM ENERGY(EV))
              CALL ZREALD(EMAX)
              WRITE(1,11)
       11     FORMAT(1H ,30HTYPE NUMBER OF REAL COLLISIONS)
       18      CALL ZINTRD(NRC)
              IF(THI.NE.0.0) GOTO 750
              IF(NRC.LE.3000) GOTO 16
              WRITE(1,710)
       710    FORMAT(1H ,45HMAXIMUM COLLISIONS IN ONE VALLEY =3000:RETYPE)
              GOTO 18
       750    IF(NRC.LE.2000) GOTO 16
              WRITE(1,17)
       17     FORMAT(1H ,46HMAXIMUM COLLISIONS IN TWO VALLEYS =2000:RETYPE)
              GOTO 18
       16     WRITE(1,12)
       12     FORMAT(1H ,30HTYPE NUMBER OF ELECTRIC FIELDS)
              CALL ZINTRD(NF)
              WRITE(10,801)NF
       801    FORMAT(I2)
              DO 30 I=1,NF
              WRITE(1,13)I
       13     FORMAT(1H ,10HTYPE FIELD,I2,10H (KV.CM-1))
              CALL ZREALD(F(I))
       30     CONTINUE
              WRITE(1,23)
       23     FORMAT(1H ,44HTYPE VALLEY FOR DISTRIBUTION FUNCTION:1 OR 2)
              CALL ZINTRD(VV)
              WRITE(1,22)
       22     FORMAT(1H ,49HTYPE DISTANCE FROM KZ AXIS OF DSTRIBUTION:1 TO 21)
              CALL ZINTRD(VR)
              CALL PAGE
       C
       C      CALCULATE PHONON FREQUENCIES AND OCCUPATION RATIOS
       C
              HWO=H*WO/(E*100.0)
              HWI=H*WI/(E*100.0)
              HWE=H*WE/(E*100.0)
              IF(WO.NE.0.0) GOTO 909
              NO=0.0
              GOTO 908
```

```
      MONTE CARLO PROGRAM
909   NO=1/(EXP(WO*H/(KB*T)*100.0)-1)
908   NI=1/(EXP(WI*H/(KB*T)*100.0)-1)
      NE=1/(EXP(WE*H/(KB*T)*100.0)-1)
C
C     CONSTANTS FOR SCATTERING RATES
C
      C1=1.0E+12*C*C*SQRT(M)*WO*(1/R1-1/R2)*(NO+1)/(1.4142*H*SQRT(E))
      C2=C1*NO/(NO+1)
      C3=1.0E+10*(2*M)**1.5*KB*T*THA*THA*E*E*SQRT(E)/(4.0*3.142*RHE
     1 *S*S*H*H*H*H)
      C4=2.0E+14*M**1.5*THE*THE*E*E*(NE+1)*SQRT(E) (1.4142*3.142
     2 *RHE*WE*H*H*H)
      C5=C4*NE/(NE+1)
      C6=1.0E+14*(EM1*M)**1.5*THI*THI*E*E*(NI+1)*SQRT(E)/(1.4142*3.142
     3 *RHE*WI*H*H*H)
      C7=3.0E+14*(EM2*M)**1.5*THI*THI*E*E*(NI+1)*SQRT(E)/(1.4142
     4 *3.142*RHE*WI*H*H*H)
      C8=C7*NI/(NI+1)
      C9=C6*NI/(NI+1)
C
C     CALCULATE K SPACE MESH ELEMENT FOR BOTH VALLEYS
C
      DK(1)=1.0E+7*SQRT(2*EM1*M*EMAX*E)/(H*20.0)
      DK(2)=1.0E+7*SQRT(2*EM2*M*EMAX*E)/(H*20.0)
C
C     CALCULATE VALUES OF KZ AT CENTRE OF EACH MESH ELEMENT
C     IN CHOSEN VALLEY AND WRITE TO FILE
C
      KZI=-DK(VV)*9.5
      DO 805 LL=1,20
      WRITE(13,807)KZI
807   FORMAT(E12.4)
      KZI=KZI+DK(VV)
805   CONTINUE
C
C     SET PARAMETERS FOR CENTRAL VALLEY,THEN CALCULATE THE
C     TOTAL SCATTERING RATE FOR REAL PROCESSES (R) FOR A NUMBER
C     OF ENERGIES UP TO THE MESH SIZE.STORE MAXIMUM VALUE OF R
C     IN GAMMA(1) TO CALCULATE PSEUDO (SELF) SCATTERING RATE.
C     TT=1 ENABLES PROGRAM TO RETURN TO LABEL 40
C
      TT=1
      EM=EM1
      V=1
31    GAMMA(V)=0.0
      EI=0.0
      J=1
35    EI=EI+EMAX/20.0
      GOTO 100
40    R=0.0
      DO 50 I=1,10
      R=R+L(I)
50    CONTINUE
      IF(R.GT.GAMMA(V)) GAMMA(V)=R
      J=J+1
      IF(J.NE.21) GOTO 35
```

COMPUTER SIMULATION USING MONTE CARLO METHODS 403

```
MONTE CARLO PROGRAM
C
C      SET PARAMETERS FOR SATELLITE VALLEY AND REPEAT PROCESS TO
C      OBTAIN GAMMA(2).
C
       IF(V.EQ.2) GOTO 71
       EM=EM2
        V=2
       GOTO 31
71     WRITE( 1,75)GAMMA(1),GAMMA(2)
       WRITE(9,75)GAMMA(1),GAMMA(2)
75     FORMAT(1H ,8HGAMMA1= ,E10.4,10H  GAMMA2= ,E10.4)
       WRITE(1,72)T,NRC,EMAX
       WRITE(9,72)T,NRC,EMAX
72     FORMAT(1H ,5HTEMP=,E10.4,12H REALCOLSNS=,I6,15H MAXENERGY(EV)=
      3  ,F4.2)
       WRITE(1,333)
       WRITE(9,333)
333    FORMAT(2X,5HFIELD,4X,8HVELOCITY,1X,9HTIME IN 1,1X
      2  ,9HTIME IN 2,1X,9HMN.ENERGY,4X,4HSELF,4X,4HMESH)
C
C      SET MESH REGISTERS TO ZERO. AND PLACE ELECTRON AT
C      STARTING POINT IN MESH.TT=0 FOR ITERATIVE PROCESS.
C
       TT=0
       J=1
80     V=1
       EM=EM1
       PSI=0.0
       KRHO=0.0
       KZF=1.0E+6
       SR=0
       SS=0
       GMAX=0
       EFIN=H*H*KZF*KZF*1.0E-14/(E*2*EM*M)
       ETOT=0.0
       KZTOT=0.0
       VTIM(1)=0.0
       VTIM(2)=0.0
       KZ(1)=0.0
       KZ(2)=0.0
       ET(1)=0.0
       ET(2)=0.0
       DO 999 K=1,20
       KMESH(VV,VR,K)=0.0
999    CONTINUE
C
C      IF NO. OF REAL PROCESSES EQUALS CHOSEN VALUE,END ITERATION
C      AND GOTO FINAL CALCULATION.
C
90     IF(SR.EQ.NRC) GOTO 470
C
C      CALL RANDOM NUMBER(NOT=0) AND CALCULATE TIME OF FLIGHT UNDER
C      ELECTRIC FIELD AND NEW POSITION OF ELECTRON IN K SPACE
C
       R=RND(B)
       IF(R.EQ.0.0) R=1.0E-20
```

MONTE CARLO PROGRAM
```
      TIME=ALOG(1/R)/GAMMA(V)
      KZI=KZF+(TIME*E*F(J)*1.0E+18)/H
      KT=SQRT(KRHO*KRHO+KZI*KZI)
      EI=H*H*KT*KT*1.0E-14/(E*2*EM*M)
C
C     IF ELECTRON LEAVES MESH PLACE IT ON EDGE OF MESH AND
C     REGISTER OCCURRENCE IN COUNTER GMAX.
C
      IF(EI.LE.EMAX) GOTO 95
      GMAX=GMAX+1
      EI=EMAX
      KT=1.0E+7*SQRT(2*EM*M*EMAX*E)/H
      IF(KZI.GT.0.0)GOTO 94
      KZI=-SQRT(ABS(KT*KT-KRHO*KRHO))
      GOTO 95
   94 KZI=SQRT(ABS(KT*KT-KRHO*KRHO))
C
C     STORE FLIGHT TIME IN TOTAL TIME REGISTER FOR APPROPRIATE
C     VALLEY,THEN REGISTER PASSAGE OF ELECTRON THROUGH ELEMENTS
C     OF K SPACE MESH.
C
   95 VTIM(V)=VTIM(V)+TIME
      KKRHO=KRHO/DK(V)+1
      IF(KKRHO.NE.VR) GOTO 890
      KKZI=KZI/DK(V)+10
      KKZF=KZF/DK(V)+10
      KF=KKZF
      KI=KKZI
      KF1=KF+1
      KI1=KI+1
      IF(KI.EQ.KF) GOTO 880
      DO 870 LL=KF1,KI1
      KMESH(V,KKRHO,LL)=KMESH(V,KKRHO,LL)+1.0
  870 CONTINUE
      KMESH(V,KKRHO,KF1)=KMESH(V,KKRHO,KF1)+KF-KKZF
      KMESH(V,KKRHO,KI1)=KMESH(V,KKRHO,KI1)+KKZI-KI1
      GOTO 890
  880 KMESH(V,KKRHO,KF1)=KMESH(V,KKRHO,KF1)+KKZI-KKZF
C
C     SUM TOTAL CHANGES IN K SPACE POSITION AND ENERGY SPACE POSITION
C     AND STORE IN KZTOT AND ETOT.SUM MEAN ENERGY CHANGE IN MNE.
C     SUM INDIVIDUAL VALUES FOR EACH VALLEY IN KZ(V) AND ET(V).
C
  890 KZTOT=KZTOT+ABS(KZI-KZF)
      ETOT=ETOT+(KZI*KZI-KZF*KZF)*10.0*H/(2*EM*M)
      MNE=MNE+(KZI*KZI*KZI-KZF*KZF*KZF)*1.0E-14*H*H/(2*EM*M*M*E)
      KZ(V)=KZ(V)+ABS(KZI-KZF)
      ET(V)=ET(V)+(KZI*KZI-KZF**2)*10.0*H/(2*EM*M)
C
C     CHECK FOR ROUNDING ERRORS LEADING TO NEGATIVE ENERGY VALUES.
C     IF THIS OCCURS,PLACE ELECTRON AT STARTING POSITION.
C
      IF(EI.GT.0.0) GOTO 100
      KRHO=0.0
      PSI=0.0
      KZF=1.0E+6
```

COMPUTER SIMULATION USING MONTE CARLO METHODS 405

```
ONTE CARLO PROGRAM
      EM=EM1
      V=1
      GOTO 90
C
C     CALCULATE FINAL ENERGY VALUE FOR EACH SCATTERING PROCESS.
C
  100 EF(1)=EI-HWO
      EF(2)=EI+HWO
      EF(3)=EI
      EF(4)=EI
      EF(5)=EI-HWE
      EF(6)=EI+HWE
      EF(7)=EI-HWI+D
      EF(8)=EI-HWI-D
      EF(9)=EI-HWI-D
      EF(10)=EI+HWI+D
C
C     SCATTERING RATES FOR REAL PROCESSES
C
      IF(EF(1).GT.0.0) GOTO 110
      L(1)=0.0
      GOTO 120
  110 L(1)=C1*SQRT(EM)*ALOG(ABS((SQRT(EI)+SQRT(EF(1)))/(SQRT(EI)-
     4  SQRT(EF(1)))))/SQRT(EI)
C
C     EMISSION OF OPTICAL PHONON
C
  120 IF(EF(2).GT.0.0) GOTO 125
      L(2)=0.0
      GOTO 130
  125 L(2)=C2*SQRT(EM)*ALOG(ABS((SQRT(EI)+SQRT(EF(2)))/(SQRT(EI)-
     5  SQRT(EF(2)))))/SQRT(EI)
C
C     ABSORPTION OF OPTICAL PHONON
C
  130 IF(EF(3).GT.0.0) GOTO 135
      L(3)=0.0
      GOTO 140
  135 L(3)=C3*EM**1.5*SQRT(EF(3))
C
C     EMISSION OF ACOUSIC PHONON
C
  140 IF(EF(4).GT.0.0) GOTO 145
      L(4)=0.0
      GOTO 150
  145 L(4)=L(3)
C
C     ABSORPTION OF ACOUSTIC PHONON
C
  150 IF(EF(5).GT.0.0) GOTO 155
      L(5)=0.0
      GOTO 170
  155 IF(V.EQ.2) GOTO 160
      L(5)=0.0
      GOTO 170
  160 L(5)=C4*EM**1.5*SQRT(EF(5))
```

```
      MONTE CARLO PROGRAM
C
C     EMISSION OF EQUIVALENT INTERVALLEY PHONON
C
170   IF(EF(6 .GT.0.0) GOTO 175
      L(6)=0.0
      GOTO 190
175   IF(V.EQ.2) GOTO 180
      L(6)=0.0
      GOTO 190
180   L(6)=C5*EM**1.5*SQRT(EF(6))
C
C     ABSORPTION OF EQUIVALENT INTERVALLEY PHONON
C
190   IF(EF(7).GT.0.0) GOTO 195
      L(7)=0.0
      GOTO 210
195   IF(V.EQ.2) GOTO 200
      L(7)=0.0
      GOTO 210
200   L(7)=C6*SQRT(EF(7))
C
C     EMISSION OF INTERVALLEY PHONON(SATELLITE TO CENTRAL)
C
210   IF(EF(8).GT.0.0) GOTO 215
      L(8)=0.0
      GOTO 230
215   IF(V.EQ.1) GOTO 220
      L(8)=0.0
      GOTO 230
220   L(8)=C7*SQRT(EF(8))
C
C     EMISSION OF INTERVALLEY PHONON(CENTRAL TO SATELLITE)
C
230   IF(EF(9).GT.0.0) GOTO 235
      L(9)=0.0
      GOTO 250
235   IF(V.EQ.1) GOTO 240
      L(9)=0.0
      GOTO 250
240   L(9)=C8*SQRT(EF(9))
C
C     ABSORPTION OF INTERVALLEY PHONON(CENTRAL TO SATELLITE)
C
250   IF(EF(10).GT.0.0) GOTO 255
      L(10)=0.0
      GOTO 270
255   IF(V.EQ.2) GOTO 260
      L(10)=0.0
      GOTO 270
260   L(10)=C9*SQRT(EF(10))
C
C     ABSORPTION OF INTERVALLEY PHONON(SATELLITE TO CENTRAL)
C
270   IF(TT.EQ.1) GOTO 40
C
C     CALCULATE SUM OF REAL PROCESS SCATTERING RATES
```

COMPUTER SIMULATION USING MONTE CARLO METHODS 407

```
MONTE CARLO PPJGRAM
C
280   SCATT(1)=L(1)/GAMMA(V)
      DO 290 K=2,10
      SCATT(K)=SCATT(K-1)+L(K)/GAMMA(V)
290   CONTINUE
C
C     CALL RANDOM NUMBER.SELECT SCATTERING CHANNEL.
C
      R=RND(B)
      IF(R.LT.SCATT(1)) GOTO 300
      IF(R.LT.SCATT(2)) GOTO 310
      IF(R.LT.SCATT(3)) GOTO 320
      IF(R.LT.SCATT(4)) GOTO 330
      IF(P.LT.SCATT(5)) GOTO 340
      IF(P.LT.SCATT(6)) GOTO 350
      IF(P.LT.SCATT(7)) GOTO 360
      IF(P.LT.SCATT(8)) GOTO 370
      IF(R.LT.SCATT(9 ) GOTO 380
      IF(R.LT.SCATT(10)) GOTO 390
      GOTO 400
C
C     SET ENERGY AFTER SCATTERING PROCESS.
C
300   EFIN=EF(1)
      GOTO 420
310   EFIN=EF(2)
      GOTO 420
320   EFIN=EF(3)
      GOTO 410
330   EFIN=EF(4)
      GOTO 410
340   EFIN=EF(5)
      GOTO 410
350   EFIN=EF(6)
      GOTO 410
360   EFIN=EF(7)
      GOTO 430
370   EFIN=EF(8)
      GOTO 430
380   EFIN=EF(9)
      GOTO 430
390   EFIN=EF(10)
      GOTO 430
400   EFIN=EI
      GOTO 450
410   SR=SR+1
C
C     REGISTER REAL COLLISION.CALCULATE K SPACE POSITION AFTER
C     ACOUSTIC, INTERVALLEY OR EQUIVALENT INTERVALLEY PHONON SCATTERING.
C
      R=RND(B)
      KT=1.0E+7*SQRT(2*EM*M*EFIN*E)/H
      KZF=KT*(1-2*R)
      KRHO=KT*SQRT(4*R*(1-R))
      GOTO 460
420   SR=SR+1
```

408 PHYSICS PROGRAMS

```
MONTE CARLO PROGRAM
C
C      REGISTER REAL COLLISION.CALCULATE K SPACE POSITION AFTER
C      OPTICAL PHONON SCATTERING.
C
       R=RND(B)
       U=RND(B)
       PHI=2*3.142*R
       EX=2*SQRT(EFIN*EI)/((SQRT(EI)-SQRT(EFIN))**2)
       BETA=(((1+EX)-(1+2*EX)**U)/EX)
       RHO=(BETA*KZI/KT-SQRT(ABS(1-BETA*BETA))*KRHO/KT*COS(PHI))
       KT=1.0E17*SQRT(2*EM*M*EFIN*E)/H
       KZF=KT*RHO
       KRHO=KT*SQRT(ABS(1-RHO*RHO))
       GOTO 460
C
C      CHANGE VALLEY PARAMETERS FOR INTERVALLEY PROCESSES.
C
430    IF(V.EQ.1) GOTO 440
       V=1
       EM=EM1
       GOTO 410
440    V=2
       EM=EM2
       GOTO 410
450    SS=SS+1
C
C      REGISTER SELF SCATTERING PROCESS.K SPACE POSITION UNCHANGED.
C
       KZF=KZI
C
C      CHECK IF ELECTRON IS SCATTERED OUT OF MESH.IF SO,REGISTER
C      PROCESS ON COUNTER GMAX,AND PLACE ELECTRON ON EDGE OF MESH.
C      LABEL 90 REPEATS ITERATIVE PROCESS STARTING WITH FREE
C      ELECTRON FLIGHT UNDER ELECTRIC FIELD.
C
460    IF(EFIN.LE.1.0) GOTO 90
       GMAX=GMAX+1
       KT=1.0E+7*SQRT(2*EM*M*EMAX*E)/H
       IF(KRHO.GT.KT) KRHO=KT
       KZF=SQRT(ABS(KT*KT-KRHO*KRHO))
       GOTO 90
C
C      FINAL CALCULATION OF DRIFT VELOCITY-VEL,TIME SPENT IN EACH
C      VALLEY-VTIM(V),MEAN ENERGY-MNE.ALSO OUTPUT NO. OF SELF
C       SCATTERING PROCESSES AND NUMBER OF TIMES MESH EXCEEDED.
C
C      SECOND TABLE OF DATA CONTAINS MEAN VELOCITY IN EACH
C      VALLEY(V1 AND V2),MOBILITY(U1 AND U2) AND FRACTIONAL TIME
C      IN EACH VALLEY(NPR1 AND NPR2).
C
470    IF(KZTOT.EQ.0.0)KZTOT=1.0E-20
       VEL(J)=ETOT/KZTOT
       MNE=MNE/KZTOT
       IF(VTIM(1).EQ.0.0) VTIM(1)=1.0E-25
       IF(KZ(1).EQ.0.0) KZ(1)=1.0E-25
       IF(KZ(2).EQ.0.0) KZ(2)=1.0E-25
```

```
MONTE CARLO PROGRAM
      U1(J)=ET(1)/KZ(1)
      U2(J)=ET(2)/KZ(2)
      U1(J)=U1(J)/(F(J)*1000.0)
      U2(J)=U2(J)/(F(J)*1000.0)
      NPR1(J)=UTIM(1)/(UTIM(1)+UTIM(2))
      NPR2(J)=UTIM(2)/(UTIM(1)+UTIM(2))
      WRITE(1,480)F(J),VEL(J),UTIM(1),UTIM(2),MNE,SS,GMAX
      WRITE(9,480)F(J),VEL(J),UTIM(1),UTIM(2),MNE,SS,GMAX
480   FORMAT(5(E9.3,1X),I7,1X,I5)
      CALL TIOU(7)
C
C     WRITE DATA TO FILES FOR GRAPHICAL DISPLAY.
C
      WRITE(10,800)F(J)
      WRITE(11,800)VEL(J)
      WRITE(12,800)MNE
800   FORMAT(E12.4)
      DO 810 LL=1,20
      WRITE(13,820)KMESH(UU,VR,LL)
820   FORMAT(E12.4)
810   CONTINUE
      J=J+1
      IF(J.NE.NF+1) GOTO 80
      WRITE(9,501)
      WRITE(1,501)
501   FORMAT(//,2X,5HFIELD,5X,4HVEL1,6X,4HVEL2,6X,4HMOB1,6X
     2 ,4HMOB2,8X,2HN1,8X,2HN2)
      DO 510 J=1,NF
      WRITE(9,500)F(J),V1(J),V2(J),U1(J),U2(J),NPR1(J),NPR2(J)
      WRITE(1,500)F(J),V1(J),V2(J),U1(J),U2(J),NPP1(J),NPR2(J)
500   FORMAT(7(E9.3,1X))
510   CONTINUE
C
C     TRUNCATE AND CLOSE ALL DATA FILES.
C
      CALL SEARCH(8,0,5)
      CALL SEARCH(8,0,6)
      CALL SEARCH(8,0,7)
      CALL SEARCH(8,0,8)
      CALL SEARCH(8,0,9)
      CALL SEARCH(8,0,10)
      CALL SEARCH(4,0,5)
      CALL SEARCH(4,0,6)
      CALL SEARCH(4,0,7)
      CALL SEARCH(4,0,8)
      CALL SEARCH(4,0,9)
      CALL SEARCH(4,0,10)
      CALL SEARCH(3,'GRAPHS',5)
      READ(9,951)Y
951   FORMAT(A3)
      CALL SEARCH(4,0,5)
      WRITE(1,952)
952   FORMAT(/,28HTO OBTAIN GRAPHS TYPE GRAPHS,//
     1 ,28HTO LEAVE PROGRAM TYPE FINISH)
      READ(1,951)X
      IF(X.NE.Y)GOTO 983
```

```
      MONTE CARLO PROGRAM
C
C     START GRAPHICAL DISPLAY
C
      CALL RESUME('*GMONTE')

  983 STOP
      END
```

Physics Programs
Edited by A. D. Boardman
© 1980 John Wiley & Sons Ltd.

CHAPTER 12

Modelling of the Thermal Conductivity of Unidirectional Composite Materials

G. S. KEEN and B. W. JAMES

1. INTRODUCTION

Composite materials, made of a host (matrix) with a number of fibres inserted into it (inserts), are now widely used because of their special mechanical properties and high strength to weight ratio. In some applications it is necessary to know the heat flow along, or through, the composite material. This heat flow will depend upon the thermal conductivities of the constituent materials. If the composite material has an anisotropic structure, the heat flow will also be anisotropic, and cannot, in general, be calculated simply on the basis of the ratios of the constituent materials since it depends upon the geometrical arrangement of the composite in the heat flow direction. It is therefore convenient to characterize the composite material by a number of effective thermal conductivities which take account of these factors.

Obviously the effective thermal conductivity of the composite material can be measured experimentally with special apparatus. However, this approach can be very time consuming if a number of different composite materials have to be considered over a wide temperature range. Also, whilst experimental measurements of the thermal conductivity of the composite material and the matrix material used in the composite are readily made, it is not possible to measure the thermal conductivity of the fibre reinforcement, due to the small physical dimensions of the fibres, although the bulk value can be found for materials which exist (occur) in bulk form. (The anisotropic structure of some fibres is not repeated in the bulk form, e.g. carbon fibres.) In contrast, an investigation of the thermal properties of composite materials, over a wide range of temperature, and for various configurations of fibres inserted into the matrix material, can be carried out with a computer program which models the real material. The advantages of this approach include obtaining values of the effective thermal conductivities much more rapidly, and at a lower cost than that involved in experimental measurement. Another advantage of computer modelling to obtain the effective thermal conductivity is that, if the thermal conductivity of the insert material is not known, and cannot be measured in the bulk form, it may be found by iterative calculations from measured values of the matrix thermal conductivity and the effective thermal conductivity of the composite.

2. HEAT CONDUCTION IN COMPOSITES

2.1 Basic ideas

The general form of the heat flow equation is

$$\mathbf{Q} = \mathbf{K} \cdot \nabla T, \qquad (1)$$

where $\mathbf{Q}$ is the thermal flux vector, whose dimensions are energy transmitted, per unit area, per unit time, $\mathbf{K}$ is the thermal conductivity tensor of the medium conducting the heat, and ∇T is the vector temperature gradient across it.

A unidirectional composite material may be idealized by a regular array of inserts, aligned in one direction in the matrix material. The basic rectangular prism, or cell, of such a composite, is a single insert, immersed in the matrix material with its axis parallel to the longest side of the prism. All cross-sections of this cell that are perpendicular to the insert axis are identical, therefore, and any sample of composite material can be simulated by using a large number of basic cells. In Figure 1 a cylindrical insert, or fibre, is shown, but any other shape is possible. If the insert is, in fact, a cylindrical fibre and is at the centre of a cell that has, in addition, a square cross-section then only a quarter of the basic cell needs to be considered because of its fourfold axis of symmetry. This has the advantage, as will be seen later, that the accuracy of the calculations improves if a fraction of the basic cell is used.

If a temperature difference is applied between any two opposite faces of a cell that is part of a large assembly then the edge effects arising from the other faces can be ignored as there will be no flow of heat across them. Since this is the case, the application of the temperature difference between faces 1 and 2, of Figure 1 causes the heat to flow in a direction parallel to the fibre axis. This is called longitudinal heat flow, whereas if the temperature difference is applied between either of the two other opposite pairs of faces, the direction of heat flow is in a direction transverse to the fibre axis and is called transverse heat flow.

In a unidirectional composite material there are three independent coefficients in the conductivity tensor, K_x, K_y, and K_z, if the insert axis is taken as the z-axis (Nye,[1] Boardman, O'Connor, and Young.[2]). Furthermore, if the

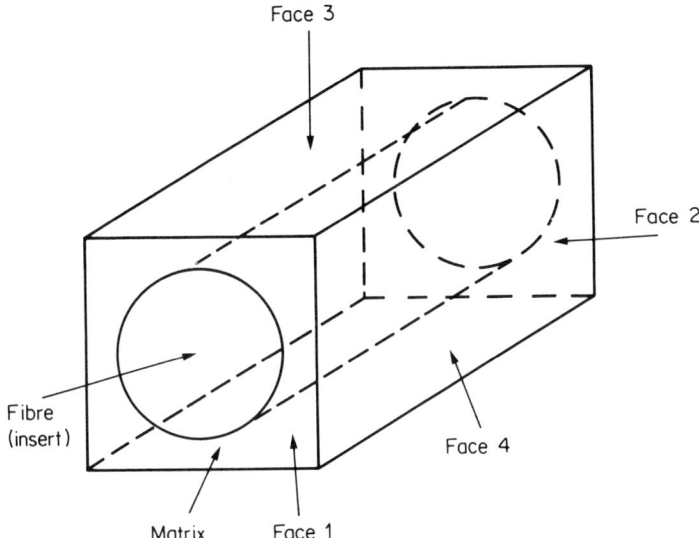

Figure 1. The basic rectangular prism or cell of a composite material containing a regular array of fibres

composite has an axis of three-, four-, or sixfold symmetry about the z-axis then $K_x = K_y$ and there is only one value for the effective transverse thermal conductivity.

2.2 Longitudinal heat flow

In Figure 1 suppose that a temperature difference is established between faces 1 and 2. By the reasoning developed above, the heat flow must be parallel to the axis of the fibre. The effective longitudinal thermal conductivity K_e of the whole cell is related to K_f the thermal conductivity of the fibre and K_m the thermal conductivity of the matrix in the following way:

$$K_e = \left(\frac{A_f}{A_f + A_m}\right) K_f + \left(\frac{A_m}{A_f + A_m}\right) K_m, \qquad (2)$$

where A_f is the area of the fibre normal to its axis and A_m is the area of the matrix not occupied by the fibre. Actually the longitudinal heat flow problem is a straightforward application of the method of mixtures, since the two materials form parallel heat flow paths.

2.3 Transverse heat flow

If the cell in Figure 1 has a temperature difference between faces 3 and 4 the heat flow is transverse. Also if there are many basic cells, and hence lots

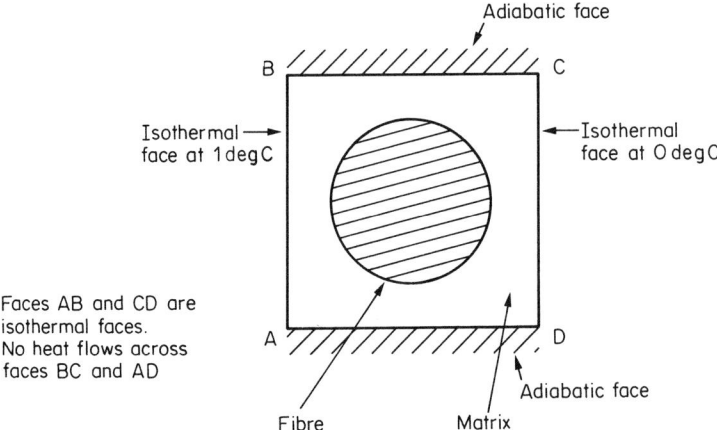

Figure 2. A perpendicular section through a basic cell of the composite

of fibres, no heat can flow across the other faces. Therefore in any section, perpendicular to the fibre, as shown in Figure 2, the temperature difference exists entirely between the faces AB and CD. One of the most informative computational pictures to construct in problems like this is the distribution of isotherms. A trivial case arises when the thermal conductivities of the fibre and matrix are equal. In this case the isotherms are parallel and, if they are drawn at constant temperature intervals, are equally spaced. The calculation of the effective thermal conductivity of the composite material is also a trivial matter.

On the other hand, if the two materials have different conductivities then the isotherms are no longer straight or parallel or evenly spaced. For a fibre with a lower conductivity than the matrix, a larger temperature gradient will exist in the fibre than in the matrix. Consequently, the majority of the heat will flow through the matrix around the fibre. This is shown in Figure 3 that shows isotherms, for a temperature difference across the cell of 1 degC, at intervals of 0.1 degC. If the fibre has a higher conductivity than the matrix, the reverse will occur, with the majority of the heat flowing through the fibre. This is illustrated in Figure 4.

The non-laminar redistribution of the flow of heat that occurs when $K_f \neq K_m$, is not a trivial matter to calculate and the remainder of this chapter is devoted to an electrical analogue technique that leads to the production of results, such as those shown in Figures 3 and 4. The electrical analogue of heat flow was considered a long time ago by Kayan.[3] It has been used both for calculations and for experimental investigations of heat flow problems. It is only necessary, with this technique, to specify the external boundary

Thermal conductivity of a composite
Thermal conductivity of matrix 125.000 watt/metre/degree kelvin
Thermal conductivity of inserts
Conductivity of insert Nos 1 was 1.000 watt/metre/degree kelvin
Effective thermal conductivity 77.117 watt/metre/degree kelvin
Sample dimensions x = 10.000 mm y = 10.000 mm
The unit cell is divided into 10 rows by 10 columns

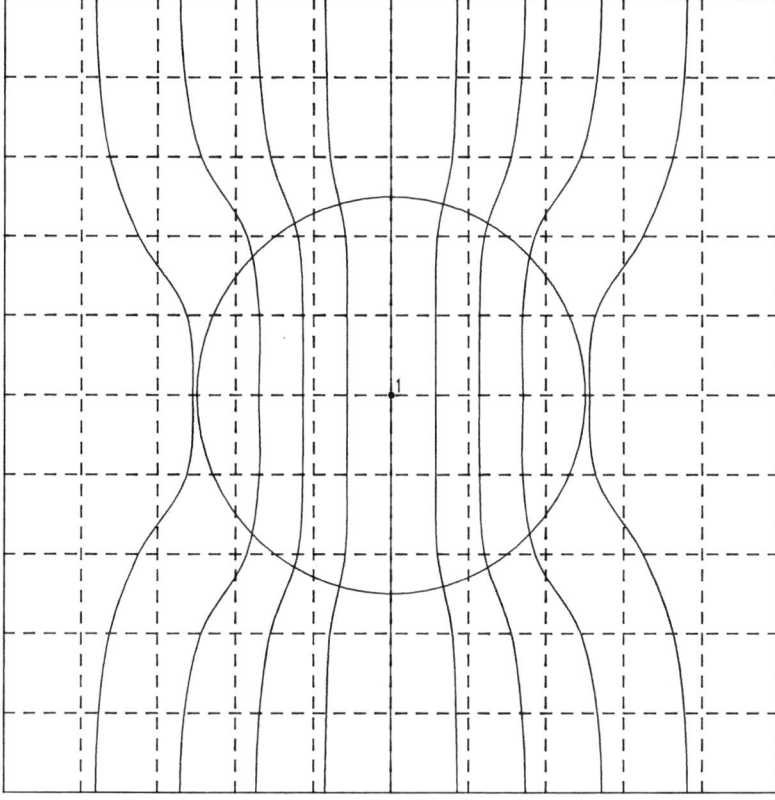

Figure 3. Section through a basic cell of a composite showing 0.1 deg C isotherms for high conductivity matrix and low conductivity fibre; n, the number of fibre inserts, is designated in the computer output as Nos n

conditions and the arrangement of the different materials of the structure to enable the temperature distribution and heat flow to be determined. An alternative approach, to the electrical analogy, is to use a relaxation method (Emmons[4]) that requires, in addition to the above information, an initial guess of the temperature distribution. It is not felt to be appropriate to the present problem, although Collier[5] has used it for problems involving complex arrangements of materials and boundary conditions.

Thermal conductivity of a composite
Thermal conductivity of matrix 1.000 watt/metre/degree kelvin
Thermal conductivity of inserts
Conductivity of insert Nos 1 was 125.000 watt/metre/degree kelvin
Effective thermal conductivity 1.471 watt/metre/degree kelvin
Sample dimensions x = 10.000 mm y = 10.000 mm
The unit cell is divided into 10 rows by 10 columns

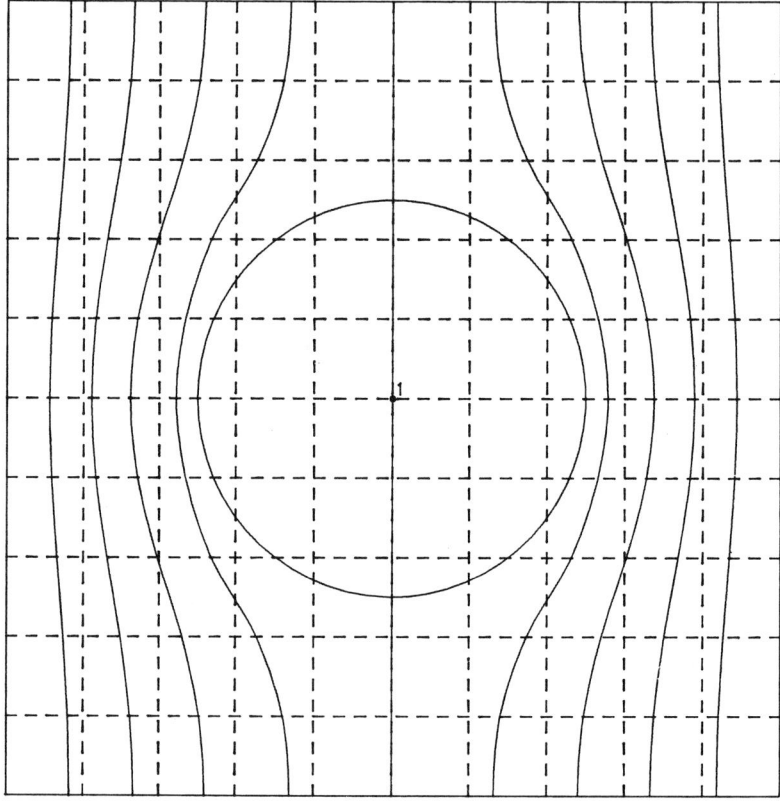

Figure 4. Section through a basic cell of a composite showing 0.1 degC isotherms for low-conductivity matrix and high-conductivity fibre; n, the number of fibre inserts is designated in the computer output as Nos n

3. THE ELECTRICAL ANALOGUE

The heat flow equation is analogous to Ohm's law of electrical resistivity. To put the matter very simply, Ohm's law for an isotropic solid, or along the longitudinal or transverse axes of a unidirectional solid, is

$$V = IR, \qquad (3)$$

where I is the electrical current flowing per unit area, V is the potential drop, and R is the resistance. The terms in equation (3) are identified, by analogy, as: (a) I with Q/A, (b) V with temperature gradient $\partial T/\partial Z$, $\partial T/\partial X$, or $\partial T/\partial Y$, (c) R with $1/K$, where K is the longitudinal or transverse thermal conductivity.

A representation of the cell of Figure 2, by a network, i.e. meshes, of resistors, is shown in Figure 5. A resistor is placed along each side of a mesh, with the exception of the two isothermal faces AB and CD, and values are assigned to them in accordance with the thermal conductivity of fibre and matrix, remembering that, in the analogy, electrical resistance is inversely proportional to thermal conductivity.

In Figure 5 the temperature difference of T degrees centigrade is replaced directly by a T volt potential difference applied across the resistor network. This network may, of course, be much larger than the 3×3 system shown here, but this is a convenient size for the present example. The calculated effective thermal conductivity changes slightly as the number of meshes changes. This is a feature that needs to be considered, in any complete

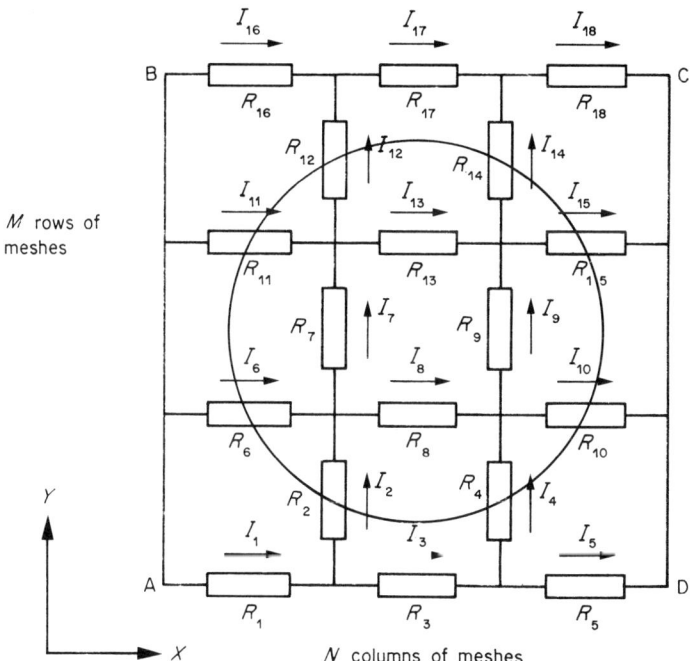

Figure 5. The resistor network to represent the composite shown in Figures 1 and 2. All points along AB are at 1.0 V with respect to points along CD, In this example $M = 3$ and $N = 3$

calculation, and is discussed later in the chapter. The method of calculating the values of the resistors R_1 to R_{18} is given below.

The thermal resistivities ρ_m and ρ_f, of the matrix and fibre, are $\rho_m = 1/K_m$ and $\rho_f = 1/K_f$ and a one-to-one correspondence between them and the electrical resistivities is taken, i.e. ρ_m and ρ_f are made numerically equal to the corresponding electrical resistivities. The individual network resistors have values that depend upon ρ_m and ρ_f, the geometry of the basic cell, and the number of rows and columns in the network.

The first stage in calculating the network resistor values is to find the apparent resistances in the X- and Y-directions of Figure 5 for a cell consisting entirely of matrix or entirely of fibre material. Suppose R_{mx} or R_{my} is the apparent matrix resistance in the X- or Y-direction and R_{fx} or R_{fy} is the apparent fibre resistance in the X- or Y-direction. If the cell dimensions are X_s and Y_s then the apparent resistances are

$$R_{mx} = \rho_m \frac{X_s}{LY_s} = \frac{1}{K_m}\frac{X_s}{LY_s},$$
$$R_{my} = \rho_m \frac{Y_s}{LX_s} = \frac{1}{K_m}\frac{Y_s}{LX_s},$$
$$R_{fx} = \rho_f \frac{X_s}{LY_s} = \frac{1}{K_f}\frac{X_s}{LY_s},$$
$$R_{fy} = \rho_f \frac{Y_s}{LX_s} = \frac{1}{K_f}\frac{Y_s}{LX_s},$$

(4)

where L is the length of the cell normal to the XY plane. This is now chosen to be 1.0 m; obviously any other value could be used but it is only a scaling factor that may as well be set equal to unity.

It is clear now that if there are M rows of meshes in the Y-direction, and N columns of meshes in the X-direction, then the individual resistor values in the network are

$$r_{mx} = \frac{1}{K_m}\frac{X_s}{Y_s}\frac{M}{N},$$
$$r_{my} = \frac{1}{K_m}\frac{Y_s}{X_s}\frac{N}{M},$$

(5)

provided they lie entirely within the matrix, and

$$r_{fx} = \frac{1}{K_f}\frac{X_s}{Y_s}\frac{M}{N},$$
$$r_{fy} = \frac{1}{K_f}\frac{Y_s}{X_s}\frac{N}{M},$$

(6)

provided they lie entirely within the fibre insert. These values, excluding those that lie partly in the matrix and partly in the fibre, are assigned to resistors $R_1 \ldots R_{18}$ in Figure 5.

Equations (5) and (6) assume that there are M lines of resistors in the X-direction between AB and CD, whereas Figure 5 shows that there are $M+1$. There are, however, effectively only M lines since the top and bottom lines, BC and AD, have a weighting factor of only $\frac{1}{2}$ since they are shared with adjacent cells. This means resistors in these network positions must have their resistance doubled. This difficulty does not arise in the Y-direction since the faces AB and CD are isothermal faces.

In Figure 5 the resistors are drawn symbolically as blocks, but this representation is possibly confusing when a resistor extends across a boundary between the fibre and the matrix. Each side of the mesh is, in fact, a resistor, as drawn in Figure 6. If a resistor happens to cross a fibre/matrix boundary then it cannot have the full $r_{mx,y}$ or $r_{fx,y}$ value. It is necessary to determine what proportion of it lies within the fibre or the matrix. For example, suppose that, as indicated in Figure 6, the boundary cuts the resistor at 0.45, measured from the junction point inside the fibre and taking the side of the mesh as 1 unit. A resistor in this network position is given the value

$$R = 0.45 r_{fx} + 0.55 r_{mx}. \tag{7}$$

All the resistance values of the resistors R_1 to R_{18} are therefore easily found and the total current flowing from side AB to side CD is used in Ohm's law to find K_e, the effective thermal conductivity of the composite.

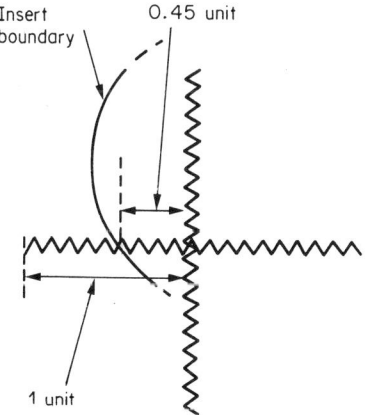

Figure 6. A resistor at the boundary between matrix and insert (fibre)

The total current is obviously the sum of the currents flowing from plane AB towards CD and to calculate these currents it is necessary to use Kirchhoff's laws.

4. KIRCHHOFF'S LAWS

The two Kirchhoff laws extend the applications of Ohm's law to a system comprising any number of branches, branch points, and electromotive force (emf) sources. They are:

(1) The junction law, that states that the sum of currents flowing in branches which form a junction is zero.
(2) The mesh law, that states that, in any closed mesh, the sum of the potential differences across the branches of which the circuit is composed, is equal to the sum of the electromotive forces in that mesh.

The application of the first law to a typical junction of four resistors in the network of Figure 5 gives

$$I_6 + I_2 - I_7 - I_8 = 0. \tag{8}$$

Obviously, similar equations for all the junctions in the network exist.

For a typical mesh consisting of three resistors in Figure 5 the application of the second law gives

$$I_6 R_6 - I_2 R_2 - I_1 R_1 = 0. \tag{9}$$

Similar equations for all the meshes of the network also exist.

In more detail an examination of Figure 5 reveals that eight junction equations like equation (8), and nine mesh equations like equation (9) exist. The fact that this makes a total of 17 equations with 18 unknown currents is a possible difficulty, but a final equation is obtained by considering any line of resistors between AB and CD and using the value of an externally applied emf. The emf can have any value, but it is simpler to use unity and scale the answers to fit any other voltage. A final equation is

$$I_1 R_1 + I_3 R_3 + I_5 R_5 = 1. \tag{10}$$

Since all points along AB are at the same voltage relative to CD, there are other choices for the final equation. It does not matter which one is used.

In a network of M by N meshes there are $(M+1)N + (N-1)M = 2MN + N - M$ resistors and currents, NM mesh equations and $MN + N - M - 1$ junction equations. Therefore the number of equations that do not involve the external potential is $(2MN + N - M) - 1$, i.e. one less than the total number of resistors.

The equations relevant to Figure 5 are, from the application of Kirchhoff's laws,

$$I_1 - I_2 - I_3 = 0,$$
$$I_3 - I_4 - I_5 = 0,$$
$$I_2 + I_6 - I_7 - I_8 = 0,$$
$$I_4 + I_8 - I_9 - I_{10} = 0,$$
$$I_7 + I_{11} - I_{12} - I_{13} = 0, \quad (11a)$$
$$I_9 + I_{13} - I_{14} - I_{15} = 0,$$
$$I_{12} + I_{16} - I_{17} = 0,$$
$$I_{14} + I_{17} - I_{18} = 0,$$

and

$$-I_1 R_1 - I_2 R_2 + I_6 R_6 = 0,$$
$$I_2 R_2 - I_3 R_3 - I_4 R_4 + I_8 R_8 = 0,$$
$$I_4 R_4 - I_5 R_5 + I_{10} R_{10} = 0,$$
$$-I_6 R_6 - I_7 R_7 + I_{11} R_{11} = 0,$$
$$I_7 R_7 - I_8 R_8 - I_9 R_9 + I_{13} R_{13} = 0,$$
$$I_9 R_9 - I_{10} R_{10} + I_{15} R_{15} = 0, \quad (11b)$$
$$-I_{11} R_{11} - I_{12} R_{12} + I_{16} R_{16} = 0,$$
$$I_{12} R_{12} - I_{13} R_{13} - I_{14} R_{14} + I_{17} R_{17} = 0,$$
$$I_{14} R_{14} - I_{15} R_{15} + I_{18} R_{18} = 0,$$
$$I_1 R_1 + I_3 R_3 + I_5 R_5 = 1.0.$$

Equation (11), in matrix form, is

$$\mathbf{R} \cdot \mathbf{I} = \mathbf{V}, \quad (12)$$

where $\mathbf{R}$ is an 18×18 square matrix, $\mathbf{I}$ and $\mathbf{V}$ are current and voltage column vectors and it is the column vector $\mathbf{I}$ that is calculated. The matrix $\mathbf{R}$ consists mainly of zeros except for 3 or 4 elements in each row. The full $\mathbf{R}$ matrix is given in Figure 7, mainly to show that it presents some mathematical problems. Technically it is known as a non-diagonal sparse matrix and it becomes more so for larger networks.

After the column vector $\mathbf{I}$ is evaluated from equation (12) the effective transverse thermal conductivity K_e is obtained from the current flow through any plane parallel to AB. The effective electrical resistance R_e of the cell, found from I_T, the sum of the currents entering through plane AB, and the applied potential (1 Volt in this case), is $R_e = 1.0/I_T$ and the effective

Figure 7. The full **R** matrix array with zeros omitted

resistivity is

$$\rho_e = R_e \frac{Y_s}{X_s} = \frac{1.0}{I_T} \frac{Y_s}{X_s}. \tag{13}$$

The one-to-one correspondence between the electrical and thermal resistivity enables the thermal conductivity to be written directly as

$$K_e = \frac{1}{\rho_e} = I_T \frac{X_s}{Y_s}. \tag{14}$$

The temperature distribution in the composite is found by calculating the potential at each network intersection. The locations of the isotherms (i.e. isopotential lines) are found by linear interpolation between the network intersections. The results are used to draw an isotherm map of the composite. Figures 3 and 4 show isotherms, plotted at intervals of 0.1 degC, for a composite, with only one fibre, in the case of a temperature difference of 1 degC.

5. THE COMPUTER PROGRAM

A modular approach is used since the program seems to split naturally into several sections, which are:

(1) Input of initial parameters that consists of the physical size of the sample, the number of meshes into which the sample is to be divided, the shape of the sample, the thermal conductivities of the constituents, and the number and sizes of the inserts.
(2) Calculation of the values of the resistances to represent inserts in the matrix, using equations (5) and (6).
(3) Setting up of the matrix and vectors of equation (12) from the above resistor values.
(4) Use of mathematical routines for calculating the current flowing in each resistor.
(5) Interpolation for isothermal values and calculation of the effective thermal conductivity of the composite from the results obtained in (4) above.
(6) Output of the temperature at each network intersection, the positions of the isothermals, and the effective thermal conductivity.

Some of these sections seem to subdivide further into many very simple modules. For example, in (3) five basic types of equations are needed to generate sufficient equations to solve for all of the currents, namely:

(a) currents flowing into and out of a four-resistor junction in the network;
(b) currents flowing into and out of a three-resistor junction in the network;
(c) currents flowing in a closed four-resistor mesh;

(d) currents flowing in a closed three-resistor mesh;
(e) currents flowing along one horizontal row of the mesh.

The main segment of the computer program calls a set of subroutines that perform nearly all these functions.

The subroutines are listed in Table 1 and their functions are outlined. The NAG library subroutines FO3AJF and FO4APF are used to evaluate equation (12). These NAG subroutines are specifically intended for use with sparse matrices. It is not absolutely necessary to use sparse matrix methods, but if they are not used the program needs a lot more storage space. Even now the program published here requires a storage of 62K to handle a 10×10 meshes network.

The format for the data required to run the program is shown below and annotated so that the numbers can be identified:

10 10	N and M
10.0 10.0	X_s and Y_s, sample cell size
1	the number of inserts in the sample
0.2041	K_m thermal conductivity of the matrix
1.998	K_f thermal conductivity of the fibre/insert
2	Insert type (1 = rectangular, 2 = circular)
10.0 0.0	the centre coordinates of the inserts
7.652	the radius of a circular insert or the length of sides of a rectangular insert in X, Y order

If the number of inserts is $n > 1$, the last four lines of the input data file occur n times.

A listing of the program, a typical data file, and the resulting output are given at the end of the chapter. There are three output channels which write to files or to the one line printer. The input data, the calculated effective thermal conductivity, and the positions of the isovoltage points, at intervals of 0.1 V, form the output on channel 2. The input data, the effective thermal conductivity, and the values of the voltage (synonymous with temperature) at the corners of the meshes form the output on channel 3. This output could be used together with a general contour plotting graphics program to produce an isotherm map of the composite. The output on channel 6 is in a form suitable for input to the accompanying graphics program and consists of the input data, the effective thermal conductivity, and the positions of the 0.1 deg C isotherms.

The graph plotting program that produces a picture of the isotherms makes use of the GINO-F software (widely available in the U.K. and elsewhere) and has been used to produce Figures 3, 4, and 8. A listing of this program is also given at the end of the chapter. Apart from drawing the isotherms the program also draws the resistor network and the position of the inserts.

Table 1. The subroutines of the main program

Subroutine name	Function of subroutine
ARRAYSIZE	Reads M and N the number of rows and columns of meshes in the network. Also reads the sample section dimensions X_s and Y_s.
CALCRESNUM	Calculates the number of resistors from given M and N.
ARRAYZERO	Sets all values of the resistance array to zero.
NUMOFINC	Reads the number of inserts or inclusions and repeats calls of ARRAYRES and INCLUSION an appropriate number of times.
ARRAYRES	Reads the conductivity of the matrix followed by the conductivity of the inserts.
INCLUSION	Reads the type of insert, coordinates of the centre of the insert, and the dimension(s) of the insert. Circular inserts are indicated by 1 and rectangular inserts by 2.
INCSC1	Called by INCLUSION for a rectangular insert to calculate the values of the resistors.
INCSC2	Called by INCLUSION for a cylindrical insert to calculate values of the resistors.
EDGERESCOMP	Doubles the values of the resistors which are at the edges of the network representing the cell of the composite.
SETVOLTARRAY	Sets **V** vector to zeros for all but one element which is set to 1.0.
MESH4I	Used for four-resistor meshes to set appropriate members of **R** equal to the appropriate values of resistance
MESH3I	Used for three-resistor meshes, otherwise the same as MESH4I
POINT4I	Used for four current junctions to set appropriate members of **R** equal to +1 or −1.
POINT3I	Used for three current junctions, otherwise the same as POINT4I.
OVERALLI	Sets up equation (9) for the network specified.
FO3AJF	Decomposes the sparse matrix **R** into triangular factors and evaluates the determinants.
FO4APF	Calculates the (approximate) solution **I** of the set of real sparse linear equations $\mathbf{R} \cdot \mathbf{I} = \mathbf{V}$.
EXITPARAM	If an error is detected in FO3AJF this provides the error exit.
EFFCOND	Calculates the effective thermal conductivity K_e from the values of **I**.
WRITE	Calculates and outputs the voltage map for the network intersections.
WRITEGRAPH	Interpolates the above voltage map to find the positions of the 0.1 V (0.1 degC) steps and outputs results for use by a graphics program.

Thermal conductivity of a composite
Thermal conductivity of matrix 0.100 watt/metre/degree kelvin
Thermal conductivity of inserts
Conductivity ot insert Nos 1 was 1.000 watt/metre/degree kelvin
Effective thermal conductivity 0.261 watt/metre/degree kelvin
Sample dimensions $x = 3.000$ mm $y = 3.000$ mm
The unit cell is divided into 3 rows by 3 columns

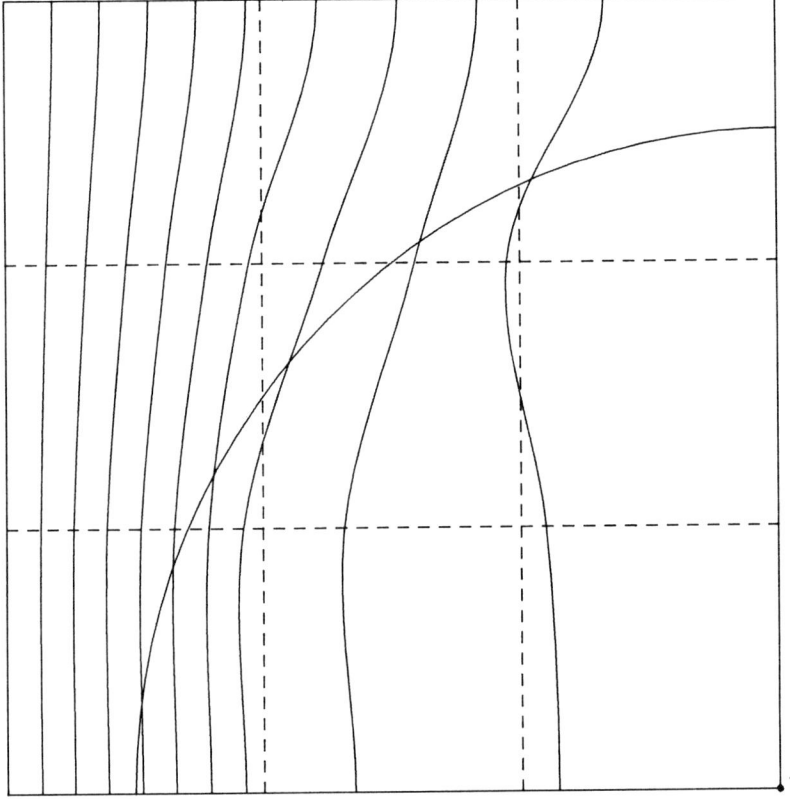

Figure 8. The output from the graphical program for the sample data given. Note that the centre of inserts is numbered as 1. These voltage contours, with 0.1 V steps, correspond to isotherms with 0.1 deg C steps

The data for the graphics program used for Figures 3, 4, and 8 is supplied from a file. Typical contents of this file are listed here under the title GRAPHICS PROGRAM INPUT FILE (p. 454). They consist of control and specification data, together with output from the subroutine WRITEGRAPH. The input file is typically.

1			Control integer
3	3	1	M N and number of inserts

3.0000 3.0000 0.1000		X_s Y_s K_m sample cell size and matrix thermal conductivity
1.000		K_f thermal conductivity of the fibre/matrix
2		Insert type (1 = rectangular, 2 = circular)
3.0000 0.0000 2.5000		The centre coordinates of the inserts and its radius
0.2614		K_e effective thermal conductivity
0.1325 .... 2.1421		
....	}	Positions of isotherms.
0.1887 .... 2.3280		
−1		Control integer to end data batch.

The same control and specification data can be used for a grid temperature plot together with the output on channel 3. An example of such a file is listed under GRID TEMPERATURES (p. 453).

6. RUNNING THE PROGRAM

The program was written for use at an interactive terminal. It is therefore provided with appropriate prompts. If interactive facilities are not available the data can be read from cards or a file of the format given in section 5. The prompts will then form part of the output and can be ignored.

The value of the effective thermal conductivity of a given composite varies slightly with the number of meshes in the resistor network and with the size of the representative cell, for a given number of meshes. The calculated value is nearest the true value for the largest number of meshes, and the approach to it as a function of the number of meshes should be investigated as it will illustrate the approximation contained in the model. As there is, inevitably, a limit to the storage available on a computer there is a consequential limit to the number of meshes that can be used. The program, listed here, can handle a maximum of 100 meshes, and requires about 62 K of storage. It is only limited to 100 meshes by the declared storage. A simple change will permit larger/smaller networks to be handled so long as they are compatible with the storage available on the computer. This feature may be useful in any development of the program beyond that given in this chapter. The greatest accuracy in the calculated value of the effective thermal conductivity is obtained by considering the smallest representative cell with the largest number of meshes possible.

In the example, given at the end of this chapter, only one-quarter of the basic cell of a square array of circular fibres is used (cf. Figure 8) since this is the minimum usable symmetry element compatible with the applied temperature gradient. (The minimum symmetry element of the cell is a sector of

45° about the centre because there are four planes of symmetry through the axis of the cell.) Apart from the square array of fibres considered in this chapter other possibilities are hexagonal and rectangular arrays of fibres. The minimum usable symmetry unit for a hexagonal array of fibres is a rectangle with side lengths in the ratio $\sqrt{3}:1$ and with a quarter fibre at opposite corners. In the case of a rectangular array of fibres one-quarter of a unit cell can be used with a quarter fibre in one corner. In this case the two orthogonal values of the effective thermal conductivity K_x and K_y have to be calculated separately. Composite systems with rectangular inserts can be considered by setting the appropriate control integer to 1 in the input data (see section 5). Other shapes of insert could also be considered by developing the appropriate subroutines.

Finally, some thermal conductivity data is given in Table 2 which will enable composites of some common materials to be investigated.

Table 2. Thermal conductivity of some common materials

Material	Thermal conductivity $(W\,m^{-1}\,(degK)^{-1})$
Silver	405
Copper	385
Aluminium	238
Brass	110
Steel	50
Concrete	1.5
Epoxy resin	0.23
Rock wool	0.042
Polystyrene (cellular)	0.035
Still air	0.024

REFERENCES

1. J. F. Nye, *Physical Properties of Crystals* (Oxford University Press, London, 1957).
2. A. D. Boardman, D. E. O'Connor, and P. A. Young, *Symmetry and its Applications in Science* (McGraw-Hill, London, 1973).
3. C. F. Kayan, *Trans. ASME*, **67,** 713 (1945).
4. H. W. Emmons, *Trans. ASME*, **65,** 607 (1943).
5. W. D. Collier, *The Solution of Heat Transfer Problems using HEATRAN*, TRG Report 2512 (R).

MODELLING OF UNIDIRECTIONAL COMPOSITE MATERIALS

```
A PROGRAM TO CALCULATE THE THERMAL CONDUCTIVITY OF COMPOSITES

      MASTER MAINPROGRAM
C
C     A PROGRAM TO CALCULATE THE THERMAL CONDUCTIVITY
C     OF COMPOSITE MATERIALS.
C
C
C     THE PROGRAM READS IN THE INSERT INFORMATION,
C     CALCULATES THE CURRENT THROUGH EACH RESISTOR,
C     AND INTERPOLATES THE 0.1 TO 0.9 VOLTAGE CONTOURS.
C
C     MAX. NUMBER OF RESISTOR VALUES IS 200 FOR STORAGE PROVIDED.
C     THIS IS OK FOR A SQUARE MESH OF SIDE UP TO & INCL. 10*10
C     OR A RECTANGULAR ONE WITH LESS THAN 200 RESISTORS.
C
      DIMENSION RNET(11000),R(210),IND(11000),IND2(11000),VOLTCURR(210)
      DIMENSION IW(210,13),IWW(22),INDW(22),W(220),VOLT(11)
      DIMENSION CONDINC(20),VOLTGRAPH(9)
C
      COMMON /NETWORK/RNET,IND,IL,IH,NUM
      COMMON /RESDATA/R,RXI,RYI,RXM,RYM
      COMMON /CURRVOLT/VOLTCURR
      COMMON /GENERAL/K
      COMMON /ANSCHECK/G,D1,ID2,IW
      COMMON /CONDDATA/CONDMAT,CONDINC,CONDEFF,XSIZE,YSIZE
C
      EQUIVALENCE (IW(1,6),W(1))
C
C     RECORD TIME OF RUN.
C
      CALL DATE(ADATE)
      CALL TIME(ATIME)
      WRITE(2,200) ADATE,ATIME
  200 FORMAT(1H1,' THIS RUN WAS DONE ON ',A8,' AT ',A8//)
C
C     READ IN DATA.
C
      CALL ARRAYSIZE(N,M)
C
C     GO TO ERROR LABEL IF THIS CONDITION IS TRUE.
C
      IF(N.EQ.1) GO TO 9
      CALL CALCRESNUM(N,M,IRNUM)
      CALL ARRAYZERO(IRNUM)
      CALL NUMOFINC(INCNUM)
C
C     CALCULATE ALL RESISTOR VALUES AND STORE THEM IN ARRAY R.
C
      DO 250 JJJ=1,INCNUM
      CALL ARRAYRES(IRNUM,JJJ,N,M)
      CALL INCLUSION(N,M,JJJ,INCNUM)
  250 CONTINUE
C
C     COMPENSATE FOR EDGE EFFECT.
C
      CALL EDGERESCOMP(N,M,IRNUM)
C
```

A PROGRAM TO CALCULATE THE THERMAL CONDUCTIVITY OF COMPOSITES

```
C     SET UP RIGHT HAND SIDES OF EQUATION.
C
      CALL SETVOLTARRAY(IRNUM)
C
      NUM=1
      IL=0
      IH=2
C
C     SET UP ALL THE SIMULTANEOUS EQUATIONS, FILLING ARRAYS
C     RNET AND IND, CONTROLLED BY THE SUBROUTINES BELOW.
C     THE RESISTOR VALUES ARE PUT INTO RNET IN COLUMN ORDER,
C     AND THE ROW INDEX IS STORED IN IND FOR THE NAG SUBROUTINES.
C     THE MESHES ARE WORKED THROUGH IN ROW ORDER,
C     LOOKING FOR THE PRESENCE OF THE PARTICULAR RESISTOR.
C
      DO 8 IC=1,IRNUM
C
      IW(IC,1)=NUM
C
      ICOUNT=0
C
C     CONCERNED WITH MESHES.
C
      DO 1 I=1,M
      DO 2 J=1,N
      ICOUNT=ICOUNT+1
      K=(I-1)*(2*N-1)
C
C     TEST FOR 3 OR 4 RESISTOR MESHES.
C
      IF(J.EQ.1.OR.J.EQ.N) GO TO 3
      CALL MESH4I(I,J,N,M,ICOUNT)
      GO TO 2
    3 CALL MESH3I(I,J,N,M,ICOUNT)
    2 CONTINUE
    1 CONTINUE
C
C     CONCERNED WITH MESH POINTS.
C
      MPLUS1=M+1
      NMINUS1=N-1
      DO 4 I=1,MPLUS1
      DO 5 J=1,NMINUS1
      ICOUNT=ICOUNT+1
      K=(I-1)*(2*N-1)
      IF(I.EQ.1.OR.I.EQ.MPLUS1) GO TO 6
C
C     TEST FOR 3 OR 4 RESISTOR POINT JUNCTIONS.
C
      CALL POINT4I(J,N,ICOUNT)
      GO TO 5
    6 CALL POINT3I(I,J,N,ICOUNT)
    5 CONTINUE
    4 CONTINUE
C
C     CONCERNED WITH ONE HORIZONTAL ROW.
```

A PROGRAM TO CALCULATE THE THERMAL CONDUCTIVITY OF COMPOSITES

```
C
      ICOUNT=ICOUNT+1
      CALL OVERALLI(N,ICOUNT)
C
      IL=IL+1
      IH=IH+1
C
    8 CONTINUE
C
      CALL TIME(ATIME)
      WRITE(2,202) ATIME
  202 FORMAT(' THE MATRIX IS NOW SET UP,TIME WAS ',A8//)
C
C     NOW SET UP THE FIRST UNOCCUPIED POSITION IN RNET.
C
      IW(IRNUM+1,1)=NUM
C
C     ERROR EXIT.
C
      IF(ICOUNT.NE.IRNUM) GO TO 7
C
C     DECREMENT NUM AS IT WAS AUTOINCREMENTED AFTER THE
C     LAST CORRECT POSITION FOUND.
C
      NUM=NUM-1
C
C     NUM NOW CONTAINS THE NUMBER OF THE LAST OCCUPIED POSITION
C     IN RNET.
C
C     SET UP PARAMETERS FOR NAG.
C
      IIW=IRNUM+10
      NW=(IRNUM/10)+1
      U=0.25
      JSCALE=0
      IFAIL=1
      IA=11000
C
C     SET IFAIL TO 1 FOR BOTH ROUTINES(SOFT FAILURE OPTION).
C
C     CALL NAG LIBRARY ROUTINES.
C
      CALL F03AJF(RNET,IND,IW,IIW,IND2,IRNUM,IWW,INDW,NW,6,U,
     CIA,D1,ID2,JSCALE,IFAIL)
      WRITE(2,100) IFAIL
  100 FORMAT(' FOR CALL TO F03A.IF,IFAIL WA` ',I2//)
      IF(IFAIL.EQ.0) GO TO 10
      WRITE(2,104) JSCALE
  104 FORMAT(' PROGRAM FAILURE,JSCALE WAS ',I4//)
C
C     PRINT ERROR INFORMATION IF FAILURE.
C
      CALL EXITPARAM
      STOP
   10 MTYPE=1
      IFAIL=1
```

A PROGRAM TO CALCULATE THE THERMAL CONDUCTIVITY OF COMPOSITES

```
C
C
      CALL F04APF(PNET,IND,IW,IIW,IRNUM,IA,W,VOLTCURR,MTYPE,IFAIL)
      WRITE(2,103) IFAIL
  103 FORMAT(' FOR CALL TO F04APF,IFAIL WAS ',I2//)
      IF(IFAIL.NE.0) STOP
C
      CALL TIME(ATIME)
      WRITE(2,203) ATIME
  203 FORMAT(' NAG ROUTINE FINISHED,TIME WAS ',A8//)
C
C
C     CALCULATE EFFECTIVE THERMAL CONDUCTIVITY.
C
      CALL EFFCOND(N,MPLUS1)
C
C     CALCULATE AND OUTPUT A VOLTAGE MAP OF MATRIX.
C
      CALL WRITE(N,MPLUS1)
C
C     CALCULATE AND OUTPUT VOLTAGE CONTOUR DATA.
C
      CALL WRITEGRAPH(N,MPLUS1)
C
C     OUTPUT EFFECTIVE THERMAL CONDUCTIVITY.
C
      WRITE(2,999) CONDEFF
  999 FORMAT(////' KEFFECTIVE = ',F12.6)
C
C
C     RECORD JOB FINISH TIME.
C
      CALL TIME(ATIME)
      WRITE(2,204) ATIME
  204 FORMAT(////' JOB FINISHED AT ',A9//)
C
C
C
C
      STOP
C
C     PROGRAM ERROR EXITS.
C
    7 WRITE(2,101) ICOUNT,IRNUM
  101 FORMAT(' THE NUMBER OF EQUATIONS IS NOT THE SAME AS THE NUMBER
     C'OF UNKNOWNS.'/
     C' THE NUMBER OF EQUATIONS = ',I4/
     C' THE NUMBER OF UNKNOWNS = ',I4/
     C' PROGRAM HALTED.'///)
    9 WRITE(2,102)
  102 FORMAT(' ERROR - THE PROGRAM DOES NOT HOLD FOR THE ONE COLUMN
     C'CASE.'/
     C' PROGRAM HALTED.'///)
      STOP
      END
C
```

A PROGRAM TO CALCULATE THE THERMAL CONDUCTIVITY OF COMPOSITES

```
C
C
      SUBROUTINE MESH4I(I,J,N,M,ICOUNT)
C
C     A PROGRAM TO SET UP IN THE MATRIX RNET,RESISTOR VALUES ASSOCIATED
C     OF A 4 RESISTANCE MESH.
C
      DIMENSION RNET(11000),R(210),IND(11000)
C
C     RNET IS THE ARRAY OF RESISTANCE VALUES.
C     R IS THE ARRAY CONTAINING THE VALUES OF EACH RESISTOR.
C
      COMMON /NETWORK/RNET,IND,IL,IH,NUM
      COMMON /RESDATA/R,RXI,RYI,RXM,RYM
      COMMON /GENERAL/K
C
C     IPOS IS THE NUMBER OF THE RESISTOR IN THE LOWEST POSITION
C     IN THE MESH.
C
      IPOS=K+2*J-1
C
C     PUT VALUES OF RESISTORS INTO ARRAY RNET.
C
C     CHECK IF RESISTOR LIES IN THE MESH.
C
      INUMROW=IPOS-1
      IF(INUMROW.GT.IL.AND.INUMROW.LT.IH) GO TO 1
      INUMROW=IPOS
      IF(INUMROW.GT.IL.AND.INUMROW.LT.IH) GO TO 1
      INUMROW=IPOS+1
      IF(INUMROW.GT.IL.AND.INUMROW.LT.IH) GO TO 2
      IF(I.EQ.M) GO TO 4
      INUMROW=IPOS+(2*N-1)
      IF(INUMROW.GT.IL.AND.INUMROW.LT.IH) GO TO 2
      RETURN
C
    4 INUMROW=K+(2*N-1)+J
      IF(INUMROW.GT.IL.AND.INUMROW.LT.IH) GO TO 2
      RETURN
C
    1 RNET(NUM)=-R(INUMROW)
      GO TO 3
C
    2 RNET(NUM)=R(INUMROW)
    3 IND(NUM)=ICOUNT
      NUM=NUM+1
      RETURN
      END
C
C
C
      SUBROUTINE MESH3I(I,J,N,M,ICOUNT)
C
C     A PROGRAM TO SET UP IN THE MATRIX RNET,RESISTOR VALUES ASSOCIATED
C     OF A 3 RESISTOR MESH.
C
```

A PROGRAM TO CALCULATE THE THERMAL CONDUCTIVITY OF COMPOSITES

```
      DIMENSION RNET(11000),R(210),IND(11000)
C
C     RNET IS THE ARRAY OF RESISTANCE VALUES.
C     R IS THE ARRAY CONTAINING THE VALUES OF EACH RESISTOR.
C
      COMMON /NETWORK/RNET,IND,IL,IH,NUM
      COMMON /RESDATA/R,RXI,RYI,RXM,RYM
      COMMON /GENERAL/K
C
C     TEST TO SEE IF THIS IS THE FIRST OR LAST MESH IN THE ROW.
C
      IF(J.GT.1) GO TO 1
C
C     IPOS IS THE NUMBER OF THE RESISTOR IN THE LOWEST POSITION
C     IN THE MESH.
C
      IPOS=K+1
C
      PUT VALUES OF RESISTORS INTO ARRAY RNET IF IN THE MESH.
C
      INUMROW=IPOS
      IF(INUMROW.GT.IL.AND.INUMROW.LT.IH) GO TO 2
      INUMROW=IPOS+1
      IF(INUMROW.GT.IL.AND.INUMROW.LT.IH) GO TO 3
      INUMROW=IPOS+(2*N-1)
      IF(INUMROW.GT.IL.AND.INUMROW.LT.IH) GO TO 3
      RETURN
C
    2 RNET(NUM)=-R(INUMROW)
      GO TO 4
C
    3 RNET(NUM)=R(INUMROW)
    4 IND(NUM)=ICOUNT
      NUM=NUM+1
      RETURN
C
C     THIS IS THE PROCESS FOR THE MESHES AT THE OTHER END
C     OF THE NETWORK.
C
    1 IPOS=K+2*J-1
      INUMROW=IPOS-1
      IF(INUMROW.GT.IL.AND.INUMROW.LT.IH) GO TO 2
      INUMROW=IPOS
      IF(INUMROW.GT.IL.AND.INUMROW.LT.IH) GO TO 2
      IF(I.EQ.M) GO TO 5
      INUMROW=IPOS+(2*N-1)
      IF(INUMROW.GT.IL.AND.INUMROW.LT.IH) GO TO 3
      RETURN
    5 INUMROW=IPOS+N
      IF(INUMROW.GT.IL.AND.INUMROW.LT.IH) GO TO 3
      RETURN
      END
C
C
C
      SUBROUTINE POINT3I(I,J,N,ICOUNT)
```

MODELLING OF UNIDIRECTIONAL COMPOSITE MATERIALS

A PROGRAM TO CALCULATE THE THERMAL CONDUCTIVITY OF COMPOSITES

```
C
C      A PROGRAM TO SET UP IN THE MATRIX RNET,RESISTOR VALUES ASSOCIATED
C      OF A 3 RESISTOR POINT LOCATION.
C
       DIMENSION RNET(11000),IND(11000)
C
C      RNET IS THE ARRAY OF RESISTANCE VALUES.
C
       COMMON /NETWORK/RNET,IND,IL,IH,NUM
       COMMON /GENERAL/K
C
C      TEST TO SEE IF THIS IS THE FIRST OR LAST ROW OF POINTS.
C
       IF(I.GT.1) GO TO 1
C
C      IPOS IS THE POSITION OF THE LOWEST AND MOST LEFTWARD
C      OF THE 3 RESISTORS USED AT EACH POINT.
C
       IPOS=2*J-1
C
C      SET APPROPRIATE ELEMENTS OF ARRAY TO 1.0
C
       INUMROW=IPOS
       IF(INUMROW.GT.IL.AND.INUMROW.LT.IH) GO TO 2
       INUMROW=IPOS+1
       IF(INUMROW.GT.IL.AND.INUMROW.LT.IH) GO TO 2
       INUMROW=IPOS+2
       IF(INUMROW.GT.IL.AND.INUMROW.LT.IH) GO TO 3
       RETURN
C
     2 RNET(NUM)=1.0
       GO TO 4
C
     3 RNET(NUM)=-1.0
     4 IND(NUM)=ICOUNT
       NUM=NUM+1
       RETURN
C
C      THIS IS THE PROCESS FOR THE POINTS ON THE TOP ROW OF THE MESH.
C
     1 IPOS=K+J
       IN=K-(2*N-1)
       INUMROW=IPOS
       IF(INUMROW.GT.IL.AND.INUMROW.LT.IH) GO TO 2
       INUMROW=IPOS+1
       IF(INUMROW.GT.IL.AND.INUMROW.LT.IH) GO TO 3
       INUMROW=IN+2*J
       IF(INUMROW.GT.IL.AND.INUMROW.LT.IH) GO TO 3
       RETURN
       END
C
C
C
       SUBROUTINE POINT4I(J,N,ICOUNT)
C
C      A PROGRAM TO SET UP IN THE ARRAY RNET,PARAMETERS ASSOCIATED
```

A PROGRAM TO CALCULATE THE THERMAL CONDUCTIVITY OF COMPOSITES

```
C     WITH A 4 RESISTOR POINT LOCATION.
C
      DIMENSION RNET(11000),IND(11000)
C
C     RNET IS THE ARRAY OF RESISTANCE VALUES.
C
      COMMON /NETWORK/RNET,IND,IL,IH,NUM
      COMMON /GENERAL/K
C
C     IPOS IS THE POSITION OF THE LOWEST AND LEFTMOST OF THE
C     4 RESISTORS USED AT EACH POINT.
C
      IPOS=K+(2*J-1)
C
C     SET APPROPRIATE COEFFICIENTS OF ARRAY TO 1.0
C
      INUMROW=IPOS
      IF(INUMROW.GT.IL.AND.INUMROW.LT.IH) GO TO 1
      INUMROW=IPOS+1
      IF(INUMROW.GT.IL.AND.INUMROW.LT.IH) GO TO 1
      INUMROW=IPOS+2
      IF(INUMROW.GT.IL.AND.INUMROW.LT.IH) GO TO 2
      INUMROW=IPOS-(2*N-2)
      IF(INUMROW.GT.IL.AND.INUMROW.LT.IH) GO TO 2
      RETURN
C
    1 RNET(NUM)=1.0
      GO TO 3
C
    2 RNET(NUM)=-1.0
    3 IND(NUM)=ICOUNT
      NUM=NUM+1
      RETURN
      END
C
C
C
      SUBROUTINE OVERALLI(N,ICOUNT)
C
C     THIS ROUTINE SETS UP ONE EQUATION USING ALL THE RESISTORS
C     ON THE BOTTOM (M=1) ROW.
C
      DIMENSION RNET(11000),R(210),IND(11000)
C
C     RNET IS THE ARRAY OF RESISTANCE VALUES.
C     R IS THE ARRAY CONTAINING THE VALUE OF EACH RESISTOR.
C
      COMMON /NETWORK/RNET,IND,IL,IH,NUM
      COMMON /RESDATA/R,RXI,RYI,RXM,RYM
C
C     CALCULATE THE NUMBER OF RESISTORS IN THE LINE AND
C     PUT VALUES OF RESISTORS INTO RNET.
C
      INUM=2*N-1
      DO 1 I=1,INUM,2
      IF(I.GT.IL.AND.I.LT.IH) GO TO 2
```

A PROGRAM TO CALCULATE THE THERMAL CONDUCTIVITY OF COMPOSITES

```
    1 CONTINUE
      RETURN
    2 RNET(NUM)=R(I)
      IND(NUM)=ICOUNT
      NUM=NUM+1
      RETURN
      END
C
C
C
      SUBROUTINE ARRAYZERO(IRNUM)
C
C     A ROUTINE TO SET TO ZERO THE RESISTANCE ARRAY.
C
      DIMENSION R(210)
      COMMON /RESDATA/R,RXI,RYI,RXM,RYM
C
C     IRNUM IS THE NUMBER OF RESISTORS AND CURRENTS.
C
      DO 1 I=1,IRNUM
      R(I)=0.0
    1 CONTINUE
      RETURN
      END
C
C
C
      SUBROUTINE CALCRESNUM(N,M,IRNUM)
C
C     A ROUTINE TO CALCULATE THE NUMBER OF RESISTORS/CURRENTS,.
C     FROM INPUT VALUES OF M AND N.
C
      IRNUM=((M+1)*N)+((N-1)*M)
      RETURN
      END
C
C
C
      SUBROUTINE ARRAYSIZE(N,M)
C
C     A ROUTINE TO READ IN THE NUMBER OF HORIZONTAL AND VERTICAL MESHES
C     AND THE OVERALL DIMENSIONS IN MILLIMETRES.
C
      DIMENSION CONDINC(20)
      COMMON /CONDDATA/CONDMAT,CONDINC,CONDEFF,XSIZE,YSIZE
      WRITE(2,101)
  101 FORMAT(' ENTER N AND M >')
      READ(1,100) N,M
  100 FORMAT(2I0)
      WRITE(2,102) N,M
  102 FORMAT(' N WAS ',I2,'   M WAS ',I2//)
      WRITE(2,103)
  103 FORMAT(' ENTER XSIZE(N-DIR) AND YSIZE(M-DIR) IN MM')
      READ(1,104) XSIZE,YSIZE
  104 FORMAT(2F0.0)
      WRITE(2,105) XSIZE,YSIZE
```

A PROGRAM TO CALCULATE THE THERMAL CONDUCTIVITY OF COMPOSITES

```
  105 FORMAT(' XSIZE(N-DIR) WAS ',F9.4,' MM   YSIZE(M-DIR) WAS ',
     CF9.4,' MM'//)
      RETURN
      END
C
C
C
      SUBROUTINE ARRAYRES(IRNUM,JJJ,N,M)
C
C     A ROUTINE TO SET UP THE VALUES OF WHOLE RESISTORS,
C     IN X AND Y DIRECTIONS,
C     FOR THE MATRIX AND EACH INCLUSION.
C
      DIMENSION R(210),CONDINC(20)
      COMMON /RESDATA/R,RXI,RYI,RXM,RYM
      COMMON /CONDDATA/CONDMAT,CONDINC,CONDEFF,XSIZE,YSIZE
C
      IF(JJJ.GT.1) GO TO 1
      WRITE(2,100)
  100 FORMAT(' ENTER THERMAL CONDUCTIVITY OF THE MESH >')
      READ(1,101) CONDMAT
  101 FORMAT(F0.0)
      WRITE(2,103) CONDMAT
  103 FORMAT(/' THE THERMAL CONDUCTIVITY OF THE MESH WAS ',F9.4//)
      CALL LINE
C
C
C
    1 WRITE(2,105) JJJ
  105 FORMAT(//' *** INFORMATION ABOUT INSERT NUMBER ',I2,' ***'//)
      WRITE(2,102)
  102 FORMAT(' ENTER THERMAL CONDUCTIVITY OF THE INCLUSION(S) >')
      READ(1,101) CONDINC(JJJ)
      WRITE(2,104) CONDINC(JJJ)
  104 FORMAT(/' THE THERMAL CONDUCTIVITY OF THE INCLUSION(S) WAS ',
     CF9.4//)
C
      IF(JJJ.GT.1) GO TO 3
      RXM=XSIZE/YSIZE/CONDMAT*M/N
      RYM=YSIZE/XSIZE/CONDMAT*N/M
    3 RXI=XSIZE/YSIZE/CONDINC(JJJ)*M/N
      RYI=YSIZE/XSIZE/CONDINC(JJJ)*N/M
      RETURN
      END
C
C
C
      SUBROUTINE EFFCOND(N,MPLUS1)
C
C     A ROUTINE TO CALCULATE THE EFFECTIVE CONDUCTIVITY,
C     BY SUMMING THE CURRENT VALUES ACROSS A PLANE
C     PERPENDICULAR TO THE APPLIED TEMPERATURE.
C
      DIMENSION CONDINC(20),VOLTCURR(210),R(210)
      COMMON /RESDATA/R,RXI,RYI,RXM,RYM
      COMMON /CONDDATA/CONDMAT,CONDINC,CONDEFF,XSIZE,YSIZE
```

A PROGRAM TO CALCULATE THE THERMAL CONDUCTIVITY OF COMPOSITES

```
      COMMON /CURRVOLT/VOLTCURR
C
      INTEGER SUMLOW
      SUMLOW=2*N-1
      SUM=0.0
      DO 1 I=1,MPLUS1
      LOC=1+(I-1)*SUMLOW
      SUM=SUM+VOLTCURR(LOC)
    1 CONTINUE
      CONDEFF=SUM/YSIZE*XSIZE
      WRITE(6,100) CONDEFF
      WRITE(3,100) CONDEFF
  100 FORMAT(1X,F10.4)
      RETURN
      END
C
C
C
      SUBROUTINE WRITE(N,MPLUS1)
C
C     A ROUTINE TO CALCULATE AND OUTPUT THE VOLTAGE MAP,.
C     OF THE MATRIX.
C
C     VOLT IS AN ARRAY TO HOLD A LINE OF OUTPUT.
C
      DIMENSION VOLTCURR(210),VOLT(11),R(210)
      COMMON /RESDATA/R,RXI,RYI,RXM,RYM
      COMMON /CURRVOLT/VOLTCURR
C
C     ZERO-ISE ARRAY VOLT.
C
      DO 5 I=1,11
      VOLT(I)=0.0
    5 CONTINUE
C
C     WRITE OUT FIRST LINE OF RESULTS,IE ALL ZERO'S.
C
      WRITE(3,100) (VOLT(I),I=1,MPLUS1)
C
C     GO THROUGH RESULTS ONE VERTICAL LINE AT A TIME.
C
      DO 1 J=1,N
      DO 2 I=1,MPLUS1
      K=(I-1)*(2*N-1)
      IF(I.EQ.MPLUS1) GO TO 3
      LOC=K+2*J-1
      GO TO 4
    3 LOC=K+J
    4 VOLT(I)=VOLT(I)+VOLTCURR(LOC)*R(LOC)
    2 CONTINUE
      WRITE(3,100) (VOLT(IM),IM=1,MPLUS1)
    1 CONTINUE
  100 FORMAT(1X,11(F7.4,4X))
C
      RETURN
      END
```

A PROGRAM TO CALCULATE THE THERMAL CONDUCTIVITY OF COMPOSITES

```
C
C
C
      SUBROUTINE SETVOLTARRAY(IRNUM)
C
C     A ROUTINE TO SET UP THE ARRAY OF VOLTAGES,.
C     FORMING THE RHS OF THE EQUATION.
C
      DIMENSION VOLTCURR(210)
C
      COMMON /CURRVOLT/VOLTCURR
C
      IEND=IPNUM-1
      DO 1 I=1,IEND
      VOLTCURR(I)=0.0
    1 CONTINUE
C
      VOLTCURR(IRNUM)=1.0
C
      RETURN
      END
C
C
C
      SUBROUTINE EXITPARAM
C
C     A ROUTINE TO WRITE OUT THE VARIOUS OUTPUT PARAMETERS OF THE NAG
C     IT IS CALLED IN ERROR CONDITION.
C
C     ROUTINES ON EXIT.
C
      DIMENSION IW(210,13)
C
      COMMON /NETWORK/RNET,IND,IL,IH,NUM
      COMMON /ANSCHECK/G,D1,ID2,IW
C
      NRELM=IW(IRNUM+1,1)-1
      WRITE(2,207) NRELM
  207 FORMAT(1H1,' THE NUMBER OF ELEMENTS IN L/U = ',I5//)
      WRITE(2,200) G
  200 FORMAT(' G = ',E15.8//)
      WRITE(2,201) D1,ID2
  201 FORMAT(' D1 = ',E15.8,'   ID2 = ',I5//)
C
C
      WRITE(2,205)
  205 FORMAT(' COLUMN PIV',18X,'COL SP',19X,'ROW SP',
     C16X,'ORDER PIV ROW'//)
      DO 203 I=1,NUM
      WRITE(2,204) IW(I,2),IW(I,3),IW(I,4),IW(I,5)
  203 CONTINUE
  204 FORMAT(1X,4(I10,15X))
      WRITE(2,206)
  206 FORMAT(//////)
C
      RETURN
```

A PROGRAM TO CALCULATE THE THERMAL CONDUCTIVITY OF COMPOSITES

```
      END
C
C
C
      SUBROUTINE WRITEGRAPH(N,MPLUS1)
C
C     A ROUTINE TO WRITE OUT DATA TO A FILE FOR USE WITH THE
C     GINO-F PROGRAM.
C     IT CALCULATES THE POSITIONS OF THE 0.1 VOLT INCREMENTS
C     ACROSS THE SAMPLE, ALONG THE ROWS OF THE MESH.
C
      DIMENSION VOLTGRAPH(9),VOLTCURR(210),R(210),CONDINC(20)
      COMMON /RESDATA/R,RXI,RYI,RXM,RYM
      COMMON /CURRVOLT/VOLTCURR
      COMMON /CONDDATA/CONDMAT,CONDINC,CONDEFF,XSIZE,YSIZE
C
C
      WRITE(2,200)
  200 FORMAT(//' POSITIONS OF 0.1 V INCPEMENTS ACROSS THE SAMPLE.'/)
C
      SUMLOW=2*N-1
      XINC=XSIZE/FLOAT(N)
      DO 1 I=1,MPLUS1
      V1=0.0
      V2=0.0
      REMAIN=0.0
      TEMP=0.1
      IJ=1
      DO 2 J=1,N
      K=(I-1)*SUMLOW
      LOC=K+2*J-1
      IF(I.EQ.MPLUS1) LOC=K+J
      V2=V1+VOLTCURR(LOC)*R(LOC)
      IF(V2.LT.TEMP.AND.ABS(TEMP-V2).GT.1.0E-6) GO TO 4
      VDIFF=V2-V1
    3 VOLTGRAPH(IJ)=(TEMP-V1)/VDIFF*XINC+REMAIN
      IJ=IJ+1
      IF(IJ.EQ.10) GO TO 5
      TEMP=TEMP+0.1
      IF(TEMP.LE.V2) GO TO 3
    4 V1=V2
      REMAIN=FLOAT(J)*XINC
    2 CONTINUE
    5 WRITE(6,100) (VOLTGRAPH(IK),IK=1,9)
      WRITE(2,100) (VOLTGRAPH(IK),IK=1,9)
  100 FORMAT(1X,9F8.4)
    1 CONTINUE
C
C     PRINT END LABEL FOR GRAPHICS PROGRAM.
C
      IC=-1
      WRITE(6,600) IC
  600 FORMAT(1X,I4)
C
      RETURN
      END
```

A PROGRAM TO CALCULATE THE THERMAL CONDUCTIVITY OF COMPOSITES

```
C
C
C
      SUBROUTINE INCLUSION(N,M,JJJ,INCNUM)
      DIMENSION CONDINC(20)
      COMMON /CONDDATA/CONDMAT,CONDINC,CONDEFF,XSIZE,YSIZE
C
C     READ IN INCLUSION DATA AND
C     OUTPUT DATA TO THE GRAPHICS DATA FILE.
C
C
      ICON=1
      IF(JJJ.EQ.1) WRITE(6,605) ICON
      IF(JJJ.EQ.1) WRITE(3,605) ICON
  605 FORMAT(1X,I5)
C
      IF(JJJ.EQ.1) WRITE(6,601) M,N,INCNUM
      IF(JJJ.EQ.1) WRITE(3,601) M,N,INCNUM
  601 FORMAT(1X,3(I3,4X))
      IF(JJJ.EQ.1) WRITE(6,606)XSIZE,YSIZE,CONDMAT
      IF(JJJ.EQ.1) WRITE(3,606)XSIZE,YSIZE,CONDMAT
  606 FORMAT(1X,3(F10.4,2X))
      WRITE(6,604) CONDINC(JJJ)
      WRITE(3,604) CONDINC(JJJ)
  604 FORMAT(1X,F10.4)
      WPITE(2,100)
  100 FORMAT(' ENTER SHAPE CODE:1=SQUARE OR RECTANGLE,2=CIRCLE')
      READ(1,101) ISC
  101 FORMAT(I0)
C
C
      WRITE(2,104) ISC
  104 FORMAT(' SHAPE CODE WAS ',I2//)
      WRITE(6,600) ISC
      WRITE(3,600) ISC
  600 FORMAT(1X,I4)
C
      WRITE(2,102)
  102 FORMAT(' ENTER CENTRE POSITION (N/X,M/Y) ')
      READ(1,103) XC,YC
  103 FORMAT(2F0.0)
      WRITE(2,110) XC,YC
  110 FORMAT(/' CENTRE POSITION IN "N/X" DIRECTION WAS ',F12.6/
     C' CENTRE POSITION IN "M/Y" DIRECTION WAS ',F12.6//)
      WRITE(2,107)
  107 FORMAT(//' ++++++ALL SIZES MUST BE REAL.++++++'//)
      GO TO (1,2) ,ISC
    1 WRITE(2,114)
  114 FORMAT(' ENTER XL & YL IN MM.'/
     C' (SIDE SIZES IN N/X & M/Y DIR.RESP.) >')
      READ(1,106) XL,YL
  106 FORMAT(2F0.0)
      WRITE(2,105) XL,YL
  105 FORMAT(' XL (N-DIR) WAS ',F12.6,'   YL (M-DIR) WAS ',F12.6//)
      HXL=XL/2.0
      HYL=YL/2.0
```

MODELLING OF UNIDIRECTIONAL COMPOSITE MATERIALS   443

A PROGRAM TO CALCULATE THE THERMAL CONDUCTIVITY OF COMPOSITES

```
      WRITE(6,602) XC,YC,HXL,HYL
      WRITE(3,602) XC,YC,HXL,HYL
  602 FORMAT(1X,4(F10.4,2X))
C
C     CALCULATE RESISTOR VALUES FOR RECTANGULAR INSERTS.
C
      CALL INCSC1(N,M,XL,YL,XC,YC)
      CALL LINE
      RETURN
C
    2 WRITE(2,108)
  108 FORMAT(' ENTER RADIUS IN MM >')
      READ(1,106) RAD
      WRITE(2,109) RAD
  109 FORMAT(' RADIUS WAS ',F12.6//)
      WRITE(6,603) XC,YC,RAD
      WRITE(3,603) XC,YC,RAD
  603 FORMAT(1X,3(F10.4,2X))
C
C
C     CALCULATE RESISTOR VALUES FOR CIRCULAR INSERTS.
C
      CALL INCSC2(N,M,RAD,XC,YL)
C
      CALL LINE
C
      RETURN
      END
C
C
C
      SUBROUTINE LINE
C
C     A ROUTINE TO DRAW A SEPARATING LINE.
C
      WRITE(2,100)
  100 FORMAT(////' +++++++++++++++++++++++++++++++++++++++++'////)
C
      RETURN
      END
C
C
C
      SUBROUTINE EDGERESCOMP(N,M,IRNUM)
C
C     A ROUTINE TO DOUBLE THE VALUE OF ALL THE EDGE RESISTORS,
C     AS THESE ARE SHARED WITH THE ADJACENT ROW OF THE MESH.
C
      DIMENSION R(210)
      COMMON /RESDATA/R,RXI,RYI,RXM,RYM
      INTEGER SUMLOW
      SUMLOW=2*N-1
      DO 1 I=1,SUMLOW,2
      R(I)=2.0*R(I)
    1 CONTINUE
      ILOW=M*SUMLOW+1
```

A PROGRAM TO CALCULATE THE THERMAL CONDUCTIVITY OF COMPOSITES

```
      DO 2 I=ILOW,IRNUM
      P(I)=2.0*R(I)
    2 CONTINUE
      RETURN
      END
C
C
C
      SUBROUTINE NUMOFINC(INCNUM)
C
C     A ROUTINE TO READ IN THE NUMBER OF INCLUSIONS.
C
      WRITE(2,100)
  100 FORMAT(' ENTER NUMBER OF INCLUSIONS')
      READ(1,101) INCNUM
  101 FORMAT(I0)
      WRITE(2,102) INCNUM
  102 FORMAT(' THE NUMBER OF INCLUSIONS WAS ',I2//)
      RETURN
      END
C
C
C
      SUBROUTINE INCSC1(N,M,XL,YL,XC,YC)
C
C     A ROUTINE TO CALCULATE THE RESISTOR VALUES FOR A
C     SQUARE/RECTANGULAR INSERT.
C
      DIMENSION R(210),CONDINC(20)
      COMMON /RESDATA/R,RXI,RYI,RXM,RYM
      COMMON /CONDDATA/CONDMAT,CONDINC,CONDEFF,XSIZE,YSIZE
      INTEGER SUMLOW
C
C
      HXL=XL/2.0
      HYL=YL/2.0
      XINC=XSIZE/N
      YINC=YSIZE/M
      SUMLOW=2*N-1
C
C     ICROSS1/2=1 IF MESH POINT IS IN INSERT ELSE =0.
C
C
C     WORK ALONG HORIXONTAL MESH LINES.
C
      RMV=RXM
      RIV=RXI
C
      DO 2 J=1,M+1
      DO 1 I=1,N+1
C
C     CALCULATE PARTICULAR RESISTOP NUMBER.
C
      NUM=(J-1)*SUMLOW+(I-1)*2-1
C
C     SPECIAL CASE FOR TOP ROW OF RESISTORS.
```

MODELLING OF UNIDIRECTIONAL COMPOSITE MATERIALS

```
A PROGRAM TO CALCULATE THE THERMAL CONDUCTIVITY OF COMPOSITES
C
      IF(J.EQ.M+1) NUM=(J-1)*SUMLOW+(I-1)
C     CALCULATE OFFSETS FROM INSERT CENTRE TO RESISTOR.
C
      XPOS=FLOAT(I-1)*XINC
      YPOS=FLOAT(J-1)*YINC
      DX=ABS(XC-XPOS)
      DY=ABS(YC-YPOS)
C
C     CHECK IF RESISTOR IS INSIDE INSERT.
C     GO TO 9 = INSIDE.
C     GO TO 3 = OUTSIDE.
C
      IF((ABS(DX-HXL).LT.1.0E-6.AND.DY.LE.HYL).OR.
     C(ABS(DY-HYL).LT.1.0E-6.AND.DX.LE.HXL).OR.
     C(ABS(DX-HXL).LT.1.0E-6.AND.ABS(DY-HYL).LT.1.0E-6)) GO TO 9
      IF(DX.GT.HXL.OR.DY.GT.HYL) GO TO 3
C
C     CHECK FOR FIRST COLUMN OF MESH POINTS.
C
    9 IF(I.EQ.1) GO TO 5
C
C     INSIDE INSERT.
C
      ICROSS2=1
C
C     IF ALL RESISTOR LENGTH IS IN INSERT, GO TO 8.
C
      IF(ICROSS1.EQ.1) GO TO 8
C
C     CALCULATE WHAT PART OF RESISTOR LIES INSIDE INSERT.
C
      DIST=DX+XINC
      RPART=DIST-HXL
C
C     CALCULATE VALUE OF THAT RESISTOR.
C
      RVAR=RPART/XINC*RMV+(XINC-RPART)/XINC*RIV
C
C     CHECK IF RESISTOR HAS BEEN ASSIGNED A VALUE ALREADY.
C
      IF(R(NUM).GT.1.0E-6.AND.ABS(R(NUM)-RMV).GT.1.0E-6) GO TO 7
C
C     SET VALUE OF RESISTOR.
C
      R(NUM)=RVAR
      GO TO 7
C
C     CHECK IF RESISTOR HAS BEEN ASSIGNED A VALUE ALREADY.
C
    8 IF(R(NUM).GT.1.0E-6.AND.ABS(R(NUM)-RMV).GT.1.0E-6) GO TO 7
C
C     SET RESISTOR TO VALUE OF RINSERT.
C
      R(NUM)=RIV
      GO TO 7
```

A PROGRAM TO CALCULATE THE THERMAL CONDUCTIVITY OF COMPOSITES

```
C
C      SHOW MESH POINT IS IN INSERT.
C
     5 ICROSS1=1
       GO TO 1
C
C      CHECK FOR FIRST COLUMN OF MESH POINTS.
C
     3 IF(I.EQ.1) GO TO 4
C
C      SHOW POINT IS IN MATRIX.
C
       ICROSS2=0
C
C      IF ALL RESISTOR LENGTH IS IN MATRIX, GO TO 6.
C
       IF(ICPOSS1.EQ.0) GO TO 6
C
C      CALCULATE WHAT PART OF RESISTOR LIES IN THE MATPIX.
C
       RPART=DX-HXL
       RVAR=RPART/XINC*RMV+(XINC-RPART)/XINC*RIV
C
C      CHECK IF RESISTOR HAS BEEN ASSIGNED A VALUE ALREADY.
C
       IF(R(NUM).GT.1.0E-6.AND.ABS(R(NUM)-RMV).GT.1.0E-6) GO TO 7
C
C      NO, SO SET VALUE TO THAT CALCULATED.
C
       R(NUM)=RVAR
       GO TO 7
C
C      SHOW MESH POINT IS IN THE MATRIX.
C
     4 ICROSS1=0
       GO TO 1
C
C      IF RESISTOR VALUE ALREADY SET, THEN SKIP.
C
     6 IF(R(NUM).GT.1.0E-6.AND.ABS(R(NUM)-RMV).GT.1.0E-6) GO TO 7
C
C      SET RESISTOR TO VALUE OF RMATRIX.
C
       R(NUM)=RMV
C
C      MAKE LATEST MESH POINT "PREVIOUS MESH POINT".
C
     7 ICROSS1=ICROSS2
     1 CONTINUE
     2 CONTINUE
C
C      REPEAT AND WORK ALONG VERTICAL MESH LINES.
C
       PMV=RYM
       RIV=RYI
C
```

MODELLING OF UNIDIRECTIONAL COMPOSITE MATERIALS   447

A PROGRAM TO CALCULATE THE THERMAL CONDUCTIVITY OF COMPOSITES

```
      DO 12 I=1,N-1
      DO 11 J=1,M+1
      NUM=2*I+(J-2)*SUMLOW
      XPOS=FLOAT(I)*XINC
      YPOS=FLOAT(J-1)*YINC
      DX=ABS(XC-XPOS)
      DY=ABS(YC-YPOS)
      IF((ABS(DX-HXL).LT.1.0E-6.AND.DY.LE.HYL).OR.
     C(ABS(DY-HYL).LT.1.0E-6.AND.DX.LE.HXL).OR.
     C(ABS(DX-HXL).LT.1.0E-6.AND.ABS(DY-HYL).LT.1.0E-6)) GO TO 19
      IF(DX.GT.HXL.OR.DY.GT.HYL) GO TO 13
   19 IF(J.EQ.1) GO TO 15
      ICROSS2=1
      IF(ICROSS1.EQ.1) GO TO 18
      DIST=DY+YINC
      RPART=DIST-HYL
      RVAR=RPART/YINC*RMV+(YINC-RPART)/YINC*PIV
      IF(R(NUM).GT.1.0E-6.AND.ABS(R(NUM)-RMV).GT.1.0E-6) GO TO 17
      R(NUM)=RVAR
      GO TO 17
   18 IF(R(NUM).GT.1.0E-6.AND.ABS(R(NUM)-RMV).GT.1.0E-6) GO TO 17
      R(NUM)=RIV
      GO TO 17
   15 ICROSS1=1
      GO TO 11
   13 IF(J.EQ.1) GO TO 14
      ICROSS2=0
      IF(ICROSS1.EQ.0) GO TO 16
      RPART=DY-HYL
      RVAR=RPART/YINC*RMV+(YINC-RPART)/YINC*RIV
      IF(R(NUM).GT.1.0E-6.AND.ABS(R(NUM)-RMV).GT.1.0E-6) GO TO 17
      R(NUM)=RVAR
      GO TO 17
   14 ICROSS1=0
      GO TO 11
   16 IF(R(NUM).GT.1.0E-6.AND.ABS(R(NUM)-RMV).GT.1.0E-6) GO TO 17
      R(NUM)=RMV
   17 ICROSS1=ICROSS2
   11 CONTINUE
   12 CONTINUE
      RETURN
      END
C
C
C
      SUBROUTINE INCSC2(N,M,RAD,XC,YC)
C
C     A ROUTINE TO CALCULATE THE RESISTOR VALUES FOR A
C     CIRCULAR INSERT.
C
      DIMENSION R(210),CONDINC(20)
      COMMON /RESDATA/R,RXI,RYI,RXM,RYM
      COMMON /CONDDATA/CONDMAT,CONDINC,CONDEFF,XSIZE,YSIZE
      INTEGER SUMLOW
C
C
```

A PROGRAM TO CALCULATE THE THERMAL CONDUCTIVITY OF COMPOSITES

```
      XINC=XSIZE/N
      YINC=YSIZE/M
      SUMLOW=2*N-1
C
C     ICROSS1/2=1 IF MESH POINT IS IN INSERT ELSE =0.
C
C
C     WORK ALONG HORIZONTAL MESH LINES.
C
      RMV=RXM
      RIV=RXI
C
      DO 2 J=1,M+1
      DO 1 I=1,N+1
      NUM=(J-1)*SUMLOW+(I-1)*2-1
      IF(J.EQ.M+1) NUM=(J-1)*SUMLOW+(I-1)
      XPOS=FLOAT(I-1)*XINC
      YPOS=FLOAT(J-1)*YINC
      DIAG=SQRT(ABS((XC-XPOS)**2)+ABS((YC-YPOS)**2))
      IF(ABS(DIAG-RAD).LT.1.0E-6) GO TO 9
      IF(DIAG.GT.RAD) GO TO 3
    9 IF(I.EQ.1) GO TO 5
      ICROSS2=1
      IF(ICROSS1.EQ.1) GO TO 8
      DIST1=XC-XPOS+XINC
      DIST2=SQRT(RAD**2-ABS((YC-YPOS)**2))
      RPART=DIST1-DIST2
      RVAR=RPART/XINC*RMV+(XINC-RPART)/XINC*RIV
      IF(R(NUM).GT.1.0E-6.AND.ABS(R(NUM)-RMV).GT.1.0E-6) GO TO 7
      R(NUM)=RVAR
      GO TO 7
    8 IF(R(NUM).GT.1.0E-6.AND.ABS(R(NUM)-RMV).GT.1.0E-6) GO TO 7
      R(NUM)=RIV
      GO TO 7
    5 ICROSS1=1
      GO TO 1
    3 IF(I.EQ.1) GO TO 4
      ICROSS2=0
      IF(ICROSS1.EQ.0) GO TO 6
      DIST1=XPOS-XC
      DIST2=SQRT(RAD**2-ABS((YC-YPOS)**2))
      RPART=DIST1-DIST2
      RVAR=RPART/XINC*RMV+(XINC-RPART)/XINC*RIV
      IF(R(NUM).GT.1.0E-6.AND.ABS(R(NUM)-RMV).GT.1.0E-6) GO TO 7
      R(NUM)=RVAR
      GO TO 7
    4 ICROSS1=0
      GO TO 1
    6 IF(R(NUM).GT.1.0E-6.AND.ABS(R(NUM)-RMV).GT.1.0E-6) GO TO 7
      R(NUM)=RMV
    7 ICROSS1=ICROSS2
    1 CONTINUE
    2 CONTINUE
C
C     WORK ALONG VERTICAL MESH LINES.
C
```

A PROGRAM TO CALCULATE THE THERMAL CONDUCTIVITY OF COMPOSITES

```
      RMV=RYM
       RIV=RYI
C
      DO 12 I=1,N-1
      DO 11 J=1,M+1
      NUM=2*I+(J-2)*SUMLOW
      XPOS=FLOAT(I)*XINC
      YPOS=FLOAT(J-1)*YINC
      DIAG=SQRT(ABS((XC-XPOS)**2)+ABS((YC-YPOS)**2))
      IF(ABS(DIAG-RAD).LT.1.0E-6) GO TO 19
      IF(DIAG.GT.RAD) GO TO 13
   19 IF(J.EQ.1) GO TO 15
      ICROSS2=1
      IF(ICROSS1.EQ.1) GO TO 18
       DIST1=YC-YPOS+YINC
      DIST2=SQRT(RAD**2-ABS((XC-XPOS)**2))
      RPART=DIST1-DIST2
      RVAP=RPART/YINC*RMV+(YINC-RPART)/YINC*RIV
      IF(R(NUM).GT.1.0E-6.AND.ABS(R(NUM)-RMV).GT.1.0E-6) GO TO 17
      R(NUM)=RVAR
      GO TO 17
   18 IF(R(NUM).GT.1.0E-6.AND.ABS(R(NUM)-RMV).GT.1.0E-6) GO TO 17
      R(NUM)=RIV
      GO TO 17
   15 ICROSS1=1
      GO TO 11
   13 IF(J.EQ.1) GO TO 14
      ICROSS2=0
      IF(ICROSS1.EQ.0) GO TO 16
      DIST1=YPOS-YC
      DIST2=SQRT(RAD**2-ABS((XC-XPOS)**2))
      RPART=DIST1-DIST2
      RVAR=RPART/YINC*RMV+(YINC-RPART)/YINC*RIV
      IF(R(NUM).GT.1.0E-6.AND.ABS(R(NUM)-RMV).GT.1.0E-6) GO TO 17
      R(NUM)=RVAR
      GO TO 17
   14 ICROSS1=0
      GO TO 11
   16 IF(R(NUM).GT.1.0E-6.AND.ABS(R(NUM)-RMV).GT.1.0E-6) GO TO 17
      R(NUM)=RMV
   17 ICROSS1=ICROSS2
   11 CONTINUE
   12 CONTINUE
      RETURN
      END
C
C
C
      FINISH
C
C
C
```

MAIN CALCULATION PROGRAM INPUT FILE

```
3 3
3.0 3.0
1
0.1
1.0
2
3.0 0.0
2.5
```

MAIN PROGRAM OUTPUT

THIS RUN WAS DONE ON 20/10/78 AT 20/49/47

ENTER N AND M >
N WAS 3   M WAS 3

ENTER XSIZE(N-DIR) AND YSIZE(M-DIR) IN MM
XSIZE(N-DIR) WAS   3.0000 MM   YSIZE(M-DIR) WAS   3.0000 MM

ENTER NUMBER OF INCLUSIONS
THE NUMBER OF INCLUSIONS WAS 1

ENTER THERMAL CONDUCTIVITY OF THE MESH >

THE THERMAL CONDUCTIVITY OF THE MESH WAS   0.1000

++

*** INFORMATION ABOUT INSERT NUMBER 1 ***

ENTER THERMAL CONDUCTIVITY OF THE INCLUSION(S) >

THE THERMAL CONDUCTIVITY OF THE INCLUSION(S) WAS   1.0000

ENTER SHAPE CODE:1=SQUARE OR RECTANGLE,2=CIRCLE
SHAPE CODE WAS 2

ENTER CENTRE POSITION (N/X,M/Y)

CENTRE POSITION IN "N/X" DIRECTION WAS   3.000000
CENTRE POSITION IN "M/Y" DIRECTION WAS   0.000000

++++++ALL SIZES MUST BE REAL.++++++

ENTER RADIUS IN MM >
RADIUS WAS   2.500000

MAIN PROGRAM OUTPUT

++

THE MATRIX IS NOW SET UP,TIME WAS 20/49/49

FOR CALL TO F03AJF,IFAIL WAS   0

FOR CALL TO F04APF,IFAIL WAS   0

NAG ROUTINE FINISHED,TIME WAS 20/49/50

POSITIONS OF 0.1 V INCREMENTS ACROSS THE SAMPLE.

```
 0.1325  0.2649  0.3974  0.5298  0.6623  0.7948  0.9272  1.3506  2.1421
 0.1317  0.2634  0.3951  0.5268  0.6585  0.7902  0.9219  1.3126  2.0958
 0.1577  0.3154  0.4731  0.6308  0.7886  0.9463  1.2348  1.5909  1.9471
 0.1887  0.3775  0.5662  0.7550  0.9437  1.2183  1.5295  1.8407  2.3280
```

KEFFECTIVE =   0.261443

JOB FINISHED AT 20/49/50

GRID TEMPERATURES

```
    1
 3      3      1
    3.0000     3.0000     0.1000
    1.0000
    2
    3.0000     0.0000     2.5000
    0.2614
0.0000    0.0000    0.0000    0.0000
0.7550    0.7593    0.6341    0.5298
0.8834    0.8894    0.9149    0.8512
1.0000    1.0000    1.0000    1.0000
```

GRAPHICS PROGRAM INPUT FILE

```
     1
  3     3     1
    3.0000        3.0000        0.1000
    1.0000
    2
    3.0000        0.0000        2.5000
    0.2614
  0.1325  0.2649  0.3974  0.5298  0.6623  0.7948  0.9272  1.3506  2.1421
  0.1317  0.2634  0.3951  0.5268  0.6585  0.7902  0.9219  1.3126  2.0958
  0.1577  0.3154  0.4731  0.6308  0.7886  0.9463  1.2348  1.5909  1.9471
  0.1887  0.3775  0.5662  0.7550  0.9437  1.2183  1.5295  1.8407  2.3280
 -1
```

A PROGRAM TO PLOT AN ISOTHERM CONTOUR MAP

```
      MASTER CONTOURMAP
C
C     A PROGRAM TO PLOT AN ISOTHERM CONTOUR MAP.
C
      COMMON/GFCURV/CB,SB,CF,SF,XB,YB,XF,YF
      DIMENSION V(9,101),D(2,20),VX(101),YY(101)
      DIMENSION POS(2,20),G(20),IC(20),B(20)
C
C     THE PROGRAM PRODUCES AN ISOTHERM CONTOUR MAP
C     FOR A COMPOSITE MATERIAL FROM DATA CALCULATED
C     BY THE MAIN ANALYSIS PROGRAM.
C     THIS PROGRAM USES SUBROUTINES FROM THE GINO-F
C     GRAPHICS PACKAGE.
C
      CALL PLOTNA(7)
      CALL DEVPAP(270.,270.,IT)
      CALL WINDOW(2)
C
      CALL SCALE(0.75)
      CALL SHIFT2(46.0,46.0)
C
C     READ DATA IN.
C
   14 READ(5,110)ICON
  110 FORMAT(1I0)
      IF(ICON.EQ.-1) GO TO 25
      READ(5,101)M,N,IP
  101 FORMAT(3I0)
      READ(5,102)XS,YS,A
  102 FORMAT(3F0.0)
      DO 1 I=1,IP
      READ(5,107) B(I)
  107 FORMAT(F0.0)
      READ(5,105) IC(I)
  105 FORMAT(I0)
      IF(IC(I).EQ.1) READ(5,103) D(1,I),D(2,I),POS(1,I),POS(2,I)
  103 FORMAT(4F0.0)
      IF(IC(I).EQ.2) READ(5,106) D(1,I),D(2,I),G(I)
  106 FORMAT(3F0.0)
    1 CONTINUE
      READ(5,108) SI
  108 FORMAT(F0.0)
      K=M+1
      DO 2 I=1,K
    2 READ(5,104)(V(J,I),J=1,9)
  104 FORMAT(9F0.0)
C
C     CALCULATE SCALING FACTOR FOR MESH.
C
      IF(YS-XS) 15,15,16
   15 FAC=200.0/XS
      GOTO 17
   16 FAC=200.0/YS
   17 CALL SCALE(FAC)
      CALL PENSEL(1,0.1,2)
C
```

A PROGRAM TO PLOT AN ISOTHERM CONTOUR MAP

```
C
C      DRAW CELL BOUNDS IN A SOLID LINE.
C
       X=XS
       Y=YS
       CALL MOVTO2(0.0,0.0)
       CALL LINBY2(X,0.0)
       CALL LINBY2(0.0,Y)
       X=-X
       Y=-Y
       CALL LINBY2(X,0.0)
       CALL LINBY2(0.0,Y)
C
C
C
C      DRAW IN HORIZONTAL MESH LINES.
C
       CALL DASHED(1,4.0,2.0,0.0)
C
C
       DY=YS/M
       X=XS
       CALL MOVTO2(0.,DY)
       DO 3 I=2,M
       CALL LINBY2(X,0.0)
       X=-X
     3 CALL MOVBY2(0.0,DY)
C
C      DRAW IN VERTICAL LINES.
C
       DX=XS/N
       CALL MOVTO2(DX,0.0)
       Y=YS
       L=N+1
       DO 4 I=2,N
       CALL LINBY2(0.0,Y)
       Y=-Y
     4 CALL MOVBY2(DX,0.0)
C
       CALL DASHED(0,1.0,1.0,0.0)
C
C
C      RESET WINDOW TO ACTUAL SIZE OF MESH.
C
       XW=184.5
       YW=184.5
       CALL WINDO2(10.,XW,10.,YW)
       FACT=1.0/FAC
       XWT=200.0
       YWT=200.0
       XWTT=186.5
       YWTT=186.5
C
C      DRAW IN INSERT OUTLINES.
C
       DO 5 I=1,IP
```

MODELLING OF UNIDIRECTIONAL COMPOSITE MATERIALS

```
A PROGRAM TO PLOT AN ISOTHERM CONTOUR MAP

      GG=2*G(I)
      CX=D(1,I)
      CY=D(2,I)
      CALL MOVTO2(CX,CY)
      CALL WINDO2(32.5,XWTT,32.5,YWTT)
      CALL DOT(0.5)
      CALL WINDO2(34.5,XW,34.5,YW)
      IF(IC(I).EQ.1) GO TO 13
C
C     CIRCULAR INSERTS.
C
      CALL MOVBY2(-G(I),0.0)
      CALL ARCBY2(G(I),0.0,-GG,0.0,1)
      GOTO 35
C
C     SQUARE/RECTANGULAR INSERTS.
C
   13 CALL MOVBY2(-POS(1,I),-POS(2,I))
      XT=2.0*POS(1,I)
      YT=2.0*POS(2,I)
      CALL LINBY2(0.0,YT)
      CALL LINBY2(XT,0.0)
      CALL LINBY2(0.0,-YT)
      CALL LINBY2(-XT,0.0)
C
C     CHECK IF INSERT IS COMPLETELY WITHIN MESH BOUNDS.
C
   35 IF(CX.LT.XS.AND.
     CCX.GT.0.0.AND.
     CCY.LT.YS.AND.
     CCY.GT.0.0) GO TO 34
      GO TO 32
C
C     SECTION TO PUT INSERT NUMBER BY CENTRE POINT
C     AS INSERT DOES NOT CROSS MESH BOUNDARIES.
C
   34 CALL MOVTO2(CX,CY)
      CALL SCALE(FACT)
      CALL MOVBY2(1.0,1.0)
      CALL CHAINT(I,-2)
      CALL SCALE(FAC)
      GO TO 5
C
C     SECTION TO PUT THE NUMBER OF THE INSERT BY THE MESH
C     SIDE NEAREST TO THE INSERT CENTRE.
C
   32 XCENT=XS/2.0
      YCENT=YS/2.0
      XL=CX-XCENT
      YL=CY-YCENT
      AHT=YCENT/XCENT*ABS(XL)
      Y1=AHT+YCENT
      Y2=AHT-YCENT
C
      IF(CX.LE.XCENT.AND.CY.LE.Y1.AND.CY.GE.-Y2) GO TO 41
C
```

A PROGRAM TO PLOT AN ISOTHERM CONTOUR MAP

```
      IF(CX.LE.XCENT.AND.CY.GT.Y1) GO TO 42
C
      IF(CX.LE.XCENT.AND.CY.LT.-Y2) GO TO 44
C
      IF(CX.GT.XCENT.AND.CY.LE.Y1.AND.CY.GE.-Y2) GO TO 43
C
      IF(CX.GT.XCENT.AND.CY.GT.Y1) GO TO 42
C
      IF(CX.GT.XCENT.AND.CY.LT.-Y2) GO TO 44
C
C
      WRITE(6,400)
  400 FORMAT(' NUMBERING EXERCISE FAILED'//)
      GO TO 5
C
   41 AMULT=ABS(XCENT/XL)
      YL=YL*AMULT
      YCENT=YCENT+YL
      CALL MOVTO2(-5.0/FAC,YCENT)
      GO TO 33
C
   42 AMULT=ABS(YCENT/YL)
      XL=XL*AMULT
      XCENT=XCENT+XL
      CALL MOVTO2(XCENT,YS+2.0/FAC)
      GO TO 33
C
   43 AMULT=ABS(XCENT/XL)
      YL=YL*AMULT
      YCENT=YCENT+YL
      CALL MOVTO2(XS+2.0/FAC,YCENT)
      GO TO 33
C
   44 AMULT=ABS(YCENT/YL)
      XL=XL*AMULT
      XCENT=XCENT+XL
      CALL MOVTO2(XCENT,-7.0/FAC)
C
   33 CALL WINDO2(0.0,XWT,0.0,YWT)
      CALL SCALE(FACT)
      CALL CHAINT(I,-2)
      CALL SCALE(FAC)
      CALL WINDO2(34.5,XW,34.5,YW)
    5 CONTINUE
C
C     SET UP CONTOUR VERTICAL STEPS.
C
      S=0.0
      DO 6 I=1,K
      YY(I)=S
    6 S=S+DY
      CB=0.0
      SB=1.0
      CF=0.0
      SF=1.0
C
```

A PROGRAM TO PLOT AN ISOTHERM CONTOUR MAP

```
C     DRAW IN CONTOURS.
C
      DO 7 I=1,9
      DO 8 J=1,K
    8 VX(J)=V(I,J)
      CALL MOVTO2(VX(1),0.0)
    7 CALL CURTO2(VX,YY,K,1,1)
C
C     RESET SCALE TO UNITY.
C
      CALL WINDOW(2)
      FAC=1.0/FAC
      CALL SCALE(FAC)
      NUMLINE=3+IP
C
C     OUTPUT ALL TITLES.
C
      FAC=1.0
      YPOS=220.0+FLOAT(NUMLINE)*6.0
      IF(IP.GT.3) YPOS=256.0
      CALL MOVTO2(0.0,YPOS)
      CALL CHASWI(1)
      IF(IP.LE.3) GO TO 31
      IHT=IP-3
      FAC=9.0/(9.0+FLOAT(IHT))
      CALL SCALE(FAC)
   31 CALL CHAHOL('T*LHERMAL CONDUCTIVITY OF A COMPOSITE*.')
      YPOS=(YPOS-6.0*FAC)
      CALL MOVTO2(0.0,YPOS/FAC)
      CALL CHAHOL('T*LHERMAL CONDUCTIVITY OF MATRIX  *.')
      CALL CHAFIX(A,8,3)
      CALL CHAHOL('*L WATT/METRE/DEGREE KELVIN*.')
      YPOS=(YPOS-6.0*FAC)
      CALL MOVTO2(0.0,YPOS/FAC)
      CALL CHAHOL('T*LHERMAL CONDUCTIVITY OF INSERTS *.')
      DO 30 I=1,IP
      YPOS=(YPOS-6.0*FAC)
      J=I+2
      CALL MOVTO2(0.0,YPOS/FAC)
      CALL CHAHOL('C*LONDUCTIVITY OF INSERT *UN*LOS *.')
      CALL CHAINT(I,-3)
      CALL CHAHOL('*L WAS *.')
      CALL CHAFIX(B(I),10,3)
   30 CALL CHAHOL('*L WATT/METRE/DEGREE KELVIN*.')
      YPOS=(YPOS-6.0*FAC)
      CALL MOVTO2(0.0,YPOS/FAC)
      CALL CHAHOL('E*LFFECTIVE THERMAL CONDUCTIVITY  *.')
      CALL CHAFIX(SI,8,3)
      CALL CHAHOL('*L WATT/METRE/DEGREE KELVIN*.')
      YPOS=(YPOS-6.0*FAC)
      CALL MOVTO2(0.0,YPOS/FAC)
      CALL CHAHOL('S*LAMPLE DIMENSIONS X =*.')
      CALL CHAFIX(XS,10,3)
      CALL CHAHOL('*L MM   Y = *.')
      CALL CHAFIX(YS,10,3)
      CALL CHAHOL('*L MM*.')
```

A PROGRAM TO PLOT AN ISOTHERM CONTOUR MAP

```
      IP=M*N
      YPOS=(YPOS-6.0*FAC)
      CALL MOVTO2(0.0,YPOS/FAC)
      CALL CHAHOL('T*LHE UNIT CELL IS DIVIDED INTO *.')
      CALL CHAINT(M,-5)
      CALL CHAHOL('*L ROWS BY *.')
      CALL CHAINT(N,-5)
      CALL CHAHOL('*L COLUMNS*.')
      CALL PICCLE
      GOTO 14
   25 CALL DEVFIN
      STOP
      END
      FINISH
C
C
C
```

Physics Programs
Edited by A. D. Boardman
© 1980 John Wiley & Sons Ltd

CHAPTER 13

Computational Study of Diffraction by Microcrystalline and Amorphous Bodies

P. J. Grundy

1. INTRODUCTION

One of the areas in science that has benefited greatly from the development of the large computers is concerned with the study of atomic and molecular structural arrangements. The successful unravelling of the atomic arrangements in DNA, proteins, and other large biological molecules and crystals by X-ray diffraction is perhaps the most well-known example of what is, essentially, a multidisciplinary field of study. Whilst some progress can be made by a combination of experience, intuition, and skill in such investigations, it is the ability of the large computer to perform many calculation steps in a reasonable time that has made complete characterization possible. This chapter is not concerned with biophysical subjects as such, but with the application of essentially similar, and simpler, techniques to crystalline and non-crystalline atomic arrangements found in metals and alloys. The examples used here to illustrate the computational exercises have been developed in connection with electron diffraction studies of thin metal films; however, the principles, theories, and subject matter apply equally well, with minor modifications, to X-ray and neutron diffraction.

The results of a diffraction experiment are usually obtained in the form of a diffraction 'pattern' in analogue or digital form. This information, often in a subsequently refined and corrected form, is determined by the time-averaged atomic arrangement in the test specimen and it can be 'inverted' mathematically to obtain the detail of these arrangements. However, this procedure is only approximate in many cases (particularly for non-crystalline structures and small crystals and atomic clusters) and an alternative procedure is to calculate a diffraction pattern from a model for comparison with the experimental pattern. This last exercise is found to be very useful, not only in diffraction applications but in any experimental technique, such as electron microscopy, that is concerned with information recorded as intensity variations in two dimensions as in a micrograph. The excellent graphical facilities available with modern computers are obviously of great benefit here.

Below we give a computer program designed to carry out these operations for simple systems. However, before doing this it is helpful to outline a simple theory of diffraction[1,2] that leads to the concepts and equations we shall use in this program.

2. DIFFRACTION THEORY

In a diffraction experiment the angular distribution of scattered intensity is given in terms of a differential scattering cross-section, $D(\theta)$, for the sample. In a convenient approximation for electrons, due to Born, where the scattering is assumed to be weak and dynamic events, such as multiple scattering, do not occur, $D(\theta)$ can be factorized into two terms:

$$D(\theta) = P(\mathbf{S}) \cdot F(\mathbf{S}, \omega).$$

Here 2θ is the scattering angle, S is the scattering parameter ($4\pi \sin \theta/\lambda$, often called k or q) corresponding to a change in momentum on scattering, and ω is a frequency dependence corresponding to a change in energy, $\hbar\omega$, on scattering. $P(\mathbf{S})$ depends on the type of interaction between the incident radiation (e.g. electron, X-ray, neutron) and the scattering centres and is calculable and fairly well known.

For elastic scattering F depends only on $\mathbf{S}$, i.e. only a spatial dependence exists, and $D(\theta)$ is essentially a function of $F(\mathbf{S})$. $F(|\mathbf{S}|)$ is isotropic for scattering from a single atom or from a randomly related set of scatterers. However, if some spatial correlation exists between the scattering centres $F(\mathbf{S})$ will vary with $\mathbf{S}$. These are, of course, the two extreme cases of a truly amorphous atomic arrangement and an ordered crystal.

$F(\mathbf{S})$ is variously called the interference function, the structure-sensitive intensity, or the structure factor for the system and, notwithstanding this confusion, is a very useful function in defining the structure and, through this, in deducing some physical properties of the system. In general $F(\mathbf{S})$ is related to the distribution of atoms in the system through a Fourier transform of a radial distribution function $G(\mathbf{r})$ which defines the deviation from an average atomic density ρ_0, $\rho(\mathbf{r}) - \rho_0$, at some point $\mathbf{r}$. Clearly, by use of the Fourier transform in an inverse sense it is possible to obtain structural information from experimental diffraction patterns, as pointed out above. In inverse relation to $F(\omega)$, G in general has a time dependence $G(\tau)$ but here we are only concerned with a time average distribution or stationary distribution of scattering centres.

To outline a calculation of $F(\mathbf{S})$, consider a plane electron wave of unit amplitude and wave vector $\mathbf{K}_0$ ($K_0 = 2\pi/\lambda$) incident on a distribution of N scattering centres or atoms at positions $\mathbf{r}_n$ from some origin atom (Figure 1). The total amplitude A_d diffracted into some angle 2θ is the sum of all the individual contributions from each atom modified by the phase factor ϕ which is determined by the relative positions of the atoms. Hence A_d is given by

$$A_d(2\theta) = \sum_n^N f_n(2\theta)\exp(i\phi_n).$$

Here $f_n(2\theta)$ is the scattering factor of the nth atom and is the fraction of the amplitude incident on that atom that is elastically scattered at 2θ. The phase difference for the two atoms at O and P is given by

$$\phi = \frac{2\pi}{\lambda} \cdot \text{(the path difference, } OA - PB\text{)}$$

$$= \frac{2\pi}{\lambda} \cdot \frac{\mathbf{r}_1 \cdot \mathbf{K}_d - \mathbf{r}_1 \cdot \mathbf{K}_0}{|\mathbf{K}_d|} = \mathbf{r}_1 \cdot (\mathbf{K}_d - \mathbf{K}_0)$$

$$= \mathbf{r}_1 \cdot \mathbf{S}_1,$$

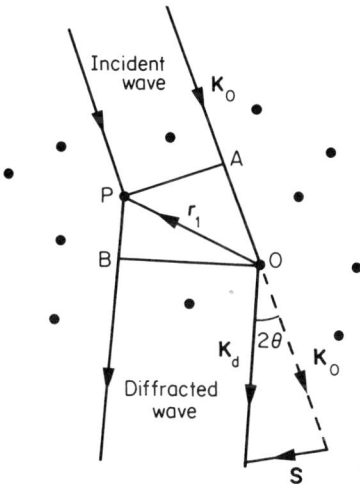

Figure 1. A schematic diagram illustrating the scattering of electrons by a group of atoms. $\mathbf{K}_0$ is the incident wave, $\mathbf{K}_d$ is the wave diffracted through 2θ, and $\mathbf{S}$ is the scattering vector

where $\mathbf{K}_d$ is the diffracted wave. For this elastic scattering $|\mathbf{K}_d| = |\mathbf{K}_0|$, and $\mathbf{S} = \mathbf{K}_d - \mathbf{K}_0$ and has the modulus $4\pi \sin \theta/\lambda$. Hence, replacing 2θ by $\mathbf{S}$ and summing for all the phase terms,

$$A_d(\mathbf{S}) = \sum_n^N f_n(S) \exp i(\mathbf{r}_n \cdot \mathbf{S}). \qquad (1)$$

The diffracted intensity is given by the real part of $A_d(\mathbf{S}) \cdot A_d^*(\mathbf{S})$ and is written as

$$I(\mathbf{S}) = \sum_{m,n}^{N} \sum^{N} f_n(S) f_m(S) \cos(\mathbf{r}_{nm} \cdot \mathbf{S}), \qquad (2)$$

where $\mathbf{r}_{nm} = \mathbf{r}_n - \mathbf{r}_m$ is the separation of the nth and mth atoms and each atom is considered as an origin in turn. This equation gives the diffraction pattern in the Fraunhofer approximation and is applicable to all atomic arrangements, ordered or disordered. For a single component specimen, i.e. one composed of like atoms, equation (2) reduces to

$$\begin{aligned} I(\mathbf{S}) &= Nf^2(S) + f^2(S) \sum_{m \neq n}^{n} \cos(\mathbf{r}_{nm} \cdot \mathbf{S}) \\ &= Nf^2(S)[1 + F(\mathbf{S})] \end{aligned} \qquad (3)$$

for $f_n(S) = f_m(S) = f(S)$, say. Here $F(\mathbf{S})$, the Debye function, is the part of $I(\mathbf{S})$ that is dependent only on the particular arrangement of the atoms and not their identity. Clearly, if we wish to abstract $F(\mathbf{S})$ from an experimental diffraction pattern and hence from an experimental trace of $I(\mathbf{S})$, then

$$F(\mathbf{S}) = \frac{I(\mathbf{S})}{Nf^2(S)} - 1 \qquad (4)$$

and to a good approximation for a multicomponent arrangement with n atomic species of atomic concentration C_i (e.g. for $i = 1$ $C_1 = $ (no. of atoms of species 1)$/N$)

$$F(\mathbf{S}) = \frac{I(\mathbf{S}) - N \sum_i^n C_i f_i^2(S)}{N\left[\sum_i^n C_i f(\mathbf{S})\right]^2}. \qquad (4a)$$

In systems containing many atoms it can be assumed that any particular $\mathbf{r}$ lies at all azimuthal angles to the incident wave and $\cos(\mathbf{r} \cdot \mathbf{S})$ can be replaced by an average value $\sin(Sr)/Sr$. This average is obtained as

$$\overline{\cos(\mathbf{r}_{nm} \cdot \mathbf{S})} = \int_{\alpha=0}^{\pi} \int_{\phi=0}^{2\pi} \cos(Sr_{nm} \cos \alpha) 2\pi \sin \alpha \, d\alpha \frac{d\phi}{2\pi},$$

where α is the angle between the moduli of **r** and **S** and ϕ is here the azimuthal angle formed by the plane containing **r** and **S** with an arbitrary plane. This simplification can also be applied to large assemblies of uncorrelated small crystals or atomic clusters if the above assumption is valid. Clearly, any intercrystalline interference effects are neglected and the interference function is then

$$F(S) = \frac{1}{N} \sum_{m \neq n}^{NN} \frac{\sin(Sr_{nm})}{Sr_{nm}}, \qquad (5)$$

and can be further written as

$$F(S) = \frac{2}{N} \sum_n B_n \frac{\sin(Sr_n)}{Sr_n} \qquad (6)$$

if the particular interatomic spacing r_n occurs B_n times but is summed over once in view of the factor 2. Here, of course, $F(S)$ is isotropic in S and a calculated $I(S)$ via equation (3) would have circular symmetry in the Fraunhofer diffraction plane, as would the experimental $I(S)$ from a specimen composed of many small crystals or a disordered 'glassy' atomic arrangement.

The radial distribution function of an arrangement of identical atoms describes the average number of atoms at distances between r and $r+dr$ from some chosen atom as origin, further averaged by taking each atom in turn as the origin. The average number of atoms is given by $4\pi r^2 \rho(r)$ and the radial distribution function $R(r)$ is just $4\pi r^2 \rho(r)$ or in a 'reduced' form $G(r) = 4\pi r \rho(r)$. Here $\rho(r)$ is equal to the density of atoms at r and for fluctuations about a mean density ρ_0, i.e. for local atomic arrangements, $G(r)$ can be written in an integral form as

$$G(r) = 4\pi r(\rho(r) - \rho_0) = \frac{2}{\pi} \int_0^{S_{max}, \infty} SF(S) \sin(Sr) \, dS \qquad (7)$$

with limits for an interference function obtained in principle out to $S = \infty$ or, as more practically possible, to some cut-off value S_{max}. Maxima in $G(r)$ represent the most commonly occurring values of r in any atomic arrangement. The Fourier transformation of known spatial detail in a structure can be used in principle, of course, to calculate a spatial frequency spectrum or diffraction pattern, $I(S)$, in the Fraunhofer diffraction plane via

$$SF(S) = \int_0^{r_{max}, \infty} 4\pi r(\rho(r) - \rho_0) \sin(Sr) \, dr \qquad (8)$$

and then equation (3).

Computation of $G(r)$ via equation (7), and indeed $R(r)$, from experimentally obtained values of $I(S)$ and then $F(S)$ are often used to obtain structural information on a specimen. However, the accuracy of the calculation can suffer from the fact that experimental measurements are usually limited to some maximum value of S and certainly the use of the 'reverse' procedure of equation (8) to calculate a diffraction pattern from a known model is unwise. Recourse to the Debye equation followed by equation (3), or its equivalent for a multicomponent structure, is advised. For such systems separate interference and radial distribution functions can, of course, be calculated for the separate correlations, e.g. for an alloy AB the values of r_{A-A}, r_{A-B}, and r_{B-B}.

Calculation of these functions for systems of significant size involves many repeated steps, i.e. N atoms involve $N(N-1)/2$ interatomic distances, and rapid and automatic computation is clearly essential. For polycrystalline assemblies of larger crystals, and for single crystals, it is obviously more convenient to take advantage of the periodicity and symmetry properties of the atomic arrangement and to calculate the structure factor and the diffracted intensity[2] via the reciprocal lattice concept.

3. STRUCTURES AND MODELS

As examples of the calculation of $F(S)$, and $I(S)$ if required, we consider particular models that have been used in investigations[3,4] of the atomic arrangements in thin metal films formed by deposition from the vapour phase on to substrates usually cooled to low temperatures. Figure 2 shows the $F(S)$ and $G(r)$ functions determined for nickel films, using equations (4) and (7), from *experimentally* determined $I(S)$ functions. These are interpreted as typical of those obtained from non-crystalline atomic arrangements, the broad maxima and 'split' nature of the second peak (at $S = 5.4$ and 5.9 Å^{-1}) being characteristic of these structures.[5]

To corroborate such conclusions interference functions are often calculated for a variety of microcrystalline models to compare with the experimental function. Table 1 gives typical information for small FCC (face centred cubic) and HCP (hexagonal close packed) microcrystals of single elements and for a hexagonal microcrystal of one particular compound. Changes in 'packing' caused by the presence of crystal defects and the presence of strains[3] can easily be incorporated in such lists.

As an example that could be used in the program at the end of this chapter consider the case of strain. The complexity of calculating the effect of an anisotropic distribution of strains on a crystal is prohibitive and here only a Gaussian distribution is considered. If, in the unstrained crystal, there are B_n atoms separated by a distance r_n then in the strained crystal there will be B_n atoms in a Gaussian distribution about r_n, and the number of

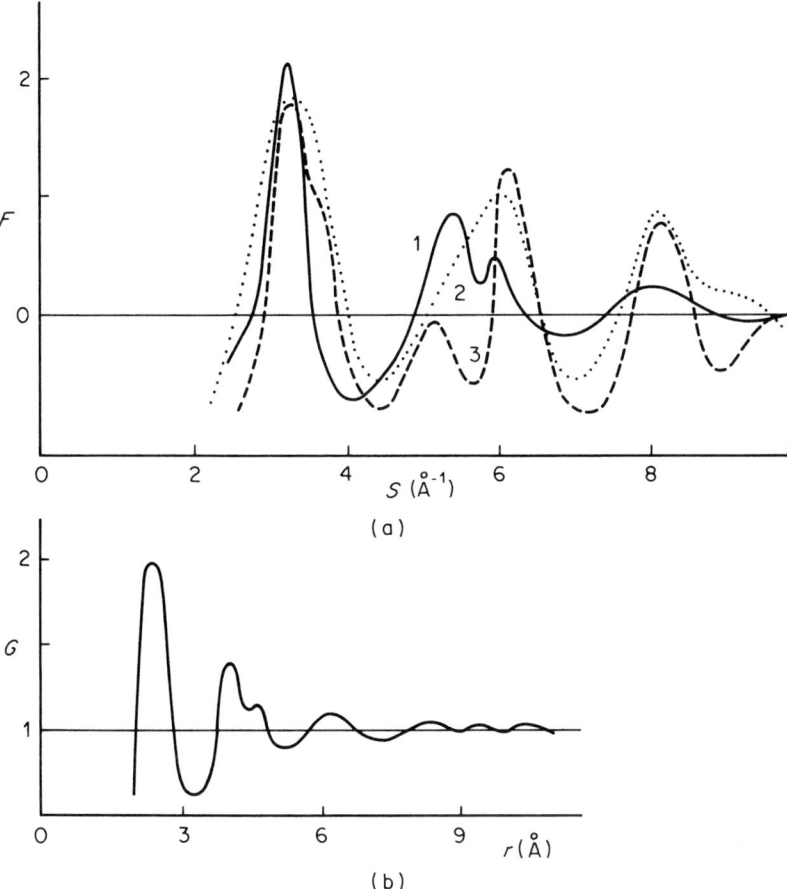

Figure 2. (a) Interference functions from: (1) a diffraction experiment on thin film (~100 nm thick) nickel evaporated on to a cold (77 K) substrate; (2) a calculation for a random assembly of 13 atom FCC nickel microcrystallites; and (3) a calculation for a similar assembly of 55-atom microcrystallites. (b) The radial distribution junction G derived from the data of curve (1)

atoms separated by r is Y_n where

$$Y_n = B_n [2\pi\sigma^2 r_n^2]^{-\frac{1}{2}} \exp[-(r-r_n)^2/2\sigma^2 r_n^2] \quad (9)$$

and σ is the rms strain. The interference function is then calculated by substituting Y_n into equation (6) and integrating over r from, say, $r_{min} = r_n - 4\sigma r_n$ to $r_{max} = r_n + 4\sigma r_n$ for each value of r_n and B_n encountered in the unstrained crystal as

$$F(S) = \frac{2}{N} \sum_{n=1}^{n_{max}} \int_{r_{min}}^{r_{max}} Y_n \frac{\sin(Sr)}{Sr} dr. \quad (10)$$

Table 1. Structural data for some microcrystalline assemblies

(a) FCC crystallite (13 and 55 atoms and lattice parameter a)

n	r_n	B_n 13 atoms	B_n 55 atoms
1	$a/\sqrt{2}$	36	216
2	a	12	90
3	$a\sqrt{3}/2$	24	264
4	$a\sqrt{2}$	6	114
5	$a\sqrt{5}/2$		192
6	$a\sqrt{3}$		48
7	$a\sqrt{7}/2$		240
8	$2a$		27
9	$a\sqrt{9}/2$		108
10	$a\sqrt{5}$		48
11	$a\sqrt{11}/2$		48
12	$a\sqrt{6}$		36
13	$a\sqrt{13}/2$		48
14	$a\sqrt{8}$		6

(b) HCP crystallite (with 9 atoms and lattice parameters c and a)

n	r_n	B_n
1	a	16
2	$a\sqrt{2}$	2
3	$(a\sqrt{8/3})c$	4
4	$a\sqrt{3}$	2
5	$a\sqrt{11/3}$	10
6	$a\sqrt{17/3}$	2

(c) Hexagonal crystallite (17 atoms for a particular AB_5 compound)

Atom	X	Y	Z
A	0	0	0
A	0	a	0
A	$0.866a$	$0.5a$	0
A	$0.866a$	$1.5a$	0
A	0	0	0
A	0	a	c
A	$0.866a$	$1.5a$	c
A	$0.866a$	$0.5a$	c
B	$0.289a$	$0.5a$	0
B	$0.577a$	a	0
B	$0.289a$	$0.5a$	c
B	$0.577a$	a	c
B	0	$0.5a$	$0.5c$
B	$0.433a$	$1.25a$	$0.5c$
B	$0.866a$	a	$0.5c$
B	$0.433a$	$0.25a$	$0.5c$
B	$0.433a$	$0.75a$	$0.5c$

$$\begin{bmatrix} \text{GdCo}_5 (a = 0.4976 \text{ nm}) \\ \text{and } c - 0.3973 \text{ nm}) \text{ is an example} \\ \text{of such an AB}_5 \text{ compound.} \end{bmatrix}$$

Table 2. Coordinates for a 66-atom polytetrahedral model

COORDINATES FOR A 66 ATOM POLYTETRAHEDRAL MODEL

ATOM	X	Y	Z
1	-.4863886E+00	-.2953945E+00	-.1922949E+00
2	.4906244E+00	-.2953474E+00	-.1990868E+00
3	-.1810224E-01	.4804142E+00	-.1818506E+00
4	-.3077860E-01	-.4364057E-01	.6651296E+00
5	-.1707290E-01	.3927625E-01	-.1013845E+01
6	-.1325838E-01	-.1029454E+01	.1871708E+00
7	-.8125635E+00	.4543660E+00	.3479783E+00
8	.7429578E+00	.4805490E+00	.3714333E+00
9	-.6418542E-02	-.8680445E+00	-.7689570E+00
10	.7867370E+00	.4776615E+00	-.6765297E+00
11	-.8519302E+00	.4572374E+00	-.6611606E+00
12	-.8202376E+00	-.6429253E+00	.6133292E+00
13	-.4281090E-01	.9934240E+00	.5788397E+00
14	.7676529E+00	-.6121813E+00	.6405042E+00
15	-.8404372E+00	-.4399824E+00	-.1081307E+01
16	.8118558E+00	-.4280836E+00	-.1108470E+01
17	-.4384662E-01	.9861227E+00	-.9754062E+00
18	-.8246563E+00	-.1144300E+01	-.3679007E+00
19	.7941500E+00	-.1183392E+01	-.3620953E+00
20	-.2423443E-01	-.1073528E+01	.1121757E+01
21	-.1411787E+01	-.1564587E+00	-.8360812E-01
22	-.5362258E+00	.1255099E+01	-.1896423E+00
23	-.5561840E+00	.6312257E+00	.1274361E+01
24	.1397363E+01	-.1015376E+00	.1769679E-01
25	.4677623E+00	.1277841E+01	-.1839424E+00
26	.4584101E+00	.6472854E+00	.1290835E+01
27	-.4313629E-01	-.6519132E+00	-.1699498E+01
28	.5415432E+00	.5603805E+00	-.1620441E+01
29	-.6312472E+00	.4897407E+00	-.1604143E+01
30	-.3894207E-01	-.1772164E+01	-.4349459E+00
31	-.7865489E+00	-.1607124E+01	.4596921E+00
32	.7303235E+00	-.1573403E+01	.5219583E+00
33	-.1147469E+01	.9814941E+00	-.1378851E+00
34	-.1348411E+01	.5837886E-01	.1042099E+01
35	-.9207767E+00	.1386698E+01	.6899666E+00
36	.1428558E+01	.9393049E+00	-.1382706E+00
37	.8190813E+00	.1413906E+01	.7087494E+00
38	.1271128E+01	.9794284E-01	.1078886E+01
39	.5236714E+00	-.1434411E+01	-.1300074E+01
40	-.5891278E+00	-.1421382E+01	-.1278153E+01
41	.1666675E+01	.1756494E+00	-.8879475E+00
42	.8683543E+00	.1380016E+01	-.1048241E+01
43	-.1683674E+01	.7033924E-01	-.9951138E+00
44	-.9411843E+00	.1363316E+01	-.1053054E+01
45	-.1614555E+01	-.1062371E+01	.2200583E+00
46	-.5294313E+00	-.3193201E+00	.1478330E+01
47	-.6843712E-01	.1924308E+01	.3454938E+00
48	-.6616315E-01	.1527842E+01	.1351182E+01
49	.1555146E+01	-.1045730E+01	.2503108E+00
50	.4487337E+00	-.3046936E+00	.1496514E+01
51	-.1656540E+01	-.8929604E+00	-.7631010E+00

52	.1613467E+01	-.7804677E+00	-.6883520E+00
53	-.5327672E-01	.1919442E+01	-.7122008E+00
54	-.5660612E-01	.3315453E+00	.2063794E+01
55	-.1363024E+01	-.9579605E+00	-.1705629E+01
56	-.1483518E+01	-.1832305E+01	-.4003913E+00
57	-.2324718E+01	-.4477878E+00	-.2060256E+00
58	.1378434E+01	-.1038785E+01	-.1612607E+01
59	.1393848E+01	-.1726983E+01	-.8897992E+00
60	.2316008E+01	-.2730557E+00	-.2304989E+00
61	.1172531E+00	.1594587E+01	-.1682998E+01
62	-.1160266E+01	.1958527E+01	-.1603868E+00
63	.1175806E+01	.1907548E+01	-.1674481E+01
64	-.3516758E-01	-.2040049E+01	.8872391E+00
65	-.108751E+01	.2836724E+00	.2011218E+01
66	.9802335E+00	.3116941E+00	.2039919E+01

In Table 1 structural information is given in terms of B_n and r_n (for lists (a) and (b)) and in terms of coordinate positions (for the particular compound in (c)). The lattice parameters a for the FCC microcrystal and c and a for the hexagonal microcrystallites can be given any reasonable values from the literature or particular values for definite materials. The example given

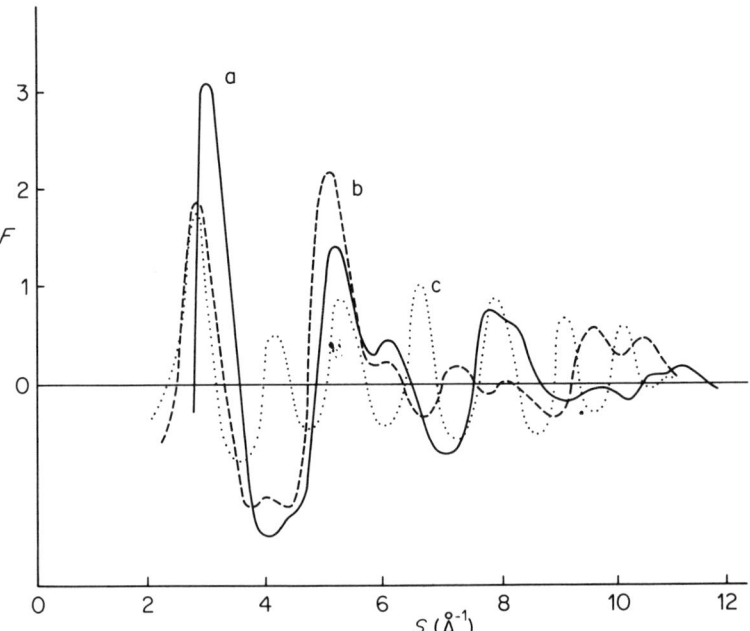

Figure 3. Interference functions for: (a) an experimental $Gd_{10}Co_{90}$ alloy thin film; (b) a calculation for a random assembly of 77 atom Gd_2Co_{17} microcrystallites; and (c) a calculation for a similar assembly of 17 atom $GdCo_5$ microcrystallites

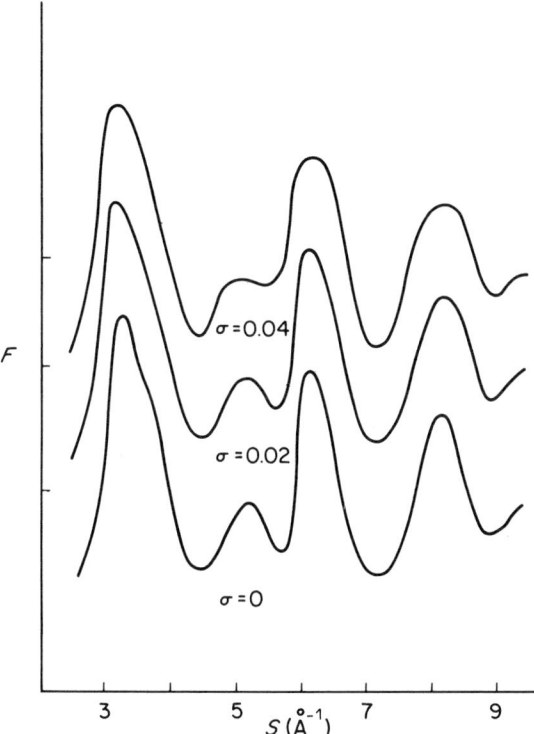

Figure 4. Curves showing the calculated effect of different strain levels on the interference function for 55-atom FCC microcrystallites

in (c) is for $GdCo_5$. Such structural details can be obtained from drawings, constructed models (e.g. 'ball and spoke' models) as in Table 2 and tabulated crystallographic data.[6,7]

Calculated $F(S)$ functions using these and other data are superimposed on those obtained from the experimental $I(S)$ (using equations (4) and (4a)) for particular thin films in Figure 3. The effect of strain on interference functions calculated via equations (9) and (10) are illustrated in Figure 4. Inspection of these curves and the experimental $F(S)$ curve of Figure 2a suggests that strained microcrystalline assemblies are not responsible for the observed diffraction pattern from the specimen.

Although the agreement, or lack of it, between experimental and model functions is not of importance here, computational exercises of this kind are of significance in showing that these materials are not composed of simple microcrystalline atomic arrangements.

4. A BREAKDOWN OF THE PROGRAM AND SUGGESTED PROBLEMS

A specimen outline of the stages of the program is given in Figure 5. The first operation of the main program is to calculate interatomic distances from values of atomic coordinates given as input data or generated in a subroutine (e.g. MODEL). However, these may be read in, for example from Table 1, in a 'sorted' form as B_n, r_n. If not in this form, the values of r_n may be then sorted into such a population count or frequency spectrum. If required such a spectrum or pair distribution function, ρ of B_n against r_n can be plotted. The sorting operation can usually be done by a standard routine available in most statistical program packages.

The interference function $F(S)$ is then calculated by equations (5) or (6) (depending on the data) or by some modification of these, such as equation (10), incorporating a particular defect or strain distribution. For multicomponent systems these functions can be calculated separately for the different correlations, e.g. Gd–Gd, Gd–Co and Co–Co distances in Table 1c, and then summed. This separate treatment is essential if the relevant scattering factors are to be included and $I(S)$ calculated. As pointed out earlier, such curves can be compared with the results of relevant experimental determinations.

A computation of the interference functions and diffracted intensities for the data given in Table 3 will serve to show the sensitivity of the calculation to small structural changes. Both sets of data are for random assemblies of clusters containing 13 atoms. The set in (a) is for atoms packed in a close-packed structure of 12 atoms, showing 6 membered rings, round a central atom. That in (b) is for a 13-atom icosahedron in which 12 atoms are evenly spaced round and in contact with the centre atom. These atoms do not touch each other and form 5 membered rings in section. The first model can fill space to form an ordered crystalline structure, the second cannot and would produce an extremely disordered structure. In the table d is the atomic diameter (this can usually be taken as the Goldschmidt diameter for 12-fold close packing[6]). For the crystalline packing d could be taken as the closest distance of approach which is usually very close to the Goldschmidt diameter,[6] e.g. for FCC nickel the lattice parameter a is 3.524 Å and the closest distance of approach is 2.49 Å which is very near to the Goldschmidt diameter of 2.50 Å. In the comparative calculation the data of Table 1a could be used for the crystal or that of Table 3 given in terms of the lattice parameter or atomic diameter in hard packing (for which $a \simeq \sqrt{2d}$). The two interference functions of Figure 6 are clearly different and this distinction is reflected fairly clearly in the diffracted intensity. Hence it is evident that with careful measurement such packings in experimental specimens could be distinguished with interference functions calculated from experimental $I(S)$ data obtained out to large S values.

COMPUTATIONAL STUDY OF MICROCRYSTALLINE AND AMORPHOUS BODIES 473

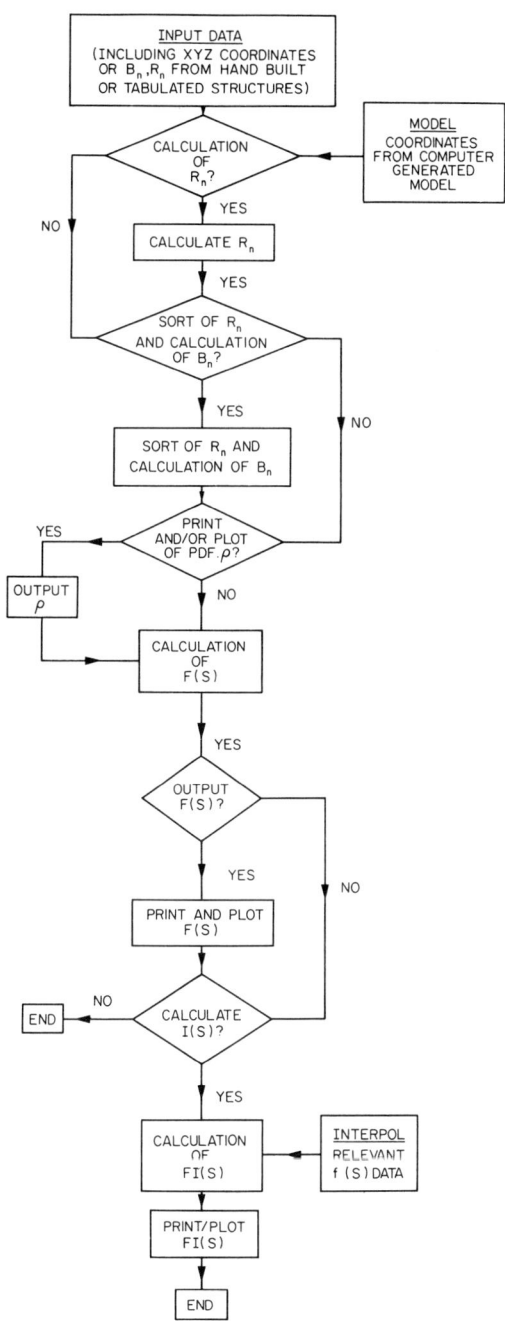

Figure 5. A schematic layout of the program given in section 5

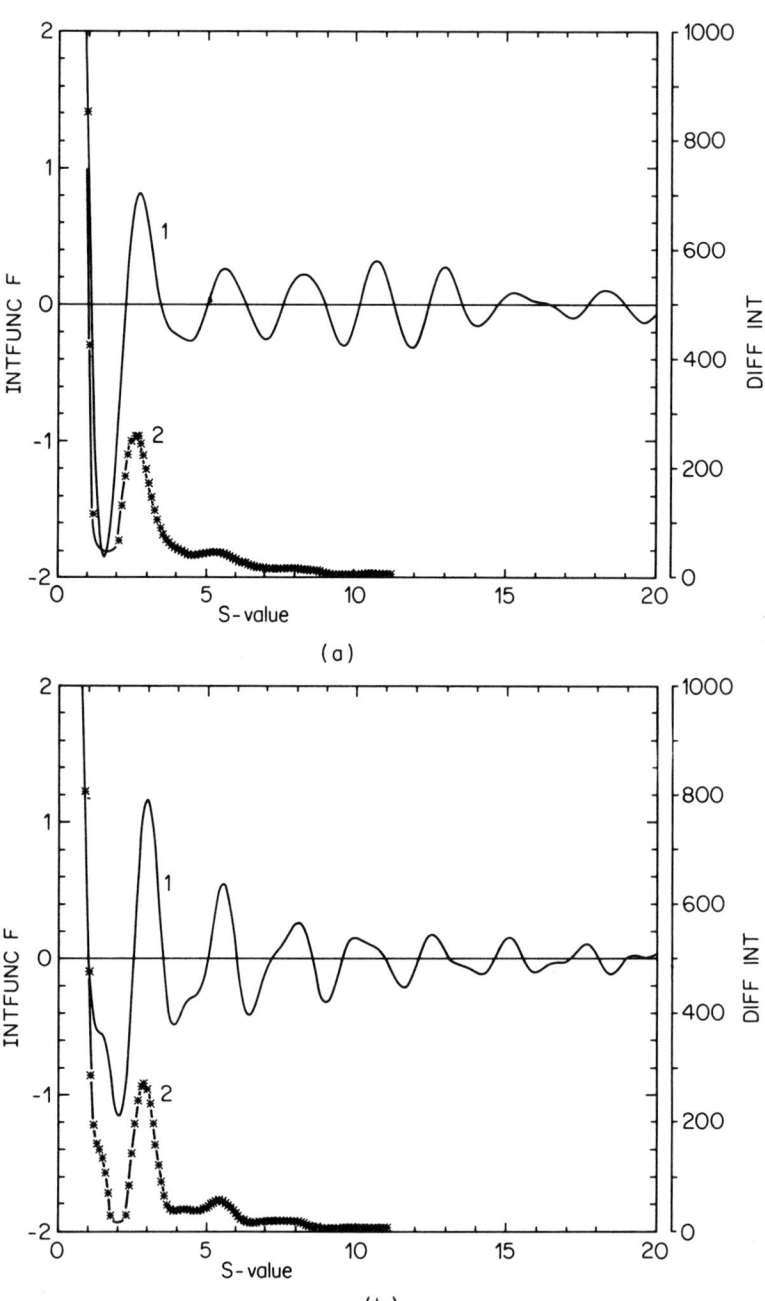

Figure 6. Typical output for the structural data given in Table 3. Curves (1) are the interference functions and curves (2) are the diffracted intensity (neglecting any inelastic contribution); (a) is for the FCC cluster and (b) is for the icosahedral packing.

Table 3. Structural data for two similar atomic arrangements

(a)	13-atom FCC packing (a = lattice parameter)		(b)	13-atom icosahedron (d = atomic diameter)	
n	B_n	r_n	n	B_n	r_n
1	36	$a/\sqrt{2}$ or d	1	12	$1.0d$
2	12	a or $d\sqrt{2}$	2	6	$2.0d$
3	24	$a\sqrt{3}/2$ or $d\sqrt{3}/2$	3	20	$1.868d$
4	6	$a\sqrt{2}$ or $2d$	4	30	$1.052d$
			5	10	$1.902d$

Similar exercises can be carried out with other packings and models, and if diffraction information is available pair distribution functions could be calculated. Electron-scattering factors have the same form for all atoms, only differing in value over the range of S (details for use in the program can be obtained from the literature[8]). Of course, these calculations can be carried out for X-ray and neutron scattering; in these cases the scattering factors, or form factors as they are sometimes called, have different forms.

REFERENCES

1. For a detailed review of general diffraction theory see D. H. Martin, *Contemporary Physics*, **18,** 81 and 193 (1977) and reference 2.
2. A. Guinier, *X-ray Diffraction in Crystals, Imperfect Crystals and Amorphous Bodies* (Freeman, San Francisco and London, 1963).
3. L. B. Davies and P. J. Grundy, *Physica Status Solidi* (a), **8,** 189 (1971).
4. S. S. Nandra and P. J. Grundy, *J. Phys. F*, **7,** 207 (1977).
5. G. S. Cargill III, *Solid State Physics*, **30,** 227 (Academic Press, London, 1975).
6. M. Hansen, *Constitution of Binary Alloys* (McGraw-Hill, New York, 1958).
7. R. W. G. Wyckoff, *Crystal Structures* (Interscience/Wiley, New York, London, Sydney, 1960).
8. P. A. Doyle and P. S. Turner, *Acta Cryst.*, **A24,** 390 (1968).

DIFFRACTION PATTERN SIMULATION

```
C THIS PROGRAM CALCULATES THE INTERFERENCE FUNCTION,A PAIR DISTRIBUTION
C FUNCTION (IF REQUIRED) AND THE COHERENTLY DIFFRACTED INTENSITY FOR AN
C ATOMIC ARRANGEMENT GIVEN EITHER THE COORDINATES X Y Z OF THE ATOMS OR
C A FREQUENCY COUNT BN OF ATOMIC SPACINGS RN AS INPUT DATA.
C ALTERNATIVELY DATA MAY BE INPUT FROM A ROUTINE GENERATING A COMPUTED MODEL
C
C IN THIS PROGRAM F IS THE INTERFERENCE FUNCTION. S AND SS ARE USED FOR
C THE SCATTERING PARAMETER. RHO IS THE PAIR DISTRIBUTION FUNCTION. R IS
C THE SCALE OF THE INTERATOMIC DISTANCES. FS IS THE ELECTRON SCATTERING
C FACTOR, X Y AND Z ARE ATOMIC COORDINATES. FI IS THE COHERENTLY
C DIFFRACTED INTENSITY. NB REPRESENTS THE FREQUENCY OF SORTED CORRELAT-
C IONS RN. NA AND NS ARE POINTERS TO THE ARRAYS X,Y,Z,NB AND RN. RNEW IS
C USED AS A WORKING ARRAY FOR ATOMIC CORRELATIONS. IARR AND IST ARE USED
C FOR POINTERS AND WORK SPACE. ITF,ITP AND ITI ARE ARRAYS FOR GRAPHS
C
      DIMENSION F(200),FI(200),S(200),RHO(100),R(100),RNEW(100),FS(200)
      DIMENSION NA(100),X(100),Y(100),Z(100),IARR(5000),IST(5000)
      DIMENSION SS(200)
      DIMENSION NS(100),NB(100),RN(100)
      DIMENSION A1(20),A2(20),B1(20),B2(20),FSS(200),SIS(200),AM(20)
      DIMENSION D(20)
      COMMON/SCDTA/RV(2000)
      DIMENSION ITF(13),ITI(12),ITP(13)
C
C HERE WE HAVE CHOSEN CONVENIENT SIZES FOR THE VARIABLE ARRAYS
C
      DATA ITF/21,21HINTERFERENCE FUNCTION,7,7HS-VALUE,7,7HINTFUNC/
      DATA ITI/20,20HDIFFRACTED INTENSITY,7,7HS-VALUE,8,8HDIFF INT/
      DATA ITP/26,26HPAIR DISTRIBUTION FUNCTION,8,8HRN-VALUE,3,3HPDF/
C
C THESE ENTRIES GIVE THE GRAPH TITLES AND ORDINATE AND ABSCISSA
C DESCRIPTIONS FOR GRAPHICAL OUTPUT OF THE CHOSEN EXAMPLES
C
      READ(1,100)N,M
      WRITE(2,44)
      WRITE(2,101)N,M
C
C N IS THE NUMBER OF ATOMS IN THE MODEL
C M/10 IS THE UPPER LIMIT TO S (THIS IS JUST A CONVENIENT CHOICE IN
C THIS EXAMPLE BUT IF FI IS SUBSEQUENTLY CALCULATED IT WILL DEPEND ON
C THE AVAILABLE DATA ON FS WHICH SHOULD BE KNOWN UP TO THE UPPER LIMIT
C CHOSEN FOR S)
C
      NN=(N*(N-1))/2
C
C NN IS THE TOTAL NUMBER OF INTERATOMIC DISTANCES
C
      READ(1,102)IA,IC
      WRITE(2,55)
      WRITE(2,103)IA,IC
      IF(IA-2)1,2,3
C
C IA ACCOUNTS FOR THE DIFFERENT DATA INPUTS. IC CHOOSES THE ALTERNATIVE
C CALCULATION FOR F IF A PAIR DISTRIBUTION FUNCTION IS NOT REQUIRED
C FOR THE CHOICE IA=1 READ IN ATOMIC COORDINATES
C
```

COMPUTATIONAL STUDY OF MICROCRYSTALLINE AND AMORPHOUS BODIES

```
DIFFRACTION PATTERN SIMULATION

    1 READ(1,104)(NA(I),X(I),Y(I),Z(I),I=1,N)
      WRITE(2,66)
      WRITE(2,104)(NA(I),X(I),Y(I),Z(I),I=1,N)
C
C NOW CALCULATE THE INTERATOMIC DISTANCES R FROM THIS DATA
C
      K=0
      IN=N-1
      DO 4 I=1,IN
      JN=I+1
      DO 5 J=JN,N
      F1=(X(J)-X(I))**2
      F2=(Y(J)-Y(I))**2
      F3=(Z(J)-Z(I))**2
      K=K+1
    5 R(K)=SQRT(F1+F2+F3)
    4 CONTINUE
      IF(IC.GT.1)GO TO 9
C
C R(K) MAY BE SORTED INTO A !FREQUENCY SPECTRUM! NB OF CORRELATIONS RN
C VIA SOME SUITABLE SORTING ROUTINE. THESE ARE USUALLY AVAILABLE IN
C STATISTICAL LIBRARY ROUTINES. THIS SORTING IS ASKED FOR OF IC=1 AND
C SHOULD BE IN A FORM SUITABLE FOR PLOTTING A PAIR DISTRIBUTION FCTN
C
C CALL !SORT!
C
C
C HERE IS AN ALTERNATIVE INPUT AF DATA - READ IN NB AND RN VALUES FROM
C TABULATED STRUCTURE OR HAND BUILT MODEL AND PLOT THE PDF
C L IS THE NUMBER OF VALUES OF NB
C
    2 READ(1,107)L
      WRITE(2,107)L
      READ(1,105)(NB(I),RN(I),I=1,L)
      WRITE(2,77)
      WRITE(2,105)(NB(I),RN(I),I=1,L)
C IF THE PDF IS NOT REQUIRED THEN
      GO TO 6
      CALL NARROW
      CALL FGPLT(RN,NB,L,2,0,1,0,ITP,2)
C
C NARROW REQUESTS A PARTICULAR OUTPUTMEDIUM AND FGPLT IS A LIBRARY GRAPH
C PLOTTING ROUTINE
C
C ANOTHER ALTERNATIVE INPUT - VALUES OF X,Y,Z OP NB AND RN GIVEN BY A
C ROUTINE GENERATING A COMPUTER BUILT STRUCTURAL MODEL MAY REPLACE THE
C FOLLOWING CONTINUE COMMAND
C
    3 CONTINUE
C
C CALCULATE, WRITE AND PLOT THE INTERFERENCE FUNCTION. OF COURSE,
C SEPERATE FUNCTIONS MAY BE CALCULATED FOR A-A, A-B,AND B-B CORRELATIONS
C IF A BINARY MODEL COMPOSED OF A AND B ATOMS IS CONSIDERED
C
    6 DO 7 IX=1,M
      SX=FLOAT(IX)/10.0
```

DIFFRACTION PATTERN SIMULATION

```
      SUM=0.0
      DO 8 IK=1,L
      IF(RN(IK).EQ.0.0)GO TO 8
      PROD=SIN(SX*RN(IK))/(SX*RN(IK))
      RES=PROD*NB(IK)
      SUM=SUM+RES
    8 CONTINUE
      F(IX)=(2.0*SUM)/N
      S(IX)=SX
    7 CONTINUE
      GO TO 12
    9 DO 10 IX=1,M
      SX=FLOAT(IX)/10.0
      SUM=0.0
      DO 11 IK=1,NN
      IF(R(IK).EQ.0.0)GO TO  1
      PROD=SIN(SX*R(IK))/(SX*R(IK))
      SUM=SUM+PROD
   11 CONTINUE
      F(IX)=SUM/N
      S(IX)=SX
   10 CONTINUE
C
C NOW WRITE THE INTERFERENCE FUNCTION
C
   12 WRITE(2,88)
      WRITE(2,106)(F(IX),S(IX),IX=1,M)
C
C IF THE COHERENTLY DIFFRACTED INTENSITY IS REQUIRED VALUES OF THE
C ELECTRON SCATTERING FACTOR AT THE S VALUES FOR WHICH F HAS BEEN
C CALCULATED ARE NEEDED. THESE CAN BE OBTAINED BY INTERPOLATING FACTORS
C OBTAINED FROM THE PUBLISHED LITERATURE. IF WANTED THE INTERPOLATED
C FACTORS MAY BE PRINTED
C
      CALL INTERPOL(FS,SS)
      WRITE(2,99)
      WRITE(2,106)(FS(I),SS(I),I=1,M)
C
C USE THESE VALUES OF S TO CALCULATE WHAT IS IN EFFECT THE DIFFRACTION
C PATTERN EXPECTED FROM THE MODEL
C
      DO 13 I=1,M
      SUM=1.0+F(I)
      FACT=FS(I)**2
      FI(I)=N*FACT*SUM
      SUM=0.0
      FACT=0.0
   13 CONTINUE
      WRITE(2,144)
      WRITE(2,106)(FI(I),S(I),I=1,M)
C PLOT THE INTERFERENCE FUNCTION AND THE DIFFRACTED INTENSITY
      CALL NARROW
      CALL FIXAXS(0.0,20.0,-2.0,2.0)
C FIXAXS IS A LIBRARY ROUTINE FOR SETTING THE LIMITS OF THE
C ORDINATE AND ABSCISSA
      CALL FGPLT(S,F,M,2,0,1,0,ITF,2)
```

COMPUTATIONAL STUDY OF MICROCRYSTALLINE AND AMORPHOUS BODIES   479

```
DIFFRACTION PATTERN SIMULATION
      CALL FIXAXS(0.0,20.0,0.0 1000.0)
      CALL FGPLT(S,FI,M,1,8,1,2,ITI,2)
      CALL DEVFIN
  999 CONTINUE
C
C DEVFIN CLOSES DOWN THE GRAPHICAL OUTPUT DEVICE
C
   44 FORMAT(1X,15HNUMBER OF ATOMS,5X,13HLIMIT TO SX10)
   55 FORMAT(1X,18HWHICH INPUT SCHEME,5X,16HWHICH CALC FOR F)
   66 FORMAT(5X,11HATOM NUMBER,10X,5HX VAL,10X,5HY VAL,10X,5HZ VAL)
   77 FORMAT(1X,18HBN FREQUENCY OF RN,5X,11HVALUE OF RN)
   88 FORMAT(1X,21HINTERFERENCE FUNCTION,15X,7HS VALUE)
   99 FORMAT(1X,17HSCATTERING FACTOR,15X,7HS VALUE)
  144 FORMAT(1X,20HDIFFRACTED INTENSITY,15X,7HS VALUE)
  100 FORMAT(2I4)
  101 FORMAT(6X,I4,15X,I4)
  102 FORMAT(2I3)
  103 FORMAT(8X,I3,20X,I3)
  104 FORMAT(10X,I5,5X,E14.7,5X,E14.7,5X,E14.7)
  105 FORMAT(10X,I4,10X,F14.7)
  106 FORMAT(5X,E14.7,10X,E14.7)
  107 FORMAT(1I4)
      STOP
      END
C
      SUBROUTINE INTERPOL(FSS,SIS)
      DIMENSION A1(20),A2(20),B1(20),B2(20),FSS(200),SIS(200),AM(20)
      DIMENSION D(20)
C
C THIS ROUTINE RETURNS VALUES OF SCATTERING FACTOR FS AND THE RELEVANT
C VALUES OF SCATTERING PARAMETER SS TO THE MAIN PROGRAM INTERPOLATED
C FROM TABULATED INPUT VALUES OF FSS (AS B2) AND SIS (AS A2). N1 IS THE
C NUMBER OF STARTING VALUES OF FSS AND SIS. IR1 IS THE
C STARTING VALUE(*1000) OFSIS. IH IS THE INCREMENT(*1000) IN SIS AT
C WHICH INTERPOLATION IS REQUIRED. IN THE USE OF THE PARTICULAR
C INTERPOLATION ROUTINE E01ADF THE INPUT ARRAYS ARE REASSIGNED FOR EACH
C ITERATION. AM AND D APE WORKING SPACE.
C
C FIRST READ IN VALUES OF NV,IR1 AND IH. CONVENIENT VALUES OF IR1 AND
C IH ARE 100.
C
      READ(1,200)NV,IR1,IH
      WRITE(2,200)NV,IR1,IH
      N1=NV+1
C
C NOW READ IN STARTING VALUES OF FSS AND SIS;CONVENIENTLY 20 VALUES IF
C KNOWN.
C
      READ(1,201)A2
      WRITE(2,202)A2
      READ(1,201)B2
      WRITE(2,202)B2
      IX=1
   40 X=FLOAT(IR1)/1000
C
C REASSIGN THE ARRAYS AND CALL E01ADF
```

DIFFRACTION PATTERN SIMULATION

```
C
   30 DO 16 I=1,20
      B1(I)=B2(I)
   16 A1(I)=A2(I)
      CALL E01ADF(NV,X,A1 B1,AM,D,N1,VAL)
C
C BUILD THE OUTPUT ARRAYS SIS AND FSS AND INCREMENT AND LIMIT THEIR SIZE.
C
      SIS(IX)=X
      FSS(IX)=VAL
      IX=IX+1
      IR1=IR1+IH
      IF(IX.GT.200)GO TO 20
      GO TO 40
   20 CONTINUE
  200 FORMAT(3I4)
  201 FORMAT(20F0.0)
  202 FORMAT(20F7.4)
      RETURN
      END
      FINISH
C
C
C ENTER THE DATA FOR THE PROGRAM IN THIS SPACE
C
C
ENDJOB
```

Index

Aberrations, 1, 10
 chromatic, 31
 spherical, 11
 transverse, 29
Acoustic branch, 274
Adams–Moulton method, 158
Angle selections, 372
Approximation
 Born, 462
 core state, 311–314
 harmonic, 225, 270
 nearly-free electron, 314, 319, 321
 single wire, 150, 151
 tight binding, 321, 325
Astigmatism, 33
Attenuated total reflection, 47, 51, 54, 56
Atomic arrangements
 glassy, 465
 microcrystalline, 471
 non-crystalline, 466
Atomic units 244, 303, 329

Back focal length, 8
Back focal plane, 81
Bending, 15, 24
Bloch functions, 303, 321
Bohr magnetons, 196
Bohr radii, 304
Boundary conditions
 electromagnetic, 49, 56
 in lattice dynamics, 271
 on magnetostatic potential, 126
 periodic, 304
Bragg reflections, 302, 320, 329, 340
Brewster modes, 58
de Broglie wavelength, 301

Capture radius, 151, 166
Clipping amplitude, 95

Clipping factor, 96
Coma, 16, 33
Concave mirror, 33
Conduction band, 342
Critical angle, 54
Critical sampling, 87
Critical trajectory, 166
Crystal field, 188, 190, 192, 199
Crystal lattice, 302
Crystal structure
 BCC, 278
 CsCl, 278, 281
 dhcp, 189, 197
 FCC, 291
 FCC crystallite, 468
 hexagonal crystallite, 468
 simple cubic, 290
Cumulative probability, 359
Cyclotron frequency, 389

Data for GaAs, 368
Debye equation, 466
Debye function, 464
Debye temperature, 230
Degeneracy, 316
Demagnetizing field, 128, 140
Density of states, 271, 282, 283, 287, 288
Diagonalization, 310
Dielectric tensor, 47
Dielectric function, 49
Differential mobility, 355, 357
Diffraction, 1, 79
 amplitude, 463
 electron, 461
 Fraunhofer, 80, 464, 465
 intensity, 464, 466, 472
 neutron, 461, 475
 pattern, 461, 464–466, 471
 theory of, 462

481

Dipole moment, 126
Dispersion curve, 50, 52, 54, 269, 271–276, 285
Dispersion equation, 48, 49, 357
Distribution function
  electron, 360, 381
  pair, 472, 475
  radial, 462, 465, 466
Dynamical matrix, 277, 278, 284, 291

Effective mass, 326, 329, 332, 341, 360
Eigenvalues, 194, 198, 284, 331
Eigenvectors, 194, 198, 199, 331
Elastic constants
  effective, 228
  second-order, 225
  third-order, 225
Electron density, 363
Electron distribution function, 360, 381
Electron flight, 362, 369
Electron plasma, 47
Electron population ratio, 376
Electron wave, 463
Energy
  first-order correction to, 243, 250, 253
  of deformation, 224
  second-order correction to, 243, 247, 250, 253
Energy bands, 301, 332, 355, 363
Energy conduction band, 363
Energy gap, 318, 338
Energy valence band, 363
Equations of motion
  atoms in a lattice, 270, 273, 277, 281
  particle in magnetic separator, 154
Euler method, 157

Fano modes, 58
Fermat's principle, 1
Fermi–Dirac gas, 301
Fermi energy, 342
Fermi surface, 342
Filter performance, 150
Finite difference equations, 129
Fluid impedance, 150
Fluid susceptibility, 149
Focal length, 8, 15
Force constants
  interatomic, 270, 275, 278, 284, 290
  interplanar, 282
Fourier analysing, 310

Fourier coefficients, 331
Fourier inverse transform, 92
Fourier plane, 80
Fourier series, 308, 331, 341
Fourier transforms, 80, 462, 465
Fresnel coefficients, 57, 62
Functions
  atomic wave, 321
  basis, 306, 308, 310, 341
  Bloch, 303, 321
  comb, 87
  Dirac delta, 84
  energy conserving delta, 369
  normalized wave, 242
  orthogonal, 310
  orthonormal, 306, 308
  rectangle, 91

Gauss–Seidel scheme, 131
Gaussian approximation, 2
Gaussian image, 9
Geometrical optics, 1
GINO-F, 104, 113, 116, 119, 169, 393, 424
Goldschmidt diameter, 472
Gunn effect, 355, 357

Hall effect simulation, 392
Hamiltonian, 240, 309, 318, 330, 331
  ionic, 190
  matrix, 193
Harmonics
  cylindrical, 152
  spherical, 341
Heat
  conduction in composites, 412
  electrical analogue of flow, 414
  flow equations, 412
  non-laminar flow, 414
Higher order spectra, 86
Histograms
  amplitude in hologram, 95, 100
  k-space, 385
  phase in hologram, 96, 100
  phonon frequencies, 282, 286
Holes, 329
Holograms, 79
  binary, 90
  computer-generated, 79
  digital, 79
  drawing of, 98

INDEX 483

Fourier transform, 109
Fraunhofer, 80
Holography, 79
Hooke's law, 221
Hot electron behaviour, 355, 362
Image
  brightness, 94
  intensifiers, 79
  twin, 109
Interactions
  Coulomb, 310
  cubic–cubic, 190
  exchange, 340
  hexagonal–hexagonal, 190
  hyperfine, 191, 200
  nearest-neighbour, 278
  RKKY, 188
  second-nearest-neighbour, 278
Interference function, 462, 467, 471, 472
Inversion centre, 309, 310, 330
Iterative formulae, 131

Kirchhoff's laws, 420
k-space, 272, 340, 355, 361
  distribution, 381, 383
  histogram, 385

Laplace's equation, 128, 152
Laser, 79
Lens
  achromatic, 30
  single, 30
Light line, 51, 52 ,
Linear atomic chain, 269
Linear chain of identical atoms, 270
Linear extrapolation, 248
Linear interpolation, 248
Locus
  of real angles, 54
  of real frequencies, 54

Magnetic charge density, 126, 128
Magnetic energy density, 155
Magnetic field, 126, 138, 140
Magnetic high gradient separation, 149
Magnetic induction, 125, 140
Magnetic interaction, 191
Magnetic moment operator, 191, 196
Magnetic moments, 187, 189, 200
Magnetic orderings, 187, 188

Magnetic traction force, 149, 154, 155
Magnetization, 126, 140, 188
  anomalies, 188
  curves, 201
  experimental, 196, 197
  in praseodymium, 187
Magnetostatic equations, 125
Magnetostatic potential, 125
Maxwell's equation, 49
Mean energy, 372, 375
Mean energy: computer-generated plot, 385
Mean velocity, 355, 372, 375
Mean velocity: computer-generated plot, 384
Methods
  Adams–Moulton, 158
  augmented plane wave, 341
  Euler, 157
  Monte Carlo, 358
  orthogonalized plane wave, 341
  phase shift, 342
  predictor–corrector, 156
  Runge–Kutta, 158
  scattering, 342
  variational, 306, 330
Microwave amplifiers, 355
Microwave current oscillations, 355
Microwave sources, 355
Models
  free electron, 60, 64
  Kronig–Penney, 301
  microcrystalline, 466
Molecular field
  parameters, 198
  Weiss, 189, 190
Monte Carlo flow diagram, 380
Monte Carlo method, 358
  results, 384, 385, 393

NAG library, 105, 204, 205, 228, 284, 293, 331, 332, 424
Non-parabolicity, 364
Norm, 306, 316
Notation
  Dirac, 191
  matrix, 223
  two-suffix, 223

Ohm's law, 355, 416

Operators
  energy, 190
  Stevens, 190
Optical character-recognition, 79
Optical path, 2, 10
Optical surface testing, 79
Optic branch, 274
Orbital angular momentum, 188
Overlap integrals, 324, 326
Overprinting, 105, 111
Over-relaxation method, 131

Paraxial equations, 2
Partial differential equations, 125
Pauli exclusion principle, 363, 392
Permanent magnet, 125
Perturbation theory, 239, 241, 311, 326, 324
Plasma frequency, 50
Poisson's equation, 126, 356
Polariton
  plasmon-, 47
  phonon-, 47
  surface, 47, 53
  surface plasmon-, 49
Potential
  arbitrary, 311
  Brooks–Herring, 387
  cosine, 311, 313, 325
  Coulomb, 321
  finite, 313, 326
  harmonic, 311, 313, 325
  infinite square well, 240, 311, 326
  periodic, 301, 310
  rectangular, 310, 311, 326, 332
  sawtooth, 311
Praseodymium, 187, 188, 191
Predictor–corrector method, 156
Probability density, 240, 304, 314, 330, 332, 358, 370
Pseudo-random distribution, 358
Pupil
  entrance, 7
  exit, 7

Quantum mechanics, 239

Radial forces, 279
Random flight, 359
Random numbers, 98, 359, 370, 372
Random phasing, 95, 98, 111

Random uniformly distributed numbers, 359, 370, 372
Ray
  finite tracing, 16
  principal, 8, 29
  tracing, 2
Rayleigh criterion, 23, 95
Reciprocal
  lattice vector, 282, 341, 466
  space, 340
Reflected intensity, 53
Refractive index, 2
Relativistic correction, 342
Root-sampling method, 282
Runge–Kutta method, 158
Rydbergs, 303

Sampling, 86
  critical, 87, 95
  over-, 87, 94, 110
  under-, 87, 110
Scattering
  by lattice vibrations, 360
  channel selection, 371
  Coulombic, 374
  differential cross-section, 462
  dipole, 374
  electron–phonon, 362
  electron–acoustic phonon, 364
  electron–polar optical phonon, 364, 374
  factor, 463, 472, 475
  intervalley, 364
  intravalley, 364
  ionized impurity, 387
  mechanisms in GaAs, 366
  parameter, 462
  real process, 371
  total rate, 362, 369
  virtual process, 370, 371
Schrodinger equation, 239, 303, 306, 317
Screening, 340, 387
Secular determinant, 277, 284
Single wire approximation, 150, 151
Snell's law, 2
Soft mode, 286, 290
Spatial damping, 50
Spatial frequency, 86, 89, 465
Spatial frequency bandwidths, 89
Stark effect, 190
Stokes equation, 152

Structure factor, 462, 466
Susceptibility
  fluid, 149
  particle, 149
Symmetry
  cubic, 189
  hexagonal, 189
  uniaxial, 192, 413

TE mode, 48
Temporal damping, 50
Tensor
  strain, 222
  stress, 223
Theorem
  convolution, 87
  shift, 84
Thermal conductivity
  effective, 411
  of common materials, 428
Thermal resistivities, 418
Thin metal films, 461, 466
Threshold bias, 355
TM mode, 47
Total internal reflection, 52
Transferred electron mechanism, 357
Transforms
  Fourier, 80
  letter group, 82
Trapezoidal integration, 331

Unit cell, 273, 310
  magnetic, 187
  two-atom, 280

Valley
  central, 363–365
  equivalent, 365
  non-equivalent, 365
  satellite, 363–365
Van Hove singularities, 283, 287, 290
Variational method 306, 330
Variational parameters, 306
Vector
  displacement, 227
  energy flow, 227
  particle motion, 227
Viscous drag, 154
V-value, 31

Wavefront, 1
Waves
  damped, 50
  electromagnetic, 47, 52
  equation, 225
  function, 240
  packet, 327
  surface, 49

Zone
  boundaries, 302, 316, 318, 319, 329, 338
  Brillouin, 271, 275, 282, 291, 302
  extended, 306
  reduced, 306
  repeated, 306

/530.028P578>C1/